THEORETICAL AND APPLIED MECHANICS

INTERNATIONAL UNION OF THEORETICAL
AND APPLIED MECHANICS

THEORETICAL AND APPLIED MECHANICS

*Proceedings of the XVIIth International Congress of Theoretical and Applied Mechanics
held in Grenoble, France, 21–27 August, 1988*

Edited by
Paul GERMAIN
Monique PIAU and Denis CAILLERIE

*Institut de Mécanique de Grenoble
Grenoble, France*

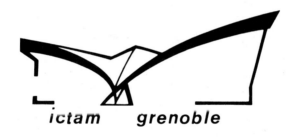

1989

NORTH-HOLLAND
AMSTERDAM · NEW YORK · OXFORD · TOKYO

ISBN: 0 444 87302 3

Published by:
ELSEVIER SCIENCE PUBLISHERS B.V.
P.O. Box 1991
1000 BZ Amsterdam
The Netherlands

Sole distributors for the U.S.A. and Canada:
ELSEVIER SCIENCE PUBLISHING COMPANY, INC.
655 Avenue of the Americas
New York, N.Y. 10010
U.S.A.

Library of Congress Cataloging-in-Publication Data

International Congress of Theoretical and Applied
    Mechanics (17th: 1988: Grenoble, France)
    Theoretical and applied mechanics.

    Bibliography: p.
    1. Mechanics, Analytic--Congresses. I. Germain,
Paul, 1920-   . II. Piau, Monique. III. Caillerie,
Denis. IV Title.
QA801.I39   1988   531   88-32948
ISBN 0-444-87302-3

PRINTED IN THE NETHERLANDS

PREFACE

This book contains the Proceedings of the XVIIth International Congress of Theoretical and Applied Mechanics, held on the Saint-Martin d'Hères Campus, near Grenoble, France, August 21-27, 1988. The Congress was organized by the International Union of Theoretical and Applied Mechanics (IUTAM) and held in Grenoble by invitation from the Comité National Français de Mécanique and from the Université Joseph Fourier and the Institut National Polytechnique de Grenoble through their joint research laboratory the Institut de Mécanique de Grenoble.

The full text of the two General Lectures, of introductory lectures of the three minisymposia and of sectional lectures, according to the list on page XVII, are included in this volume and will not appear elsewere.The contributed papers presented at the congress are listed by author and title; most of them will be published in appropriate scientific journals.

The publication of these Proceedings has been handled promptly and capably by North Holland and their editors, to whom we are very grateful

Grenoble
October 1988   Paul Germain   Monique Piau   Denis Caillerie

1 - Entrance to the Reception Hall at the Bibliothèque Universitaire de Sciences during the week of the Congress.

2 - Arrival of the French Minister of Research, Prof. Hubert CURIEN prior to the Opening of the Congress on Monday morning.

3 - The Minister of Research, Prof. GERMAIN, Prof. LIGHTHILL and Prof. PAYAN talking in front of Amphitheatre Weil.

4 - Amphitheatre Weil was filled to capacity with one thousand mechanicians and accompanying persons during the Opening Ceremony.

5/6 - The Opening Ceremony, on the podium, from the left :
Prof. Denis CAILLERIE, Executive Secretary of ICTAM Grenoble - Prof. Georges LESPINARD, President of INP Grenoble - M. Charles DESCOURS, Sénateur - M. Joël GADBIN, Sous-Préfet - Prof. Hubert CURIEN, Minister of Research - Prof. Paul GERMAIN, Chairman of ICTAM Grenoble - Sir James LIGHTHILL, President of IUTAM - Dr Monique PIAU Cochairman of ICTAM Grenoble - Prof. Jean-Jacques PAYAN, President of UJF Grenoble - Prof. Michel COMBARNOUS, Director of Research at the Ministry of Education - Prof. Keith MOFFATT, Secretary of the Congress Committee of IUTAM.

7 - Prof. LIGHTHILL giving his speach, on his right and left sides are Prof.
GERMAIN and Dr Monique PIAU.

8 - At the Opening Ceremony - First person, on the left is Prof. ARNOLD who
gave the Opening Lecture.

9   - A moment of rest on the parvis in front of the Amphitheatre Weil.

10 - Charming hostesses welcoming participants at the Bibliothèque de
      Sciences throughout the Congress.

11 - Presentation of the "History of IUTAM" by Prof. Stephen JUHASZ.

12 - A Poster-Session.

SPONSORING ORGANIZATIONS AND COMPANIES

The following organizations and companies have provided financial support:

- Université Joseph Fourier.
- Institut National Polytechnique de Grenoble.
- Centre National de la Recherche Scientifique (Département Sciences Physiques pour l'Ingénieur).
- Ministère de la Recherche et de l'Enseignement Supérieur.
- Ministère de la Défense (Direction des Recherches et Etudes Techniques).
- Agence Française pour la Maîtrise de l'Energie.
- Association Universitaire de Mécanique.
- Fédération des Industries Mécaniques et Transformatrices des Métaux.
- Société Française des Mécaniciens.
- Société Française de Tribologie.
- Société Hydrotechnique de France.
- G.A.M.A.C..
- Electricité De France.
- Centre d'Etudes Nucléaires (C.E.A.).
- Régie Nationale des Usines RENAULT.
- Institut Français du Pétrole.
- Centre National d'Etudes Spatiales.
- Société Nationale d'Etude et de Construction de Moteurs d'Avions.
- Office National d'Etudes et de Recherches Aérospatiales.
- Peugeot S.A.
- Aérospatiale (SNIAS).
- Total C.F.P.
- S.A.G.E.M.
- Société Européenne de Propulsion.
- Matra S.A.
- Société Bertin.
- Association Aéronautique et Astronautique de France (A.A.A.F.)

- The International Center of Scientific Unions (ICSU).
- The Ministère des Affaires Etrangères.
- The Direction de la Coopération et des Relations Internationales of the Ministère de l'Education Nationale.
- The Office of Naval Research-London (US Navy).
- The Foundation Robert Maxwell.

gave subsidies to welcome foreign participants.

CONTENTS

OPENING AND CLOSING LECTURES

INTRODUCTORY LECTURES OF MINISYMPOSIA

SECTIONAL LECTURES

CONTRIBUTED PAPERS

CONGRESS COMMITTEE OF IUTAM

President : M.J. Lighthill (UK)*

J. Achenbach (USA)
A. Acrivos (USA)
E.R. de 'Arantes e Oliveira (Portugal)
L. Bevilacqua (Brazil)
B.A. Boley (USA)*
J. Brilla (CSSR)
G.G. Chernyi (USSR)
I.F. Collins (New Zealand)
D.C. Drucker (USA)
W. Fiszdon (Poland)*
N.J. Hoff (USA)
J. Hult (Sweden)
I. Imai (Japan)
A. Yu. Ishlinsky (USSR)
S. Kaliszky (Hungary)
Y.H. Ku (USA)

J. Lemaitre (France)
T.C. Lin (China)
G. Maier (Italy)
H.K. Moffatt, Secretary (UK)*
R. Moreau (France)
E.A. Müller (FRG)
R. Narasimha (India)
F.I. Niordson (Denmark)*
J.R. Philip (Australia)
Lingxi Qian (China)
F.P.J. Rimrott (Canada)
W. Schiehlen (FRG)
W. Schneider (Austria)
J. Singer (Israel)
L. van Wijngaarden (Netherlands)*
J.R. Willis (UK)

* Members of Executive Committee

LOCAL ORGANIZING COMMITTEE

Co-chairmen :   P. Germain* and Monique Piau*

D. Caillerie *   Secretary
J.M. Piau*   Treasurer

A. Alemany
J.L. Auriault*
J. Baudoin
G. Biguenet*
G. Binder
D. Bouvard*
D. Cordary
L. Debove

J. Desrues
E. Flavigny
J.P. Gaudet*
Odile Lantz*
M. Lesieur
G. Lespinard
R. Moreau*
Colette Morel

* Members of Executive Committee

XVII

CONGRES INTERNATIONAL
MECANIQUE
THEORIQUE APPLIQUEE
GRENOBLE
88

LIST OF AUTHORS OF PAPERS IN THIS VOLUME

Prof. **V. Arnold** - Steklov Mathematical Institute, Vavilova 42, 117966 GSP-1 MOCKBA, USSR.

Prof. **G. Batchelor** - DAMTP, University of Cambridge, Silver Street, CAMBRIDGE CB3 9EW, UNITED KINGDOM.

Dr. **J.L. Chaboche** - ONERA, 29 avenue de la Division Leclerc, 92320 CHATILLON, FRANCE.

Prof. **J.F. Davidson** - Dept. of Chemical Engineering, Cambridge University, Pembroke Street, CAMBRIDGE CB2 3RA, UNITED KINGDOM.

Prof. **J. P. Gollub** - Haverford College, HAVERFORD PA 19041, USA.

Prof. **P. Hagedorn** - Institut für Mechanik, Technische Hochschule Darmstadt, Hochschulstr. 1, D-6100 DARMSTADT, FRG (WEST GERMANY).

Prof. **J. R. Herring** - National Center for Atmospheric Research, BOULDER Colorado 80307, USA.

Dr. **E. Hopfinger** - Institut de Mécanique de Grenoble, Domaine Universitaire BP 53X, 38041 GRENOBLE, FRANCE.

Prof. **J. W. Hutchinson** - Division of Applied Sciences, Harvard University, Pierce Hall, 29 Oxford St., CAMBRIDGE MA 02138, USA.

Dr. **S. Kieffer** - U.S. Geological Survey, 2255 N. Gemini drive, FLAGSTAFF Arizona 86001, USA.

Prof. **M. Kiya** - Department of Mechanical Engineering, Hokkaido University, 060 SAPPORO, JAPAN.

Prof. **J.W. Miles** - University of California, San Diego IGPP, A-025, LA JOLLA CA 92093, USA.

Prof. **A. Needleman** - Engineering Div. Box D, Brown University, PROVIDENCE 02912 RI, USA.

Prof. **S. B. Savage** - Dept. of Civil Engineering & Applied Mechanics, McGill University, H3A 2K6 MONTREAL, QUEBEC CANADA.

Prof. **F. Simonelli** - Physics Dept., The University of Pennsylvania, PHILADELPHIA, PA 19104, USA.

Prof. **F.T. Smith** - Maths. Dept., University College, Gower St., LONDON WCIE 6BI, UNITED KINGDOM.

Prof. **K. Sobczyk** - IPPT, Polish Academy of Sciences, Ul. Swietokrzyska 21, 00-049 WARSZAWA, POLAND.

Prof. **D.A. Spence** - Mathematics Dept., Imperial College, 80 Queen's Gate, South Kensington, LONDON SW7 2BZ, UNITED KINGDOM.

Prof. **B. Storakers** - Dept. of Strength Materials and Solid Mechanics, Royal Institute of Technology, 10044 STOCKHOLM, SWEDEN.

Prof. **D. L. Turcotte** - Cornell University, Snee Hall, ITHACA 14853, USA.

Dr. **V. Tvergaard** - Tech. University of Denmark, 2800 LYNGBY, DENMARK.

Dr. **J.H.J. van der Meulen** - Maritime Research Institute, Haagsteeg 2, POB 28, 6700 AA WAGENINGEN, THE NETHERLANDS.

Prof. **L. van Wijngaarden** - University Twente, POB 217, 7500 AE ENSCHEDE, THE NETHERLANDS.

Prof. **Ren Wang** - Department of Mechanics, Peking University, 100871 BEIJING, CHINA (PRC).

LIST OF PARTICIPANTS

AUSTRALIA (4 participants)
Ansourian P.
De Vahldavis G.
Karihaloo B.L.
Philip J.R.

AUSTRIA (5 participants)
Fischer F.D.
Rammerstorfer F.G.
Schneider W.
Troger H.
Ziegler F.

BELGIUM (6 participants)
Janssen H.J.M.
Kestens J.
Lamy A.
Lhost O.
Sarlet W.
Sergysels R.

BRAZIL (7 participants)
Bevilacqua L.
Frota M.N.
Pamplona D.
Steffen V.
Tenenbaum R.
Vargas A.
Zindeluk M.

BULGARIA (8 participants)
Baltov A.
Brankov G.
Christov C.I.
Karagiozova D.
Markov K.

Popov K.
Yamboliev K.
Zapryanov Z.

CAMEROUN (1 participant)
Tayou Simo J.

CANADA (25 participants)
Axelrad D.R.
Aydemir N.U.
Bertrand F.
Blackwell J.H.
Bourassa P.A.
Cleghorn W.L.
Dickinson S.M.
Dost S.
Ellyin F.
Ethier R.
Gladwell G.M.L.
Goldak J.A.
Jeffrey D.J.
Lardner R.W.
Marandi S.
Maslowe S.A.
Neale K.W.
Ranalli G.
Rimrott F.
Robillard L.
Savage S.B.
Swanson S.R.
Tabarrok B.
Tanguy P.A.
Tennyson R.C.

CHILE (1 participant)
Letelier

CHINA (PRC) (28 participants)
    Chen Y.N.
    Cheng C.J.
    Dai S.
    Fan T.Y.
    Gao Y.C.
    Guo Z.H.
    Hu Y.
    Jin Z.M.
    LI C.x.
    Li L.
    Li J.
    Lin T.C.
    Ling F.
    Liu Y.
    Lu Y.
    Shao C.X.
    Shi G.Y.
    Sun Y.D.
    Wang R.
    Wang M.
    Xu J.
    Yang W.
    Yin X.C.
    Zhang R.J.
    Zheng Z.
    Zheng Z.C.
    Zhou M.
    Zhu M.M.

CZECHOSLOVAKIA (6 participants)
    Bartak J.
    Curev A.G.
    Hyca M.
    Jira J.
    Krupka V.
    Markus S.

DANEMARK (10 participants)
    Byskov E.
    Dalgaard-Jensen K.
    Goltermann P.
    Hansen E.B.
    Mikkelsen B.
    Niordson F.
    Olhoff N.

    Pedersen P.T.
    Pyrz R.
    Tvergaard V.

FINLAND (11 participants)
    Aliranta J.P.
    Autio M.
    Kangaspuoskari M.
    Lahtinen H.T.
    Mikkola M.J.
    Parland H.
    Ranta M.A.
    Raty R.
    Sjolind S.G.
    Tervonen M.
    Valiheikki O.T.

FRANCE (340 participants)
    Abdelmoula R.
    Adjedj G.
    Adler M.
    Adou K.J.
    Akel S.
    Alaoui Soulihani A.
    Alemany A.
    Allain C.
    Allani M.N.
    Allix O.
    Anselmet F.
    Anthoine A.
    Argoul P.
    Armand J.L.
    Atten P.
    Aubry E.
    Auriault J.L.
    Baetz J.P.M.
    Bamberger Y.
    Baptiste D.
    Barbi C.
    Barrere M.
    Barthes-Biesel D.
    Batoz J.L.
    Beaumont N.
    Belorgey M.
    Benallal A.
    Benyettou F.
    Bernadou M.

Bernet C.

Berthaud Y.

Berveiller M.

Biguenet G.

Binder G.

Blanc-Benon P.

Blanchard D.

Boehler J.P.

Bohineust

Bois P.A.

Bongrain H.

Bonnet M.

Bonnet G.

Bonthoux C.

Boucherit A.

Boulon M.

Bouré J.A.

Bourquin F.

Bousgarbiès J.L.

Boussaa D.

Bouvard D.

Brancher J.P.

Bremand F.

Brière P.

Brousse P.

Brun L.

Bruneau M.

Bui H.D.

Buisine D.

Burlet H.

Cabannes H.

Caillerie D.

Cailletaud G.

Caltagirone J.P.

Cambou B.

Campbell A.

Canova G.R.

Cao H.L.

Caperan P.

Caron J.F.

Catheline G.

Caulliez G.

Cerdan J.P.

Chabert d'Hières G.

Chaboche J.L.

Chaissé F.

Chambon R.

Charlez P.

Chau T.H.

Chedmail P.

Cheret R.

Chesneau C.

Chevalier Y.

Chevalier J.F.

Chollet J.P.

Chollet H.

Chomaz J.M.

Chpoun A.

Chrysochoos A.

Cimetiere A.

Cochelin B.

Coffignal G.

Cognet G.

Cointe R.

Comte-Bellot G.

Cormier R.

Coulouvrat F.

Coussy O.

Dahan M.

Daignières M.

Damamme G.

Dambrine B.

Dangla P.

Danho E.

Darve F.

Daube O.

Dautray R.

Davaille A.

de Buhan P.

de Fouquet J.

Debailleux C.

Delhaye J.M.

Deriat E.

Desoyer T.

Desrues J.

Di Benedetto H.

Djeran T.

Doan-Kim S.

Dodu J.

Dogui A.

Domaszewski M.

Dragon A.

Drouot R.

Dubois M.

Dubois P.

Dudeck M.

Dumontet H.

Dunand A.

Duperray B.
Duvaut G.
Dyment A.
Ehrlacher A.
El Hawa E.
El Mouatassim M.
Etay J.
Fabrie P.
Farge M.
Faure R.
Favre A.
Fedelich B.
Ferraris G.
Feuillebois F.
Fleitout L.
Fortuné D.
Fournier J.D.
Franc J.P.
François D.
Frémond M.
Frene J.
Frisch U.
Gachon H.
Gagne Y.
Gardin C.
Gasc C.
Gatignol R.
Gatignol P.
Gaudet J.P.
Gautherin M.T.
Gelin J.C.
Gentile D.
Georges J.M.
Gérard A.
Germain P.
Germain Y.M.
Géron M.
Gilibert Y.
Gilletta de S Joseph D.
Grediac M.
Grellier J.P.
Gremillard J.
Grolade D.
Halphen B.
Hopfinger E.
Hounkanlin M.
Iooss G.
Jardon A.
Jayaraman V.

Jean M.
Jemmali M.
Joubert F.
Julliard P.
Jullien J.
Kapsa P.
Karimi K.
Karray M.A.
Klepaczko J.
Krasucki F.
Labbé P.
Lacomme M.
Ladeveze P.
Ladeveze J.
Lagarde A.
Lalanne M.
Lambelin J.P.
Lan Q.
Lanchon H.
Lanier J.
Lascaux P.
Lasek A.
Laure P.
Le Houédec D.
Le Provost C.
Le Ray M.
Leboeuf F.
Leborgne P.
Lecertisseur C.
Lega J.
Leguillon D.
Lemaître J.
Léné F.
Lesieur M.
Leuchter O.
Levy T.
Leyrat J.P.
Linder R.
Liu Q.
Loehr
Loubet J.L.
Lusseyran F.
Luu T.S.
Madani K.
Maigre H.
Maingre E.
Maisonneuve O.
Marcelin J.L.
Maresca C.

Marichal D.

Marle C.

Marquis D.

Masbernat L.

Mathia T.

Maugin G.A.

Meance R.

Meister E.

Merlen A.

Métais O.

Meziere Y.

Michel J.M.

Millard A.

Mizzi J.P.

Molinari A.

Molodtsof Y.

Moreau R.

Moreau J.J.

Morilhat P.

Morlet D.

Morvan D.

Muhé H.

Muller P.

Naciri T.

NGuyen Q.S.

Nicolas P.

Niemann E.

Nizou P.Y.

Oddou C.

Ohayon R.

Orsero P.

Oudin J.

Palgen L.

Pascal M.

Patoor E.

Peerhossaini H.

Pegoraro P.

Pelissier R.

Peube J.L.

Piau J.M.

Piau M.

Pierrard J.M.

Pierre M.

Polack J.D.

Potier-Ferry M.

Predeleanu M.

Putot C.

Qian Y.H.

Rahier C.

Rebourcet B.

Remond Y.

Renard J.

Renner M.

Rey C.

Rey C.

Richer J.P.

Rivet J.P.

Robert R.

Rosant J.M.

Roseau M.

Rossi M.

Roth J.C.

Roucous R.

Rougee P.

Rousseau M.

Saanouni K.

Sabir M.

Sabot J.

Saint Jean Paulin J.

Salençon J.

Santon L.

Sauvage G.

Scain J.

Scibilia M.F.

Sero-Guillaume O.

Shi G.

Sidoroff F.

Sirieys P.

Sommeria J.

Sornette D.

Staquet C.

Sulem P.L.

Suo X.Z.

Suquet P.

Surrel Y.

Ta Phuoc L.

Taghite M.B.

Taheri S.

Tamagny P.

Tardivel F.

Teodosiu C.

Tocquet B.

Touratier M.

Tournier C.

Touzot G.

Trad A.

Trompette P.

Truong Dinh Tien J.

Valentin G.
Valid R.
van Huffel A.
Verchery G.
Vilotte J.P.
Vincourt M.C.
Viviand H.
Voldoire F.
Wack B.
Waldura H.
Wesfreid J.E.
Woillez J.
Yin J.
Yin H.P.
Zaoui A.
Zarka J.
Zenouda J.C.

Najar J.
Nastase A.
Pfeiffer F.
Popp K.
Raszillier H.
Rozvany G.
Schiehlen W.
Stein E.
Stumpf H.
Viehl M.
Vogel H.
Wallaschek J.
Wang Y.
Wedig W.
Weichert D.
Windrich H.
Wu C.C.

FRG (WEST GERMANY) (45 participants)

Assenheimer M.
Bartlma F.
Bertram A.
Bohme G.
De Boer R.
Detemple E.
Ehlers W.
Eschenauer H.A.
Fernholz H.H.
Gersten K.
Göz M.
Hagedorn P.
Haupt P.
Herrmann K.P.
Hiller W.J.
Holtfort J.
Kienzler R.
Kleczka M.
Kowalewski T.A.
Kreuzer E.J.
Labisch F.K.
Lehmann T.
Lorenzen A.
Meier G.E.A.
Mielke A.
Muller E.A.
Muller I.
Muschik W.

GRD (EAST GERMANY) (2 participants)

Altenbach J.
Günther H.

GREECE (5 participants)

Baniotopoulos C.C.
Kounadis A.N.
Palassopoulos G.V.
Sapountzakis E.
Soldatos K.

HONG-KONG (1 participant)

Ko N.W.M.

HUNGARY (5 participants)

Bosznay A.
Kaliszky S.
Pomazi L.
Szabo F.J.
Tarnai T.

INDIA (3 participants)

Narasimha R.
Nigam N.
Roy A.

IRELAND (1 participant)
Hayes M.A.

ISRAEL (23 participants)
Alperovitch H.
Banks-Sills L.
Beltzer A.I.
Ben-Haim Y.
Bodner S.R.
Durban D.
Elishakoff I.
Greenberg J.B.
Hashin Z.
Hauser S.
Ishai O.
Kitron A.
Libai A.
Ophir E.
Parnes R.
Sabbag M.
Shapiro M.
Shemer L.
Shilkrut D.
Singer J.
Toren M.
Tsinober A.
Zewi I.G.

ITALY (15 participants)
Bergamaschi S.
Capecchi D.
Cazzani A.
Cercignani C.
Cinquini C.
Contro R.
Ferrari M.
Galletto D.
Germano M.
Maier G.
Rega G.
Rieutord M.
Rovati M.
Sinopoli A.
Stamm H.

JAPAN (33 participants)
Ashida F.
Fukumoto Y.
Funakoshi M.
Hasimoto H.
Himeno R.
Horii H.
Imai I.
Ishii K.
Ishii Y.
Izutsu N.
Jousselin F.
Kambe T.
Kawata K.
Kiya M.
Makinouchi A.
Maruo H.
Matsumoto Y.
Matsumoto N.
Miyamoto H.
Miyazaki T.
Mizushima J.
Monji H.
Nakanishi S.
Noda N.
Oshima Y.
Sano O.
Sato H.
Sumi Y.
Takahashi K.
Tsuji T.
Umeki M.
Umemura A.
Yamamoto Y.

KOWAIT (1 participant)
Megahed S.

MEXICO (2 participants)
Cruz J.
Soto L.

NETHERLANDS (34 participants)
Arbocz J.
Bakker P.G.
Besseling J.F.

Coene R.
De Bruin G.J.
Dieterman H.
Ernst R.Y.
Geerlings J.J.P.
Geurst J.A.
Janssen J.
Kalker J.J.
Kapteyn C.
Klever F.J.
Koiter W.T.
Kok J.B.W.
Kuijpers W.
Leroy
Meijaard J.P.
Meijers P.
Menken C.M.
Reynen J.
van Beek P.
van Campen D.
van de Coevering G.F.M.
van de Ven A.
van der Giessen E.
van der Heyden A.
van der Meulen J.H.J.
van Heugten P.C.M.
van Keulen F.
van Wijngaarden L.
van Zwol B.
Velthuizen H.G.M.
Vreenegoor A.J.

NEW ZEALAND (3 participants)
Collins I.
Segedin R.H.
Woods B.A.

NIGERIA (1 participant)
Aderogba K.

NORWAY (3 participants)
Dysthe K.
Irgens F.
Tyvand P.A.

POLAND (23 participants)
Bielski W.R.
Bogacz R.
Dems K.
Duszek M.K.
Gawecki A.
Grabacki J.,
Gutkowski W.
Kapitaniak T.
Kosinski W.
Lewinski T.
Majorkowska-Knap K.
Makowski J.
Mroz Z.
Pecherski R.B.
Perzyna P.
Petryk H.
Rogalska E.
Sobczyk K.
Szemplinska-Stupnika W.
Szuwalski K.
Telega J.J.
Wittbrodt E.
Zyczkowski M.

PORTUGAL (5 participants)
Arantes e Oliveira E.R.d.
Bairrao R.
Camotim D.
Trabucho L.
Veiga M.F.

RUMANIA (1 participant)
Chiriacescu S.

SAUDI ARABIA (1 participant)
Al Athel S.

SPAIN (1 participant)
Antoranz J.C.

SWEDEN (44 participants)
Alavyoon F.
Amberg G.

Andersson B.
Bark F.
Berndt S.
Bjorkman G.
Borgstrom L.
Carlsson C.G.
Dahlkild A.A.
Digby P.J.
Edlund U.
Faleskog J.
Gudmundson P.
Gustavsson H.
Hedner G.
Hsieh R.K.T.
Hult J.
Inge C.
Johnson E.
Jonsson M.
Kao H.
Karlsson L.
Klarbring A.
Lagerstedt T.
Larsson P.L.
Levenstam M.
Lofqvist T.
Lundh H.
Melin S.
Moberg H.
Nabo O.
Nasstrom M.
Nilsson K.F.
Ostlund S.
Rajaratnam S.
Ristinmar M.
Stensson A.
Storakers B.
Troive L.
Tryding J.
Yang Q.X.
Zahrai S.
Zang W.
Zhou S.A.

SWITZERLAND (14 participants)
Bargmann H.W.
Borne L.
Brauchli H.
Chyou Y.P.

Dual J.
Gyr A.
Haas R.
Lefrançois A.
Meyer M.
Ryhming I.
Sayir M.
Sofer M.
Thomann H.
Wuthrich C.

TAIWAN (8 participants)
Chang C.o.
Chao C.C.
Chiang C.R.
Chien L.K.
Kam T.Y.
Lee C.J.
Sun F.T.
Yau J.D.

TUNISIA (1 participant)
Ben Ouezdou M.

TURKEY (7 participants)
Alici E.
Erbay S.
Erbay H.A.
Inan E.
Meriç R.A.
Suhubi E.S.
Tuncer C.

UNITED KINGDOM (60 participants)
Batchelor G.
Berger M.A.
Byatt-Smith J.G.B.
Calladine C.R.
Chambers L.G.
Coelho S.L.V.
Couët B.
Craik A.D.D.
Craine R.E.
Craven A.

Dalziel S.B.

Davidson J.F.

Drath L.J.

Drazin P.G.

Dunwoody J.

Feit D.

Fleck N.A.

Gribben R.J.

Hansen J.

Hawkes T.D.

Hunt G.W.

Hunt J.C.R.

Huppert H.E.

Johnson K.L.

Keane A.J.

Kerr R.C.

King A.

Lawrence R.

Lighthill S.J.

Moffatt H.K.

Nagata M.

Parker D.F.

Pellegrino S.

Ponter A.

Popat N.R.

Price W.G.

Proctor M.R.E.

Roberts J.W.

Sellin R.H.J.

Simoes da Silva L.A.P.

Smith F.T.

Smith F.I.P.

Sneddon I.

Spence D.A.

Spencer A.J.M.

Stronge W.J.

Stuart J.T.

Teles da Silva A.F.

Thompson J.M.T.

Tranah D.

Tritton D.J.

Ursell F.

Virgin L.N.

Webster P.

Whittaker A.

Wickham G.R.

Willis J.R.

Worster M.G.

Yu T.X.

Zhang W.

URSS (28 participants)

Aristov V.V.

Arnold V.

Babeshko V.

Burchuladze T.V.

Cherenkov A.P.

Chernyi G.C.

Derzhavina A.

Dobrovol'skaya Z.N.

Dulov V.

Glaznev S.E.

Godunov S.

Gorodeiski A.

Guz A.N.

Ishlinsky A.

Khusnutdinova N.V.

Krasheninnikova N.L.

Mikhailov G.K.

Obraztsov I.P.

Parton V.Z.

Rumjantsev V.V.

Sabelnikov V.A.

Sedov L.I.

Stepanov V.A.

Sulikashvili R.S.

Terentjev A.G.

Teshukov V.M.

Vasil'ev D.G.

Yavorskaya I.M.

USA (123 participants)

Abeyaratne R.

Abramson H.N.

Achenbach J.D.

Acrivos A.

Antar B.

Aref H.

Atluri S.N.

Bammann D.J.

Batra R.C.

Bazant Z.P.

Blume J.A.

Bogy D.B.

Boley B.A.

Botsis J.

Brock L.M.
Browand F.
Buckmaster J.
Budiansky B.
Chandra A.
Chen Y.
Chen C.F.
Chona R.
Christensen R.M.
Chung K.
Cohen D.S.
Corona E.
Crandall S.H.
Drucker D.C.
Dundurs J.
Dvorak G.
Engblom J.J.
Erdogan F.
Folias E.S.
Freund L.B.
Gao Y.
Gharib M.
Gilbert R.P.
Givoli D.
Gollub J.P.
Havner K.S.
Herring J.R.
Herrmann G.
Hetnarski R.B.
Hodge P.
Hoff N.J.
Hogan H.A.
Holt M.
Hsu C.S.
Huang T.C.
Hussain F.
Hutchinson J.W.
Ibrahim R.A.
Jasiuk I.
Johnson R.E.
Johnson G.C.
Juhasz S.
Kant R.
Katsube N.
Kieffer S.
Kleinman R.
Knauss W.G.
Knowles J.K.
Krajcinovic D.

Kyriakides S.
Lawrence C.J.
Lebovitz N.
Lee K.D.
Leibovich S.
Leissa A.
Liechti K.M.
Lin T.H.
Markenscoff X.
Marshall J.S.
McCoy J.
Melville W.K.
Metzner A.W.K.
Miksad R.W.
Miles J.W.
Moon F.C.
Mura T.
Naghdi P.M.
Nash W.A.
Needleman A.
Pandey B.D.
Payne E.R.
Pence T.J.
Pian T.H.H.
Pierre C.
Plaut R.
Ponte-Castaneda P.
Powell R.L.
Richmond O.
Rubinstein A.A.
Sacks S.
Schmeelk J.
Schreyer H.L.
Shih F.
Shirron J.
Shukla A.
Siginer D.A.
Simmonds J.
Snyder J.M.
Sridharan S.
Steele C.R.
Symonds P.S.
Synolakis
Szewczyk A.A.
Ting T.C.T.
Traugott S.
Trevino G.
Triantafyllidis N.
Tulin M.

Turcotte D.L.
van den Broeck J.M.
Wagoner R.H.
Wang K.C.
Weitsman Y.
Weng G.J.
Wu T.
Youngdahl C.K.
Zhang K.
Zhang J.
Zureick A.H.

VENEZUELA (1 participant)
   Power H.

YUGOSLAVIA (5 participants)
   Atanackovic T.N.
   Djukic D.
   Lazic J.D.
   Lubarda V.
   Vujicic V.

REPORT ON THE CONGRESS

Paul GERMAIN - Monique PIAU

The decision to accept the invitation from France to hold the XVIIth International Congress of Theoretical and Applied Mechanics in Grenoble was made by the Congress Committee of IUTAM during its second meeting in August 84 in Lyngby. The french proposal was submitted jointly by the Comité National Français de Mécanique and by the Institut de Mécanique de Grenoble, a joint laboratory of the University Joseph Fourier (UJF) and of the Institut National Polytechnique de Grenoble (INPG).

It was decided that the format of the Congress would follow the one successfully adopted in Lyngby. The Congress would cover the entire field of Mechanics, including analytical, solid and fluid mechanics, and their applications. Two general lectures - the first just after the opening ceremony and the second just before the closing ceremony - and fifteen invited sectional lectures would be given. Moreover special emphasis would be placed on three selected topics. For each topic a convenor would be appointed to be responsible for the organization of a half day introductory session devoted to key-note lectures giving an overview of the field. Finally the Congress Committee decided to organize poster/discussion sessions in parallel with the usual oral presentations as in Lyngby where this initiative met an unquestionable success.

The decisions of the Congress Committee were implemented by its Executive Committee which worked in close cooperation with the Local Organizing Committee. Among a number of proposals, the Executive Committee finally selected the following three topics constituting the so-called three mini-symposia.

(i)   Mechanics of large deformation and damage
(ii)   The Dynamics of two-phase flows
(iii)   Mechanics of the Earth's Crust

It was decided that in addition to half-day sessions each mini-symposium would have one sectional lecture and contributed papers grouped in lecture and poster sessions. Professor V.I. Arnold from USSR and Professor J.W. Miles from U.S.A. were invited to give the two general lectures. All the fifteen authors invited to give a sectional lecture were present at Grenoble except Professor G. Milton who was unable to attend because of illness.

The number of contributed papers submitted to the congress was 1354. Finally 573 were selected on the basis of summaries and abstracts. The distribution according to countries was as follows:

ICTAM 88 Grenoble
Submitted and presented papers (S/P)
Participants at the Congress (A)

	S	P	A		S	P	A
Algeria	2	0	0	Japan	36	22	33
Australia	6	4	4	Jordan	1	0	0
Austria	7	5	5	Kowait	3	1	1
Bahrain	1	0	0	Lybia	2	0	0
Belgium	8	3	6	Morocco	1	0	0
Brazil	15	6	7	Mexico	2	2	2
Bulgaria	12	4	8	The Netherlands	20	11	34
Cameroun	0	0	1	New Zealand	2	1	3
Canada	36	16	25	Nigeria	2	1	1
Chile	2	1	1	Norway	3	2	3
China (PRC)	337	20	28	Poland	37	16	23
Czechoslovakia	16	4	6	Portugal	7	2	5
Denmark	3	3	10	Rumania	3	1	1
Egypt	6	0	0	Saudi Arabia	3	0	1
Finland	5	3	11	South Africa	1	0	0
France	275	103	340	Spain	6	1	1
FRG (West Germany)	55	25	45	Sweden	14	9	44
GRD (East Germany)	3	1	2	Switzerland	7	5	14
Greece	6	4	5	Taiwan (ROC)	25	7	8
Hong-Kong	3	1	1	Tunisia	2	2	1
Hungary	4	3	5	Turkey	17	7	7
India	18	2	3	United Kingdom	48	27	60
Irak	1	0	0	URSS	36	17	28
Iran	2	0	0	USA	202	84	123
Ireland	0	0	1	Venezuela	2	1	1
Israel	16	11	23	West Indies	1	0	0
Italy	21	8	15	Yugoslavia	10	4	5
				TOTAL	1354	450	951

Among the selected papers, 250 were designated as lectures and 323 as posters, with equal status and requirement of quality.

In Canada, China, France, Germany (Fed. Rep.), Japan, Poland, UK, USA, USSR, National Committees reviewed papers from within their own countries and made recommendations to the International Committee authorized by the Congress Committee for final selection.

One thousand scientists registered before and during the meeting. But late cancellations reduced the actual number of participants to 951 and 450 papers were presented. There were also 310 accompanying persons. In all, 44 countries were represented and the distribution by country can be seen from the preceeding list. The list of authors and papers presented at the Congress is given on pages 427 to 451.

On the Sunday evening preceeding the Congress there was a reception at the Palais des Sports, downtown Grenoble, where most of participants met and received their registration material.

The Congress was held on the Saint-Martin-d'Hères campus, near Grenoble. The entire ground floor of the Bibliothèque Universitaire des Sciences was devoted to the Registration desk and to various exhibitions - Scientific, Technical and Artistic - and enabled the participants to meet and to rest. The opening and closing ceremonies, the General Lectures and most of the Sectional Lectures took place in the Amphitheatre Weil, (with one thousand seats). Other sessions were held in four buildings located around the large and spacious esplanade in front of the Bibliothèque Universitaire des Sciences.

The official opening ceremony started Monday 22 at 9.15. Seated on the podium were :

- Professor Hubert Curien, Ministre de la Recherche et de la Technologie.
- Sir James Lighthill, President of IUTAM, President of the Congress.
- Professor Jean Jacques Payan, Président de l'Université Joseph Fourier and Professor Georges Lespinard, Président de l'Institut National Polytechnique de Grenoble.
- Professor H.K. Moffatt, Secretary of the Congress Committee.
- Professor Michel Combarnous, Directeur de la Recherche au Ministère de l'Education Nationale.
- Le Sénateur Charles DESCOURS
- Le Sous Préfet Joël GABDIN

and three members of the Local Organizing Committee, the two co-presidents:

- Doctor Monique Piau, Directeur de l'Institut de Mécanique de Grenoble.

- Professor Paul Germain, Président du Comité National Français de Mécanique.

and the Secretary-General: Professor Denis Caillerie.

All the speakers on the podium spoke in French (a simultaneous translation was available). Only a brief account of their addresses is given below.

Professor Paul Germain who chaired this opening ceremony welcome the participants on behalf of the Comité National Français de Mécanique and explained why the Institut de Mécanique de Grenoble was the best place to be chosen in France for the International Congress of Mechanics.

Doctor Monique Piau, head of this Institut gave a brief presentation of this large laboratory where nearly 200 people are involved in many fields of research in fluid and solid mechanics, theoretical as well as applied. Then she gave a short account of the work of the Local Organizing Committee. She acknowledged in particular the various administrations, organizations and societies for their financial support which contributed to the necessary expenses, hence kept the registration fees below a limit and increased the number of grants for scientists from abroad. The list of these sponsoring organizations may be found on page XII.

The next speaker was President Jean-Jacques Payan on behalf of the two Scientific Universities of Grenoble. He expressed pleasure in welcoming the International Congress of Mechanics to this campus. The President mentioned that the Amphitheatre Weil was completely renovated for the occasion. He emphasized some of the special feature and the unique strengths of the Universities of Grenoble. He pointed out that the Institut de Mecanique is surely one of the strengths and expressed great pride in its many achievements, especially in connection with the prosperity of the industries of the "Region Rhône-Alpes".

Monsieur le Sénateur Charles Descours spoke in the absence of Monsieur Alain Carignon, Député Maire de Grenoble, Président du Conseil Régional, who was also Ministre de l'Environnement during the period 1986-1988. He extended to all participants the warmest greetings of the city and of the region of Grenoble. He underlined all the recent achievements of this dynamic region and especially the involvement of Grenoble in the most fascinating scientific and technical advanced realizations. He noted that the city of Grenoble has considerably increased the amount of subsidies devoted to research and to the development of educational establishments especially on the University level.

After all these testimonies, Sir James Lighthill who is the Provost of University College of London, the President of IUTAM and the President of the Congress presented his congratulations to the participants gathered in this prestigious amphitheatre, and gave his thanks to the French dignitaries for their friendly welcome. Recalling the already long tradition of IUTAM and of the International Congress of Theoretical and Applied Mechanics, the President emphasized the importance of such a worldwide gathering for a clear view of the progresses of mechanics. He pointed out that one of the most recent achievement may be described as the marriage of theoretical mechanics which began by Newton three hundred years ago, and of applied mechanics, which started much earlier. Only recently have these two mechanics joined each other in order to extend considerably the field of their applications which may be attacked by mathematics and treated with modern computing facilities. In particular our view of the world of Mechanics has been profoundly modified from the one held a century ago, by the theory of bifurcations and the discovery of strange attractors and of chaos. He emphasized the importance of an International Congress in providing the best opportunity to follow the rapid evolution of our discipline.

In accordance with the French traditions, Professor Hubert Curien, Ministre de la Recherche et de la Technologie was the last to speak, and brought the greetings of the French Government. He remarked that for the success of the ambitious programs of France, Mechanics has to fullfill an essential role. Recalling his personal experience as President of the Centre National d'Etudes Spatiales, he cited as an example the crucial role of Mechanics in space technology. He presented to the President of the Congress and to the participants his best wishes for a fruitfull and a pleasant week in Grenoble.

Sir James Lighthill expressed his thanks to the Ministre Hubert Curien and declared open the XVIIth International Congress of Theoretical and Applied Mechanics. A few minutes later Professor V.I. Arnold who is a founder of the modern theory of dynamical systems presented his general lecture.

* * * * *

xxxvi

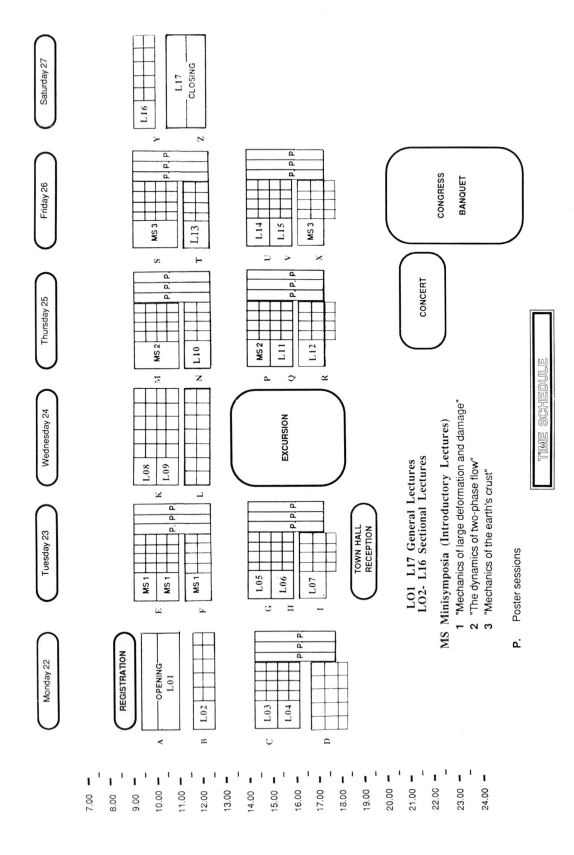

TIME SCHEDULE

LO1 L17 General Lectures
LO2- L16 Sectional Lectures

MS Minisymposia (Introductory Lectures)
1 "Mechanics of large deformation and damage"
2 "The dynamics of two-phase flow"
3 "Mechanics of the earth's crust"

P.   Poster sessions

A survey of the programme of the week is given on the opposite page.

The Lord Mayor of Grenoble graciously received all participants at the City Hall of Grenoble on Tuesday, 23 August at 7 p.m.. A garden party was organized for this occasion.

Excursions were arranged for participants and accompanying persons on Wednesday afternoon.They could choose between excursions to the Chartreuse Mountains or Bourget lake, and Mountain hiking on scenic trails.

During the period of the congress, a scientific and technical exhibition, as well as an Art Exhibition, were held in the Bibliothèque Universitaire de Sciences, on the Reception floor. Participants were invited to a private viewing of the Art Exhibition on Thursday evening.

A piano concert was held in Olivier Messiaen hall later on Thursday evening. François-René DUCHABLE was the featured pianist who presented an outstanding program of several classical works.

The Congress banquet was held in the Palais des Congrès, a modern building situated in the centre of La Villeneuve on Friday, 26 August. About 600 participants were present at the dinner. The CACHIMBO orchestra performed traditionnal and modern jazz, and dancing music until well after midnight.

* * * * *

The official closing ceremony followed immediately the general lecture of Professor J.W. Miles. It was chaired by Doctor Monique Piau.

Professor Paul Germain gave the statistics (see above) on the number of people participating in this congress. He thanked all participants who, by their presence and their work made this congress a real success. Then he thanked all those who arranged for the technical, recreationnal and artistic environment of the Congress which contributed to make the stay of the participants and of the accompanying persons a real pleasure. On behalf of the French communities of mechanics, he thanked all the members of the Local Organizing Committee. Special thanks were given to those who had the hardest tasks ; Denis Caillerie the Secretary General, Jean-Michel Piau the treasurer and Monique Piau the President of this Committee, who, with her colleagues at IMG, was in charge of the whole effort of organization and was able to solve all the small problems arising during the week. Finally Prof. Germain extended his gratitude to the Congress Committee and especially to Secretary General Professor Keith Moffatt, for their unvaluable contribution to the preparation of the Congress and their help to the Local Organizing Committee, to IUTAM for all what it has done for the development of Mechanics in the world, and to the General Assembly for his own election as future President of the Union. He emphasized how difficult it will be for him to be the successor of the outgoing President Sir James, who is not only one of the most distinguished Scientist of our century but also a very clever and perspicuous diplomat.

Professor Keith Moffatt called this week of Congress quite successful. He expressed his personal gratitude to all those who helped him in the preparation of this Congress, the members of the Congress Committee and especially the members of the International Papers Committee who have the hard job to select the best papers for presentation. He also acknowledged the importance of the work done by the Local Organizing Committee and thanked especially Monique Piau, Denis Caillerie and all their coworkers for their kind and fruitful cooperation. He commented that the general format appeared very adequate. In particular, the poster sessions allow quite extensive presentations and discussions. He suggested that the Congress Committee would begin immediately to think of the improvements to ensure the continued success of the next Congress.

Sir James Lighthill, President of IUTAM and President of this Congress was the last speaker. First of all, he described the difficulties faced by the Congress Committee to choose the site of the next Congress between two very attractive candidates : China and Israël. Advantages and disadvantages of these two possibilities were carefully examined and discussed. Finally, the Congress Committee voted for Israel by a very slight margin. Consequently the next Congress will be held in Haïfa.

Professor Lighthill, who will become next Vice-President according to the statutes, also confirmed the decision of the General Assembly that Paul Germain will be the next President ; Professor Leen Van Wijngaarden and Werner Schiehlen will remain respectively Treasurer and Secretary General. Professor Imaï will remain member of the Bureau. Professor Boley, Professor Chernyi and Professor Ziegler were new elected as members. The General Assembly also elected as members of the General Assembly Professor Drucker former President of IUTAM and Professor Ishlinsky and Professsor Hult, former Secretary General of IUTAM, in recognition of all their important contributions as members of IUTAM Bureau.

The President recalled also the initiative of the IUTAM Bureau which recently created two "Bureau Prizes" to honour an outstanding presentation of an excellent paper. He explained the process of selection by a jury. The decision shows that lecture sessions as well as poster sessions are taken into consideration and that prizes may be given to young scientists whose native language is not English. The three authors who have awarded the prizes are :

Prof. H. Horii, from the Department of Civil Engineering of the University of Tokyo for the paper : <u>Mechanics of cracked solids and models of fracture process zone</u> presented at the lecture session Damage III : Micromechanics.

Dr. R.E. Johnson from Talbot Laboratory, University of Illinois, Urbana, for the paper : <u>Coating flows in rotational molding</u> presented at the poster/discussion session on <u>Low Reynolds number and non-newtonian fluid mechanics.</u>

Dr. A. Lorenzen from Max-Planck Institute for Strommungsforchung
       Gottingen for the paper : <u>Observation of hysteresis in
       baroclinic unstability</u> presented at the lecture session
       <u>Stability (fluids) II</u>.

Finally, before closing the Congress, the President stressed again how
successfull and pleasant was the meeting, thanked all those who have prepared and
organized this excellent Congress and hoped that all the participants will meet again
within four years for the XVIIIth ICTAM 1992 in Haïfa.

Theoretical and Applied Mechanics
P. Germain, M. Piau and D. Caillerie (Editors)
Elsevier Science Publishers B.V. (North-Holland)
  IUTAM, 1989

1

BIFURCATIONS AND SINGULARITIES IN MATHEMATICS AND MECHANICS

V.I. Arnold

Steklov Mathematical Institute
MOSCOW, USSR

The role of bifurcation theory in the analysis of the dependence of mechanical and other systems on the parameters is similar to the role of the qualitative theory of phase portraits of dynamical systems in oscillation theory. The foundations of bifurcation theory, like those of the qualitative theory of dynamical systems, go back to the works of Poincaré (while some important problems were studied by Huygens). However the solution of these problems has become possible only in the last few decades, due to the progress of singularity theory and to some new mathematical techniques which have connected these problems of mechanics to domains of abstract mathematics, such as algebraic geometry and topology, the theories of Lie algebras and crystallographical reflection groups.

The modern bifurcation theory has become *multiparametrical* . The general scheme of this theory involves the *space of all the systems* or objects we are interrested in (say, the space of differentiel equations of some class - usually this is an infinite - dimensional functional space). Degenerate systems (or objects) form a *hypersurface* in the space of all the systems (objects), that is a subvariety of codimension one. This hypersurface is the *boundary*, separating the domains of systems (objects) of our functional space, formed by the systems (objects) with qualitatively different behaviour - say, the domains formed by the stable and unstable systems.

In most cases the degenerate systems' (objects') hypersurface itself has *singularities*. To more degenerate systems (objects) there correspond subvarieties of singularities, whose codimensions in the functional space grow with the degree of degeneracy.

A *multiparameter family* of systems (of objects) is a mapping A from the parameter space to the functional space of all sytems (fig. 1). If A is a generic mapping, its image intersects the hypersurface of degenerate systems transversally. The counterimage of the surface of degenerate systems in the parameter space is called the *bifurcation diagram* of the family. If the family is sufficiently general (technically speaking, *versal*), then the singularities of the bifurcation diagram represent exactly the singularities of the hypersurface of the degenerate systems . For instance, in fig. 1 one sees a 2-parameter versal family, containing, for some value of the parameter, a "codimension two bifurcation" corresponding to a singular point at the bifurcation diagram.

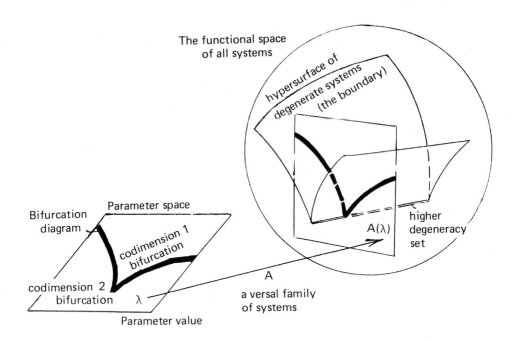

Fig. 1

In fig. 2 one sees three one-parameter families. The first one is versal, the two others are not. In the first case, the curve representing the family intersects the hypersurface of degenerate systems transversally. Such an intersection is a stable event and corresponds to a typical, stable bifurcation: any neighbouring curve intersects the hypersurface of degenerate systems at a neighbouring point, hence every neighbouring family displays a similar bifurcation.

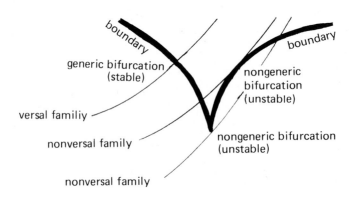

Fig. 2

This is not the case for the two other families in fig. 2. Indeed, one of those curves is tangent to the hypersurface of degenerate systems. Such an intersection point may disappear or may be transformed into two intersection points under a small deformation of the curve. Hence the bifurcation, corresponding to the tangency point is unstable.

The last curve contains a singular point of the hypersurface of degenerate systems. This event is also nontypical, it disappears under generic small deformations of the curve. Such a bifurcation is impossible in singular one-parameter versal families.

The versal families were used by Poincaré (see Lemma 4 in his Thesis [1], 1879). But the systematic study of the bifurcations occuring in versal families has become available only recently, when the mathematical difficulties of this approach have been overcome with the help of new mathematical technique. In this survey I shall try to describe some of the concrete results, obtained in the realization of the general programme, described above, applied to mechanical systems (both with a finite number of degrees of freedom and with an infinite one, describing continuous media). Naturally, neither the mathematical difficulties nor the technique used to overcome them can be described here: I shall only describe the results themselves.

1.   Stability Boundary in Linear Evolutionary Systems

Let us consider, following L.V. Levantovskii [2], the *stability boundary* in the space of parameters of a k-parameter family of linear systems. This boundary has, for a typical family, only a finite (but growing with k) number of different singularities (we call two singularities different, if one cannot be transformed to the other by a smooth change of variables). The same finite list describes the typical singularities of the graphs of the increment as a function of the parameters in typical families depending on k-1 parameters.

For k = 3, the list contains 4 singularities (see [3]), represented in fig. 3. The singularity, corresponding to a pair of 2 x 2 Jordan cells is described for the formula $y^2 = z\, x^2$, defining a surface, called the Witney-Cayley *umbrella* (this excentric umbrella contains also a handle, corresponding to z < 0). The stability boundary is one half ot the umbrella surface. This is one of the standard patterns one encounters in all the branches of modern bifurcation theory.

It is interesting to note that the stability domain is always wedged inside the instability domain, displaying the general principle of fragility of all good things, according to which the good things (such as stability) are more vulnerable than the bad ones. Indeed, a small shift of the parameter value corresponding to a singular point of the stability boundary makes the system unstable with a higher probability than stable.

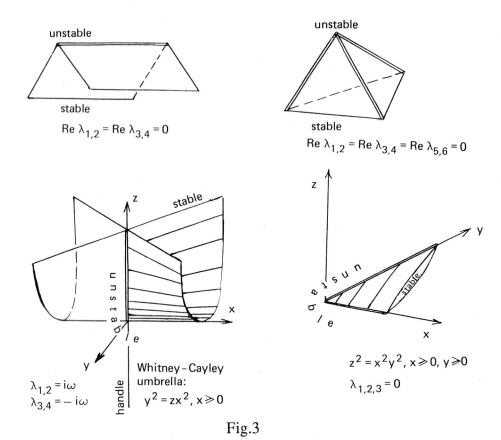

Fig.3

The results described above can be applied to the systems with a finite number of degree of freedom as well as to those describing continuous media (in nonsymmetric domains).

Before leaving the discussion of the loss of stability theory, I shall mention one important phenomenon, discovered recently in the one-parameter *dynamical bifurcation* theory - the *delay* in the loss of stability manifested by the systems depending on a slowly changing parameter.

Everybody knows the typical scenario of the mild loss of stability in one-parameter families of evolutionary systems (fig. 4). The amplitude of the selfoscillations generated by such a bifurcation grows like the square root of the distance of the parameter value from the critical one (this bifurcation has been studied by Andronov [4] in 1931 and by E. Hopf in 1942, and hence is usually called the Hopf bifurcation).

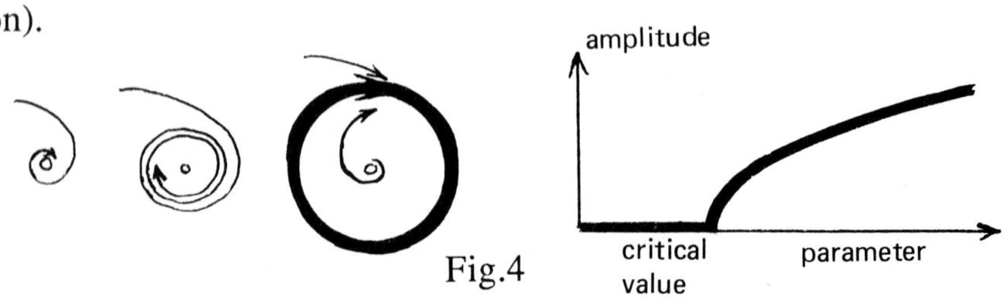

Fig.4

It happens that when the parameter really (while slowly) changes with time, a typical system's behaviour is very different from that described by the quasistationary theory, which assumes that everything goes as in the case where the parameter and the time are independent variables.

Namely, for a long period of time after the moment when the parameter passes the critical value, the system does not leave the neighbourhood of the equilibrium point, which has become unstable; in fact, so long, that the slowly varying parameter increment becomes finite. Only then the system leaves the heighbourhood of the equilibrium point. It leaves it with a jump, going to the periodic attractor born at the critical moment (fig. 5). This attractor has grown up to a finite amplitude selfoscillation regime.Thus the transition from the stationary attractor to the periodic one looks like a sudden jump at a parameter value very different from the theoretical critical value of the loss of stability.

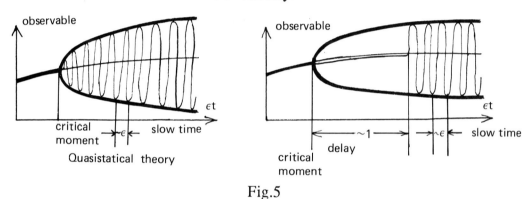

Fig.5

It is interesting that this happens only in *analytic* systems. In finitely smooth and even in infinitely smooth nonanalytic systems the delay in the stability loss, measured by the difference between the critical value of the parameter and its value at the jump moment, tends to zero with the velocity of the variation of the parameter.

The delay phenomenon has been first described by Shishkova [5] in a model example (1973). The general theory, due to A.I. Neistadt [6], [7], [8], appeared only in 1985.

2.   Wavefronts and their Perestroikas

The phenomena I shall discuss are very general and common to all problems of wave propagation in nonhomogeneous ans nonisotropic media. But to indicate the nature of the results I shall decribe them in the simplest case of waves on the Euclidean plane, propagating with velocity 1.

Suppose that the initial disturbance is bounded by some smooth curve (the initial *wavefront*). After time t the wavefront will propagate at a distance t along every ray orthogonal to the initial wavefront (fig. 6). For small t, this new wavefront is smooth, but for larger t, it may acquire singularities.

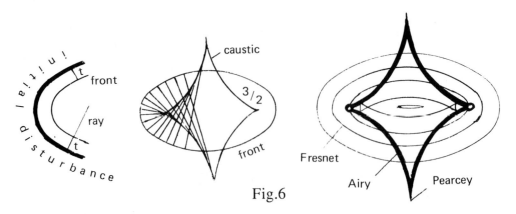

Fig.6

The bifurcation, leading to the formation of such singularities, has been studies by Huygens [9] in 1654. Let us consider, for instance, the propagation of the disturbances inside an ellipse. The family of rays (of normals to the ellipse) has an envelope (the *caustic*) with 4 cusps. When a moving front comes to the cusp of the caustic it experiences a *perestroika*. Two cusps are formed at the moving wavefront. These cusp points move along the caustic (the actual boundary of the domain of disturbances is only a part of the moving wavefront; this part countains no cusps). The perestroika of the front at the cusp point of the caustic is stable: for any curve close to the initial ellipse a similar perestroika occurs.

The above description (at the level of geometrical optics) of the generation of singularities at the moving wavefront has a "physical optics" counterpart. In this theory, to each kind of front singularities there corresponds a type of special functions, defined by oscillating integral. For a generic point of the front these are the Fresnel integrals, for the cusp points of the front (that is, for the generic points of the caustic) - the Airy function, for the cusp points of the caustic (that is, for the point of perestroikas of the wavefronts) - the Pearcy integral.

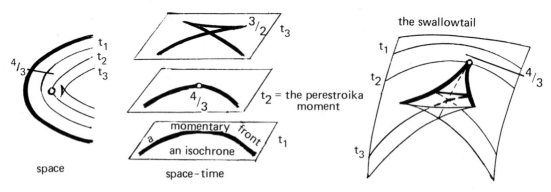

Fig. 7

The momentary wavefronts form a surface in space-time (fig. 7). At the perestroika moment this surface has a singular point. Such a singular point has been studied by Kroneker [10] in 1878 and is called the *swallowtail*. The swallowtail surface is the set of polynomials $x^4 + ax^2 + bx + c$, having a multiple roots. The isochrone surfaces t = const intersect this bifurcation diagram along the momentary fronts. These momentary front can be transformed to the plane section of the swallowtail surface, a = const by a diffeomorphism (a smooth change of variables in the space-time).

Singularities of a generic wavefront in the physical 3-space cannot be more complicated than swallowtails. A moving wavefront experiences the perestroikas at some discrete time moments. The list of typical perestroikas of wavefronts is presented in fig. 8. These perestroikas are classified by the Weyl groups or by the simple Lie algebras, also presented in fig. 8. I refer to [11], [12] for the description of the relation of wavefront perestroikas to the Weyl and Lie algebras - it is too technical to be discussed here.

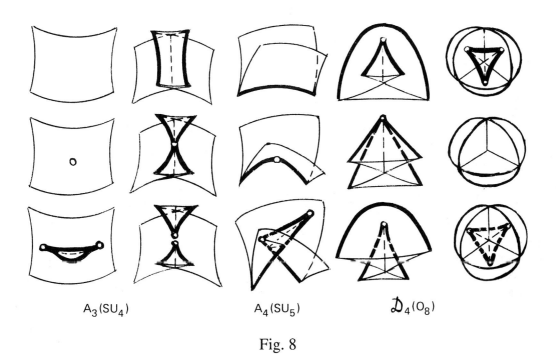

$A_3 (SU_4)$ $A_4 (SU_5)$ $D_4 (O_8)$

Fig. 8

3. Singularities and Bifurcations of Caustics

The typical singularities of caustics in the 3-space - the *swallowtail*, the *pyramide* and the *purse* - are depicted in fig. 9.

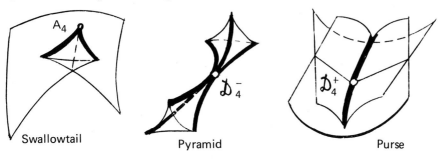

Swallowtail   Pyramid   Purse

Fig. 9

A caustic may be described in terms of classical mechanics, as the set of critical values of the projections of a Lagrangean submanifold from the phase space to the configuration space, (p, q) ↦ q (see fig. 10). A *Lagrangean submanifold* is a submanifold of the phase space, of dimension one half the dimension of the phase space, along which the integral ∫ pdq does not depend (locally) on the integration path. The graph of a potential velocity field p = ∂S/∂q is a Lagrangean submanifold. A general Lagrangean submanifold may be viewed as a graph of the multivalued potential velocity field. The Kolmogorov tori of the integrable or almost integrable Hamiltonian systems are also Lagrangean submanifolds.

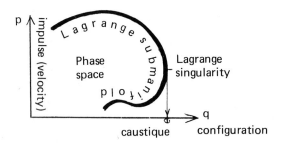

Fig. 10

The theory of singularities and of bifurcation of caustics is the base of the theory of the large scale structure of the Universe, due to Ya. B. Zeldovich [13]. The evolution of a Lagrangean submanifold under any Hamiltonian phase flow leaves it a Lagrangean submanifold (because of the conservation of the Poincaré integral invariant). But even in the case, when the initial Lagrangean submanifold was a graph of a potential velocity field and hence had no caustic (fig. 11), the time running, the slowest particles are overrun by the fastest ones and the caustic appears. The density of the particles has a singularity of order $1/\sqrt{r}$ at distance r from the caustic.

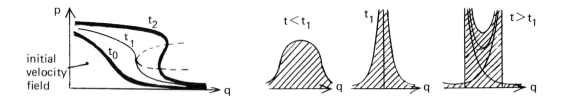

Fig. 11

After the formation of the caustics most of the particles are concentrated in the neighbourhood of the surfaces of the caustics. The density is even larger at the singular lines of the caustics, and maximal at the most singular points where these lines intersect each other.

Thus a complicated large scale *cellular structure* appears, with large *voids* inside the cells and a net of surfaces and lines of concentration of matter. This unusual structure is consistent with the astronomical observations.

The newborn caustics have the form of small plates (or *pancakes* in Zeldovich's words). The birth of a pancake is one of the typical bifurcations of the caustics. The complete list of all the typical bifurcations of caustics in the 3-space, depending on one parameter, is given in fig. 12 (borrowed from [14]).

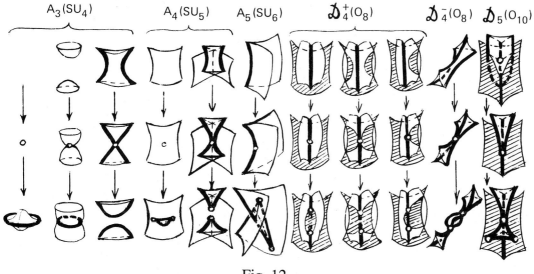

Fig. 12

4. Bifurcations of Shock Waves

The theory described below is very general, but I shall restrict myself here to the simplest (while quite relevant) example of the Burgers equation with vanishing viscosity $u_t + uu_x = \varepsilon\Delta u$

and of its potential solutions $u = \partial s/\partial x$.

Everybody knows that for $\varepsilon \to 0$ the solution acquires singularities, namely the *shock waves*. The *Florin formula* [15]*, reducing the Burgers equation to the heat equation (and usually called the Coula-Hopf transformation) implies the description of the shock wave in the space-time as of the set of nonsmoothness of the function

$$F(t, x) = \min_y f(y, t, x), \quad f = \frac{(x-y)^2}{2t} - u_0(y).$$

---

* This formula is present at page 101 of the volume 1 of the classical Forsyth treatise of differential equations (I am greatfull for Prof. Woods who have communicated to me this information at the Congress, it seems that the discovery of this reference is due to Prof. Drazin).

Geometrically (fig. 13) this formula means that the graph of f(y) is a parabola, whose axis has the abscissa x and whose width grows with t. One pushes the parabola down till it touches the graph of the initial function u_o (q). The ordinate f of the touching parabola vertex depends smoothly on x and t, if the parabola touches the graph at exactly one point.

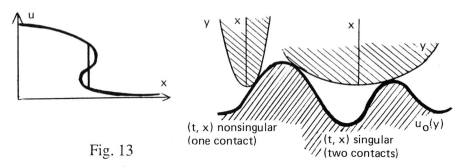

Fig. 13

If there are more points of tangency, the function F is, generically, nonsmooth. Such values of t and x form the *shock wave hypersurface* in the space-time.

The typical singularities of the minima of generic families of smooth functions, depending on the parameters, have been classified in the theory of singularities, in the case where the number of parameters does not exceed 5. In the case of 3 parameters (which corresponds to the space-time of dimension 3 = 2 + 1, that is to the shock waves propagating in 2-plane) there exist only 5 typical singularities of the minima function. The sets of discontinuity of the derivatives of these minima functions are represented in fig. 14 (the same picture describes the typical singularities of a momentary shock wave in the 3-space).

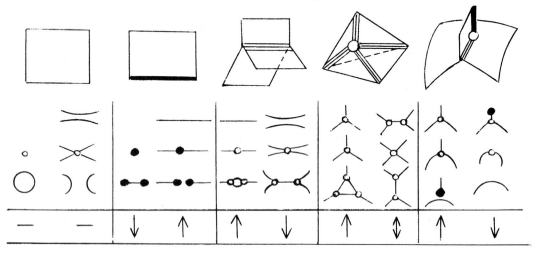

Fig. 14

To investigate the bifurcations of the singularities of the shock waves in the plane, occurring at some discrete time moments, one should intersect the surface of fig.14 by the *isochrones* t = const, where t is a generic smooth function having no critical points.

Doing so, we obtain a list containing 10 perestroikas of plane curves. This list is also presented in fig. 14. These perestroikas (as well as their higher-dimensional analogues) were studied by I.A. Bogaevskii in 1983. In 1984 Ourbatov and Saichev [16] have tried to observe these perestroikas experimentally. They discovered, that only some of these perestroikas are observable.

In some cases such as the vanishing of a triangle formed by the shock waves, the perestroika is possible only in one direction: the triangle may not reappear. In other cases both directions are admissible, in yet other case - none. In fig. 14 the *admissible perestroika directions* are indicated by the arrows.

The admissible directions of the shock wave pererstroikas depend on the convexity properties of the Hamilton function defining the corresponding Hamilton-Jacobi equation, restricted to the space of impulses (for the case of the Burgers equation this Hamilton function is $p^2 / 2$). Namely, the necessary and sufficient conditions for a typical perestroika of singularities of a minima function to be a perestroika of (convex Hamiltonian) shock waves are:

1. The local shock wave, born after the perestroika, is contractible in a neighbourhood of the perestroika point (I.A. Bogaevskii).

2. The homotopy type of the complement of the shock wave immediately after the perestroika moment is the same as at this moment (Ju. Baryshnikov).

Every one of these: two conditions is necessary and sufficient for the admissibility of a perestroika of shock waves in a plane or in a 3-space. But it is unknown whether the same is true in higher-dimensional spaces - it is even unknown whether conditions 1 and 2 are equivalent for higher-dimensional shock waves.

5.   Relaxatory Oscillations and Implicit Differential Equations

Consider a system having one *fast* (vertical) and two *slow* variables.

$$\left\{ \begin{array}{l} \dot{x} = v\,(x, y) \\ \dot{y} = \varepsilon w\,(x, y) \qquad y = (y_1, y_2), \quad \varepsilon \to 0 \end{array} \right\}$$

Equation $v(x, y) = 0$ defines a surface M in the phase 3-space, called the *slow surface* (fig. 15).

First the phase curve comes fast to the slow surface along a vertical direction ("the fast variable relaxes"). Then the phase point moves slowly along the slow surface.

This motion is defined by a plane field generated by the vertical direction of the fast motion v and the direction of the perturbation w. This plane field defines, generically, a contact structure in the 3-space of the fast and slow variables. Hence the study of the slow motion is mathematically equivalent to the investigation of an *implicit differential equation*

$$F(x, y, p) = 0, \qquad p = dy / dx.$$

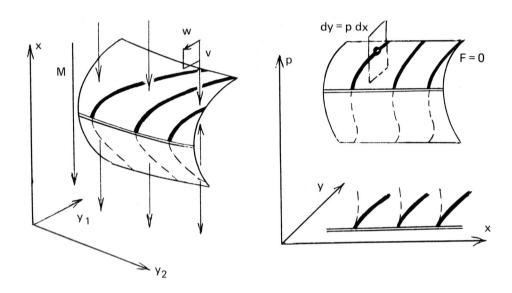

Fig. 15

The latter problem was one of the 4 problems proposed for the prize of the Swedish king Oscar II in 1885 (one of the other problems on the list was the 3-body problem).

The standard *contact structure* (a completely nonintegrable field of the planes) is given in the case by the equation dy = pdx, defining a plane at each point of the (x, y, p) space. The traces of these planes at the surface F(x, y, p) = 0 define a direction field on the surface. Its integral curves (which are similar to the orbits of the slow motion along the slow surface) should be studied at the neighbourhoods of the points, where one cannot solve the equation with respect to the derivative (that is, of the points where the tangent plane to the surface F = 0 contains the vertical direction of the p-axis, or, in terms of relaxatory oscillations, of the jump points).

This problem has been solved for the generic jump points by M. Chibrario [17] in 1932. The equation is reducible to the normal form $p^2 = x$ (fig. 15).

The theory of singular points was settled only in 1984. The *normal form*

$$(p + kx)^2 = y$$

to which the equation can be reduced by an infinitely smooth (analytic) change of variables (A.A. Davydov [18]) describes, according to the value of the parameter k, one of the 3 singularities of fig. 16 - *the folded saddle*, the *folded node* or the *folded focus*. The topological study of these pictures goes back to the work [19] of Phakadze and Shestakov (1959). Their results were rediscovered by R. Thom, L. Dara, F. Takens in 1972-1976. However even the topological equivalence of the implicit equation to its normal form was considered by these authors as a conjecture.

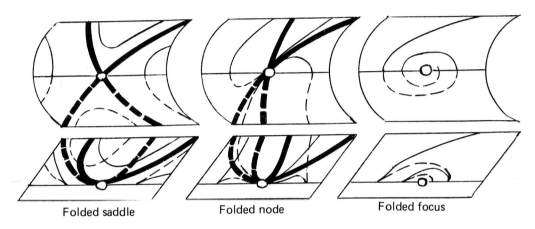

| Folded saddle | Folded node | Folded focus |

Fig. 16

6.   Interior Scattering of Waves in Nonhomogeneous Media

The waves of different kinds governed by a linear equation - say, the longitudinal and the transverse waves - usually propagate independently. I shall describe here an unusual bifurcation - a kind of transformation of the waves of one type into the waves of the other type at some interior points of a linear nonhomogeneous medium[*].

Consider a hyperbolic system described by a variational principle $\delta \int L \, dt = 0$, where

$$L = T - U \qquad T = \frac{1}{2} \int \left(\partial u \, / \partial t \right)^2 dx, \quad U = \frac{1}{2} \int a_{ijkl} \frac{\partial u_i}{\partial u_j} \frac{\partial u_k}{\partial u_l} \, dx .$$

The Euler-Lagrange equation has the form

$$\square \, \mathbf{u} = 0, \qquad\qquad \square = \partial^2 / \partial t^2 + A$$

---

[*] This transformation seems to be similar to the "wave conversion" in plasma physics. I am greatfull to the Prof. Dysthe for this remark.

The principal part A_2 of A, for fixed values of t and x, is an operator with constant coefficients, related to its *principal symbol* by the formula $A_2 = \sigma$ (D/i). Here the principal symbol is defined by the action of A_2 on the plane waves,

$$A_2\, e^{i(k,x)}\, (\cdots)\, \mathbf{w} = e^{i\,(k,x)}\, (\cdots)\, \sigma(k)\, \mathbf{w}.$$

In this formula $\mathbf{w}$ is an n-vector, $\sigma(k)$ - a symmetric m x m matrix, whose elements are homogeneous polynomials in the components of the wave vector k.

The equation of the *light hypersurface* Γ (in other terms, the *dispersion relation*) has the form

$$\det (\sigma(k) - c^2\, I) = 0,$$

where I is the identity matrix and where σ depends on t and on x as on the parameters.

The light hypersurface defines the *rays* and the *wavefronts* as the characteristics and the Legendre submanifolds of the hypersurface Γ^{2D} in the space $K^{2D+1} = PT^*\, R^{D+1}$ of contact elements of the space-time R^{D+1}, endowed with its natural contact stucture (this is the *Huygens' principle*).

If follows from the general singularity theory that the light hypersurface Γ^{2D} of a generic hyperbolic variational system has singular points, which form a set of codimension two in Γ^{2D}. In a neighbourhood of a typical singular point Γ^{2D} is diffeomorphic to the quadratic cone $u^2 + v^2 = w^2$. There exist local *Darboux coordinates* in the contact space K^{2D+1} (that is, coordinates in which the contact structure acquires the Darboux normal form $\alpha = 0$, where

$$\alpha = dz + (pdq - qdp)\,/\,2, \quad p = (p_1, \ldots, p_D), \quad q = (q_1, \ldots, q_D)$$

such that the equation of a generic hypersurface with conical singularity is formally reduced to the normal form H = 0, where

$$H = p^2 \pm q^2 - z + cz^3 \quad (D = 1), \qquad\qquad H = p_1^2 \pm q_1^2 - q_2^2 \qquad (D > 1)$$

For D = 1 (that is for the case of 1 space variable) the sign $\pm$ is - for Γ describing a hyperbolic system. The corresponding characteristics of Γ^2 are represented in fig. 17 in K^3 where their projections to the space-time are also indicated.

The conclusion of all this machinery is that some particular rays, propagating in a generic 1-dimensional medium, at some particular moment in time display a bifur-

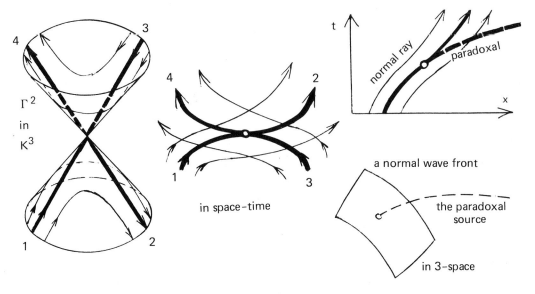

Fig. 17

cation, giving birth to a *paradoxical, transformed wave*. In fig. 17 the curves 13 and 24 are two smooth curves which are quadratically tangent at the origin. The characteristic curves, however, are the curves 12 and 34. Hence at the bifurcation moment both characteristics in the space-time are tangent, and they loose their second derivatives. One may obtain more details on the scattering of the neighbouring rays from the normal forms given above.

A typical wavefront in 3 dimensions (D = 3) acquires some isolated singular points, where it is adjacent to the transformed rays.

The normal from of a surface with a conical singularity in the contact 3-space provides also a description of one of the generic bifurcations in the theory of relaxation oscillations (or of implicit differential equations): it corresponds to the simplect(Morse) degeneration of the slow motion surface (or of the surface $F(x, y, p) = 0$).

7. Bifurcations of Periodic Solutions in Reversible Systems

The simplest example comes from the *paradoxical bifurcations* of the stationary waves defined by the equation

$$U_t = U_{xxx} + 2 A \ U_{xx} + U + U^2_x \tag{1}$$

The numerical experiments (Demechin, Shkadov [20]) show that the corresponding stationary equation ($u_t = 0$) has a family of periodic solutions (of *stationary waves*) of a fixed period 2π in x, descending on the parameter A and bifurcating at A = 1.

This is paradoxical because the Poincaré bifurcation theory predicts periodic solutions of a fixed period in a generic system only for some isolated values of the parameter.

An asymptotic formula for these stationary waves was obtained for $A = 1 - \varepsilon, \varepsilon \rightarrow 0$, by Malomed and Tribelski [21].

The explanation of this paradoxical bifurcation depends on the special symmetry of the system: the system is reversed by the reversal of the x-axis.

A vector field in a phase 4-space is called *reversible*, if it is reversed by an involution of the 4-space that change the signs of two of the coordinates. In our case the involution σ transforms the point (u, u_x, u_{xx}, u_{xxx}) of the phase 4-space into the point ($u, -u_x, u_{xx}, -u_{xxx}$). It transforms the vector field V, corresponding to (1), to the field -V in the 4-space.

The plane Fix σ of fixed points of the involution σ has dimension 2. The phase curves, starting at Fix σ, form a 3-manifold. Its next intersection with Fix σ is, generically, a curve, since 2 + 3 - 4 = 1. The reversibility implies that a phase curve γ, starting at a point of the intersection curve Γ, is closed. Hence a generic reversible vector field has non-isolated closed orbits, forming continuous families (fig.18), in contrast to the generic vector fields, which have only isolated closed orbits.

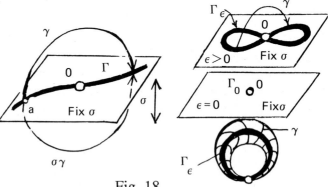

Fig. 18

Let us vary the parameter, say, $A = 1 - \varepsilon$ in equation (1). The curve Γ_ε, formed by the traces of the periodic orbits at Fix σ, bifurcates (for $\varepsilon = 0$ in the case of equation (1)). The typical bifurcations in the one-parameter families of the reversible systems are described by the Whitney umbrella we have discussed in § 1.

Place each curve Γ_ε with its horizontal plane Fix σ at height ε. We obtain a surface in 3-space. This surface has a singularity at the origin (fig. 19). The singularity is a

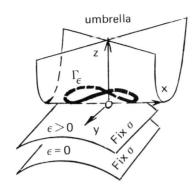

Fig. 19

Whitney umbrella: the surface is diffeomorphic to that given by the equation $y^2 = zx^2$, and the diffeomorphism reduces the plane $\varepsilon = $ const to the form $z = \varepsilon \mp x^2$ (see M.B. Sevryuk [22]).

For instance, consider a reversible vector field having an equilibrium point with two pairs of imaginary roots of characteristic numbers. Suppose that the field depends on a parameter and that the two pairs collide for some parameter value.

Before the collision there exist two Lyapunov surfaces, formed by the closed phase curves and tangent to the invariant planes of the linearized equation at the equilibrium point. The normal form given above (for the "-" case) describes what happens at the collision moment. Namely, the figure 8 curve in fig. 19 corresponds to an invariant surface in the phase space, covered by closed orbits. This surface is homeomorphic to a 2-sphere, with the poles glued together and decomposed into parallels.

The polar regions of the sphere are the Lyapunov surfaces. Hence the above normal form implies that both Lyapunov surfaces in a neighbourhood of the collision moment continue each other, collapsing for the critical value of the parameter to the equilibrium point. The equation (1) corresponds to the "-" sign in the normal form. Thus the surface formed by the stationary waves of period 2π of the equation (1) collapses to the origin for $A = 1$.

The case of the sign "+", as well as the bifurcations at other resonances (where the ratio of the eigenfrequencies in an arbitrary integer), are also studied in [22].

8. Bifurcations in Symmetric Systems

The reversible systems form a very special case of symmetric systems. The bifurcations in generic symmetric systems represent a very interesting subject, which is very vast but almost unexplored (there exist so many different symmetry group actions, giving rise to many works at different levels on the classification of singularities and bifurcations that the theory becomes rather larger than deeper - see, for instance, the book [23] by Golubitsky and Schaeffer).

I shall only mention the classical (and deep) theory of the bifurcation leading to the *hexagonal convection patterns* [24], [25] and the (hard) theory of the loss of stability of the selfoscillations at resonances, which leads to the bifurcation theory of plane vector fields that are transformed into themselves under rotation through an angle $2\pi / n$ [26].

In the latter theory a mathematical challenge is still alive: a 50-years old problem of the *loss of stability at a resonance of order* 4 is still unsettled. The problem is to study the perestroikas of the phase portraits of the equations

$$\dot{z} = \varepsilon z + A z \left| z \right|^2 + \bar{z}^3 ,$$

where ε is a complex small parameter, and where A is a generic complex number.

One conjectures that, depending on the value of A, there exist a *finite* number of perestroika patterns (namely, that the A-plane is divided into 48 domains, indicated in fig. 20, such that in each domain the perestroika pattern is topologically independent of A).

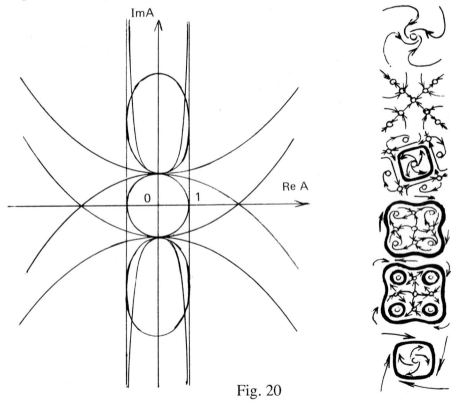

Fig. 20

While both the numerical works and the asymptotical developments ([27], [28]) agree with this conjecture, even the finiteness is still unproved.

It is interesting to note that the *"strong"* resonance cases (n ≤ 4) are the most difficult ones in this theory while the higher order "weak" resonance theory is much easier. Rational numbers with denominators greater than 4 behave in this theory rather similarly to the irrational numbers, while the case n = 4 is the only one exceeding the modern mathematical technical possibilities.

One may consider as a particular case of the theory of bifurcations in symmetrical systems the classical problem of the bifurcations in the so-called ecological *generalized Lotka-Volterra systems* (fig. 21).

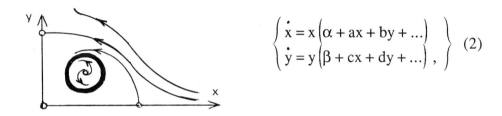

$$\left\{ \begin{aligned} \dot{x} &= x\left(\alpha + ax + by + ...\right) \\ \dot{y} &= y\left(\beta + cx + dy + ...\right), \end{aligned} \right\} \quad (2)$$

Fig. 21

depending on two parameters α, β (to understand the relation to the symmetrical systems, put $x = X^2$, $y = Y^2$).

The difficult majoration of the number of limit cycles for such systems is obtained in a recent paper by H. Žoladek [29]. System (2) is also important for the general theory of equilibria stability loss, where it appears as the *normal form for the amplitude variations* for a generic two-frequency system under the codimension 2 bifurcation occurring at the simultaneous transition of two pairs of eigenvalues to the right complex halfplane.

The codimension two bifurcation corresponding to a zero eigenvalue and a pair of eigenvalues on the imaginary axis has also been studied by Žoladek [30]. Hence the codimension two cases in the theory of loss of stability of equilibria in generic systems are now solved (at the level of the averaged or amplitude equations).

9.   Doubling Cascades and Feigenbaum-Coullet-Tresser Universality

The *doubling* of the period of a cycle is one of the typical bifurcations occurring in generic one - parameter families. Its simplest example is provided by the ecological model, described by the line mapping $x \mapsto A\, xe^{-x}$. When the parameter A grows, this mapping first has a stable fixed point 0, (fig. 22) and later a stable positive fixed point x > 0 (a stable population). Then this point looses its stability, forming a period

2 attracting cycle, which in its turn looses its stability, transferring it to a doubled attracting cycle (of period 4), and so on (this scenario has been described in a 1974 paper of A.P. Shapiro [31] in connection with the dynamics of the populations of some fishes at the Far East of the USSR).

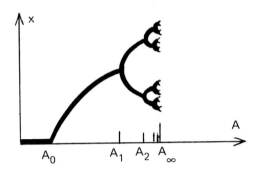

Fig. 22

After the paper by May [32], M. Feigenbaum [33] and Coullet with Tresser have discovered the *universality phenomenon* of the whole sequence of doublings. The infinite sequences of doublings are encountered in generic one-parameter families of mappings, and the subsequent doubling parameter values A_n form asymptotically a geometrical progression, whose denominator is a universal constant, independent of any system

$$\frac{A_n - A_{n-1}}{A_{n-1} - A_n} \to 4.6692 \ldots$$

The computer-assisted proof of this "renormgroup" theorem doubling corresponds to the stability loss of a limit cycle, is published by Landford (see also [34]).

In mechanics and in the theory of differential equations a doubling corresponds to the stability loss of a limit cycle forming the axis of a Möbius band, while the newborn attractor is the border of the band (fig. 23).

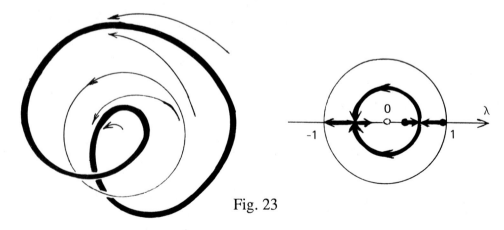

Fig. 23

At the moment of stability loss one of the multipliers (that is of the eigenvalues of the Poincaré mapping of the cycle) goes outside the unit disc through the point -1. Hence the newborn stable cycle of an approximately double period has a multiplier close to 1. Before the next doubling takes place it must travel from 1 to -1. But it is impossible to travel from 1 to -1 along the real axis because the path is obstructed by the origin which cannot be a multiplier for any parameter value. Hence the multiplier has to leave the real axis. To do this, it has to collide with another real multiplier. Then the two multipliers follow two conjugate paths contourning the origin from both sides, meet one another at the negative part of the real axis and only then one of the multipliers goes to the -1 point (fig. 23, on the right).

This whole complicated processus is also universal: the trajectories of the multipliers in the complex plane become more and more circular with each doubling and their diameters, as well as the collision moments the real axis, have a universal asymptotic, described by M.V. Jacobson (see [35]).

This behaviour of the multipliers implies the following picture of the doublings on the surface of section of Poincaré (see a - e, fig. 24).

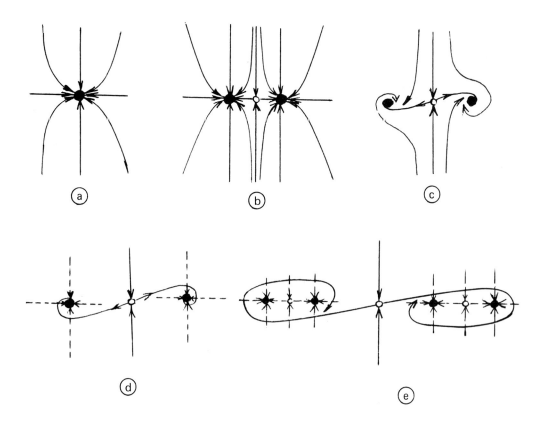

Fig. 24

At the first doubling the attracting node - point of the Poincaré mapping becomes a saddle-point and the stability is transferred to the pair of fixed points of the square of the mapping, lying at the outcoming separatrice of the saddle fixed point (b). The node points become foci (c), and later, still remaining on the outgoing separatrices (which come to the foci in a spiral) the foci become the nodal fixed points of the square of the Poincaré mapping, reversing the orientation of one of the eigenlines (d). The next doubling (e) repeats the first one (b), transforming each node to a saddle, and so on. Thus, the picture of doublings in the real theory of differential equations is very different from that of the model problem of the mappings of a line into itself.

The stability loss at a transition of a multiplier through a root of 1 of degree higher than 2 is a codimension 2 bifurcation, and it becomes generic only in two-parameter families. In such families one also encounters the asymtotically geometric progressions of bifurcations, similar to the doubling cascades, say, the *tripling cascades*. But the universal denominators of the corresponding progressions are complex numbers. Hence, for instance, a tripling cascade corresponds to a sequence of bifurcation points, lying asymtotically on a logarithmic spiral (transformed by an affine transformation).

The doubling cascades are generic also in hamiltonian systems. Here the universal denominator is greater than 8. In hamiltonian systems the multipliers equal to all the roots of unity are encountered in generic one parameter families (not in the 2-parameter ones, as for generic, nonhamiltonian systems). Indeed, the multiplier 1 in a hamiltonian system has multiplicity 2 and travels to the point -1 along the unit circle and not along the real axis.

This leads to many bifurcation sequences, each one corresponding to a particular universal denominator.

For instance, between any two consecutive doublings the cycle generates a cycle of a triple (quintuple …) period. Hence, the sequences (2, 2, 2, …) (3, 6, 12, …), (5, 10, 20, …) … all have the same universal denominator. But one may leave the initial cycle at the moment of a tripling and then one may follow the new cycle till its doubling, then leave it at the next tripling and so on - one obtains in this way the sequence (2, 3, 2, 3…) to which there should correspond a new universal denominator. Other periodical sequences, such (2, 2, 3, 2, 2, 3,…) or (3, 5, 3, 5, …), generate other denominators, and one may consider even some nonperiodical sequences.

These phenomena at present are not well understood even at the level of the numerical experiments (as far as I know), while, it seems, they explain the strange form of the islands that one can see in the very first computer experiments by Henon and Heils [36] and by Chirikov in their studies of the stochastization of the area-preserving mappings.

I hope these examples show how in the modern fast developing bifurcation theory the most abstract constructions of modern theoretical mathematics directly imply mechanical phenomena observable both in the numerical and experimental works while the questions, arising in the applied domains of mechanics, lead to deep mathematical problems and to unexpected relations between apparently unrelated domains of mathematics and physics.

REFERENCES

[1] Poincaré, H., Sur les propriétés des fonctions définies par des équations aux différences partielles. Thèse, Paris : Gautier-Villars (1879) 93.

[2] Levantovskii, L.V., Singularities of Stability Boundary. Funkt. Anal and Appl. 16 NI (1982) 44-48.

[3] Arnold, V.I., Lectures on Bifurcations in Versal Families. Russian Math. Surveys 27: 5 (1972) 119-184.

[4] Andronov, A.A., Mathematical problems of the selfoscillations theory. Report at the 1931 all-Union conference on oscillations, first published in 1933: I Vsesojusnaia Konferenzia po kolebanijam. M.L. GTTI 1933, pp. 32-72. Reprinted in: Andronov, A.A. Collected Works. Ac. Sc. USSR, M., (1956) 85-124.

[5] Shishkova, M.A., Study of One System of Differential Equations with Small Parameter at Higher Derivative. Doklady Ac. Sc. USSR 209: 3 (1973) 576-579.

[6] Neistadt, A.I., Asymptotical Study of Stability Loss of Equilibrium under Slow Transition of Two Eigenvalues through Imaginary Axe. Uspehi Math. Nayk 40: 5 (1985) 300-301.

[7] Neistadt, A.I., On Delayed Stability Loss under Dynamical Bifurcations I. Differential Equations 23: 12 (1987) 2060-2067.

[8] Neistadt, A.I., On Delayed Stability Loss under Dynamical Bifurcations II. Differential Equations 24: 2 (1988) 226-233.

[9] Huygens, C., Horlogium Oscillatorium. Paris (1673).

[10] Kroneker, L., Über Sturm'sche Funktionen Monatsber. Dtsh. Acad. Wiss. Berlin 14 Fevrier 1878, 95-121.

[11] Arnold, V.I., Wave Fronts Evolution and Equivariant Morse Lemma. Comm. Pure and Appl. Math. 29: 6 (1976) 557-582.

[12] Arnold, V.I., Varchenko, A.N., Gusein-Zade, S.M., Singularities of differentiable maps. vol. I M.: Nauka (1982) 304, vol. II M.: Nauka (1984) 336. (English translation by Birkhäuser, vol. I (1985), vol. II (1988)).

[13] Zeldovich, Ja. B., Decomposition of Uniform Matter into Pieces due to Gravitation. Astrophysics 6: 2 (1970), 319-335.

[14] Arnold, V.I., Perestroikas of Singularities of Potential Flows in Collisionless Media and Metamorphoses of Caustics in 3-Space. Trudy Sem. Im. I.G. Petrovski 8 (1982) 21-58 (Engl. Transl.: Selecta Mat.).

[15] Florin, V.A., Some Simplest Nonlinear Problems of Consolidation of Watersaturated Soils. Isvestia Ac. Sci. USSR, Otd. Techn. Nauk, (1948) N 9 1389-1397.

[16] Gurbatov, S.N., Saichev, A.I., Shandarin, S.F., Large Scale Structures of Universe in Model Nonlinear Diffusion Equation. Preprint 152 IPM im. M.V. Keldysh, M. (1984).

[17] Cibrazio, M., Sulla reduzione a forma canonica delle equationi lineari alle derivata parziali di secondo ordine di tipo misto. Accademia di science e lettere. Instituto Lombardo Rendiconti, 65 (1932) 889-906.

[18] Davydov, A.A., Normal form of Differential Equation Unresolved in Derivative Funktion. Analysis and Appl. 19: 2 (1985) 1-10.

[19] Phadakze, A.V., Shestakov, A.A., On Classification of Singular Points of First Order Differential Equations Unresolved in Derivative. Mathem. Sbornik 49: 1 (1959) 3-12.

[20] Demechin, E.A., Branching of Solutions of Stationary Running Waves Problem in Viscous Fluid Layer at Inclined Plane. Isvestia Acad. Sc. USSR MGH (1983) N 5, 36-44.

[21] Malomed, B.A., Tribelski, M.I., Bifurcations in Distributed Kinetic Systems with Aperiodic Instability. Physica D, 14 (1984) 67-87.

[22] Sevryuk, M.B., Reversible Systems. Lecture Notes in Math. Springer-Verlag 1211 (1986) 320.

[23] Golubitsky, M., Schaeffer, D.G., Singularities and Groups in Bifurcation Theory. Springer-Verlag (1985) 464.

[24] Busse, F.H., Transition to Turbulence in Rayleigh-Benard Convection. In: Hydrodynamical Instabilities and Transition to Turbulence. M. MIR (1984) 124-168.

[25] Malomed, B.A., Tribelski, M.I., On Stability of Stationary Periodic Structures at Weakly supercritical Convection and Related Problems, JETP 92: 2 (1987) 539-548.

[26] Arnold, V.I., Geometrical Methods of the Theory of Ordinary Differential Equations Springer (1983) (Engl. Transl. of: Dopolnitelnye Glavy Theorü Obyknovennych Differentialnych Uravnenii, Moscow, Nauka (1978) 304.

[27] Beres ov skaia, F.S., Hibnik, A.I., On Separatrices Bifurcations in Selfoscillations Stability Loss Problem of Resonance I/ 4. Appl. Math. and Mech. 44: 5 (1980) 938-943.

[28] Neistadt, A.I., Phase Portraits Bifurcations for some Equations of Stability Loss Problem at Resonance I 4. Appl. Math. and Mech. 42: 4 (1978) 830-840.

[29] Žoladek, H., Bifurcations of a Certain Family of Planar Vector Fields Tangent to Axes. Jour. of Different. Equs. 67: 1 (1987) 1-55.

[30] Žoladek, H., Versality of a Certain Family of Symmetric Planar Vector Fields. Mathem. Sbornik 120 (162) (1983) 473-499.

[31] Shapiro, A.P., Mathematical Models of Concurrence. In the book: Control and Information. Vladivostok, DVNZ Acad. Sci. USSR, 10 (1974) 5-75.

[32] May, R.M., Biological Populations Obeying Difference Equations: Stable Points, Stable Cycles and Chaos. J. Theor. Biol. 51 (1975) 511-524.

[33] Feigenbaum, M., Quantitative Universality for a Class of Nonlinear Transformations. J. Stat. Phys. 19: 1 (1978) 25-32.

[34] Collet, P., Eckman, J.P., Iterated Maps of the Interval as Dynamical Systems. Boston: Birchäuser (1980) 248.

[35] Arnold, V.I., Afraimovich, V.S., Ilyashenko, Ju. S., Shilnikov, L.P., Bifurcation Theory. Itogi Nauki i Tekhniki. Sovremennye Problemy Mathematiki. Fundamentalnye Napravlenia, v. 5. VINITI, Moscow (1986) 5-217 (English translation: Enciclopaedia of Mathematical Sciences, Dynamical Systems 5; Springer (1988)).

[36] Henon, M., Heils, C., The Applicability of the Third Integral of Motion: Some Numerical Experiments. Astron. J. 69 (1964) 73.

Theoretical and Applied Mechanics
P. Germain, M. Piau and D. Caillerie (Editors)
Elsevier Science Publishers B.V. (North-Holland)
© IUTAM, 1989

A BRIEF GUIDE TO TWO-PHASE FLOW

G.K. BATCHELOR

Department of Applied Mathematics and Theoretical Physics
University of Cambridge
Cambridge CB3 9EW, U.K.

1. INTRODUCTION

Understanding the dynamics of a mixture of two materials in mobile form is one of the great challenges in fluid mechanics today. Many intriguing phenomena and strange effects have been observed, but we do not know how to account for them quantitatively. This is unfortunate, in view of the great practical importance of two-phase flow. Engineering needs cannot wait for the growth of physical understanding, and many empirical formulae, ad hoc models, and calculational procedures have been developed for use with particular types of two-phase flow. These working procedures have practical value, but they do not teach us much about the principles of two-phase flow and they will play little part in this mini-symposium. The focus in this and the two later introductory lectures by John Davidson and Leen van Wijngaarden will be on fundamentals, and we shall endeavour in particular to identify the relevant physical processes. Part of the difficulty here is that two-phase flow encompasses an extremely wide variety of conditions and processes; it is a family of topics, rather than a single topic. My lecture is intended as a very brief guide to this variety of problems, and the other two introductory lectures will each consider an important particular type of two-phase flow.

It will be useful at the outset to say what we mean by 'two-phase flow' and to specify the nature of the problems to be discussed. The words indicate that two different kinds of matter are present. The difference may lie in the states (gas, liquid, solid) of the two kinds of matter and/or in their molecular compositions. The words also imply that each of the two phases is in motion. A single drop of one liquid falling under gravity through a second immiscible liquid, a gas bubble being pushed by liquid through a capillary tube, and wind-generated gravity waves on the sea surface are thus all cases of two-phase flow in the literal sense. However, the methods of single-phase theory are adequate for the investigation of such problems, and it seems preferable to reserve the term two-phase flow for more intimate mixtures of the two moving phases requiring different techniques and concepts. Novel methods are required, for example, for the case of numerous dispersed particles (a word which I use to mean discrete lumps of matter in the gaseous, liquid or solid state) moving through a second, fluid, phase, and this is in fact the typical situation in present-day studies of two-phase flow. The table attached to section 3 indicates the various possible configurations with a complex interface between the two phases, most of them being particle dispersions.

Dispersions of many particles in a second fluid phase are thus central to the topic of two-phase flow. 'Many' here usually means a large number, and so if there is any randomness in the system, perhaps resulting from the fact that the positions of the particles at the initial instant are not closely controllable, a deterministic approach is not feasible. It is appropriate instead to regard the positions as random variables to be characterized statistically. The particle velocities and other parameters of the system are usually also random variables. We have here a parallel with turbulent flow of a pure fluid: the detailed state of the system at an initial instant varies from one realization to another subject to the same macroscopic conditions, and we seek the values of various ensemble averages describing the system, using the deterministic laws governing the motion of each phase to relate the values of these averages at different instants.

Another common and important feature of two-phase flow is relative motion of the two phases. This relative motion might arise from acceleration of a mixture of two phases with different densities or from the action of a body force like gravity on two phases which respond differently. There are of course cases in which the relative (translational) velocity of dispersed particles and the ambient fluid is negligibly small, as may happen when the particles are of very small size. The problem may be simplified in such cases, and it is certainly changed in character, because a mixture of fluid and many small particles which are not in translational motion relative to the fluid behaves in certain respects as a continuum, with rheological properties which depend on the statistics of the particle configuration and whose determination has features in common with composite-materials theory. However, in the general case to be discussed here in which the two phases are in relative motion it is essential to consider the two phases as separate dynamical systems.

It is often permissible to assume that the stress at a point in a fluid phase is of the Newtonian form, and we shall make that assumption here (while recognizing that non-Newtonian fluids do occur in two-phase flow systems in the biological and polymer processing industries). The heart of the problem of two-phase flow is that the interaction between the two phases occurs at an interface whose shape and motion fluctuate randomly. Determination of the statistical properties of the stress at this moving interface in a systematic way is seldom possible. Faced with the difficulty, familiar from turbulence theory, of closing a set of equations relating mean quantities of different kinds, many authors have chosen to regard each phase as a space-filling fluid continuum and to make ad hoc assumptions about the properties of the two overlapping continua and the way in which they interact (see Ishii 1975 and Drew 1983). This does not seem to be a realistic concept, and as a model it has the weakness that it fails to make use of the information represented by the positions, shapes and velocities of the particles comprising one of the phases.

In many of the industrial occurrences of two-phase flow there is a difference between the temperatures of the two phases and an associated transfer of heat and perhaps also mass between the phases. Transfer of mass between the phases can have a strong effect on the flow in gas-liquid systems in particular in view of the large value of the ratio of the two specific volumes. Despite the great importance of thermal and phase-change effects in liquid-based cooling systems used in the power industry in particular (Butterworth & Hewitt 1977, Hetsroni 1982), lack of time obliges us to set them aside and to concentrate on dynamical processes in this mini-symposium.

2.   SOME GOVERNING DIMENSIONLESS PARAMETERS

The complexity of a problem is sometimes measured by the number of relevant dimensionless parameters. The number in most cases of two-phase flow appears to be rather modest, but this is deceptive because it is not always possible to make clear and precise use of the principle of dynamical similarity in a problem with stochastic variables. It is worthwhile none-the-less to list the most important dimensionless parameters as a way of indicating some of the relevant dynamical processes.

The flow pattern in a (Newtonian) fluid phase depends on the relative magnitude of inertia and viscous forces, so one relevant parameter is the Reynolds number

$$R = \rho_f L U / \mu_f,$$

where ρ_f and μ_f are the density and viscosity of the fluid phase, L is a measure of the length over which the fluid velocity varies appreciably and U is a measure of the magnitude of the velocity variations. If one phase is a connected ('continuous' is the word usually employed in the literature) fluid phase and the other consists of dispersed spherical particles all of the same size, then the best choice for L is the diameter of the particles, although lengths involved in the statistical distribution of particle positions also have some relevance. The appropriate choice of U in the case of a dispersion of particles in fluid will normally be the magnitude of the mean relative velocity of the two phases. One gets an impression of the circumstances in which viscous forces in the fluid phase are negligible or dominant from the fact that $R \approx 1$ in the cases of (1) an isolated spherical rigid particle of density twice that of water and diameter 130 μm falling under gravity through water and (2) an isolated drop of water of diameter 80 μm falling in air.

Inertia and gravity forces are both significant in many two-phase flow systems, and the Froude number

$$F = U^2 / gL$$

provides a measure of the ratio of their magnitudes. Here U^2/L is an estimate of the acceleration of one of the phases, and the most appropriate choice of the length L and the velocity variation U will not necessarily be the same as in the Reynolds number of the flow in a fluid phase. The Froude number is normally of primary importance, and the whole range of values from zero to infinity is relevant to practical situations.

When one phase is in the form of dispersed discrete particles, the mean relative velocity of the two phases will be determined by the balance between the effects of gravity, acceleration of the mixture, and fluid stresses at the particle surfaces. The nature of this balance depends on the Stokes number, that is, the ratio of the 'relaxation time' of the particles (the time in which the relative velocity of a force-free particle decreases by a factor e^{-1}) and the imposed time scale of variation of the fluid-phase velocity at the position of a moving particle (T, say). The particle relaxation time depends in turn on the Reynolds number of the local flow, and if for definiteness we assume this to be small the relaxation time is measured by $\rho_p L^2 / \mu_f$ (where L is here the particle diameter) and the Stokes number becomes

$$S = \rho_p L^2 / \mu_f T.$$

(For the other extreme case of high Reynolds number of the flow past a solid particle, there is a relaxation <u>distance</u> of order $L\rho_p/\rho_f$.) If $S \ll 1$ the particles adjust rapidly to changes in the velocity of the surrounding fluid and the relative velocity of the particles and the fluid is always small (although not necessarily negligible), whereas if $S \gg 1$ the particle velocity is slow to respond to changes in the fluid motion. The latter situation is representative of many practical problems of impact of particles on the boundary of the flow of gas-particle mixtures.

Another dimensionless parameter whose value affects the fluid flow in the case of a disperse phase of freely-moving particles of density ρ_p is the density ratio ρ_p/ρ_f. The direct effect is probably small in most cases, but ρ_p/ρ_f may have a role as a modifier of the Froude number. In the case of a fluid disperse phase with viscosity μ_p, in the form of either liquid drops or gas bubbles, there is also the viscosity ratio μ_p/μ_f. Although the value of this ratio may have a marked effect on the shape and integrity of a fluid particle immersed in fluid undergoing deformation, it seems unlikely that it influences strongly the bulk motion of the system when the size distribution of the particles is stable. These two parameters we may regard as secondary. Similar remarks apply to the parameter $\gamma/\mu_p U$, where γ is the surface tension at the interface of two fluid phases (perhaps non-uniform, owing to its dependence on the local surfactant concentration), though this parameter becomes of primary importance in a case of a gas-liquid or liquid-liquid interface moving through a porous medium.

3.   THE IMPORTANT TYPES OF TWO-PHASE FLOW

The variety of types of two-phase flow, each with its distinctive phenomena and processes and applications, provides part of the fascination of the subject. The table shows the four different combinations of two phases of which at least one is a fluid, and, for each of these combinations, various possible geometrical arrangements which permit movement of both phases (rather limited movement of one of the phases in a couple of cases) and some practical or natural manifestations of two-phase flow for each of these configurations.

Three important types of two-phase flow in which one of the phases is in the form of discrete particles (solid particles in a gas, solid particles in a liquid, and gas bubbles in a liquid) will now be considered briefly. There will not be time in this introductory talk for a description of the associated theoretical work. However, this is not as serious an omission as it might seem to be because, although there are numerous calculations of specific flow fields based on assumed models, there are few fundamental developments which aid our general understanding of two-phase flow. In these circumstances it is more important to devote the available time to making the physical picture clear. The references given at the end are almost all books or review articles on a particular type of two-phase flow. General text-books have nothing to say on the subject, with the exception of two pithy sections in the book by Prandtl (1952, §§V4,5), but there are some massive handbooks with a wealth of specialized information about applications and practical problems (Hetsroni 1982, Cheremisinoff 1986).

4.   A DISPERSION OF SOLID PARTICLES IN A GAS

This is the type of two-phase mixture for which the variety of cases occurring in nature or industry is perhaps the greatest. If we order the different cases by the particle size, which is roughly equivalent to considering different Stokes numbers for a given characteristic time scale of the flow field, in the first case the particle diameter is less than a micron, the particles are colloidal (meaning that they settle out by gravity extremely slowly) and the mixture is now an aerosol (see Friedlander 1977 for a review of the relevant mechanical and physical processes). The particles normally move with the gas, and there is diffusion due to Brownian motion in the case of very small particles. However, if the particles carry an electrical charge (and they may acquire one by collision with molecules in an ionized gas) they respond to an applied electric field, as in an electrostatic precipitator, and they are also attracted to solid surfaces on which there is a charge of opposite sign, which enables

CASES OF TWO-PHASE FLOW

Phase combination	Configuration of the two phases	Practical or natural occurrence
Gas-solid	Dispersion of solid particles in gas	Aerosols Filters and particle-collection devices Compression waves in a dusty gas Explosive volcanic jets Pneumatic transport of particles Saltation and formation of sand dunes Gas-fluidized beds Powder snow avalanches Flow of granular materials
Liquid-Solid	Dispersion of solid particles in liquid	Hydrosols Pattern formation by swimming micro-organisms Filtration and dewatering of slurries Hydraulic transport of particles Alluvial channels, bed forms and meanders Liquid-fluidized beds Turbidity currents
	Porous solid filled with liquid	Elastic waves in saturated rock
Liquid-Gas	Dispersion of liquid drops in gas	Mist, fog and cloud Coalescence of drops and formation of rain High-speed flow of vapour with condensation Droplet removal devices Jet pumps Drop break-up by acceleration of gas Industrial spray column
	Dispersion of gas bubbles in liquid	Compression waves in bubbly liquid Flow of boiling liquid in pipes Vertical flow of oil with exsolving gas Bubble or slug regime of vertical gas-liquid flow Transport of oil and gas in horizontal pipe Bubble-curtain break-water Connected bubbles forming a foam
	Both liquid and gas connected	Annular regime of vertical gas-liquid flow Movement of air-water interface in soil
Liquid-Liquid	Dispersion of liquid drops in liquid	Emulsions and creams Hydrodynamic break-up of immiscible drops Coalescence of drops and phase separation
	Both bodies of liquid connected	Transport of two liquids in horizontal pipe Movement of oil-water interface in porous rock

the efficiency of collection of particles on the fibres of fabric filters to be increased. Another force which may act on an aerosol particle is the thermophoretic force causing it to migrate down a temperature gradient. The thermophoretic force is usually small but it may lead to significant rates of deposition on cold surfaces when it acts across the gas streamlines, as in thermal precipitators and in flow past gas turbine blades and heat-exchanger tubes, and it causes long-term staining of interior ceilings with non-uniform temperature over which there is little or no flow. Aerosols are almost always so dilute that the effect of the particles on the gas flow is insignificant, and this allows many of the problems of deposition due to the action of a force on the particles to be solved by analysing the trajectories of isolated particles in a given gas flow.

For larger or denser particles and values of the Stokes number of order unity or larger, the effects of particle inertia are important. There is a regime here, often referred to as 'dusty-gas flow', in which inertial impaction or deposition of unwanted particles on solid surfaces may be achieved by clever design of the gas flow. Inertial effects on the trajectories of large heavy molecules in a carrier gas of light molecules may similarly be significant, and have been used as a basis for the aerodynamic separation of isotopes. The particle volume fraction is normally very small in practical cases of dusty-gas flow, so the aerodynamic resistance exerted on a particle is approximately the same as if it were isolated. However, the particle mass fraction may not be small, in which case the particles exert a significant body force on the gas. If now in a case of laminar flow of the gas the (identical) particles in a small volume all have the same velocity at an initial instant, or far upstream in a steady-flow problem, the particle velocity will remain locally uniform. In these circumstances a particle moves under the action of an aerodynamic resistance force $\mathbf{F}$ exerted by the fluid (and perhaps gravity), and the gas flow is modified by the action of a body force $-n\mathbf{F}$ where n is the local number density of particles; and $\mathbf{F}$ depends on the local relative velocity in a way which is determined by the particle Reynolds number alone. These two relatively simple coupled equations of motion have been used for the analysis of a number of problems of laminar dusty-gas flow with small particle Reynolds number, many of them allowing examination of the two limits $S \rightarrow 0$, when the particles are 'frozen' into the mixture, and $S \rightarrow \infty$, when the particles move independently of the gas (see Saffman 1962, Marble 1970). Note that the tractability of the equations for (laminar) dusty-gas flow is a consequence of the absence of fluctuations in the particle velocity.

Later today Susan Kieffer will be describing one of the most spectacular natural illustrations of two-phase flow, namely, a jet of particle-laden gas which emerges explosively from an erupting volcano. There is some evidence, from her previous laboratory studies of such jets and comparison of the qualitative results with observations of volcanoes (Kieffer & Sturtevant 1984), to suggest that for some volcanic jets S is so small that the sound speed in these jets is small (owing to the contribution of the particles to the inertia of the medium), implying that the jets are effectively supersonic, with interesting consequences for their structure.

Another group of practical problems characterized by values of the Stokes number of order unity or larger is connected with pneumatic transport of particles in a horizontal direction. These problems include conveyance of granular materials like wheat by air flow in pipes and movement of sand by wind. Effects of gravity are significant. The particle volume fraction is again usually very small, and the path of an airborne particle is determined by the balance between the inertia force, gravity, and the aerodynamic resistance to relative motion of the particle and the gas. However, the dusty-gas equations are no longer applicable because the particle velocity fluctuates and is no longer simply a function of time and position in the gas. These velocity fluctuations arise from the random nature of the pick-up of particles from the bed over which the gas is flowing, and since the gas flow is usually turbulent there is another source of randomness in the aerodynamic resistance to the particle motion. Our ability to analyse quantitatively such two-phase flow systems is limited, but there is some interesting data about the statistically steady state in pipe flow (Owen 1969) and about the average trajectories of sand particles being blown by turbulent wind. The classical work by Bagnold (1941) on the saltation process and the interaction of the moving gas-sand mixture with the form of the particle bed has been extended, and has recently been found useful for the interpretation of observations of markings on the surface of Mars made from spacecraft.

A conceptually simple type of two-phase flow in which particle interactions and velocity fluctuations are intrinsic is a fluidized bed, that is to say, a dispersion of particles in a vertical cylinder through which gas flows upwards steadily at a speed somewhere between that for which the particles are just lifted off the porous base-plate and that for which they would all be carried out the top of the cylinder (see Hetsroni 1982 and Davidson, Harrison & Clift 1985 for collections of articles reviewing work on different aspects of fluidization). This flexible and convenient device in which the particle concentration adjusts itself to suit the given gas flow speed (whereas in a sedimenting dispersion in a closed cylinder the mean fall speed adjusts itself to suit the given concentration, being smaller for larger concentrations) is widely used in chemical engineering as a means of achieving high rates of heat and mass transfer and chemical reaction between the particles and the gas and between the particles and the cylinder wall, and there is an enormous technical literature on the operation of fluidized beds. A remarkable feature of most gas-fluidized beds is the spontaneous formation of bubbles of clear fluid which rise through the bed with shapes resembling those of gas bubbles in liquid. We shall hear more about this and other properties of fluidized beds from John Davidson, one of the pioneers in the study of this type of two-phase flow.

The origin of these bubbles in gas-fluidized beds is not clear, but a common speculation is that they are an outcome of the nonlinear change of form of an exponentially growing concentration wave of small amplitude. There is indeed some evidence to suggest that under certain conditions the state of uniform concentration of particles is unstable to small wavy disturbances with horizontal wave-fronts. The instability is evidently promoted by particle inertia (Jackson 1985), but, unusually in a stability problem, it has proved difficult to identify a significant physical process which opposes the instability and so determines a criterion for instability; in a typical stability problem in fluid mechanics, perturbation analysis reveals the relevant physical process, but here there is not yet an agreed established equation of (one-dimensional) mean motion of the particles which can be perturbed. I have argued recently (Batchelor 1988) that the opposition to growth of a small disturbance comes from two terms in this equation of motion having the form of a stress gradient. In one of these terms the stress represents the flux of particle momentum associated with fluctuations in particle velocity, and in the other, which appears to be numerically larger and which has not previously been included in the equation of motion, the stress represents the effective short-range repulsive force between randomly-moving particles with excluded volume; and the two together determine an effective bulk modulus of elasticity of the configuration of particles, the relevance of which to the criterion for instability was pointed out some years ago by Wallis (1969). The non-dimensional form of the resulting criterion for instability corresponds to the particle Froude number being greater than a critical value of order unity, in agreement with the observation that a uniform fluidized bed is more unstable for larger and denser particles.

Gravity may also generate an important collective motion of particles when the particle concentration varies over a horizontal plane. A powder snow avalanche is believed to be a mobile dispersion with small volume fraction of the snow particles which is flowing down-hill rapidly as a 'density current', although the way in which the snow picked up by the advancing avalanche becomes aerated or fluidized is not clear (Hopfinger 1983). A gravity-driven mean circulation in a fluidized bed may also arise if the flow of gas through the distributor plate is not uniform.

Finally, in the limit $S \rightarrow \infty$ the particles are not affected by the presence of gas and can be considered separately. A flow of this type occurs when compact solid particles are 'poured' down a chute. It is found that the mean flow here resembles that of a liquid with a free surface, even to the extent of exhibiting a hydraulic jump under appropriate conditions. It is natural to analyse such a high-speed flow of granular material in terms of the concepts and methods of the kinetic theory of gases, with the vital difference that the kinetic energy associated with fluctuations of the particle velocity about the mean is here not an independent variable of state, as it is in the case of a gas composed of molecules, but is determined by the gravitationally-driven shear flow itself. On the other hand, when grains of sand are crowded together in flow through the neck of an hour-glass or hopper, frictional forces between particles in contact are dominant, particle inertia forces are small and the problem is effectively one of statics (Wieghardt 1975). The Sectional Lecture on the flow of granular materials that Stuart Savage will be giving later today is being included in the mini-symposium in view of the connection of this topic with general gas-particle flow systems.

5. A DISPERSION OF SOLID PARTICLES IN A LIQUID

Many of the dynamical properties of a dispersion of solid particles in a liquid are similar to those of a dispersion in a gas, and it will be sufficient to point out the main differences, most of which arise from the much larger fluid density.

In the case of a colloidal dispersion of very small particles in liquid, electrostatic forces play a more important role because most liquids contain ions of both signs and there are usually adsorbed ions of one sign at the surfaces of particles (Russel 1980). According to the classical double-layer theory each charged particle is surrounded by a diffuse layer in which counter-ions are proponderant and whose thickness is determined by the balance between electrostatic attraction to the particle surface and Brownian diffusion away from it. When the particles move relative to the fluid under the action of gravity (sedimentation) or an applied electric field (electrophoresis), the diffuse layer of counter-ions is deformed and the resistance to the particle motion is increased (Saville 1977). The effects of electrical charges at particle surfaces are especially important in concentrated dispersions (which are not as uncommon as they are in the case of aerosols), and are relevant for example to the tendency for clay particles to form aggregates which settle out and form jelly-like mud at the bottom of a river estuary at slack water.

A fascinating example of the effect of internal mechanical forces on the motion of a dispersion of very small particles in liquid is provided by swimming micro-organisms (Kessler 1986). Algal cells with roughly spheroidal shape and linear dimensions of several microns swim in a direction related to their body orientation by waving two flagella. A cell is slightly denser than water, and the density distribution is non-uniform, with the consequence that in stationary liquid a cell takes up an orientation such that swimming produces an upward motion. Cells thus become concentrated near an upper horizontal boundary, and the gravitational instability of this layer of liquid with excess density leads to convection patterns. Furthermore, if the liquid containing the cells is in motion, the local velocity gradient exerts an orienting effect on a cell body and so on the swimming velocity. For instance, in liquid in Poiseuille flow in a vertical tube, the cells have a radial component of swimming velocity which is towards the axis in down-flow, generating a cell-rich column near the axis, and away from it in up-flow. In general, a non-uniform cell concentration results in a bulk motion driven by gravity, and this too affects the orientation of the cells and the direction of their swimming. Improved understanding of these dynamical processes could be useful for the harvesting of cells.

For particles with diameters above 10 μm, on which electrostatic forces are normally negligible, collection by inertial impaction on solid surfaces is not as readily achieved as in the case of an aerosol, because the Stokes number is a good deal smaller, for given values of L and T, in consequence of the larger values of the viscosity (at small R) or the density (at large R) for liquids than for gases. The greater resistance to relative particle motion in a liquid also implies smaller values of the Froude number, for given particle size and density, with the consequence that instability of a uniform liquid-fluidized bed to small disturbances occurs at larger particle sizes than in the case of a gas-fluidized bed. The existence of 'bubbles' of clear liquid in a liquid-fluidized bed has not been established beyond question, so it seems that the nonlinear development of a disturbance also is affected by the greater density and viscosity of the liquid.

For obvious practical reasons the extreme case $S \gg 1$ can not be realized for a dispersion of particles in liquid, so that the flow of granular material in liquid is never independent of the presence of the liquid. On the other hand, the kind of collective gravity-driven motion represented by a powder snow avalanche, which does not require large values of the Stokes number, does have a counter-part in liquid. The accumulated loose mud and silt on a sloping part of an estuarial canyon on the ocean floor may ultimately slip and form a similar mobile fluidized mixture of mud and water which flows rapidly down the canyon as a 'turbidity current' (Simpson 1987). Breaks in trans-Atlantic cables on the ocean floor are believed to have been caused by such currents.

In the extensive and important subject of hydraulic transport of solid granular material like sand and gravel in approximately horizontal closed pipes or open streams, the Reynolds number of the bulk flow is normally large and the flow is turbulent. Since the Stokes number is order unity or less (relaxation distances of the order of the particle diameter), there is strong interaction of the particle motion and the turbulence in the liquid, with the consequences that the particle trajectories are not smooth and the particle velocities fluctuate randomly. Moreover the particle volume fraction may not be very small, in which event there are significant hydrodynamic interactions between the particles, likewise giving rise to fluctuations in the particle velocity. Thus the simple 'dusty-gas' equations are not applicable, and we do not know what should be put in their place.

The nature of the problems involved in hydraulic transport may be indicated by thinking initially of a single particle in a statistically steady turbulent stream of liquid of depth h and friction velocity u_*. The particle moves under the action of the inertia force, gravity (normal to the mean flow), and the hydrodynamic stress at the particle surface. Depending on the density ratio ρ_p/ρ_f, the particle may follow approximately the local random movements of the liquid or it may have a different trajectory with a characteristic length scale of random variation which exceeds that for an element of the liquid (when $\rho_p/\rho_f > 1$) and increases with ρ_p/ρ_f. We are especially interested in the mean vertical motion of the particle, which is a combination of a downward drift due to gravity and a tendency for the random vertical excursions to make more uniform the probability density of vertical position. It is common practice to equate the mean downward drift velocity to the steady velocity of fall of the particle in still liquid (U_0), which is a plausible approximation, and to represent the effect of the random excursions as a diffusion process, which is less obviously acceptable because the length scale of the vertical excursion of a particle due to the turbulence is not usually small compared with the total depth of the liquid. If we then assume the diffusivity to be a constant of order hu_*, as would perhaps be reasonable for most of the stream away from the lower boundary, it follows that the probability density of particle position falls off exponentially with height, with an e-folding distance hu_*/U_0. This simple calculation reveals the influence of the important parameter u_*/U_0, values of which may be either small, of order unity, or large in practical cases, and it also shows that the finer particles are distributed over the whole depth whereas the coarser particles are concentrated near the bottom, as would be expected.

If now there are a number of particles in the turbulent liquid stream, although too few for hydrodynamic interactions to be significant, the vertical distribution of mean concentration coincides with that of the probability density of position of a single particle. The arbitary constant multiplying the exponential function of height may be determined, by integration, in terms of the total number of particles dispersed in the liquid in a given length of the channel. One might think of this total number as the kind of parameter of a mathematical solution that can be regarded as 'given', but it is a peculiar feature of the hydraulic transport problem that the total number of dispersed particles is not specifiable in advance but is determined by the balance between the rate at which particles fall to the bottom of the channel and the rate at which they are picked up from the bed of stationary or slow-moving particles at the bottom. The study of the hydrodynamic lift force on particles in the surface layers of the bed below a turbulent stream is consequently a major part of this subject. The lift force on a bed particle is not steady, but fluctuates as the liquid velocity in its neighbourhood fluctuates, and we need to know the probability that the lift force exceeds the value needed to make it jump into the stream. The statistical geometry of the bed surface and the ratio of the roughness height to the thickness of the viscous sub-layer are clearly relevant.

In many practical situations the dispersed particles are not so far apart that collective effects are negligible. The vertical gradient of particle concentration is greatest at the bottom of the stream – and is much larger than the above exponential distribution suggests because the diffusivity is not constant there but tends to zero at the bottom – and the corresponding vertical gradient of mixture density averaged over horizontal planes may be large enough at the bottom of the flow to have a significant stabilizing influence on the turbulence there. This stabilizing influence seems likely to weaken the ability of the stream to pick up particles, so we have here a possible mechanism for the limitation of the total number of particles dispersed in the liquid.

These questions relating to the horizontal transport of particles dispersed in a turbulent stream of liquid in a channel or pipe are among the most difficult in the whole subject of two-phase flow, and relatively little progress has yet been made in reaching an understanding of the physical processes at work and in describing them quantitatively (Yalin 1977). More is known (empirically) about the intriguing variety of forms of a bed of particles that result from differential scouring and deposition of particles by the stream of liquid (Kennedy 1969, Engelund & Fredsøe 1982) and the meandering, in a horizontal plane , of a channel with erodible banks (Callander 1978). The formation of ripples in the bed of sand particles below an oscillating stream of water is another bed-form phenomenon which awaits a satisfactory explanation (Kaneko & Honji 1979).

6. A DISPERSION OF GAS BUBBLES IN A LIQUID

Some of the special features of this case derive from the very large ratio of the bulk moduli of elasticity of the liquid and the gas. When a sound wave propagates through liquid containing gas bubbles, it is the compliant gas phase that undergoes compression whereas the liquid provides the inertia. For such small frequencies that the gas remains approximately in thermodynamic equilibrium, the usual formula shows that the speed of the sound wave is greatly reduced by the presence of the bubbles, being no more than about 50 ms^{-1} for bubble volume fractions of a few percent in the case of air and water and so considerably smaller than the sound speed in either phase. This is the basis of the well-known acoustic screening property of a region of bubbly liquid in the ocean. At larger frequencies non-equilibrium effects associated with the oscillatory radial motion of the liquid in the neighbourhood of each (spherical) gas bubble become significant. The wave is then dispersive, and the sound speed first falls and then increases sharply as the frequency increases through the value appropriate to the natural (surface-tension controlled) oscillations of bubble volume in stationary ambient liquid. These and other interesting aspects of compression waves in bubbly liquid were reviewed by Leen van Wijngaarden (1972) some years ago, and he will no doubt bring us up-to-date in his expository lecture later in the mini-symposium.

Other novel dynamical features of a dispersion of gas bubbles in liquid are associated with the negligible density of the particles. When the velocity of an isolated particle relative to the ambient fluid changes, there is a change in the momentum of the fluid and an associated acceleration reaction is exerted on the particle by the fluid. In its simplest form this acceleration reaction is dynamically equivalent to the addition of a virtual mass to the real mass of the particle, and the virtual mass is comparable with the mass of fluid displaced by the particle. The acceleration reaction is thus negligible, relative to the inertia force, for particles dispersed in gas, it may be quantitatively significant for solid particles dispersed in liquid, and it is certainly significant for gas bubbles in liquid. If the flow due to an isolated moving particle can be assumed to be irrotational, the virtual mass may be determined without difficulty and, for example, is found to be half the mass of the displaced fluid for a sphere of constant radius. The assumption of irrotational flow is reasonable in many cases of oscillatory or approximately impulsive motion or for a short time after motion develops from rest; and so we find for example that the velocity of a spherical gas bubble in liquid through which a sound wave is passing is three times that of the ambient liquid, the inward displacement of a gas bubble in water flowing round a sharp bend in a pipe is quadratic in the angle of turn, and the initial upward acceleration of a spherical gas bubble from rest is $2g$.

It is also known that, when the acceleration of the fluid in which a solitary particle is immersed is non-zero as a consequence of spatial variation of the fluid velocity, there is a similar acceleration reaction on the particle and that the general expression for this reaction contains just one particle parameter, which is numerically equal to the virtual mass.

The effective inertia of a bubble is thus measured by the liquid density, and the definition of the Stokes number that is appropriate for a dispersion of gas bubbles in liquid is obtained from that given in section 2 by replacing ρ_p by ρ_f.

If we are to be able to analyse the (incompressible) flow of a dispersion of gas bubbles in liquid we need to extend the above classical results in certain respects. First we need to take account of the hydrodynamic interactions between gas bubbles in view of their importance in dispersions with bubble volume fractions of a few percent or larger. Specifically we need to know the mean acceleration reaction on a bubble in a statistically homogeneous dispersion in liquid which is being accelerated. It would be expedient to assume irrotational flow, since this is permissible under certain conditions and it may be the only condition under which we can hope to make progress mathematically. It is unlikely that the acceleration reaction will be found to be exactly equivalent to the addition of a virtual mass to the real mass of the particles. Some results have been obtained for the case of small bubble fraction, as Leen van Wijngaarden will report.

Second, it would be helpful to know something about the effect of the vorticity that is undoubtedly present in liquid through which gas bubbles are moving, even if it is only a single observation showing the general magnitude of the effect. This is not relevant at small bubble Reynolds numbers, because the acceleration reaction on a bubble is then in general small compared with the resultant force due to viscous stresses. At bubble Reynolds numbers between about 30 and 300 a single air bubble rises steadily in water without appreciable departure from sphericity and it is known that the wake containing non-zero vorticity is narrow. The question then is whether the accumulated vorticity due to the wakes shed from many bubbles in a statistically homogeneous dispersion affects the acceleration reaction significantly; if not, the way is clear for the use of irrotational-flow theory in a calculation of the acceleration reaction.

At bubble Reynolds numbers above about 300, corresponding to diameters of gas bubbles in water above 1 mm, the bubbles become flattened on the under-sides and rise under gravity unsteadily. There is a corresponding growth of velocity fluctuations in the liquid and the appearance of a true turbulence with strong nonlinear interaction between different scales of motion (the fluctuating motions that exist at small bubble Reynolds numbers in consequence of the random positions of bubbles being viscous dominated). There is no doubt a similar development of fluid turbulence as the size of solid particles in a gas-fluidized bed is increased, but it is more important in the case of gas bubbles in liquid owing to their strong response to local acceleration of the liquid.

There is also a shear-generated contribution to the turbulence in the liquid phase in long vertical bubble columns with up or down-flow of the liquid (a common industrial set-up). The liquid-phase turbulence is of special importance here because it has a controlling influence on the distribution of bubble concentration over the cross-section of the column. There appear to be two effects causing non-uniformity of this distribution. One is the 'lift' force of inertial origin that acts on a bubble in translational motion through fluid in simple shearing motion and that causes the bubble to move towards the wall of a pipe when the relative translational velocity of the bubble and the shear-flow velocity of the liquid are in the same direction. The other is a radial force on a bubble arising from the radial pressure gradient in the liquid due to non-uniformity of the radial normal Reynolds stress; the mean-square radial velocity fluctuation in turbulent flow through a pipe is normally largest near the wall, so the radial pressure gradient pushes bubbles towards the wall. Then, when the bubble concentration is not radially uniform, two further effects come into play. First, turbulent diffusion tends to restore the uniformity of the concentration. Second, the vertical force exerted on the liquid by the rising bubbles is now non-uniform and so causes changes in the mean liquid velocity profile and in the distribution of turbulence intensity, the changes being different for up and down-flow of the liquid. Attempts to analyse this complex flow system with neglect of the hydrodynamic interaction of bubbles and simple modelling of the liquid turbulence indicate a maximum of the bubble concentration near the wall in the case of up-flow and near the centreline in the case of down-flow, and this is also what observations of a dilute dispersion show (Lahey 1987).

Depending on the gas and liquid flow rates, other regimes of two-phase flow in a long vertical pipe may occur. The bubbles may coalesce, as a consequence of random 'collisions' or of larger bubbles overtaking smaller bubbles, and ultimately form a train of large bubbles, or 'slugs', each of which nearly fills the pipe cross-section and has a length larger than the pipe diameter. Another possibility is that the gas flux may occur mainly in a central connected region which is surrounded by an annular region of flowing liquid. Waves are generated at such a gas-liquid interface and liquid droplets may be entrained into the gas stream, and in a case of down flow of the liquid there may be 'flooding', or reversal of the direction of the liquid flow. These different regimes, which are made possible by the convertibility of the gas phase from discrete to connected form, have been studied extensively in view of their importance in chemical processing plants and the power industry (Delhaye 1983).

7. IN CONCLUSION

What summarizing remarks can usefully be made at the end of this sketchy outline of the dynamics of two-phase flow? Well, for those who like their research problems to be reasonably well defined physically and to be formulatable in mathematical terms, this is probably not a suitable field. A taste for pioneering and a preparedness to live with confusion are more appropriate qualifications. It will be clear from the fore-going, I hope, that the main obstacle to progress in two-phase flow theory is not a want of mathematical, numerical or experimental technique but is essentially that we lack a closed form of the equations governing the mean motion of the separate phases. (This is rather like the position that turbulence theory was in during the early 1930's.) Underlying this lack of governing dynamical equations is ignorance or uncertainty about the relevant physical processes. There is an acute need for observations which will tell us something about these physical processes, that is to say, for experiments planned to meet scientific needs rather than to provide engineering information. Two-phase flow is too important and interesting a subject to be left to those who are obliged to find quick answers to practical questions.

REFERENCES

Bagnold, R.A. 1941 *The Physics of Blown Sand and Desert Dunes*. Methuen.

Batchelor, G.K. 1988 A new theory of the instability of a uniform fluidized bed. *J. Fluid Mech.* **193**, 75-110.

Butterworth, D. & Hewitt, G.F. (Ed.) 1977 *Two-Phase Flow and Heat Transfer*. Oxford University Press.

Callander, R.A. 1978 River meandering. *Ann. Rev. Fluid Mech.* **10**, 129-158.

Cheremisinoff, N.P. (Ed.) 1986 *Encyclopedia of Fluid Mechanics*. Vol.3, Gas-Liquid Flows; Vol.4, Solids and Gas-Solids Flows; Vol.5, Slurry Flow Technology; Vol.6, Complex Flow Phenomena and Modeling. Gulf Publishing.

Davidson, J.F., Harrison, D. & Clift, R. (Ed.) 1985 *Fluidization*, 2nd edition. Academic Press.

Delhaye, J.M. 1983 Two-phase pipe flow. *Intern. Chem. Eng.* **22**, 385-410.

Drew, D.A.   1983   Mathematical modeling of two-phase flow. *Ann. Rev. Fluid Mech.* **15**, 261-291.

Engelund, F. & Fredsøe, J.   1982   Sediment ripples and dunes. *Ann. Rev. Fluid Mech.* **14**, 13-37.

Friedlander, S.K.   1977   *Smoke, Dust and Haze.* Wiley-Interscience.

Hetsroni, G. (Ed.)   1982   *Handbook of Multiphase Systems.* Hemisphere Publ.

Hopfinger, E.J.   1983   Snow avalanche motion and related phenomena. *Ann. Rev. Fluid Mech.* **15**, 47-76.

Ishii, M.   1975   *Thermo-Fluid Dynamic Theory of Two-Phase Flow.* Eyrolles.

Jackson, R.   1985   Hydrodynamic stability of fluid-particle systems. Chap.2 of *Fluidization*, 2nd ed., edited by Davidson, Harrison & Clift. Academic Press.

Kaneko, A. & Honji, H.   1979   Initiation of ripple marks under oscillating water. *Sedimentology* **26**, 101-113.

Kennedy, J.F.   1969   The formation of sediment ripples, dunes and antidunes. *Ann. Rev. Fluid Mech.* **1**, 147-168.

Kessler, J.O.   1986   Individual and collective fluid dynamics of swimming cells. *J. Fluid Mech.* **173**, 191-205.

Kieffer, S.W. & Sturtevant, B.   1984   Laboratory studies of volcanic jets. *J. Geophys. Res.* **89**, 8523-68.

Lahey, R.T.   1987   Turbulence and phase distribution phenomena in two-phase flow. *Proc. Conference on Transient Phenomena in Two-Phase Flow*, Dubrovnik. ICHMT.

Marble, F.E.   1970   Dynamics of dusty gases. *Ann. Rev. Fluid Mech.* **2**, 397-446.

Owen, P.R.   1969   Pneumatic transport. *J. Fluid Mech.* **39**, 407-432.

Prandtl, L.   1952   *The Essentials of Fluid Dynamics.* Blackie.

Russel, W.B.   1980   Review of the role of colloidal forces in the rheology of suspensions. *J. Rheology* **24**, 287-317.

Saffman, P.G.   1962   On the stability of laminar flow of a dusty gas. *J. Fluid Mech.* **13**, 120-128.

Saville, D.A.   1977   Electrokinetic effects with small particles. *Ann. Rev. Fluid Mech.* **9**, 321-337.

Simpson, J.E.   1987   *Gravity Currents in the Environment and the Laboratory.* Ellis Horwood.

Wallis, G.B.   1969   *One-Dimensional Two-Phase Flow.* McGraw-Hill.

Wieghardt, K.   1975   Experiments in granular flow. *Ann. Rev. Fluid Mech.* **7**, 89-114.

Wijngaarden, L. van   1972   One-dimensional flow of liquids containing small gas bubbles. *Ann. Rev. Fluid Mech.* **4**, 369-396.

Yalin, M.S.   1977   *Mechanics of Sediment Transport.* Pergamon.

Theoretical and Applied Mechanics
P. Germain, M. Piau and D. Caillerie (Editors)
Elsevier Science Publishers B.V. (North-Holland)
© IUTAM, 1989

PHENOMENOLOGICAL ASPECTS OF CONTINUUM DAMAGE MECHANICS

J.L. CHABOCHE

Office National d'Etudes et de Recherches Aérospatiales
Châtillon, France

The objective of the paper is to review the general principles and the main possibilities of CDM, considering both the experimental aspects, the theoretical concepts and some applications to structures. CDM is considered as a general method to treat the progressive deterioration of materials and structures in the framework of Continuum Mechanics.
The various definitions of damage variables and the associated measurement procedures are considered first, including the remaining life and the effective stress concept. The influence of damage on the constitutive equations is described in the framework of thermodynamics with internal variables. Some kinetic equations for damage are recalled which are based on phenomenological considerations. Both the cases of isotropic and anisotropic damage are discussed briefly.
The possibility of applying CDM concepts to the life prediction of structural components finally illustrated by some examples on smooth and notched bars and the creep crack growth is predicted in the framework of a local approach to fracture. The numerical convergence is discussed and a modification of the continuum damage variable is introduced in order to solve the problem of localization.

1. INTRODUCTION

For some twenty years, a great deal of research has been going on in the concepts of Continuum Damage Mechanics, originally introduced by Kachanov [1] then Rabotnov [2]. The deterioration processes of the material are described in the framework of Continuum Mechanics by means of one or several damage parameters. Then evolution of such mechanical variables obey to some phenomenological equations which are based on microstructure observations, homogenization results and mechanical experiments.

References [3] to [8] describe Damage Mechanics relatively extensively. The possibilities of this approach are rather broad, covering the domain going from the initial damage to the inclusion of the nonlinear cumulative effects of damage and the prediction of the initiation of a macroscopic crack.

Many developements of Damage Mechanics have been made in various fields and for many different kinds of materials, as recalled in section 2.3. The present computational possibilities allow structural application in which the material degradation and its corresponding strength decrease are taken into account during the whole calculation. Several applications have been done recently in practical situations.

The purpose of the present paper is to review briefly the main features of the phenomenological approach of Damage Mechanics, and its application to predict both crack initiation and crack propagation in structures, in the framework of local approaches of fracture. The difficulties associated with the corresponding numerical techniques will be underlined and the need for some modification of the classical definition of damage will be discussed on the basis of two examples.

2.   THE CONTINUUM DAMAGE MECHANICS

2.1. Two different approaches

The CDM shows the two different levels of development, the micromechanics approach and the phenomenological one.

- The first level, figure 1.a, treats both the present state and its evolution in terms of a micro-macroprediction. For example, in a volume element to be homogenized, the local fields (micro stresses) are deduced from the macro quantities through a localization procedure and the prior knowledge of the local defects (say d). Then the growth rate of the defect (say $\dot{d}$) is defined from physical (or phenomenological) assumptions. The new state, after damage increment, is then homogenized, giving rise to macroscopic quantities (for instance average compliance).

- The second one, figure 1.b, introduces a set of damage variables and defines their evolution equations directly at the macroscale. There are two steps in the development of such theories : (i) the nature and the general properties of the internal variables D can be determined by applying the above localization/homogenization procedure for some arbitrarily fixed defect state (arrays of penny shaped microcracks for example). (ii) the damage rate equations ($\dot{D}$) are postulated from both the general framework of thermodynamics with internal variables and macroscopic experiments. Two kinds of damage measurements play role in choosing the form of these phenomenological models: the remaining life measures, important to describe the cumulative aspects of damage in terms of life, the mechanical effects of damage, using the concept of effective stress. It is this second approach, the phenomenological one, which is considered below.

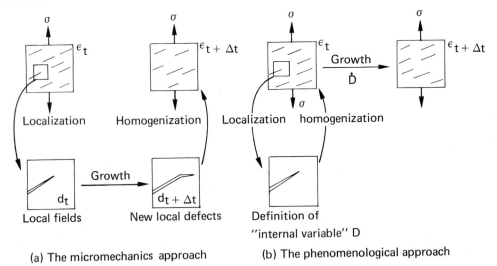

(a) The micromechanics approach   (b) The phenomenological approach

Fig. 1.

2.2. Various definitions of damage

They are many different ways to define the internal variables associated to the damage processes. Each definition must correspond to some method of measurement. In fact, such quantities are much more hidden variables than is for example the inelastic strain, generally defined through the concept of an instantaneously released configuration. Among the various posibilities, we may distinct:

- The microscructural measures of the real defects: volume fraction of cavities in ductile damage, surface fraction of cavitated grain boundaries in creep of metals, accumulated length (or surface) of microcracks in fatigue. Such measurements are difficult to integrate directly in the life prediction methods in terms of their macroscopic effects.

- Measurements of the evolution of various physical parameters such as density, acoustic emission, resistivity, may lead to some consistent macroscopic definitions of damage parameters but do not introduce the mechanical characters of damage.

- Measures of the remaining life, which are of prior importance for the engineer in order to take into account the cumulative aspects of damage processes in the life prediction methods.

- Measures of variation in the mechanical behavior. These are the most appropriate from the mechanical point of view. These measures are interpreted through the effective stress concept [4, 7] with an equivalence in strain (Fig. 2).

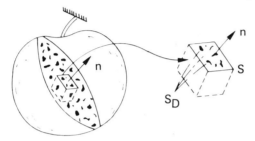

(a) The damaged element

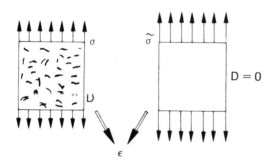

(b) The concept of damage interpreted
by the idea of effective stress.

Fig. 2 —(a) The damaged element and the net sectional area $S - S_D$.
(b) The concept of effective stress.

The effective stress σ is the one that would have to be applied to the undamaged volume element for it to deform the same way as the damage element subjected to the actual stress σ (Fig. 2.b) [9]. Taking the damage D to represent a loss in effective sectional area, considering the decohesions and local stress concentration effects, the effective stress is written:

$$\sigma = \sigma \, \frac{S}{\tilde{S}} = \frac{\sigma}{1 - D}$$

This definition is consistent with the homogenization techniques [10] which provide a thread of reasoning by which the concept can be generalized to the multiaxial case with anisotropic damage [11].

Various parameters can be used to measure the mechanical effect of damage: (i) the variation in the elastic modulus used for ductile damage , but also in fatigue (especially in composite materials); (ii) the viscoplastic behavior (during tertiary creep); (iii) the cyclic behavior in the low-cycle fatigue range. Let us point out some recent good correspondences obtained in fatigue [12] and in creep [13] between these different measures.

2.3. Phenomenological models

Since the first propositions of Kachanov and Rabotnov, an increasing number of works has been performed, using similar definition to the effective stress concept, in various fields of application. The following list, certainly non exhaustive, gives an idea of the variety of materials and loading conditions:

- the creep of metals [14-16], especially under multiaxial states of stress;

- the fatigue of metals [17-19] and of composite materials [20-21];

- the ductile fracture damage of metals [22-26];

- the damage of concrete [27-29];

- the damage effects in composite materials [30-31].

The damage theories have been initially developed within the simplified assumption of an isotropic damage. They correspond to the present level of application in actual components. However numerous other theories have been proposed to take into account the anisotropy of damage. The help of micromechanics and micro-macro approaches is quite evident in this field of research. Let us simply mention here the various types of definitions that can be used for the damage variable. References [4-8] review these more completely:

- scalar valued vectorial functions. The reference [33] shows that a decomposition into elementary measures of damage leads to even-order tensors;

- vector fields [27], each associated with a microcracking system of a given orientation;

- second-order tensor [25, 34] stemming directly from microstructural measures, with which the idea of a net stress tensor can be associated [15], used in the damage law;

- fourth-order tensor [7, 10], by way of the concept of effective stress, which expresses the effects of anisotropy simultaneously on the damage growth law and on the mechanical constitutive equation.

Let us point out a recent variant of the last of these theories, in the framework of a formulation based on the total strain and the decomposition of the effective stress [35].

In addition to the direction, it may be necessary to consider the sense of the stress: in a cyclic loading, for example, the effects of closing the microcracks may play a major role. So, for a fixed damage state, the damage may be active for a positive loading (open cracks), i.e., modify the stiffness of the material with respect to the initial stiffness, or it may be inactive for a negative loading (microcracks closed). This effect is automatically introduced in the theory based on vector fields [27]. Another method consists in considering the effect of damage on the mechanical behaviour as different when the effective maximum principal stress is in tension or in compression [36].

2.4. Correlations between microstructure and the phenomenological models

Let us point out some connections which can be made between physically based damage growth equation and the more phenomenological ones. Such correlations are based either on micromechanics studies or thermodynamic considerations [28, 37].

In the case of creep for example [37], it is possible to combine the secondary creep power law and the Kachanov-Rabotnov creep damage equation in order to obtain an equation which describes fairly well the growth of intergranular cavity fraction [38, 39]. On the other hand, the influence of oxidation processes [40] may be considered as implicitly taken into account by the classical creep damage equation [37, 41].

In the case of fatigue, several works have shown the possibility of correlating the accumulated crack length with a macroscopic damage parameter [11, 17, 42]. Such a correlation may be defined explicitly through thermodynamic consideration, by equating the power dissipated in two models of the process [28, 37]. The macroscopic model is based on CDM and the microcrack growth is written in terms of Linear Fracture Mechanics. The figure 3, taken from ref. [43], shows an application for low-cycle fatigue of a 316 stainless steel. In that case, the macroscopic damage evolution for two strain levels is given by the fatigue damage model used in references [16, 18]. The energetic equivalence leads to predicted accumulated microcracks which compare fairly well with the experimental results obtained by Levaillant [43].

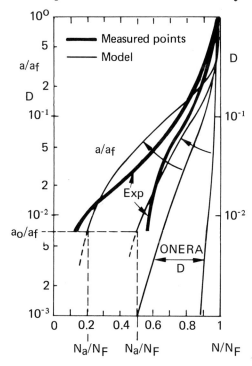

Fig. 3 — Comparison between the kinetics of fatigue damage as interpreted by metallography and by the ONERA model. Steel 316 L, 600°C.

3. LOCAL APPROACHES TO FRACTURE

3.1. The principle of local approaches

The global methods of Fracture Mechanics, based on phenomenological relationships between the crack growth rate and global parameters such as the stress intensity factor K (or J or C*, etc). are extremely practical and relatively easy to use. They make very precious tools with a broad domain of use. But in certain cases, for example when the plasticity is not confined or in the case of creep, these methods may prove defective or be more difficult to apply [44, 45].

The purpose of the local aproaches to cracking is to provide a possible replacement tool. These approaches consist of computing as precisely as possible the stress and strain fields at the crack tip (including all the history effects) and to apply a local fracture criterion (of the element of matter). In the framework of the finite element methods, two kinds of techniques can be used:

(i)   The application of the criterion at a critical distance from the crack tip (generally on the first element), then the release of the crack tip node when the criterion is reached (taking into account the stress redistributions induced by this discrete progression of the crack). This method has already been used successfully in fatigue [46, 47], ductile fracture [48] and in creep [49] computations. It has the disadvantage of depending on the fineness of the mesh.

(ii)   Incorporation of the total coupling by CDM. The continuous progression of the crack is then described by the gradual decrease in the local strength of the damaged material. The crack is then the locus of material points for which the critical damage has been reached ($D = D_c \simeq 1$) [50]. This method has been used in creep [51-53] and in ductile fracture [54, 55].

3.2. Numerical problems associated with local approaches

The second type of local approach is attractive because it corresponds to a complete prediction, using models that have been calibrated against simple experiments. Unfortunately, the numerical implementation is not straightforward and several kinds of difficulties have to be mentioned:

- The very stiff differential equations which generally govern the damage growth are difficult to integrate properly. This is especially true in the case of creep. This problem can be solved by using a self-adaptative time step control in the coupled viscoplastic integration scheme. The method used in the examples below [41, 52] is implemented as a very simple second-order explicit method [7] ,with a special automatic time step control taking into account the specificities of damage growth [41] (especially when the damage attains large values, near from the local breaking point).

- The use of CDM to describe crack growth needs the complete coupling between constitutive equations and damage, in order to describe the progressive disappearance of any material resistance in the process zone. In the case of creep crack growth for example, we have to take into account the effects of damage on both the viscoplastic strain rate (tertiary creep effect) and the elastic behavior. This second coupling is introduced as a continuous evolution of the stiffness matrix in the finite element model. In the framework of the explicit integration scheme, a very simple method has been designed in order to save most of the matrix reduction time by means of a transfer to the second member of the stiffness variations [41, 52]:

$$K^* q = F + F_p + (K^* - K)\tilde{q}$$

where F and F_p are the normal elastic and "thermoviscoplastic" loads, q the present unknowns of the problem (node displacements), K^* a constant stiffness matrix, K the present stiffness matrix depending on the present damage in each material point (Gauss integration points). The correction term in the second member uses a predictor $\tilde{q}$ of the displacements, defined from the previous time step. The reactualization of the K^* matrix takes place from an automatic error control (which parameter is fixed by the user). Due to the small time steps associated with the explicit scheme, this method has been found as very efficient and convergent [41].

- The coupling between damage and elastic behavior induces a non-monotonic material behavior (for pure tensile conditions under strain control). The tensile curve then shows damage induced softening. Such a situation may lead to non-uniqueness of the solution (bifurcations), numerical instabilities and localization. After a number of numerical and theoretical studies [56-62], it appears that uniqueness and stability is preserved for the viscoplastic material, at least when the residual elastic modulus is much lower than the stress [57]. Due to the damage induced stress redistribution, this condition always prevails. However the negative tangent stiffness is still associated to localization possibility [60-62], which introduces a mesh dependency effect in the finite element models. For example, in the case of the creep of a CT specimen [52], only the first row of elements is completely broken and global result (predicted crack growth rate) depends strongly on the height of this first row. This problem is considered in the next section.

3.2. Non-local damage

Several methods can be used to prevent the localization effects :

(i)   limiting downward the size of the finite elements by an idea of characteristic volume associated with the defect statistic [54-55];

(ii)   introducing higher-order gradient in strain and stress, in the framework of the non-local continuum mechanics;

(iii) using localization limiters [29], in particular a non-locally defined deformation;

(iv)   using a non-local approach to define the damage growth law, but conserving a local definition for the strain [63].

It is this last method which has been used successfully [41, 64, 65] in various examples. The damage law is expressed simply as:

$$\dot{D}(x) = \frac{1}{\Omega^*_d} \int_{\Omega_d} \phi(x, \xi) \, \dot{\bar{D}}(\xi) \, d\xi$$

where x and ξ designate, respectively, the point considered and any point in its vicinity Ω_d. $\bar{D}$ is the growth law as it is defined locally. We can choose any given form for the vicinity function ϕ. One practical form is:

$$\phi = e^{-\frac{d^2(x, \xi)}{d^{*2}}}$$

where $d(x, \xi)$ is the distance between x and ξ, and d^* represents a characteristic distance.

$$\Omega^*_d = \int_{\Omega_d} \phi(x, \xi) \, d\xi$$

is the corresponding characteristic volume of material.

3.3. Application to a tensile specimen

This example is interesting to show the localization effects and the ability of the above non-local damage definition to prevent the localization. The axisymmetrical specimen has a cylindrical gauge length of 8 mm and a diameter of 4.5 mm, with a toroidal zone of radius 5 mm. It is submitted to an elongation control under a constant rate. The initial low elastic stress concentration at the end of the cylindrical zone redistributes due to the viscoplastic flow. The stress concentrates more at the center of the specimen, which explains the damage initiation in this region.

The figure 4.a gives the finite element model of the quarter of the specimen and figure 4.b gives the global load-displacement curve (it indicates also the evolution of damage at the center of the specimen). The material is described by the combination of an elasto-viscoplastic constitutive equation with isotropic hardening and a Kachanov-Rabotnov creep damage equation. It is determined from creep tests and describes both primary, secondary and tertiary creep.

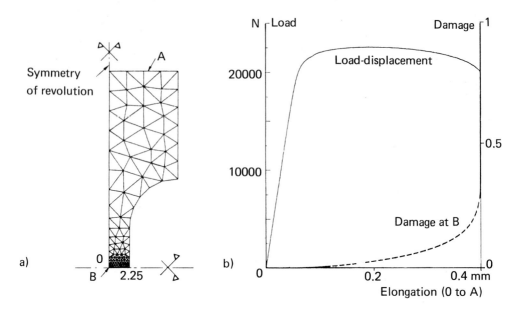

Fig. 4 — The tensile specimen — (a) mesh of one quarter,
(b) lead-elongation curve and the damage evolution.

Four meshes have been studied. With the classical local damage definiton ($d* = 0$),
for every mesh, the highly damaged zone localizes in the first row of elements. This
induces an important mesh dependency, as shown in figure 5.a. Moreover the local
stress redistributions show chaotic evolutions during the few last seconds of the
computation. During the "failure" of every Gauss point, which introduces a very
important stress decrease (to a zero value), the next one is overloaded until its own
failure takes place. The figure 3.a illustratres this behavior for 4 points in the central
section. Let us note the very precise time control of the numerical solution, which
proves the efficiency of the automatic time step control but leads, in that case, to a
very large number of time steps.

Using the non-local damage ($d* = 100$ μm in that case), the localization is prevented
and the solution is much less mesh dependent (figure 5.b). Moreover, the stress
redistributions are now smoothed as shown figure 5.b for the same points in the
central section. The trace of the spatial discretization no more appears in the time
evolution, and the rupture of every Gauss point takes place at the same instant. The
present solution saves a considerable amount of CPU time.

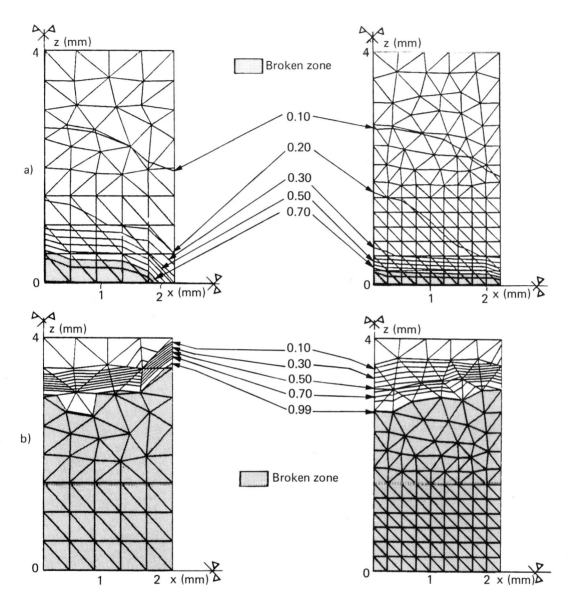

Fig. 5 — Damage zones obtained with the coarse mesh and the median mesh.

(a) with the classical local damage, (b) with the non-local damage (d* = 100 µm).

3.4. Application to a CT specimen

The above example was dealing with the homogeneous stress-field in the smooth specimen. The present one is situated at the opposite extreme side, where the initial "elastic" stress singularity is removed by the viscoplastic and damage effects.

Two materials are considered here, corresponding to very different behaviors: the INCONEL 718 which is brittle and the 316 stainless steel showing a ductile behavior. The convergence study has been made for the first material which is the most difficult case for the numerical procedure. The CT specimen is submitted to a constant load at constant temperature (600°C) for both materials). Uniform meshes have been designed in the region of cack propagation, in order to be able to study the convergence problems. The mesh sizes, for six node isoparametric triangles, are 100, 75, 50 and 20 µm. The figure 6.a shows the finite element model of the specimen and the region of crack tip in the case of the largest mesh size.

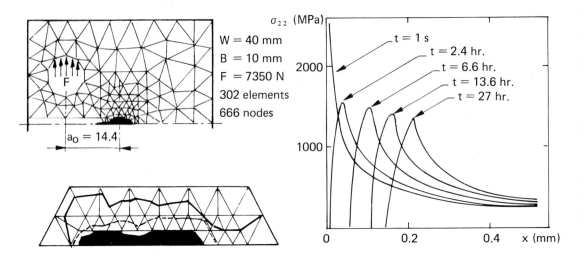

Fig. 6 — Prediction by local approach of creep crack growth of a CT test specimen
(INCO 718 alloy, 600°C).

The crack growth is simulated by the region where the damage attains its critical value $D_c = 1$. The figure 7.a gives the predicted crack growth rate as a function of the crack length when the damage is defined as a classical "local" field quantity. Clearly, no convergence is obtained as a function of the mesh size. In every case, the completely damage zone (the crack) was localized to the first row of elements.

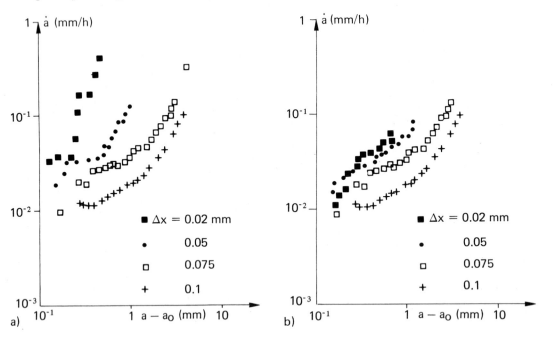

Fig. 7 — Calculated creep crack growth rates for the CT specimen (INCO 718, 600°C,
F = 7350 N), for various mesh sizes. (a) Classical local damage definition, (b) non-local
damage definition (d* = 10 μm).

The figure 6.b shows the distribution of the maximum principal stress ahead of the "crack" tip at several instants during the creep crack growth. In the damage zone (or process zone) where damage has values between 0 and 1, the stress redistributes completely, thanks to the complete and continuous coupling. The present local aproach leads to very different stress fields compared to the linear or non-linear fracture mechanics approaches.

Using the non-local definition for the damage variable (with $d^* = 20$ µm for the INCO 718) leads to a much better convergence, as shown in figure 7.b. The predicted crack growth compares fairly well with experimental results but some improvements could take place by taking into account the environmental effects [45]. The present solution has shown a relative insensititivy of the predicted growth rate to the crack length, provided the applied loading corresponds to the same stress intensity factor. This has been observed experimentally [45]. For such brittle material, the linear fracture mechanics can be applied as a good approximation.

In the case of the 316 stainless steel, the simulated crack growth [41] is extremely good compared to the experimental results of ref. [66], as shown in figure 8 for two different initial crack length. In the two cases, the loading conditions correspond to the same stress intensity factor but the crack growth rates differ by a factor between 5 and 8. The present local approach is able to reproduce exactly such differences which are not predicted by the linear or non-linear fracture mechanics. In these cases, the values of the C^* integral parameter are not sufficiently different to predict correctly the experimental results [41, 45, 66].

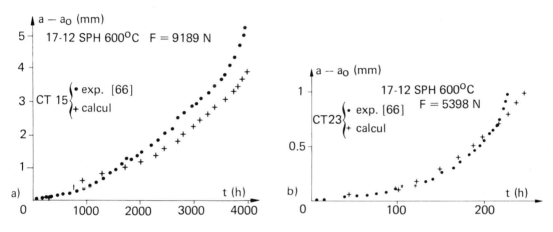

Fig. 8 — Creep crack growth prediction for two CT specimens in 316 S.S. at 600°C.
(a) CT 15, F = 9189 N, (b) CT 23, F = 5398 N.

CONCLUSION

Continuum Damage Mechanics has developed during the past twenty years in a fairly extensive way, in the frame of fundamental research studies as well as for some practical applications. Its main aspects have been discussed briefly, considering successively :

- measures and definitions of the damage variables,

- the phenomenological approach of damage growth equations,

- the effect of damage on the mechanical properties,

- the use of CDM in the development of local approaches to fracture.

Damage Mechanics is an extension of Continuum Mechanics. It offers numerous possibilities and, among other things, it is possible to relate the microstructural measures to the mechanical parameters. Moreover, this theory finds applications in a variety of situations (varied conditions of loading, varied materials and physical processes, etc).

The use of Damage Mechanics in the context of the local approaches has been largely developed in recent years to remedy certain inadequacies of Fracture Mechanics. The approach by continuous damage consists of considering the crack as the locus of totally damaged points (having reached the critical value).

This approach has been followed recently by a number of researchers. This is the case for example of the microstructurally based approach adopted by Tvergaard [67] for creep damage. He was able to predict beautiful results in several kinds of specimens. Creep crack growth is also a subject treated in the present paper, where a total coupling between creep damage and elastic-viscoplastic behavior is considered as the key to describe crack growth in terms of the extent of the completely damaged material.

The numerical problems associated to such a CDM approach of the crack propagation have been summarized and discussed with some details. The main problem concerns the localization effects which are associated to the intrinsic negative tangent modulus induced by the complete coupling between damage and the elastic behavior. Such difficulties have been illustrated in two extreme cases: the smooth tensile specimen under a constant rate elongation control and the CT specimen with creep crack growth under a constant load. In the two cases, the localization induces a great mesh dependency, even for very fine meshes (up to 20 µm mesh sizes).

The solution to this problem adopted in the present paper was independently proposed in some recent studies [63-65]. It consists in a slight modification of the local definition of the classical Continuum Mechanics, which introduces the notion of a non-local damage variable. All other internal variables (strain, temperature, hardening variables) are defined in the classical local way but the damage variable is averaged over a volume of material (by means of some influence function). This is justified by the presence, at the same scale, of both large defects (creep damage at grain boundaries) and large macroscopic stress gradients. In such a case the classical assumptions of the homogenization procedures are not verified [68] and the Continuum Mechanics approach needs the introduction of heigher order gradients or non-local variables.

The use of a non-local damage variable presents the advantage of conciliating two initially opposite points of view: the concepts of the materials science, which consider as necessary the use of a critical volume of material, the concepts of Continuum Mechanics, for which the finite element method gives a numerical solution which converges as a function of mesh refinements.

ACKNOWLEDGEMENTS

Part of this work has been realized in the framework of the GRECO "Grandes Déformations et Endommagement" and the GIS "Rupture à Chaud". The support of the french Ministry of Research M.R.E.S. is gratefully acknowledged.

REFERENCES

1.   KACHANOV, L.M., "Time of the rupture process under creep conditions". Isv. Akad. Nauk. SSR, Otd Tekh. Nauk. n° 8, (1958), p. 26-31.

2.   RABOTNOV Y.N., "Creep problems in structural members", North-Holland, (1969).

3.   HULT J., "Continuum damage mechanics". Capabilities limitations and promises. Mechanisms of Deformation and Fracture. Pergamon, Oxford, (1979), p. 233-247.

4. CHABOCHE J.L., "Continuous damage mechanics. A tool to describe phenomena before crack initiation". Nuclear Engineering and Design, vol. 64, (1981), p. 233 247.

5. MURAKAMI S. , "Notion of continuum damage mechanics and its application to anisotropic creep damage theory". J. Eng. Mat. and Techn., vol. 105, (1983), p. 99.

6. KRAJCINOVIC D., "Continuum Damage Mechanics". Applied Mechanics Reviews, vol. 37, n° 1, (1984).

7. LEMAITRE J and CHABOCHE J.L., "Mécanique des Matériaux Solides". Dunod, Paris, (1985). English edition, Cambridge Univ. Press, 1988.

8. CHABOCHE J.L., "Continuum Damage Mechanics, Parts I and II". J. of Applied Mechs., vol. 55, (1988), p. 59-72.

9. CHABOCHE J.L., "Sur l'utilisation des variables d'état interne pour la description du comportement viscoplastique et de la rupture par endommagement". Symp. Franco-Polonais de la Rhéologie et Mécanique, Cracovie, (1977).

10. DUVAUT C., "Analyse fonctionnelle-mécanique des milieux continus-homogénéisation". Theoretical and Applied Mechanics North-Holland, Amsterdam, (1976).

11. CHABOCHE J.L., " Le concept de contrainte effective, appliqué à l'élasticité et à la viscoplasticité en présence d'un endommagement anisotrope. Col. EUROMECH 115, Grenoble, (1979), CNRS, (1982).

12. PLUMTREE A. and NILSSON J.O., "Damage mechanics applied to high temperature fatigue". Journées Internationales de Printemps, Paris, (1986).

13. RIDES M., COCKS A.C.F., HAYHURST, D.R., "The elastic response of creep damaged materials". To appear in the ASME J. of Applied Mechs..

14. HAYHURST D.R., "Creep rupture under multi-axial state of stress". J. Mech. Phys. Solids, vol. 20, n° 6, (1972), p. 381-392.

15. LECKIE F.A., HAYHURST D.R., "Creep rupture of structures". Proc. Royal Soc., London, vol. 340, (1974), p. 323-347.

16. MURAKAMI S. and OHNO N., "A continuum theory of creep and creep damage". 3rd IUTAM Symp. on Creep in Structures, Leicester, (1980).

17. CHABOCHE J.L., Une loi différentielle d'endommagement de fatigue avec cumulation non linéraire". Revue Française de Mécanique, n° 50-51, (1974). English translation in "Annales de l'ITBTP, HS, (1977).

18. SOCIE D.F., FASH J.W., LECKIE F.A., "A continuum damage model for fatigue analysis of cast iron". ASME Conf. in Life Prediction, Albany, New York (1983).

19. CHABOCHE J.L., LESNE P.M., "A non-linear continuous fatigue damage model". Fatigue and Fract. Engng. Mater. and Struct., vol. 11, n° 1, (1988), p. 1-17.

20. CHAREWICZ A. and DANIEL I.M., "Fatigue damage mechanisms and residual properties of graphite/epoxy laminates". IUTAM Symp. on Mechanics of Damage and Fatigue, Haifa, (1985), Engng. Fract. Mech., 25, n° 5/6 (1986).

21.   OGIN S.L., SMITH P.A., BEAUMONT P.W.R., "Matrix cracking and stiffness reduction during the fatigue of a [0/90]$_s$ GFRP laminate. Compos. Sci. Technol. 22, p. 22-31, (1985).

22.   CORDEBOIS J.P., "Critères d'instabilité plastique et endommagement ductile en grandes déformations". Thèse, Université Paris-VI, (1983).

23.   ROUSSELIER G., "Finite deformation constitutive relations including ductile fracture damage". IUTAM Symp. on Three-Dimensional Constitutive Relations and Ductile Fracture, Dourdan (1980), Eds. Nemat-Nasser, North-Holland Publ. Comp., (1981), p. 331-355.

24.   LEMAITRE J., "A Continuum Damage Mechanics model for ductile fracture". J. Engng. Mat. and Technology, ASME, vol. 107, (1985), p. 83-89.

25.   LEE H., PENG K., WANG J., "An anisotropic damage criterion for deformation instability and its application to forming limit analysis of metal plates". Engng. Fract. Mech., 21, p. 1031-1054 (1985).

26.   CHOW C.L., JUNE WANG, "An anisotropic theory of Continuum Damage Mechanics for ductile fracture". Engng. Fract. Mechs., 27, n° 5, p. 547-558, (1987).

27.   KRAJCINOVIC D. and FONSEKA G.U., "The continuous damage theory of brittle materials, Parts 1 and 2". J. of Applied Mechanics, ASME , vol. 48, (1981), p. 809-824.

28.   MAZARS J., "A description of micro- and macroscale damage of concrete structures". Engng. Fract. Mechs., 25, n° 5/6, p. 729-737 (1986).

29.   BAZANT Z.P., BELYTSCHKO T., "Localization and size effect". 2nd Int. Conf. Constitutive Laws for Engineering Materials; Theory and Applications, Tucson, (1987), DESAI C.S. et al. eds., Elsevier.

30.   ALLIX O., GILLETTA D. and LADEVEZE P., "Non-linear mechanical behaviour of laminates". EUROMECH 204, "Structures and Crack Propagation in Brittle Matrix Composite Materials", Jablonna, Poland, (1985), BRANDT A.M. and MARSHALL H.I. eds., Elsevier App. Sc., London.

31.   TALREJA R., "A continuum mechanics characterization of damage in composite material". Proc. R. Soc. Lond. A399, (1985), p. 195-216.

32.   WANG S.S., SUEMASU H and CHIM E.S.M., "Analysis of fatigue damage evolution and associated anisotropic elastic property degradation in random short fiber composite". Engng. Fracture Mechs., vol. 25, n° 5/6, (1986), p. 829-844.

33.   LECKIE F.A. and ONAT E.T., "Tensorial nature of damage measuring internal variables". IUTAM Symp. on Physical Non-Linearities in Structural Analysis, Senlis, France, (1980) (Springer).

34.   CORDEBOIS J.P. et SIDOROFF F., "Anisotropie élastique induite par endommagement". Col. EUROMECH 115, Grenoble, (1979), CNRS, (1982).

35.   SIMO J.C., JU J.W., TAYLOR R.L. and PISTER K.S., "On strain-based continuum damage models: formulation and computational aspects". 2nd Int. Conf. on Constitutive Laws for Engineering Materials: Theory and Applications, Tucson, (1987).

36.   LADEVEZE P. and LEMAITRE J., "Damage effective stress in quasi-unilateral material condition". IUTAM Congress, Lyngby, Denmark, (1984).

37. CHABOCHE J.L., "Continuum Damage Mechanics and its application to structural lifetime predictions". La Recherche Aérospatiale, n° 1987-4, English edition.

38. LEVAILLANT C and PINEAU A., "Assessment of high temperature low-cycle fatigue life of austenitic stainless steels by using intergranular damage as a correlating parameter". Int. Symp. on Low-Cycle Fatigue and Life Prediction, Firminy, France, (1980), ASTM STP 770, (1982).

39. BEZIAT J., DIBOINE A., LEVAILLANT C. and PINEAU A.: "Creep damage in a 316 S.S. under triaxial stresses". 4th Int. Seminar on Inelastic Analysis and Life Prediction in High Temperature Environment, Chicago, Illinois, (1983).

40. DYSON B.F., GIBBONS, T.B., Tertiary creep in nickel base superalloys". Acta Met., vol. 35, n° 9, p. 2355, (1987).

41. SAANOUNI K. "Sur l'analyse de la fissuration des milieux élasto-viscoplastiques par la théorie de l'endommagement continu". Thèse d'Etat, Université de Compiègne, (1988).

42. CAILLETAUD G. and LEVAILLANT C., "Creep-fatigue life prediction: what about initiation?" Nuclear Engineering and Design, n° 83, (1984), p. 279-292.

43. LEVAILLANT C., "Approche métallographique de l'endommagement d'aciers inoxydables austénitiques sollicités en fatigue plastique ou en fluage : description et interprétation physique des interactions fatigue-fluage-oxydation". Thèse d'état, UTC, (1984).

44. PINEAU A., "Review of fracture mechanisms and local approaches to predicting crack resistance in low strength steels". ICF5, Cannes, Advances in Fracture Researches, FRANCOIS D. et al. eds., Pergamon Press, New York, (1981).

45. BENSUSSAN P., MAAS E., PELLOUX R., PINEAU A., "Creep crack initiation and propagation: Fracture Mechanics and Local Approach". 5th Int. Seminar on Inelastic Analysis on Life Prediction in High Temperature Environment, Paris, 1985.

46. ANQUEZ L. "Elastoplastic crack propagation fatigue and failure". La Recherche Aérospatiale n° 1983-2, French and English editions.

47. NEWMAN Jr. J.C., "Finite element analysis of fatigue crack propagation including the effects of crack closure. Ph. D. Thesis, VPI, Blacksburg, (1974).

48. DEVAUX J.C. et MOTTET G., " Déchirure ductile des aciers faiblement alliés : modèles numériques". Rapport n° 79-057, Framatome, (1984).

49. WALKER K.P. and WILSON D.A., "Constitutive modeling of engine materials". PWA Report FR17911 (AFWAL-TR-84-4073), (1984).

50. LEMAITRE J., "Local approach of fracture". Symp. on Mechanics of Damage and Fatigue, Haïfa, Israel, Engng. Fracture Mechs., vol. 25, n° 5/6, (1986), p. 523-527.

51. HAYHURST D.R., DIMMER P.R. and CHERNUKA M.W., "Estimates of the creep rupture lifetime of structures using the finite element method". J. Mech. Phys. of Solids, vol. 23, (1975), p. 335.

52. SAANOUNI K., CHABOCHE J.L. and BATHIAS C., "On the creep crack growth prediction by a local approach". IUTAM Symp. on Mechanics of Damage and Fatigue, Haïfa, (1985). J. of Engineering Francture Mechanics, 25, n° 5/6, (1986), p. 677-691.

53.   MURAKAMI S., "Anisotropic damage theory and its application to creep crack growth analysis". 2nd Int. Conf. on Constitutive Laws for Engng. Materials - Theory and Applications. Tucson, (1987), Publ. Elsevier, DESAI et al. eds., p. 187.

54.   ROUSSELIER G., DEVAUX J.C. and MOTTET G., "Ductile initiation and crack growth in tensile specimens. Application of Continuum Damage Mechanics". SMIRT 8, Brussels, (1985).

55.   BILLARDON R., "Fully coupled strain and damage finite element analysis of ductile fracture". Int. Seminar on Local Approaches of Fracture, Moret-sur-Loing, France, (1986). To appear in Nuclear Engng. and Design.

56.   NGUYEN Q.S., BUI H.D., "Sur les matériaux élastoplastiques à écrouissage positif ou négatif". J. de Mécanique, vol. 13, n° 2, 1974.

57.   RICE, J.R., "The localization of plastic deformation - Theoretical and Applied Mechanics". 14th Int. Cong. Theoret. Appl. Mech., ed. KOITER W.T., North-Holland, Amsterdam, (1977), p. 207-220.

58.   VALANIS K.C., On the uniqueness of solution of the initial value problem in softening materials". J. of Applied Mechanics, (1985), vol. 52, p. 649-653.

59.   TVERGAARD V., "Influence of voids on shear band instabilities under plane strain conditions". Int. J. of Fracture, 17, n° 4, (1981), p. 389-407.

60.   TVERGAARD V., "Effect of yield surface curvature and void nucleation on plastic flow localization". J. Mechs. Phys. Solids, 35, n° 1, 1987, p. 43-60.

61.   LEMONDS J., NEEDLEMAN A., "Finite element analyses of shear localization in rate and temperature dependent solids. Mechs. of Materials, (1986), p.339-361.

62.   NEEDLEMAN., "Material rate dependence and mesh sensitivity in localization problem". Computer method in Applied Mechanics and Engng., 67, 1988, p. 69-85.

63.   BAZANT Z.P. and PIJAUDIER-CABOT G., "Modeling of distributed damage by non local continuum with local strain". 4th Int. Conf. on Numerical Methods in Fracture Mechanics, San Antonio, Texas, (1987), LUXMOORE A.R. et al., Pineridge Press, Swansea, p. 411-432.

64.   SAANOUNI K., CHABOCHE J.L. and BATHIAS C., "Prévision de l'amorçage et de la propagation des fissures en élasto-viscoplasticité". Int. Seminar "High Temperature Fracture Mechanisms and Mechanics", Dourdan, France, (1987).

65.   BILLARDON R., "Strain softening and mesh sensitivity of fully coupled strain and damage finite element analysis". Int. Seminar "High Temperature Fracture Mechanisms and Mechanics", Dourdan, France, (1987).

66.   MAAS E., PINEAU A., "Creep crack growth behavior of type 316L steel". Engng. Fracture Mechanics, vol. 22, n° 2, 1985, p. 307-325.

67.   TVERGAARD V., "Analysis of creep crack growth by grain boundary cavitation". Int. J. of Fracture, 31, (1986), p. 183-209.

68.   ONAT T., LIU C.H., "On errors in homogenization". Symp. on the Mechanics of Composite Materials, Applied Mechs. and Engng. Sciences Conf., Berkeley, (1988).

Theoretical and Applied Mechanics
P. Germain, M. Piau and D. Caillerie (Editors)
Elsevier Science Publishers B.V. (North-Holland)

FLUIDISATION OF SOLID PARTICLES

John F. Davidson

Department of Chemical Engineering, University of Cambridge,
Pembroke Street, Cambridge, England, CB2 3RA.

A fluidised bed is formed when a gas or liquid passes up through a bed of particles at a rate such that the pressure drop is sufficient to support the weight of the particles. The bed of fluidised particles has many of the properties of an ordinary liquid. A primary purpose of the paper is to demonstrate the similarities between a fluidised bed and a gas-liquid system.

A gas-fluidised bed contains gas bubbles: much of the gas flow may pass through the bed as bubbles, whose behaviour is strikingly similar to that of large gas bubbles in liquid. The bubbles play a dominant role in the behaviour of the bed, governing its expansion. The bubbles carry particles into the freeboard, the gas space above the bed. In the freeboard, "ghost bubbles" are important: a ghost bubble is a zone of upward moving gas arising from a bubble in the particle bed below. Ghost bubbles increase the turbulence level in the freeboard and thereby assist diffusion of particles to the vertical containing wall where the particles form a falling film.

Particle jets from fluidised beds show remarkable coherence and their behaviour is similar to that of a liquid jet.

Falling films of particles on the wall are believed to dominate the behaviour of "fast fluidised beds". In a fast fluidised bed, the upward gas velocity in the vertical containing tube is such that there is a net upward flow of particles, recycled to the bottom of the containing tube via an external cyclone. Recent work has shown that turbulent diffusion to falling particle films on the wall appears to explain important characteristics of fast fluidised beds.

1. INTRODUCTION

The purpose of this paper is to show the analogies between the behaviour of fluidised beds and the behaviour of two-phase gas-liquid systems. Figure 1 indicates the types of fluidised bed that will be considered. Figure 1(a) shows a bubbling fluidised bed with an upward flow of gas through a bed of particles, supported on a distributor plate. The mean upward gas velocity, U, is less than the terminal free falling velocity of the majority of particles in the bed. Consequently only fine particles are carried up into the freeboard above the bed and recycled via the cyclone. For most gas-fluidised systems of this kind, the majority of the gas passes through as bubbles, and an important feature of this paper will be a discussion of the behaviour of the bubbles. Figure 1(b) shows a fast fluidised bed: the distinguishing feature of such beds is that the upward gas velocity, U, in the riser pipe is greater than the terminal free falling velocity of the majority of the particles in the bed. Consequently there is a large upward flow of particles in the riser pipe: these particles enter the cyclone and are recycled via a small dense phase bed at the foot of the cyclone. In this

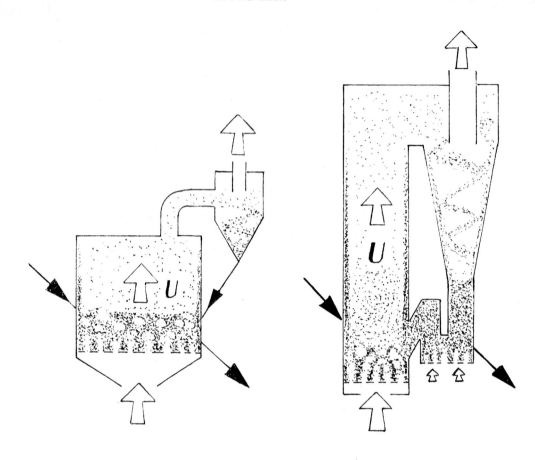

(a)   (b)

FIGURE 1   Types of fluidised bed.   Gas ⇨ Solids ➝

(a) Bubbling bed for which $U < U_t$: U_t = free-falling velocity of a
single particle.

(b) Fast or recirculating fluidised bed: $U \gg U_t$.

U = average riser gas velocity.

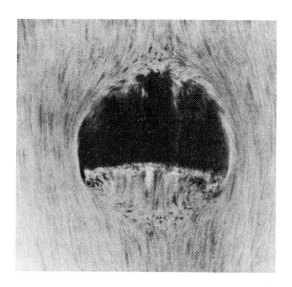

FIGURE 2   Photograph of a bubble near the wall of a fluidised bed [1].   The
            camera moved with the bubble; the exposure of 1/50 s shows the
            particle streamlines.

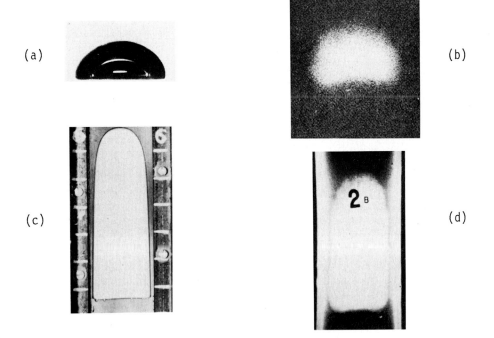

(a)

(b)

(c)

(d)

FIGURE 3   Comparison between bubbles in liquids (a) (c), and in fluidised
            beds (b) (d).   (a) Air bubble in viscous liquid [2].   (b) X-ray
            photograph of bubble in fluidised bed [3].   (c) Air bubble in
            water between parallel plates [4].   (d) X-ray photograph of bubble
            in tube [4].

dense phase bed between the bottom of the cyclone and the bottom of the riser
pipe, the dense phase particles flow rather like a liquid.   This paper
includes a discussion of the flow of a dense phase fluidised bed through an
orifice; here the dense bed again demonstrates it liquid-like nature.

2.   BUBBLES IN FLUIDISED BEDS

2.1 Steady rise of a bubble

Visual observation of fluidised particles in a transparent container shows
that much of the upward flowing gas passes through the particles as bubbles.
Figure 2 shows a photograph of such a bubble taken by a camera which moved at
the same upward velocity as the bubble.   The bubble thus appeared to the
camera as if it was held fixed by a downward flow of particles, and the
relatively long exposure time of 1/50 second revealed the particle
streamlines.   It is clear from these streamlines that the particles move
round the bubble rather than raining through the roof which is an alternative
possibility.   Figure 2 also shows that the bubble has the spherical cap shape
which is familiar for a large bubble in a liquid of low viscosity.

The similarity of shape is further shown in figure 3.   This gives a
comparison between bubbles in ordinary liquids, [figure 3(a), (c)] and the
corresponding bubbles in a fluidised bed, as seen by X-ray photography,
[figure 3(b) (d)].   Figure 3(a), (b), show bubbles which are small in
comparison with the diameter of the container and they have the characteristic
spherical cap shape.   Figure 3(c), (d), show large bubbles, the so-called
"slugs", where the bubble volume is such that the bubble extends right across
the diameter of the tube or other container so that the motion is dominated by
wall effects.   These slugs have a characteristic form: there is a rounded
nose and a flat base with a thin film of liquid draining down the walls for
the gas-liquid case and a corresponding thin film of particles in the case of
the fluidised bed bubble.

Many experimentalists have measured the rising velocities of such bubbles in
fluidised beds using a variety of methods.   They have reported the rising
velocity, u_b, in terms of the diameter, D_e, of the sphere having the same
volume as the bubble.   Also relevant is the diameter of the containing tube,
D, and the acceleration of gravity, g.   For bubbles with D_e much less than D,
the rising velocity is given by

$$u_b = 0.71 \sqrt{gD_e} \ . \qquad\qquad (1)$$

For slugs, which are formed when D_e is greater than D, the rising velocity is
given by

$$u_b = 0.35 \sqrt{gD} \ . \qquad\qquad (2)$$

The results are displayed in figure 4 and although the data are very
scattered, they agree generally with equations (1) and (2).

Equations (1) and (2) are appropriate for large gas bubbles rising in a liquid
of low surface tension and low viscosity, suggesting that the fluidised bed of
particles behaves as such a liquid.   Further evidence for this hypothesis is
given in figure 5 which shows a large number of measurements of bubble
velocity plotted against bubble height in the case where D_e is less than D.
The data in figure 5 suggest that bubble velocity is proportional to the
square root of a representative bubble dimension, which again suggests that
the fluidised bed of particles behaves like an inviscid liquid of low surface
tension.

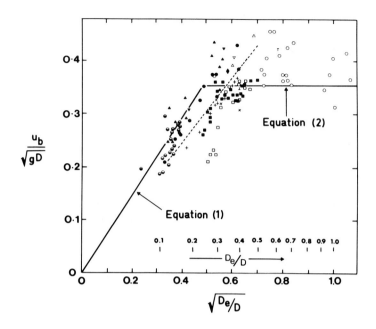

FIGURE 4   Velocity u_b of a single bubble injected into a fluidised bed of
diameter D. D_e = diameter of the sphere whose volume is that of
the bubble [5].

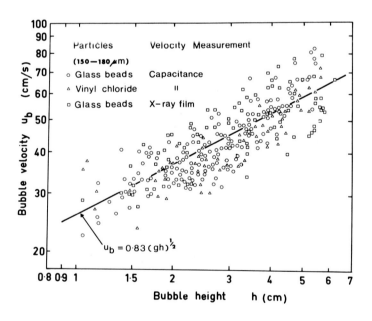

FIGURE 5   Rising velocity of a single bubble in a fluidised bed as a function
of bubble height h [5] [6].

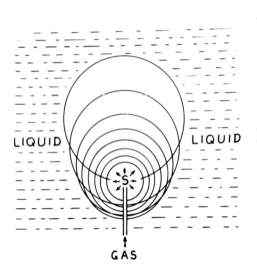

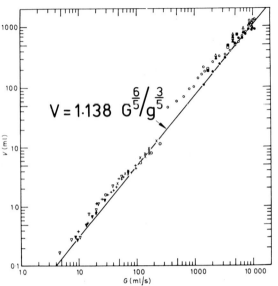

FIGURE 6   Bubble formation at an
orifice in a liquid: theoretical
development [7].

FIGURE 7   Bubble volume V as a
function of gas flow G for continuous
bubble formation in water [7].

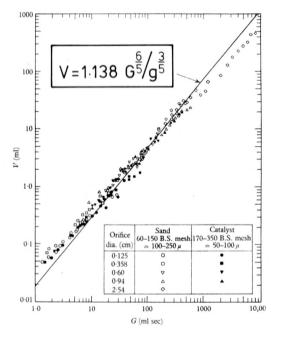

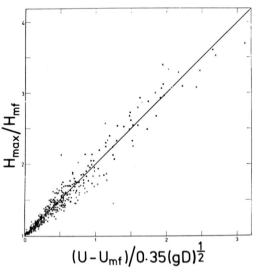

FIGURE 8   Continuous bubble
formation at an orifice in an
incipiently fluidised bed [7].

Figure 9   Maximum bed height H_{max}
as a function of fluidising velocity
U. H_{mf} and U_{mf} are at incipient
fluidisation [8].   The straight line
is from equations (2) and (4).

2.2 Bubble formation

When gas is blown steadily through a tube into a liquid, a stream of bubbles is formed with a more or less regular frequency.   Figure 6 shows an idealised picture of the bubble formation on the supposition that the growing bubble is at all times spherical.   The bubble grows as gas flows into it, expanding as a sphere whose centre starts at the end of the tube, but the centre must rise due to the buoyancy of the bubble.   Initially the base of the bubble must move downwards;   later on, the upward motion due to buoyancy overcomes the downward motion due to growth, so the base of the bubble moves upwards, eventually detaching from the orifice.   Simple theory, based on the sequence shown in figure 6, leads to the following result for a liquid of low viscosity.

$$V = (6/\pi)^{\frac{1}{5}} G^{\frac{6}{5}}/g^{\frac{3}{5}} = 1.138 \ G^{\frac{6}{5}}/g^{\frac{3}{5}} . \qquad (3)$$

Figure 7 shows the results of extensive measurements of bubble volume, V, in water, detaching from an orifice through which gas is supplied at the steady flow rate, G.   The volume of the detaching bubbles agrees approximately with equation (3) over a wide range of flow rates.   The corresponding data for bubbles formed at an orifice in an incipiently fluidised bed are shown in figure 8 and these results also agree approximately with equation (3).   This provides further evidence that the fluidised bed behaves like a liquid of low viscosity with negligible surface tension.

2.3 Bed expansion due to bubbles

When air or other gas is blown steadily upwards through a bed of particles, observation shows that the bed expands due to the presence of bubbles.   The bubble velocity being determined by equations such as (1) or (2), the bubbles have a finite residence time in the bed and must therefore occupy a certain volume, causing the bed to expand.   To make a prediction of expansion, the two-phase theory is applied as follows.   For a given bed of particles there is a minimum velocity, U_{mf}, at which the pressure drop just equals the weight of the bed;   this is the incipient fluidising velocity.   The basis of the two-phase theory is that the additional flow rate beyond U_{mf}, $(U - U_{mf})$ x (the cross-sectional area of the bed), passes through the bed as bubbles.   Using the two-phase theory, together with an assumed bubble velocity, u_b, given by equation (1) or (2), the following expression can be derived [8] for the bed height, H_{max}.

$$\frac{H_{max} - H_{mf}}{H_{mf}} = \frac{U - U_{mf}}{u_b} . \qquad (4)$$

Equation (4) gives a simple method of predicting the bed height, H_{max}: equation (1) is used when the bubble diameter D_e is much less than the bed diameter, and equation (2) when the bubble diameter D_e is comparable with, or greater than, the bed diameter.   The former situation is usually relevant for large industrial units, because the bed diameter is usually much greater than the bubble diameter;   the latter, i.e. bubble diameter comparable with bed diameter, applies for most laboratory units.   This generates scale up problems familiar to chemical engineers.

It is important to note that equation (4) predicts the maximum bed height H_{max};   particularly with slug flow, the bed height fluctuates with time as the bubbles break through the surface, so that the surface moves up and down with a reasonably regular frequency.   It is the maximum height which is predicted by the two-phase theory, because the maximum height occurs when the slug is just breaking through the surface, so each slug is wholly within the bed until its nose reaches the position of maximum bed height.   Consequently the residence time of the slug, which determines the bed expansion, relates to the maximum bed height. Confirmation of this reasoning is given by the data in figure 9 which shows H_{max}/H_{mf} plotted against the ratio $(U - U_{mf})/0.35 \sqrt{gD}$. The excellent agreement between the data and the straight line representing equations (2) and (4) shows that the simple two-phase theory gives a good prediction of maximum bed height. Note that the data in figure 9 include results for a very vigorously fluidised bed in which the maximum height is about four times the height at incipient fluidisation.

The agreement between theory and experiment in figure 9 represents a further confirmation of the hypothesis that the fluidised bed of particles behaves like a liquid of low viscosity and negligible surface tension.

2.4 Equations of particle and fluid motion

Why should the fluidised bed behave like an inviscid liquid with negligible surface tension?   Partial answers to this question may be obtained by considering the equations governing continuity and pressure distribution within the fluid and particles.

If it is assumed that the voidage fraction of the moving particulate phase round the bubble is constant, as suggested by photographs like figure 2, the continuity equations for the fluid and for the particles respectively are given by

$$\text{Fluid, div } u = 0. \quad \text{Particles, div } v = 0 . \qquad (5)$$

For small particles, the relative motion between the fluid and the particles is governed by Darcy's law so that the fluid velocity is given by

$$u = v - K\text{grad } p_f . \qquad (6)$$

Here K is a constant, provided the voidage fraction is uniform; p_f is the fluid pressure.   Combining equations (5) and (6) gives

$$\text{div grad } p_f = 0 . \qquad (7)$$

Now this is the equation governing the pressure distribution in a fixed bed of particles, and an important result;   the simple, and perhaps surprising, conclusion is that the pressure distribution, for given boundary conditions, is unaffected by the particle motion.   It is then easy to infer the form of the pressure distribution above and below the bubble, indicated in figure 10(a).   This shows the pressure distribution in the particulate phase on a vertical axis through a bubble of spherical cap form.   There must be a pressure gradient within the particulate phase, because it is everywhere fluidised, so that in regions remote from the bubble the pressure gradient is constant.   This means that there is an upward percolation of fluid through the interstices between the particles; as the fluid approaches the bubble, the streamlines will bend in towards the cavity, through which they find an easy route and consequently near the bubble there will be steeper pressure gradients as indicted in figure 10(a).   Of course the pressure within the bubble must be constant as indicated in figure 10(a).

Now considering the motion of an inviscid liquid round the spherical cap bubble, the pressure gradients may be estimated from Bernoulli's theorem, at least above the nose of the bubble where there is negligible loss of energy. This gives pressure gradients of the form shown in figure 10(b), which gives an indication of the pressure profile arising from the particle motion. There is of course a uniform pressure gradient in regions remote from the bubble corresponding to the hydrostatic change of pressure within the ideal liquid.   Likewise there must be constant pressure within the bubble because

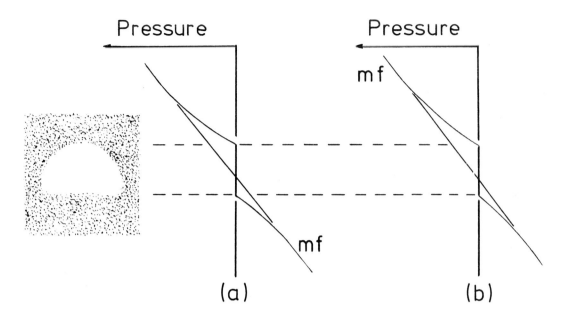

FIGURE 10   Pressure profiles round a bubble [7].

   (a) Pressure profile arising from percolation of fluid through

    the interstices between particles.

   (b) Pressure profile arising from particle motion.

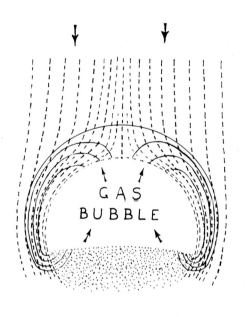

FIGURE 11   Idealised particle (----)
and fluid (——) streamlines for a
bubble in a fluidised bed as seen by
a camera moving with the bubble
(Figure 2).

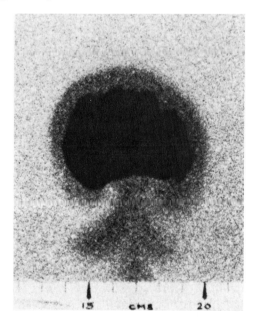

FIGURE 12   'Cloud' of NO_2 round an

NO_2 bubble in a fluidised bed [9].

it contains gas of negligible density. The probable matching between the forms of the pressure gradients in figures 10(a) and (b) gives a qualitative explanation as to why the bubble in the fluidised bed behaves as if it was in a liquid of low viscosity: the pressure gradients induced by the percolation are of the appropriate form to cause the necessary particle motion.

These considerations lead to the particle and fluid streamlines shown in figure 11 for a gas bubble held fixed by a downward flow of particles, just as seen in figure 2. The downward moving particles, shown by broken curves, keep the gas bubble fixed. But the pressure gradient forces gas upwards relative to the particles and hence there is an upward flow of gas through the bubble. When the gas enters the bubble roof, it is swept downwards by the motion of the particles; hence the gas from the bubble penetrates only a finite distance into the particulate phase, leading to the formation of the so-called "cloud" around the bubble. This cloud is indicated in figure 12, which shows a bubble containing nitric oxide rising in a bed of glass beads, the bubble being near a transparent wall. The finite penetration of the nitric oxide into the glass ballotini can be seen clearly. Measurements [9] show that the cloud diameter can be predicted by theory derived from equations (5), (6) and (7).

3.   PARTICLE JET FROM FLUIDISED BED

Another instance of liquid-like behaviour arises when there is an orifice in the side of a fluidised bed. A jet of particles emerges and if the orifice is suitably shaped, the jet has the very coherent form shown in figure 13. In this case, the jet diameter was about 3 mm and the jet retained its coherence, apart from a few particles stripped off at the surface, over a flight of 1 or 2 metres. From the jet trajectory, the exit velocity could be measured, and obeyed the equation

$$u = C_v \sqrt{2gh} \, . \qquad\qquad (8)$$

Here h is the effective head of particles above the orifice. The validity of equation (8) demonstrates liquid-like behaviour, although the velocity coefficient, C_v, is of order 0.6-0.7, rather less than would be obtained for an inviscid liquid. The reasons for this low velocity coefficient are not yet clear. The discharge coefficient, based on the use of Bernoulli's theorem, is of order 0.6-0.7; it is rather higher for shaped orifices than for orifice plates and in this respect the particulate phase does behave like a liquid.

4.   GAS MOTION IN THE FREEBOARD ABOVE A BUBBLING FLUIDISED BED

The turbulence level in the freeboard immediately above the bed is high because the gas emerges in an irregular fashion, due to the bursting bubbles. The bursting bubbles carry up fine particles which are thrown into the freeboard, some of them falling back into the bed and some being carried out of the equipment into the cyclone, indicated in figure 1(a). Particle carry over is of great importance to the fluid bed designer and hence elutriation has been intensively studied.

Pemberton [11] made a detailed study of freeboard gas motion using both gas-liquid systems and fluidised beds as follows.

4.1 Ghost bubble rising from stagnant liquid

Figure 14 shows a sequence of photographs arising when an ammonium chloride bubble generated in water rises through the water surface into the air space

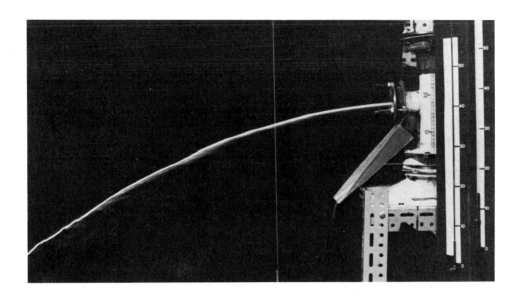

FIGURE 13   Particle jet from a shaped nozzle (exit diameter 3.1 mm) supplied by a fluidised bed [10].

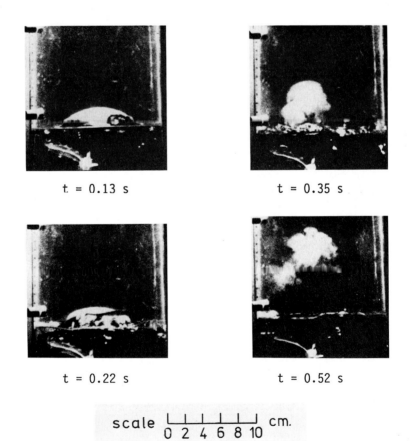

t = 0.13 s   t = 0.35 s

t = 0.22 s   t = 0.52 s

scale |_|_|_|_|_| cm.
0 2 4 6 8 10

FIGURE 14   Eruption of a 40 ml NH_4Cl bubble from water into air.   Times t are after release [11].

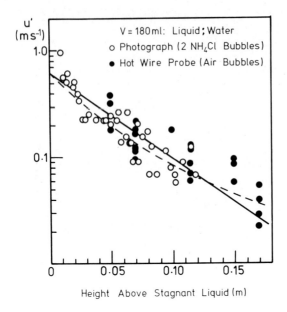

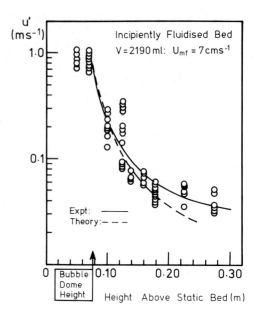

FIGURE 15   Velocity fluctuations u' above water, due to the eruption of two NH₄Cl bubbles [11].

Best fit ——.   Puff theory ----.

FIGURE 16   Velocity fluctuations u' above incipiently fluidised bed, due to the eruption of one bubble [11].

Best fit ——.   Puff theory ----.

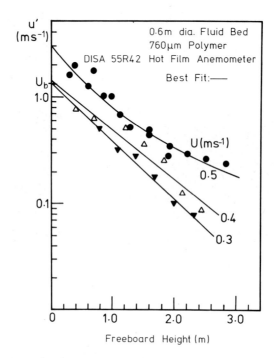

FIGURE 17   Turbulent velocity fluctuations above a fluidised bed [11].

Fluidising velocity U; U_{mf} = 0.05 m/s.

above.   The ammonium chloride particles delineate the bubble which emerges, more or less intact, to form a 'ghost' bubble in the freeboard.   The ghost bubble retains its identity, continuing to rise in the freeboard and mixing with freeboard air as it moves: the entrainment causes the upward velocity to diminish as the ghost bubble rises in the freeboard.

Figure 15 shows fluctuating velocities generated by ghost bubbles rising from water into air.   The velocity immediately above the liquid surface is higher than the bubble rise velocity, because of the complexities of the eruption process, but the decay of velocity above the liquid surface is quite rapid. A theoretical prediction of u', the fluctuating velocity, assuming the bubble behaves as a ring vortex with turbulent entrainment, the so-called puff model, is shown in figure 15.

4.2 Bubble erupting from an incipiently fluidised bed

Similar experiments were performed with an incipiently fluidised bed into which bubbles were injected one at a time, each bubble emerging into the freeboard to form a ghost bubble.   Detection of fluctuating velocities in the freeboard was by hot wire anemometer and the results are shown in figure 16. As with a gas-liquid system, the fluctuating velocities decay rapidly with distance above the surface of the fluidised bed.   The data agree approximately with the puff model, though with rather different parameters compared with those used for the gas-liquid system.

4.3 Decay of turbulence above a fluidised bed

For a vigorously fluidised bed (U much greater than U_{mf}) ghost bubbles emerge continuously into the freeboard, giving a high level of turbulence which decays with distance above the bed.   Figure 17 shows measurements of fluctuating velocity using a hot film anemometer, for a range of flow velocities.   Immediately above the bed surface, the fluctuating velocities are much higher than the mean flow velocity.   These fluctuations decay with distance above the bed surface, and with a tall freeboard, eventually reach the fluctuation characteristic of turbulent flow in a pipe, of order 0.1 U.

No satisfactory theoretical prediction of the decay of fluctuating velocities has been obtained.   But it may be noted that the decay process is slower for the continuously fluidised bed than for the single bubble experiments of figures 15 and 16: thus the fluctuations decay from a value of about 1 m/s to 0.1 m/s over a freeboard height of about 2 metres in figure 17; this implies a gas residence time of order 4-6 sec with the flow velocities used, 0.3-0.5 m/s.   By contrast, the data of figure 15 and 16 show that a single ghost bubble reduces its velocity from the emerging value of about 1 m/sec to 0.1 m/s in a height of about 0.1 m; this implies a decay time of less than 1 second.   With a continuously fluidised bed, it is clear that the turbulence energy per unit volume of emerging gas due to the fluctuations is much greater than for a single ghost bubble, and hence it is plausible that the decay process should be slower for the continuously fluidised bed.

5.   FAST-FLUIDISED BEDS

Fast fluidised beds are widely used as calciners and for combustion of coal and other solid fuels such as peat.   In the riser pipe of the fast fluidised bed, shown in figure 1(b), the upward gas velocity is typically of order 4-8 m/s.   The corresponding mean upward solids velocity is order 0.5-1 m/s, indicating a large slip velocity, far beyond the free-falling velocity of single particles.   The explanation for this paradox emerges from studies of the flow regime in the riser.

5.1 Flow regime in riser

Measurements of gas velocity and of particle flux show that the flow pattern is not uniform across the diameter of the riser. Near the wall, gas and particles are moving downwards; in the middle part of the riser, over about three quarters of its area, gas and particles move upwards at high speed, the solids content of the gas being relatively low. Recent studies [12, 13] show that there is a dense film of particles running down the wall of the riser. The total downward flow of particles in this film may be three or four times the nett throughput of particles entering the cyclone in figure 1(b). This suggests the flow pattern indicated in figure 18. At the bottom of the riser is a very vigorously bubbling fluidised bed, similar in type to the gas-liquid motion known as "churn turbulent flow", from which there is a high rate of entrainment into the "freeboard". Particles are thus carried upwards in dilute phase: but the intense turbulence generated by the ghost bubbles transfers particles rapidly to the wall where they form a downward moving film which returns particles at a high rate at the bottom of the riser. The dilute phase core has an upward flux, E, and the wall film has a downward flow rate equivalent to flux w; these fluxes are related to the distance, z, above the churn turbulent region by the following equations.

$$w = w_\infty + (w_0 - w_\infty)e^{-Kz} \ . \qquad (9)$$

$$E = E_\infty + (E_0 - E_\infty)e^{-Kz} \ . \qquad (10)$$

The nett throughput of particles which enters the cyclone in figure 1(b) is defined by a flux G, related to E and w as follows.

$$G = E - w = E_0 - w_0 = E_\infty - w_\infty \ . \qquad (11)$$

E_∞ and w_∞ are the rates corresponding to the top of a very tall riser where the rate of transfer from the dilute phase to the wall film is equal to the rate of entrainment from the wall film into the dilute core. E_0 is the upward flux at the top of the churn-turbulent region; w_0 is the corresponding rate at which particles enter the churn-turbulent region from the wall film.

The decay constant, K, can be obtained by considering diffusion from the dilute phase to the wall film: balancing the upward flow by convection in the core against mass transfer of particles to the wall film gives

$$\frac{D}{4} (U - U_t) \frac{dc}{dz} = -k_d (c - c_\infty) \ . \qquad (12)$$

Here c is the concentration of particles in the core; c_∞ is the concentration at which the rate of entrainment from the film equals the rate of deposition. The solution to equation (12) leads to equations of the same form as (9) and (10), giving

$$K = \frac{4 \, k_d}{D(U - U_t)} \ . \qquad (13)$$

The measurement of w [12, 13] give values of K and thence k_d from equation (13): these values are consistent with data for radial diffusion of droplets or of dust, for gas flowing through tubes.

5.2 Conjectural calculations for full-scale fast fluidised bed

From the work on small scale units [12, 13] it is possible to make estimates of k_d and K for large units. An important point is that K is proportional to the reciprocal of D, from equation (13). It follows that the decay constant, K, diminishes with increasing bed diameter and hence a greater bed height is needed to get an acceptably low rate of transport of particles into the

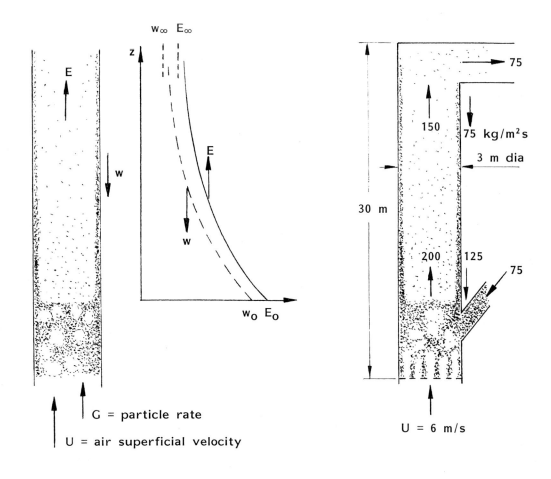

(a)   (b)

FIGURE 18   Dilute rising core/falling wall film model for fast fluidised bed

riser.

(a) Variations of core particle flux E and downward wall flux w

with z.

(b) Estimated fluxes for large unit.

cyclone at the top of the riser. Results from such a calculation are given in figure 18(b) for a bed 30 m high x 3 m diameter, the scale now being considered for large scale combustors. It is seen that the internal recycle of particles within the riser unit is extremely important and this must have a significant bearing on the gas to solid contact which is important for combustion and for sulphur removal.

The gas-liquid analogy appears to hold good even for fast fluidised beds. The fast fluidised bed can be regarded as a very large wetted wall column. The film of particles running down the wall is exactly analogous to the liquid film that may run down the wall of a vertical tube when gas transports liquid droplets up the tube.

ACKNOWLEDGEMENTS

We live in an age of 'planned research', stimulated by Peer Review Committees whose dubious deliberations lead to extravagant contracts. The work described herein was done without contracts, but with uncommitted funds: the benefactors who provided the funds are gratefully acknowledged, notably The Shell Company, Trinity College Cambridge, and the Science and Engineering Research Council through its Studentships.

REFERENCES

[1] Reuter, H., Über den Mechanismus der Blasen im Gas-Feststoff-Fliessbett (Doktor-Ingenieurs dissertation, Technischen Hochschule Aachen, 1963).

[2] Jones, D.R.M., Liquid analogies for fluidised beds (PhD dissertation, Cambridge 1965, Fig. 9).

[3] Rowe, P.N. and Partridge, A., Trans. Instn Chem. Engrs 43 (1965) T157.

[4] Ormiston, R.M., Slug flow in fluidised beds (PhD dissertation, Cambridge 1966, Figs. 4, 5).

[5] Davidson, J.F., Harrison, D. and Guedes de Carvalho, J.R.F., Ann. Rev. Fluid Mech. 9 (1977) 55.

[6] Toei, R., Matsuno, R., Kojima, H., Nagai, Y., Nakagawa, K. and Yu, S., Mem. Fac. Eng. Kyoto 27 (1965) 475.

[7] Davidson, J.F. and Harrison, D., Fluidised Particles, pp50, 53, 54, 76 (Cambridge University Press 1963).

[8] Hovmand, S. and Davidson, J.F., Fluidization, p224 (Academic Press, 1971, edited by J.F. Davidson and D. Harrison).

[9] Rowe, P.N., Fluidization, pp182, 185 (Academic Press 1971, edited by J.F. Davidson and D. Harrison).

[10] Martin, P.D. and Davidson, J.F., Chem. Eng. Res. Des. 61 (1983) 162.

[11] Pemberton, S.T. and Davidson, J.F., Chem Eng Sci 39 (1984) 829.

[12] Bolton, L.W., Circulating fluidised beds (PhD dissertation, Cambridge 1987).

[13] Bolton, L.W. and Davidson, J.F., Second International conference on circulating fluidized beds. Compiegne, 1988 (Proceedings, edited by P. Basu, to be published).

Theoretical and Applied Mechanics
P. Germain, M. Piau and D. Caillerie (Editors)
Elsevier Science Publishers B.V. (North-Holland)

BIFURCATIONS AND MODAL INTERACTIONS IN FLUID MECHANICS: SURFACE WAVES

Jerry P. GOLLUB and F. SIMONELLI

Physics Department
Haverford College
Haverford, PA 19041 and

Physics Department
The University of Pennsylvania
Philadelphia, PA 19104

An experimental approach is described for the quantitative study of bifurcations and modal interactions in spatially extended fluid systems such as surface waves. Trajectories are followed in the phase space spanned by the amplitudes of the various modes into which the patterns can be decomposed. The various stable and unstable fixed points and the bifurcations that affect their stability are determined. The experiments illustrate the importance and usefulness of symmetry considerations in determining the nature of the dynamics and the presence or absence of chaos.

1. INTRODUCTION

The experimental study of bifurcations in fluid mechanics, though extensive, has not generally been able to reveal the richness and detail that is found, for example, in the theoretical or numerical study of elementary nonlinear evolution equations. The basic reason for this difficulty lies in the complex interplay between the spatial and temporal domains. Experimental measurements based on local sampling miss the spatial features, while visual and digital image processing methods generally miss important temporal properties. Ways of bringing the spatial and temporal domains together are required to make significant progress.

One way to characterize an evolving fluid system is to decompose it (at each instant) into a suitable set of spatial modes, and then to form a phase space whose coordinate axes are the various mode amplitudes. The time evolution of the system may then be followed in phase space, at least in principle. This program is of course difficult to realize for time-dependent systems because it requires a modal decomposition (for example, using a Fourier basis) at each time step. Where it can be accomplished, however, it offers the hope of much better contact between theory and experiment.

It would be quite impossible at present to apply this methodology to a turbulent flow. On the other hand, there are interesting fluid systems in which only a few spatial modes are excited, and yet the dynamics are quite complex. A good example of such a system is the pattern of waves on the interface of a confined fluid layer driven by vertical excitation of the container. A pattern of standing waves (oscillating at half the driving frequency) appears above a dissipation-dependent threshold. This system has a long history, beginning with Faraday [1]. Many recent theoretical and experimental studies have also appeared [2]. There are several advantages of focusing attention on interfacial waves:

- Well-controlled laboratory studies are possible.
- Weak nonlinearity and symmetry considerations can be used to simplify the dynamical equations.
- There are many applications in geophysical fluid dynamics and engineering.
- This system provides an important intermediate case between dynamical systems with no spatial structure on the one hand, and those with complex structures and many degrees of freedom on the other.

In this paper, we consider the situation in which the driving frequency is relatively small, so that the excited waves are low order resonant modes with a wavelength not too much smaller than the system size. If the cell containing the fluid is square, there are geometrical degeneracies in which two spatial modes of the interface have the same resonant frequency. In that case, a fascinating set of bifurcations ensues that involves competition between several distinct spatial wave patterns. When the cell is not exactly square, the patterns can oscillate periodically or chaotically. An experimental approach is described that allows the full bifurcation structure to be determined as a function of the driving amplitude and frequency, which are the main control parameters. A more extensive description of this investigation is available elsewhere [3].

2.   EXPERIMENTAL METHODS: STUDIES IN PHASE SPACE

The phenomena of interest here involve the interaction of two degenerate modes, (3,2) and (2,3), where the indices give the number of half wavelengths in each of the horizontal cell dimensions. These modes are excited near 14 Hz in a cell 6 cm square, filled to a depth of 2.5 cm with n-butyl alcohol. The vertical oscillation was provided by an electromagnetic shaker driven by a frequency synthesizer and power amplifier under computer control. The vertical acceleration and drive amplitude were measured with an accelerometer.

The wave patterns may be easily visualized using shadowgraphs. Examples of two typical patterns are shown in Figure 1. These represent a pure (3,2) mode and a "mixed mode" that is a superposition of the (3,2) and (2,3) modes. The bright regions show areas that are elevated at some time in the wave cycle. These are the antinodes of the waves. (Since the shadowgraphs are somewhat nonlinear, only the general periodicity of the waves, but not the exact wave shapes, can be inferred from the photographs.) The examples shown are stable time-independent patterns (though the underlying waves oscillate periodically at half the driving frequency). The modal content of the patterns can vary with time, for example during their formation or when the cell is not exactly square.

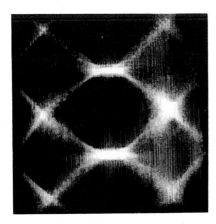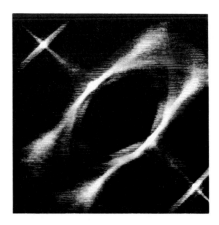

FIGURE 1. Shadowgraph images of two surface wave patterns: (a) a pure (3,2) mode and (b) a "mixed mode" containing equal proportions of the (3,2) and (2,3) modes. The images have been averaged over one period of the forcing.

Each mode may be described by two amplitudes $A_{mn}(t)$ and $B_{mn}(t)$ that give the contributions in-phase and out-of-phase with respect to the driving frequency. The contribution of the mode (m,n) to the surface displacement $Z_{mn}(x,y,t)$ may be written

$$Z_{mn}(x,y,t) = [A_{mn}(t)\cos(\omega t/2) + B_{mn}(t)\sin(\omega t/2)] \times [\cos(m\pi x/L)\cos(n\pi y/L)] \quad,$$

where ω is the driving angular frequency and L is the horizontal dimension of the cell. If there are two active modes, then four mode amplitudes must be determined at each instant to characterize the fluid interface. This can be accomplished without Fourier analysis using only two local probes (photodiodes) of the image intensity. The output of each probe is sent to a double-phase lock-in amplifier that gives signals proportional to the in- and out-of-phase components of the surface displacement at the position of the probe. Then, predetermined linear combinations of the various signals will yield the individual mode amplitudes A_{mn} and B_{mn}. For the case of interest here, there are four amplitudes, so the resulting phase space is four-dimensional.

The power of this approach is illustrated in Figure 2, which shows a two-dimensional projection of the full phase space onto the subspace of the in-phase amplitudes of the two modes. The driving amplitude and frequency are chosen to be in a regime for which mixed modes are the only stable patterns. The system is repeatedly started either from a driving amplitude below threshold, or from one that is larger than the desired final value. The transient evolution of the system is then determined by measurement of the mode amplitudes. The resulting trajectories give a wealth of information about the dynamics of the system.

For example, the origin (the flat surface) is unstable and is called a "source" in the language of dynamical systems. There are four "sinks" (solid dots) that are located off the coordinate

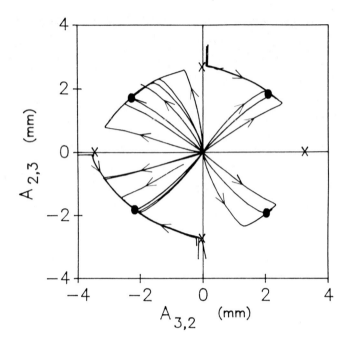

FIGURE 2.   A set of experimental transient trajectories in phase
space.   In this case, the mixed states (solid dots) are stable
and the pure states (x) are unstable.

axes.   Therefore, these states are superpositions of the individual modes.   Each sink is surrounded by a basin of attraction, so
they are not simply linear superpositions, but rather definite
stable states of the system.   There are four of them, a fact that
is related to the square symmetry of the cell.   The various sinks
correspond   to   distinct   phase   relationships   between   the   mode
amplitudes and the driving frequency.

It is also possible to detect unstable fixed points (denoted by
the symbol x in Figure 2).   These are the pure states formed from
only a single mode amplitude.   They are apparently repelling in
(at least) one direction and attracting in another direction.   In
the language of dynamical systems, they are called "saddles."
Finally, we note that the system evolves quickly to the vicinity
of a circle centered on the origin, and then more slowly to the
attracting fixed points.   This technique of observing the transient behavior starting from a variety of initial conditions, and
in each case determining the evolution in phase space, allows
much more information to be obtained than the usual approach of
observing only the final states.

3.   BIFURCATION STRUCTURE

The interactions between the two pure modes gives rise to a variety of pure and mixed states.   Their stability is a strong function of the control parameters, which are the driving amplitude
and driving frequency.   Therefore,   complex sequences of bifurcations can occur for even small changes in the parameters.   The

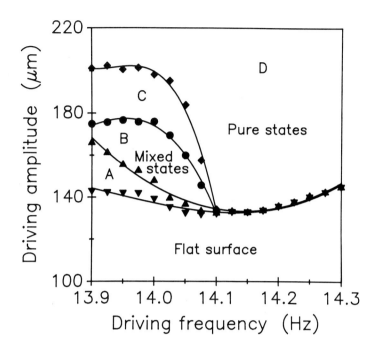

FIGURE 3.   Parameter space diagram for a square cell showing the various dynamical regimes in the neighborhood of the resonance.

various dynamical regimes occurring near the resonance are shown in Figure 3.   For low driving amplitudes, only the flat surface is stable.   For high driving amplitudes, only pure states are stable (region D).   In region B only mixed states are stable. Finally, regions A and C have several distinct types of patterns, depending on initial conditions.   In region A, the flat surface coexists with mixed states, and in B, the mixed and pure states coexist.

The nature of each of these distinct dynamical regimes is shown schematically in Figure 4.   Each of these diagrams was obtained from experiments similar to those used to obtain Figure 2.   In each diagram, the stable fixed points (sinks) are shown as solid circles, and the unstable fixed points are shown as x's (saddles) or open circles (sources).   The latter are unstable in all coordinate directions.   The pure states (those on the coordinate axes) are stable in regions C and D only, while the mixed states are stable in regions A, B, and C.   Arrows are used to show the local directions of trajectories in the vicinity of the various fixed points.   It is apparent that there a great many fixed points, up to 17 (region C).   Most were detected experimentally; a few were determined by a mixture of theoretical and experimental considerations.

Finally, in Figure 5 we show "bifurcation diagrams" illustrating the bifurcation sequences that are obtained as the driving amplitude is varied for frequencies slightly below (Figure 5a) and slightly above (Figure 5b) the resonance.   The diagrams represent two-dimensional projections of the five-dimensional

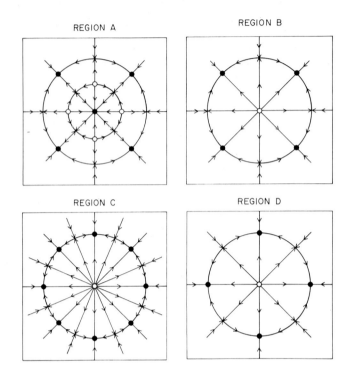

FIGURE 4. Schematic phase space structure (determined experimentally) showing the stable and unstable fixed points in the various dynamical regimes. Sinks, saddles, and sources are shown as solid circles, crosses, and open circles, respectively.

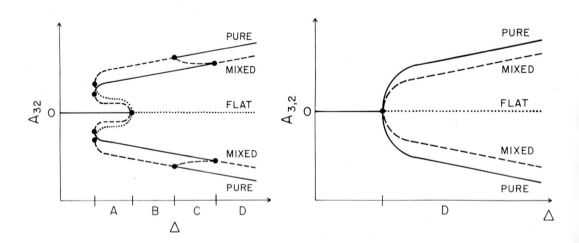

FIGURE 5. Schematic bifurcation diagrams showing one of the wave amplitudes as a function of driving amplitude. The various bifurcations are indicated by solid dots. (a) Driving frequency slightly below the resonance; (b) Driving frequency slightly above the resonance.

space spanned by the four mode amplitudes and the driving amplitude Δ.   One can think of the mixed modes as having non-zero
components perpendicular to the plane of the diagram, so that
they appear to have smaller amplitudes when projected onto the
plane of the diagram.   Sinks are represented by solid lines,
saddles by dashed lines, and sources by dotted lines.

Figure 5b is the simpler of the two diagrams since there is no
hysteresis and the only bifurcation is a forward one.   Note that
both sinks and saddles are generated at this bifurcation.   In
Figure 5a there are several bifurcations of the saddle/node type
in which saddles and sinks are generated at finite amplitude.
These are "multiple bifurcations" in the sense that several
distinct fixed points are generated at each one.   (These fixed
points were illustrated in Figure 4.)   The "backward" bifurcation
on the axis is the one that eliminates the stability of the flat
surface state.   Note that the flat surface is stable in region A,
the mixed states are stable in regions A-C, and the pure states
are stable in regions C and D.

4.   DISCUSSION AND CONCLUSION

Figure 5 summarizes a wealth of experimental data, and this
degree of detail in mapping out bifurcation sequences has not
previously been obtained from any experimental system, to the
best of our knowledge.   The determination of the trajectories in
phase space and the use of repetitive transients from various
initial conditions were essential in obtaining this information.
Furthermore, the data needed to unravel this bifurcation structure required extensive automation [4].

For example, the parameter space diagram of Figure 3 was
constructed as follows.   At each driving frequency, the driving
amplitude was incremented under computer control, and several
transients were observed to determine the dynamical regime.   When
a change in dynamics was detected, the computer was programmed to
make smaller changes of amplitude to determine the stability
boundaries accurately.   The entire process was automated and
required approximately 1-2 weeks without operator intervention.
It would have been quite difficult to construct the parameter
space diagram manually.

Beyond demonstrating the power of computer control, what is the
significance of these results?   The reader might suppose that one
would need to start from the full hydrodynamic equations for this
problem and then utilize perturbation theory to determine the
form of the coupled ordinary differential equations governing the
time evolution of the various mode amplitudes.   This approach has
in fact been carried out by Feng and Sethna [5].

Remarkably, however, the essential observations (Figures 3-5) do
not depend on detailed hydrodynamic considerations.   Rather, the
bifurcation structure is qualitatively determined by elementary
considerations involving invariances and symmetries that determine the form of the amplitude equations.   For example, a translation in time equal to one period of the forcing leaves the
physical system unaffected.   This leads to the condition that
the dynamical equations must be invariant under a sign change of
all the amplitudes.   Similarly , the square symmetry of the cell

implies that the coefficients of corresponding terms in the equations for the two competing mode amplitudes are identical.
Though the coefficients in the equations cannot be obtained by
symmetry considerations, the structure of the equations can be
determined to be as follows:

$$dC_{32}/dt = \alpha C_{32} + \beta C^*_{32} + \gamma |C_{32}|^2 C_{32} + \mu C_{32}|C_{23}|^2 + \nu C^*_{32} C_{23}^2 ,$$

where the complex amplitude $C_{32} = A_{32} - iB_{32}$. An identical equation holds for the rate of change of C_{23}. The coefficients of
the nonlinear terms are required to be imaginary so that the
system of equations is Hamiltonian in the absence of damping, as
is generally assumed for surface waves. In that case, the equations lead to predictions that are qualitatively consistent with
the experimental results of Figures 3-5. A more extensive
discussion of this method is contained in Reference 3 and is
based on theoretical work by various authors. [6]

Finally, we point out that since symmetry considerations are so
important in this problem, a slight change in the geometry can
give rise to qualitatively new phenomena. If the square symmetry
of the cell is broken by making the two horizontal dimensions
differ by a few percent, the parameter space of Figure 3 is
dramatically changed. The mixed modes are destabilized, and both
periodic and chaotic oscillations of the patterns occur in a
significant domain in driving amplitude and frequency.

An example of the chaotic regime is shown in Figure 6, which
contains a time series of one of the amplitudes, and two different projections of the four-dimensional phase space. The
attractor has the qualitative appearance of the Rossler strange
attractor, and it has the requisite properties of fractional
dimension and a positive Lyapunov exponent. The power spectrum
of the time series is broadband. In Reference 3, we discuss the
sequence of bifurcations leading to this interesting chaotic
state. There is presently no theory of this interesting
phenomenon.

ACKNOWLEDGEMENTS

We appreciate helpful discussions and an exchange of preliminary
results with P.R. Sethna and J. Guckenheimer. We have also bene-
fited from discussions with M. Golubitsky and E. Meron. This
work was supported by the DARPA Applied and Computational Mathematics Program through the University Research Initiative
Program, Contract Number N00014-85-K-0759.

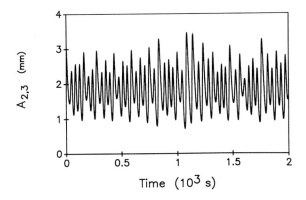

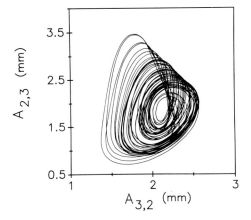 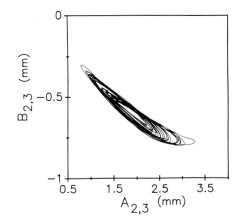

FIGURE 6.   Example of a chaotic state in a cell that is not
exactly square.   (a) Time series of one of the mode amplitudes.
(b) and (c) Two projections of the trajectories in phase space.

REFERENCES

[1]   Faraday, M., Phil. Trans. R. Soc. Lond. **121** (1831) 319.

[2]   Ciliberto, S. and Gollub, J.P., J. Fluid Mech. **158** (1985)
      381; Miles, J.W., J. Fluid Mech. **146** (1984) 285; Meron, E.
      and Procaccia, I., Phys. Rev. A **35** (1987) 4008; Holmes,
      P.J., J. Fluid Mech. **162** (1986) 365; Gu, X.M. and Sethna,
      P.R., J. Fluid Mech. **183** (1987) 543; Douady, S., and Fauve,
      S., Europhys. Lett. **6** (1988) 221; Ezerskii, A.B.,
      Rabinovich, M.I., Reutov, V.P., and Starobinets, I.M., Sov.
      Phys. JETP **64** (1987) 1228.

[3]   Simonelli, F. and Gollub, J.P., J. Fluid Mech. (1989) in
      press.

[4]   Simonelli F. and Gollub, J.P., Rev. Sci. Instrum. **59** (1988)
      280.

[5]   Feng, Z.C. and Sethna, P.R., J. Fluid Mech. (1989) in press.

[6]   Meron, E., Phys. Rev. A. **35** (1987) 4892; Guckenheimer, J.
      and Holmes, P., Nonlinear Oscillations, Dynamical Systems,
      and Bifurcation of Vector Fields (Springer-Verlag, Berlin,
      1983); Golubitsky, M. and Shaeffer, D.G., Singularities and
      Groups in Bifurcation Theory (Springer-Verlag, Berlin,
      1985).

Theoretical and Applied Mechanics
P. Germain, M. Piau and D. Caillerie (Editors)
Elsevier Science Publishers B.V. (North-Holland)
© IUTAM, 1989

ACTIVE VIBRATION DAMPING IN LARGE FLEXIBLE STRUCTURES

Peter HAGEDORN

Institut für Mechanik
Technische Hochschule Darmstadt
Hochschulstraße 1, D-6100 Darmstadt, W.-Germany

ABSTRACT

Active vibration damping is being seriously considered for applications in large flexible space structures, in robotics, and to some extent also in the more traditional branches of engineering. In the traditional approach of designing the control, the structure is discretized, via finite elements, modal analysis or some other technique, and the problem of controlling a distributed parameter system is thus transformed into a new problem, described by a system of ordinary differential equations. The powerful techniques developed for discrete control problems are then applied, methods such as pole placement or optimal control being used. Important additional questions in large structures concern the placement of actuators and sensors.

Recently, a new approach has been proposed. In this new technique the structure is not dicretized but split up into structural elements, such as beams, cables, etc., each of these elements being a simple continuous system. The boundary and transition conditions are then dropped in the first stage of the control design. The motion of the individual structural elements described by hyperbolic equations are represented by traveling waves and the controls are designed with intent to absorb locally as much of the energy of these traveling waves as possible. In this lecture, the relative merits of these different approaches – discretization vs. traveling waves – are dicussed and some new ideas are developed.

1. INTRODUCTION

In the past, structural vibrations have usually been supressed by passive means, either by increasing the structural damping, for example by artificially enlarging friction in the joints, or by adding new damping elements. In this manner, it was possible in many engineering problems to reduce vibrations to acceptable levels. Recently, in some engineering applications the damping mechanism can be adjusted to the actual working conditions, which may vary in time. In vehicle dynamics, for example, in certain automobiles the damping and the stiffness of the suspension can be adjusted to the actual road conditions and to the speed of the car. This can be done either manually by the driver or automatically. In both cases the time scale on which these stiffness and damping coefficients are adjusted is usually large with respect to the time scale of the oscillations of the car. Particularly, if this adjustment is done manually, the driver may typically choose certain dynamic characteristics for the suspension and then drive possibly for hours without resetting it. This type of controlling vibrations (either manually or automatically) by adjusting coefficients on a large time scale is usually termed "semi-active", it requires comparatively little energy and very small forces.

A different philosophy of controlling the motion of a vehicle or of other systems is related to the use of true "active" elements, producing control forces varying on the same time scale as the vibrations themselves. In an automobile, for example, pneumatic and/or hydraulic elements could be used as "active springs" and "active dampers". These active elements would be controlled by a computer having continuous access to information on the actual vibrations of the car, for example through accelerometers placed at different points of the car body, and possibly also on the road profile ahead of the car, which could eventually be picked up by optical sensors. In addition, some "slowly" changing parameters such as load of the car, tire pressure, etc. would also be taken into account. Probably all the major car manufacturers are presently experimenting with some kind of "active vehicle suspension", and particularly racing cars with active suspension have already been built. One problem of this "active vibration control" in vehicle dynamics is the fact that both the active forces and the energies envolved in controlling the vehicle's motion may be quite large.

In the above remarks on vehicle dynamics, the term "semi-active control" was used to describe the "slow" adjustment of spring and damper parameters, and the expression "active control" for the case in which actuators apply rapidly varying control forces to the car body. The case of "fast" adjustment of the stiffness and damping coefficients - i.e. on the same time scale as the vibrations themselves - was left out of this classification. This last approach, which may be the most promising one in vehicle dynamics, is sometimes called "active", sometimes "semi-active", depending on the author.

In studying the active or semi-active control of the suspension, the vehicle can usually be modelled by means of a few rigid bodies, the elastic tires as well as springs and dampers, so that the vehicle system to be controlled is mathematically described by a system of ordinary differential equations of moderately high order (maybe up to 30). The design of the control law is therefore a classical control problem and the powerful methods such as pole placement, optimal control, etc., which have been available for a long time, can be used to treat this problem.

Also in other types of machines and even in structures, active vibration damping is being considered or already applied. Magnetic bearings are for example being used to support rotating shafts and to control their vibrations (/13/, /14/) and base-isolated structures with active control are being studied in connection with tall buildings under earthquake excitation (/15/). As the operational speed of robots is increased, they can no longer be modelled by means of rigid bodies only, but the flexibility of their arms has to be taken into account. Presently, in some experimental robots the driving motors at the joints are already being used not only to produce the large overall motion of the arms, but also to compensate for their vibrations. Fortunately only the lowest elastic modes of the arms of a robot are usually of importance, so that the dynamics can be described by a relatively small system of ordinary differential equations and the classical tools from control theory can again be applied.

The situation is different if a very large number of elastic modes has to be controlled, as is the case in the large flexible space structures, planned for the coming decades. For the determination of the eigenfrequencies and eigenmodes these structures are discretized to a high degree, and models of thousands or hundreds of thousands of degrees of freedom are used; such models are of course much too large to solve for example the matrix Riccatti equations involved in the solutions of the related optimal control problems, and they are also too large for the application of the other classical techniques. A much simpler model with a small number of degrees of freedom is therefore used to formulate the control problem and for the choice of control law. In a further step the performance of the vibration control designed with the low-order model is tested in simulations with a higher order "validation" model.

One of the difficulties arising in this context is "control spill-over". By this expression one designates the fact that the vibration control designed with the low-order model may cause instabilities in the higher-order unmodelled modes, i.e. energy may spill over from the low-order controlled modes into the higher modes. This goes hand in hand with "observation spill-over" – the analogous phenomenon in the related observation problem. These spill-over effects have led to strategies such as the "co-location principle", helping to avoid spill-over. "Co-location" means that sensors and actuators are located at exactly the same points of a continuous structure.

Spill-over is however not only related to discretization. Even without discretization the mathematical model used to design the control law usually differs in some aspects from an adequate mathematical description of the actual physical systems. In particular the boundary conditions, i.e. the ways in which different parts of the structure are interconnected, are often not well defined in engineering problems. This implies that no matter how fine a discretization is used, the computed eigenmodes always differ to some extent from the real modes (if the structure behaves linearly at all) and this is an additional source of spill-over.

Recently, a new approach has been proposed for the design of active vibration control (/3-6/, /12/, /15/). In this new technique the structure is not discretized but split up into structural elements, such as beams, cables, etc., each of these elements being a simple continuous system. The boundary and transition conditions are then dropped in the first stage of the control design. The motion of the individual structural elements described by hyperbolic equations are represented by traveling waves and the controls are designed with the intent to absorb locally as much of the energy of these traveling waves as possible. Instead of solving a control problem for a large system of ordinary differential equations, the local control of an unbounded medium described by partial differential equations is thus examined and a number of local and uncorrelated control problems has to be solved. It turns out that the solution of these local problems is relatively simple, at least for bars, cables and Timoshenko beams, although it requires a new measurement technique to divide the vibrations into contributions coming from waves traveling into different directions. This measurement technique is closely related to the intensity measurements used in acoustics. It will not be further discussed here. The controls designed in this manner for a unbounded medium also work with good performance for a finite medium with arbitrary boundary conditions as well as for complex systems.

Many of the planned space structures contain spatial beamlike lattices, which macroscopically can very well be modelled as Timoshenko beams. In these cases a preliminary design of the active vibration control can be done using the Timoshenko beam model. In a further step the "beam" can then be substituted by a truss, which is then used for a more detailed analysis of the vibration control.

The present paper describes the main results obtained in the recent past at the Institut für Mechanik in Darmstadt concerning the control of bars, strings and Timoshenko beams. In model problems it is shown how the active damping devices can be designed using traveling waves. Most of the material has been published previously in /4-6/ and in particular in the doctoral dissertation /15/ of J. Schmidt.

2. CONTROLLER DESIGN FOR A TAUT STRING

2.1. The traveling wave approach

In nondimensional form the forced transverse vibrations of a taut string are described by

$$\ddot{w}(x,t) - w''(x,t) = q(x,t); \tag{1}$$

in active vibration damping the forcing term $q(x,t)$ is used to control the string's vibrations. Here $q(x,t)$ is assumed to be of the form

$$q(x,t) = \sum_{l=1}^{2} \delta(x-a_l)\, q_l(t), \tag{2}$$

i.e. it contains the control functions $q_1(t)$, $q_2(t)$, which correspond to concentrated control forces at $x = a_1$ and $x = a_2$ (see Fig. 1).

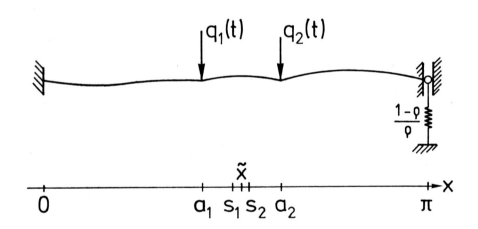

FIGURE 1
String with control forces

In the design of the control laws the measurements of the transverse displacements

$$m_k(t) := w(s_k,t), \quad k = 1,2 \tag{3}$$

are supposed to be available. These measurements are taken at points located in the neighborhood of the locations of the actuators. Using d'Alembert's solution

$$w(x,t) = f(x-t) + g(x+t), \tag{4}$$

control forces can be found such that all the incoming waves arriving at the actuator locations from the exterior of the interval $[a_1, a_2]$ are completely absorbed, no energy being reflected to the exterior (see /15/ for details). The corresponding control laws are

$$q_1(t) = - \left[\frac{m_2 - m_1}{s_2 - s_1} + \frac{\dot{m}_2 + \dot{m}_1}{2} \right] \Bigg|_{t + (a_1 - \tilde{x})} , \qquad (5a)$$

$$q_2(t) = \left[\frac{m_2 - m_1}{s_2 - s_1} - \frac{\dot{m}_2 + \dot{m}_1}{2} \right] \Bigg|_{t - (a_2 - \tilde{x})} , \qquad (5b)$$

where $\tilde{x} := (s_1 + s_2)/2$.

It is easy to see that these controls have a delay character. The simple form of the solution is due to the fact that in the case of the wave equation (1), waves travel with constant speed and without changing their profile, since there is no dispersion. The waves are measured and then compensated by the control forces, and the distance between the points of measurement and the actuators causes the delay in the control. Note that the boundary conditions are not used in the determination of this control law.

It can be shown that the design of the control law with the objective to maximize the absorption of the energy arriving from the exterior of the interval $[a_1, a_2]$ is equivalent to solving two independent and simpler problems. In each of these, a simple control element is designed such that energy travels through it in one direction only. This is symbolized by the "diode" in Fig.2a. This element can be understood as an active "valve" which permits the passage of vibration energy in one direction only. If two such basic elements with distinct preferential directions are combined, as in Fig.2b, the same optimal control as described by (5) is again obtained.

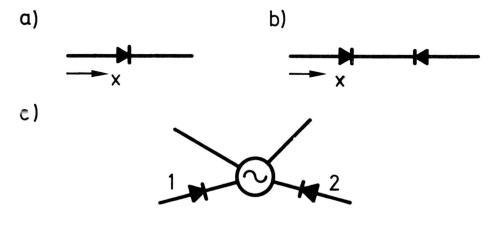

FIGURE 2
Energy "valves" : a) unidirectional energy valve,
b) combination of two valves, c) valves in a network-like structure

In a large network-like structure the mechanical vibrations are often generated at certain locations of the structure only. In a space station for example, the astronauts may represent the main source of "noise" and the mechanical vibrations generated by them should be kept away from the sensitive parts such as antennas, telescopes, etc. If these substructures are interconnected by onedimensional structures, the sensitive subsystems can be shielded from the vibrations by means of "energy valves". This is symbolized by Fig.2c, where these elements protect the lower part of the structure from the vibrations generated at the given node.

In section 3 a similar active vibration control will be developed for beam vibrations. In order to clarify the analogy between the string and the beam vibrations, it is convenient to rewrite the wave equation (1) as a first order system of the form

$$\mathbf{u}_t + \tilde{\Gamma}\,\mathbf{u}_x = \mathbf{0} \tag{6a}$$

with

$$\tilde{\Gamma} := \begin{bmatrix} 0 & -1 \\ -1 & 0 \end{bmatrix} \in \mathbb{R}^{2\times2}, \tag{6b}$$

where the new variables are defined as

$$u_1(x,t) := w_t(x,t), \qquad u_2(x,t) := w_x(x,t), \tag{7a}$$

$$\mathbf{u}\,(x,t) := \left[u_1(x,t),\ u_2(x,t)\right]^T \in \mathbb{R}^2. \tag{7b}$$

The hyperbolicity of the system is reflected in the fact that the matrix $\tilde{\Gamma}$ is real symmetric and the system can therefore be transformed into *normal form* by means of the linear transformation

$$\mathbf{v}(x,t) := \mathbf{T}\,\mathbf{u}(x,t) \tag{8}$$

with

$$\mathbf{T} := \frac{1}{\sqrt{2}} \begin{bmatrix} -1 & 1 \\ 1 & 1 \end{bmatrix} \tag{9}$$

where the matrix $\mathbf{T}$ is orthonormal, i.e. $\mathbf{T}^{-1} = \mathbf{T}^T = \mathbf{T}$. The linear transformation (8) results in the normal form

$$\mathbf{v}_t(x,t) + \Gamma\,\mathbf{v}_x(x,t) = \mathbf{0} \tag{10}$$

with

$$\Gamma = \mathbf{T}\,\tilde{\Gamma}\,\mathbf{T}^{-1} = \text{diag}\,(1,-1) \in \mathbb{R}^{2\times2}. \tag{11}$$

2.2. The "classical" approach with modal discretization

A vibration controller using the "classical" approach with modal discretization was designed in /5/ for the boundary conditions

$$w(0,t) = 0, \tag{12a}$$

$$(1-\rho)\ w(\pi,t) + \rho\ w'(\pi,t) = 0 \tag{12b}$$

with $0 \leq \rho \leq 1$ (see Fig.1). The controls were designed for the nominal boundary conditions corresponding to $\rho = 0$, i.e. the string fixed at both ends. The modes are then simply sine functions, and with a modal discretization using N modes, the wave equations (1) can be discretized as

$$w(x,t) = \sum_{k=1}^{N} z_k(t)\ \frac{1}{\sqrt{2}}\ \sin\ kx \tag{13}$$

giving

$$\ddot{z}_k(t) + k^2\ z_k(t) = \frac{1}{\sqrt{2}}\ \left[\ q_1(t)\ \sin\ ka_1 + q_2(t)\ \sin\ ka_2\ \right]\ . \tag{14}$$

With $\mathbf{y} = (z,\dot{z})^T$ this can be written as

$$\dot{\mathbf{y}}(t) = \mathbf{A}\ \mathbf{y}(t) + \mathbf{B}\ \mathbf{q}(t)\ , \tag{15}$$

with $\mathbf{y} \in \mathbb{R}^{2N}$, $\mathbf{A} \in \mathbb{R}^{2N\times2N}$, $\mathbf{B} \in \mathbb{R}^{2N\times2}$, $\mathbf{q} \in \mathbb{R}^2$.

If the controls are assumed to be linear functions of the state variables

$$\mathbf{q}(t) = -\ \mathbf{F}\ \mathbf{y}(t),\quad \mathbf{F} \in \mathbb{R}^{2\times2N}\ . \tag{16}$$

a suitable matrix $\mathbf{F}$ can be determined e.g. solving a matrix Riccatti equation, in which of course the weighting matrices have to be chosen.

Since the state $\mathbf{y}(t)$ is not being measured directly (only the two measurements $w(s_1,t)$, $w(s_2,t)$ are available), a linear optimal observer was used in /5/ to find the estimate $\hat{\mathbf{y}}(t)$ of the state, and $\mathbf{y}$ is replaced by $\hat{\mathbf{y}}$ in (16). Also, as location was used, i.e. $a_k = s_k$, $k = 1, 2$.

2.3. Numerical simulation, comparison

The performance of the two types of control was compared via numerical simulations in /5/ using a "validation model" with a high order of discretization. In this model a finite difference scheme was used to discretize the original wave equation (1), which was divided into one hundred elements. In both cases the true state vector was not used for the feedback law, but rather the measurements at the points $x = s_1$, s_2 and the estimate

$\hat{y}$ in the case of the classical controller. In the approach with modal discretization the number of modes taken into account was N = 8.

For certain initial conditions (see /5/) the equations of the
controlled systems were then integrated and the results are shown in Fig.3
and Fig.4 . While both controllers work with the nominal system ($\rho = 0$),
the modal controller ceases to work for $\rho \neq 0$, since the modes used in the
determination of the control law no longer are identical to the actual
modes in the system. In Fig.4b this becomes very clear for $\rho = 0.5$.

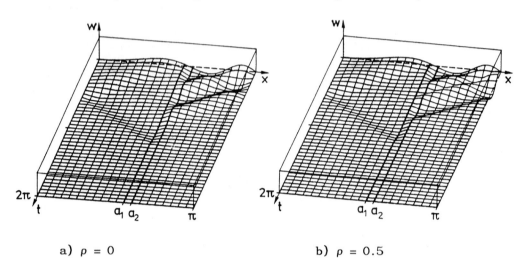

a) $\rho = 0$ b) $\rho = 0.5$

FIGURE 3
Performance of the traveling wave controller for the string

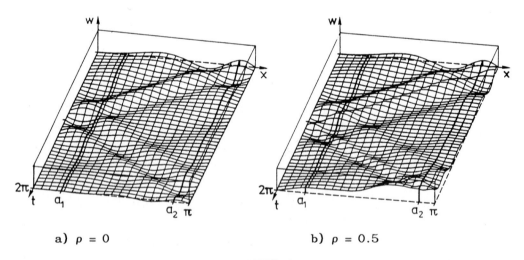

a) $\rho = 0$ b) $\rho = 0.5$

FIGURE 4
Performance of the controller designed with modal discretization

The controller designed with the traveling wave approach is of course
insensitive to changes in the boundary conditions. Since in most technical

problems the boundary conditions are not well known, the control
spill-over, due not only to discretization but also to deviations in the
boundary conditions, is almost omnipresent with the controller designed via
discretization. It is completely avoided with the traveling wave design.

The actuators in the two cases were located at different positions:
for the modal controller they were chosen such as to give best performance
for the first eight modes. The points of measurement were identical to the
locations of the actuators. In the traveling wave controller the actuators
are closely spaced, their relative distance being small with respect to the
shortest wavelength still of interest in the problem. Their absolute location on the string is of no importance in this case. The measurements in
the traveling wave controller were taken in the neighborhood of the
actuators.

3.   THE TIMOSHENKO BEAM: A HYPERBOLIC SYSTEM

In this section the problem previously solved for the string will be studied for beam vibrations. We therefore consider the beam of Fig.5 with
concentrated control forces and moments at the locations $x = a_1$ and $x = a_2$.
As in the case of the string, the goal is to find control laws such that
all vibration energy arriving from the outside of the interval $[a_1, a_2]$ is
absorbed in this interval and nothing is reflected to the exterior. Only
measurements in the immediate neighborhood of the actuator locations are
assumed, so that the boundary conditions and the beam motion far away from
$x = a_1$ and $x = a_2$ need not to be known. It can be shown that this problem
can again be reduced to the combination of two simple elements as shown in
Fig.2. In what follows we therefore only give the control law for these
"energy valves".

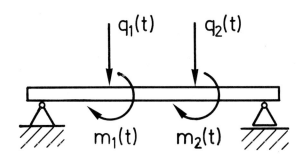

FIGURE 5
Beam with control forces and moments

Since bending waves are dispersive – as opposed to the waves in a
prestressed string – the control laws will be more complicated than (5).
The Euler-Bernoulli beam model, which in most technical problems describes
the beam vibrations effectively, does not lead to a hyperbolic system. It
turns out that the Timoshenko beam, which is usually regarded as a more
complicated beam model, is more convenient for solving the control problem,
since it leads to hyperbolic equations. The Timoshenko beam is therefore
the simpler beam model for the solution of the active vibration control
problem via the traveling wave approach.

Its equations of motion are

$$\rho A\, w_{tt}(x,t) - GA_s\big[w_x(x,t) - \psi(x,t)\big]_x = 0, \tag{17a}$$

$$\rho I\, \psi_{tt}(x,t) - EI\, \psi_{xx}(x,t) + GA_s\big[\psi(x,t) - w_x(x,t)\big] = 0, \tag{17b}$$

with the mass density ρA, the shear stiffness GA_s, the rotatory inertia ρI and EI as the beam's bending stiffness. The functions $w(\cdot,\cdot)$ and $\psi(\cdot,\cdot)$ represent the transverse displacement of the middle line and the rotation of the cross section, respectively. It is convenient to introduce the parameters

$$c_1^2 := \frac{EI}{\rho I} = \frac{E}{\rho}\,, \tag{18}$$

$$c_s^2 := \frac{GA_s}{\rho A}\,, \tag{19}$$

$$i^2 := \frac{\rho I}{\rho A} = \frac{I}{A} \tag{20}$$

and the nondimensional variables

$$\left.\begin{aligned}
\tau &:= \frac{c_s}{i}\, t, \\[6pt]
\xi &:= \frac{1}{i}\, x, \\[6pt]
\tilde{w}(\xi,\tau) &:= \frac{1}{i}\, w(x(\xi),t(\tau)), \\[6pt]
\tilde{\psi}(\xi,\tau) &:= \psi(x(\xi),t(\tau)).
\end{aligned}\right\} \tag{21}$$

In these new variables the equations (17) are now given by

$$\tilde{w}_{\tau\tau}(\xi,\tau) - \big[\tilde{w}_\xi(\xi,\tau) - \tilde{\psi}(\xi,\tau)\big]_\xi = 0, \tag{22a}$$

$$\tilde{\psi}_{\tau\tau}(\xi,\tau) - \gamma^2\, \tilde{\psi}_{\xi\xi}(\xi,\tau) + \big[\tilde{\psi}(\xi,\tau) - \tilde{w}_\xi(\xi,\tau)\big] = 0 \tag{22b}$$

with

$$\gamma^2 := \left[\frac{c_1}{c_s}\right]^2 = \frac{EI}{\rho I}\,\frac{\rho A}{GA_s} = \frac{EA}{GA_s}\,. \tag{23}$$

If for simplicity the symbols w, ψ, x and t are again used for the nondimensional variables, the equations (22) can be rewritten as

$$w_{tt}(x,t) - \big[w_x(x,t) - \psi(x,t)\big]_x = 0, \tag{24a}$$

$$\psi_{tt}(x,t) - \gamma^2\, \psi_{xx}(x,t) + \big[\psi(x,t) - w_x(x,t)\big] = 0. \tag{24b}$$

The equations (24) are now transformed into a first order system in the new variables

$$u_1(x,t) := w_t(x,t), \qquad u_2(x,t) := w_x(x,t) - \psi(x,t), \qquad (25a)$$

$$u_3(x,t) := \psi_t(x,t), \qquad u_4(x,t) := \varsigma\, \psi_x(x,t), \qquad (25b)$$

$$u(x,t) := \left[u_1(x,t), u_2(x,t), u_3(x,t), u_4(x,t)\right]^T \in \mathbb{R}^4 \qquad (26)$$

and one obtains

$$u_t + \tilde{\Gamma}\, u_x + \tilde{B}\, u = 0 \qquad (27)$$

with

$$\tilde{\Gamma} = \begin{bmatrix} 0 & -1 & 0 & 0 \\ -1 & 0 & 0 & 0 \\ 0 & 0 & 0 & -\varsigma \\ 0 & 0 & -\varsigma & 0 \end{bmatrix}, \qquad \tilde{B} = \begin{bmatrix} 0 & 0 & 0 & 0 \\ 0 & 0 & 1 & 0 \\ 0 & -1 & 0 & 0 \\ 0 & 0 & 0 & 0 \end{bmatrix}. \qquad (28)$$

Note that the only parameter still appearing in the equations is the ratio ς of the speeds of the longitudinal and the isovolumetric waves (shear waves in an unbounded elastic medium).

Similarly to the wave equation (6), also (27) can be written in the normal form

$$v_t + \Gamma\, v_x + B\, v = 0, \qquad (29)$$

with the diagonal matrix

$$\Gamma = T\, \tilde{\Gamma}\, T^{-1} = \mathrm{diag}(\varsigma, 1, -1, -\varsigma) \in \mathbb{R}^{4\times4} \qquad (30)$$

and the skew-symmetric matrix

$$B = T\, \tilde{B}\, T^{-1} = \frac{1}{2} \begin{bmatrix} 0 & 1 & 1 & 0 \\ -1 & 0 & 0 & 1 \\ -1 & 0 & 0 & 1 \\ 0 & -1 & -1 & 0 \end{bmatrix}. \qquad (31)$$

The transformation matrix of the linear transformation $v = T\, u$ is now

$$T := \frac{1}{\sqrt{2}} \begin{bmatrix} 0 & 0 & -1 & 1 \\ -1 & 1 & 0 & 0 \\ 1 & 1 & 0 & 0 \\ 0 & 0 & 1 & 1 \end{bmatrix}. \qquad (32)$$

The equilibrium conditions for the elementary beam section of Fig.6 give

$$u_2(a^-,t) - u_2(a^+,t) = q(t),\tag{33a}$$

$$u_4(a^-,t) - u_4(a^+,t) = m(t)/\varepsilon,\tag{33b}$$

with $q(t)$, $m(t)$ as control force and control moment respectively, while continuity leads to

$$u_1(a^-,t) - u_1(a^+,t) = 0,\tag{34a}$$

$$u_3(a^-,t) - u_3(a^+,t) = 0.\tag{34b}$$

In the new variables this results in

$$v_1(a^-,t) - v_1(a^+,t) = \frac{1}{\sqrt{2}}\,\frac{1}{\varepsilon}\,m(t),\tag{35a}$$

$$v_2(a^-,t) - v_2(a^+,t) = \frac{1}{\sqrt{2}}\,q(t),\tag{35b}$$

$$v_3(a^-,t) - v_3(a^+,t) = \frac{1}{\sqrt{2}}\,q(t),\tag{35c}$$

$$v_4(a^-,t) - v_4(a^+,t) = \frac{1}{\sqrt{2}}\,\frac{1}{\varepsilon}\,m(t).\tag{35d}$$

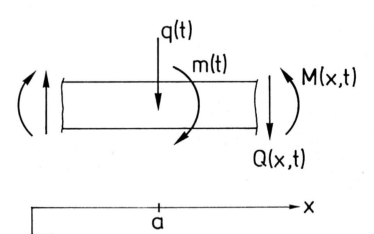

FIGURE 6
Equilibrium of a beam section

The control law for the "energy valve" of Fig.2a should maximize the energy flow. In /15/ it was shown that the energy flow is simply given by

$$p(x,t) = \frac{1}{2}\,v^T(x,t)\,\Gamma\,v(x,t)\tag{36}$$

or

$$p(x,t) = \frac{1}{2} \left[\gamma\, v_1^2(x,t) + v_2^2(x,t) - v_3^2(x,t) - \gamma\, v_4^2(x,t) \right].$$ (37)

If the energy flow $p(a^-,t)$ at $x = a^-$ it to be maximized, (37) obviously implies

$$v_3(a^-,t) = 0, \quad v_4(a^-,t) = 0$$ (38)

and with (35) one obtains the control laws

$$q(t) = - \sqrt{2}\, v_3(a^+,t),$$ (39)

$$m(t) = - \gamma \sqrt{2}\, v_4(a^+,t).$$ (40)

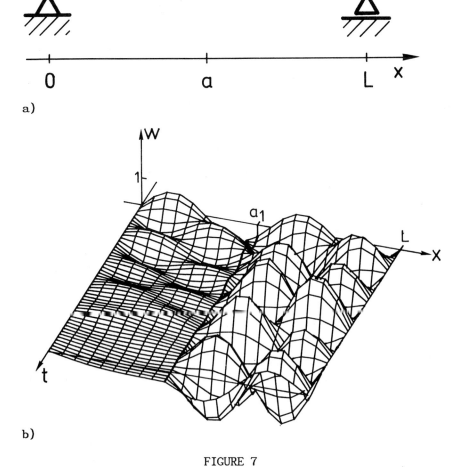

a)

b)

FIGURE 7
Timoshenko beam with one control element
a) beam with one energy valve, b) motion of the controlled beam

Similarly, for a "valve" maximizing the energy flow in the negative direction the control laws

$$q(t) = \sqrt{2}\, v_2(a^-,t),$$ (41)

$$m(t) = \gamma\, \sqrt{2}\, v_1(a^-,t)$$ (42)

are obtained. The control laws (39)–(42) imply the measurement of $v_3(a^+,t)$, $v_4(a^+,t)$ and $v_2(a^-,t)$, $v_1(a^-,t)$, respectively. This measurement problem will not be discussed here.

Numerical simulations were carried out for a simply supported Timoshenko beam and for different types of control. First, the beam of Fig.7a with a single control element at x = a was examined. For certain initial conditions (initial displacement given by the fourth eigenfunction, zero initial velocity) the equations of motion were numerically integrated again using a finite difference scheme with 100 elements. Fig.7b shows the decay of the vibrations in the region $0 \le x \le a$. Since there is an active control element, the mechanical vibration energy in the complete beam does not necessarily decrease monotonically. (It is in fact *not monotonic* in t).

In a second example, the motion of the beam of Fig.8a with two "energy valves" was simulated for the same initial conditions as before. In this case the total energy contained in $\{x|x < a_1\} \cup \{x|x > a_2\}$ decreases monotonically, while the energy stored in the region $\{x|a_1 < x < a_2\}$ is not monotonic in t. Moreover, the control obtained with the two "energy valves" is identical to the one resulting from the solution of the optimal control problem in which the time derivative of the energy outside the interval $[a_1, a_2]$ is minimized. In addition, it can be shown that the active vibration control is more efficient than the damping obtainable with passive linear and rotatory dampers at the same locations.

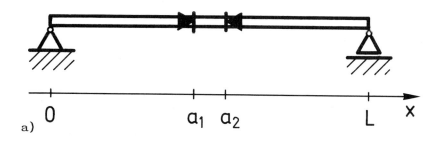

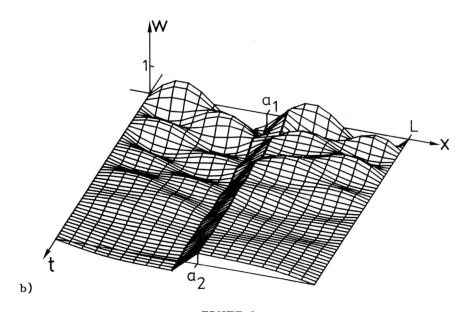

FIGURE 8
Timoshenko beam with a pair of control elements
a) beam with two energy valves, b) motion of the controlled beam

4.   SOME OPEN QUESTIONS

So far in this paper simple control laws for the vibration control in
one-dimensional continuous systems described by hyperbolic equations were
discussed. Normally in engineering applications beam vibrations are described by the Euler–Bernoulli theory, the Timoshenko theory only being used in
exceptional cases. Since the equations of the Euler–Bernoulli beam do not
form a hyperbolic system, the present ideas cannot immediately be applied
to it. In /15/ Schmidt studied the robustness of the vibration controls of
a Timoshenko beam with respect to changes in the system's parameters. It
remains to be studied whether the control designed for the Timoshenko beam
can also be applied successfully to an Euler–Bernoulli beam in some way.
One of the characteristics of the Euler–Bernoulli beam is that there is no
finite speed of propagation for the signals. On the other hand, in any
discretized mathematical model of the Euler–Bernoulli beam this difference
between finite or infinite speed of propagation is smoothened out. Similar-

ly, in any physical system this speed will obviously be limited. Therefore the vibration control designed for a Timoshenko beam with very high shear stiffness may possibly work well if implemented in a real physical beam, which normally would be described by the Euler-Bernoulli theory. The energy transport associated to the dispersion mode involving mainly shear may possibly be disregarded in the control law. These phenomena need further analytical and experimental clarification.

A different question which is of considerable interest with regard to technical applications concerns the nature of the control forces and moments. In this paper so far the control forces were always *external forces*. This will usually be a disadvantage in any technical implementation, since in the case of space structures, for example, it means that the control forces will have to be produced by reaction jets. These require fuel, i.e. mass to be carried along, and are therefore of a limited life. If the control forces on the other hand were *inner forces*, i.e. forces acting *between* parts of the same structure, they could be produced by electromechanical actuators recquiring *energy* but no mass expenditure. Energy is however available in virtually unlimited amounts (but obviously with limited power), since it can be obtained through the solar pannels. In addition, vibration control via inner forces and moments would leave invariant the total angular and linear momentum of the structue, thus decoupling the orbit and attitude problem from the vibration problem. This is a highly desirable feature in space applications.

Also in more down-to-earth applications, such as in robotics, for example, the active control forces will usually not be external forces acting between an element of the robot and an external base, but internal forces (and moments) acting between different parts of the robot. The problem of designing active vibration controls via the travelling wave approach should therefore be readdressed with view to the usage of internal forces and moments. The energy valves depicted in Fig.2 should then be produced by a group of forces (and moments) forming equilibrium groups.

The vibration controls designed via the traveling wave approach have a purely local character. In a given structure their performance should nevertheless be tested using a large discretized model and in these tests the actuator location will be important. The questions raised before by other authors with regard to the optimal actuator and sensor locations (see /8/, /11/) will therefore have to be reexamined in the light of this new approach.

5.   FINAL REMARKS

In the present paper it was shown how vibration controls can be designed for one-dimensional continua described by hyperbolic equations. Rather than discretizing the boundary value problem and designing the control for the discretized system described by ordinary differential equations, here the controller design is carried out directly with the partial differential equations, with view to maximum energy absorbtion. The ideas were applied to the transverse vibrations of a string and to a Timoshenko beam. This was possible in spite of the dispersiveness of the bending waves. No such simple solution is available for the Euler-Bernoulli beam, since the equations of motion are not hyperbolic in this case. The Timoshenko beam is therefore the simpler beam model for the active vibration control problem!

It turned out to be convenient to define an elementary controller corresponding to an "energy valve", giving way to energy flow in one direction only. Vibration energy can be efficiently extracted from the structure combining such elements, regardless of the boundary conditions.

The decentralized character of the control makes it extremely robust, as was shown by numerical experiments. This robustness also holds for changes in the system's parameters, such as stiffness and mass. Due to this fact, the author believes the approach to be useful in the design of active vibration controls for large structures containing beamlike elements, as is the case for the planned space station.

Additional open questions to be addressed in the future concern the vibration control using internal forces and the application of the controls computed for the Timoshenko beam to the Euler beam problem. The difficulties encountered by other authors, with regard to the destabilizing effect of time delays in the controls, were not found here, due to the robustness of the solution (see /1/). The given solution is represented entirely in the time domain which simpifies its implementation as compared to solutions given in the frequency domain (see /3/).

ACKNOWLEDGEMENTS

The author gratefully acknowledges the support of the Stiftung Volkswagenwerk.

REFERENCES

[1]   Datko, R.; Lagnese, J. & Polis, M.P.; An Example on the Effect of Time Delays in Boundary Feedback Stabilization of Wave Equations, SIAM J. Control and Optimization, Vol.24, No.1, 1986, 152-156

[2]   Delfour, M.C.; Lagnese, J. & Polis, M.P.; Stabilization of Hyperbolic Systems Using Concentrated Sensors and Actuators, IEEE Trans. Automatic Control, Vol.AC-31, No.12, 1986, 1091-1096

[3]   Flotow, A.H. von; Travelling Wave Approach to the Dynamic Analysis of Large Space Structures, Paper 83-0964 presented at the 24th AIAA Structures, Structural Dynamics and Materials Conference, Lake Tahoe, May 2-4, 1988

[4]   Hagedorn, P.; On a New Concept of Active Vibration Damping of Elastic Structures, Proceedings of the 2nd International Symposium on Structural Control, Waterloo, Canada, 1985

[5]   Hagedorn, P. & Schmidt, J.T.; Active Vibration Damping of Flexible Structures Using the Travelling Wave Approach, Proceedings of the Second International Symposium on Spacecraft Flight Dynamics, Darmstadt, October 20-23, 1986 (ESA SP-255, Dec 1986)

[6]   Hagedorn, P. & Schmidt J.; On the Active Vibration Control of Distributed Parameter Systems, in: Dynamics and Control of Large Structures (Editor: L. Meirovitch), Proc. of the Sixth VPI & SU/AIAA Symposium held in Blacksburg, Va., June 29 July 1, 1987, 359-373

[7]   van Horsen, W.T.; An Asymptotic Analysis of a Class of Nonlinear
        Hyperbolic Equations, Doctoral Diss., TH Delft, 1988

[8]   Hu, A.; Shelton, R.E. & Yang, T.Y.; Modelling an Control of a
        Beam-like Structure, J. Sound & Vibration (1987) 117 (3),
        475–496

[9]   Kelly, J.M.; Leitmann, G. & Soldatos, A.G.; Robust Control of
        Base-Isolated Structures under Earthquake Exitation, JOTA,
        Vol.53, No.2, 1987, 159–180

[10]   Meirovitch, L.; Analytical Methods in Vibrations, Macmillan,
        London, 1967

[11]   Müller, P.C.; Optimale Positionierung von Dämpfern in Schwingungs-
        systemen, to appear in ZAMM 1988

[12]   Noor, A.K. & Nemeth, M.P.; Analysis of Spatial Beamlike Lattices
        with Rigid Joints, Computer Methods in Appl. Mechs. and
        Engineering, 24 (1980) 35–59

[13]   Salm, J. & Schweitzer, G.; Modelling and Control of a flexible
        rotor with magnetic bearings, Proc. of the Third Int. Conf. on
        Vibrations in Rotating Machinery, held at the University of
        York, England, September 11–13, 1984 (Mechanical Eng.
        Publication, Ltd., London), 553–562

[14]   Salm, J.R.; Active Electromagnetic Suspension of an Elastic Rotor:
        Modelling, Control and Experimental Results, Proc. of the 1987
        ASME Design Technology Cont., held at Boston, Ma., September
        27–30, 1987, Vol.2, 141–149

[15]   Schmidt, J.; Entwurf von Reglern zur aktiven Schwingungsdämpfung an
        flexiblen mechanischen Strukturen, Doctoral Diss., TH Darmstadt
        1987

Theoretical and Applied Mechanics
P. Germain, M. Piau and D. Caillerie (Editors)
Elsevier Science Publishers B.V. (North-Holland)
© IUTAM, 1989

NUMERICAL SIMULATION OF TURBULENCE

Jackson R. HERRING and Robert M. KERR
National Center for Atmospheric Research*
Boulder, CO, U.S.A.

The role of direct numerical simulations (DNS) in providing insight into homogeneous turbulent flows are described. We examine both two-dimensional turbulence and isotropic three-dimensional turbulence convecting a passive scalar. Of particular concern in all these problems, is the role of DNS in revealing coherent structures, and the problems they create in vitiating traditional turbulence notions, such as found in two-point closures.

First, some dynamical elements are reviewed that are essential for understanding two-dimensional turbulence, which numerical results (McWilliams [1] Basdevant *et al.* [2]) for 2-D turbulence indicate the presence of persistent isolated vortices, which — in certain cases — severely modify the picture of energy inverse cascade and forward enstrophy cascade. Next the elements missing from closure are indicated, and suggestions on how these elements may be incorporated, so as to make the physics of the closure correspond to Navier-Stokes dynamics more closely.

Results drawn from decaying isotropic turbulence in three-dimensions are next discussed, with special stress on the role of scalar structures in determining the covariance-level dynamics. We review briefly recent results of Kerr [3], whose DNS findings shed new light on the validity of low Prandtl number spectra as, for example, the classical findings of Batchelor, Howells, and Townsend [4]. We then examine systematically, the fourth-moment statistics of decaying turbulence convecting a passive scalar, and indicate to what extent these depart from Gaussian values. This is done for a range of Prandtl numbers. We also present information on the single, point-distribution function, $P(x)$ for turbulence, and the passive scalar. Results indicate $P(x) \sim exp(-c \mid x \mid)$ (see also Yamamoto and Hosokawa, [5]) for the distribution of $x = \partial_z u$, with similar results for the scalar. We discuss the possible connection of these results to the intermittency of the flow, and for the scalar. Next, we note to what extent the classical two-point closure scheme (the Direct Interaction Approximation, Kraichnan [6]) is able to reproduce the tendency for the nonlinear and pressure terms in the Navier-Stokes equations to evolve to nominally small values, as compared to that produced by a Gaussian, incompressible flow field (this is done at the fourth-moment level).

* The National Center for Atmospheric Research is sponsored by the National Science Foundation.

1. INTRODUCTION

Numerical simulation of turbulence has emerged as a principal tool in the contribution to our understanding of this difficult subject. The reasons are not only the rapid increase in raw, "number crunching", computing power, but also in the development of three-dimensional color graphics. What has in fact happened, is the development of a new experimental tool, which is in much more intimate contact with the equations of motion than are real experiments and observations. The disadvantages of most numerical experiments (which we shall call direct numerical simulations [DNS]) is the modest Reynolds number to which they are frequently limited.

In this paper, we focus on certain recent numerical results in which DNS have modified, in a fundamental way, our perception or prejudice of what is turbulence. DNS are frequently undertaken to fix coefficients, for example, spectral power laws, or rate coefficients. We are here more interested in computations, that do not have power-law regimes. The common prejudice is that turbulent flows have statistics as close to "Gaussian" as permitted by the overall constraints that follow from the governing equations of motion. Another assumption is that the statistical symmetry towards which the flow evolves is unique, not a function of initial conditions. What DNS has shown is that such randomness, although pervasive, is interrupted by ordered structures, which play a vital, significant role despite their possibly rare occurrence. These findings have, of course, a counterpart in experiments (Hussain [7]). Our examples here are largely drawn from homogeneous turbulence, in both two and three dimensions.

2. SOME GENERAL COMMENTS ON NUMERICAL TECHNIQUES

Before presenting detailed results on turbulent flows, several current numerical methods that have proved useful are discussed. The presentation will serve as a backdrop to help focus on certain issues in both DNA and statistical theories upon which we wish to make comments. We leave out reference to "leading-edge" methods such as *Cellular Automata* or the *piece-wise parabolic method*. We also will have little to say about Lagrangian methods, which are essential in the study of flows at finite Mach numbers.

Let us symbolically denote the Boussinesq limit of the Navier-Stokes (NS) equations as,

$$\mathcal{F}\{\boldsymbol{u}, T, \boldsymbol{x}, t\} \equiv (\partial_t + L)\psi - N(\psi, \psi) = 0 \ . \tag{2.1}$$

Here ψ stands for the collection $(\boldsymbol{u}(\boldsymbol{x}, t), T(\boldsymbol{x}, t))$ (velocity and temperature fields), L is the linear operator including dissipation effects $(\nu\nabla^2\boldsymbol{u}, \kappa\nabla^2 T)$ and possible spacial means of derivatives of ($\boldsymbol{u}$ and T), and N represents the non-linear terms $(\boldsymbol{u} \cdot \nabla\boldsymbol{u}, \boldsymbol{u} \cdot \nabla T)$. N also contains the pressure gradient, the latter being eliminated through the use of continuity (*i.e*: $\nabla \cdot \boldsymbol{u} = 0$). We approximate ψ in (2.1) by:

$$\psi_N(\boldsymbol{x}, t) \equiv \sum_{n=1}^{N} a_n(t)\phi_n(\boldsymbol{x}), \tag{2.2}$$

(to fix ideas, we may imagine ϕ_n as box-normalized, complex exponential $\exp(i\boldsymbol{k} \cdot \boldsymbol{x})$, appropriate for homogeneous flow. For this case, the index n is a mapping of $\boldsymbol{k}$, whose cartesian components are $2\pi\ell_j/L_J$, with L_J the box dimension in direction J, and $\ell_j = \pm 2\pi L_J(0, 1, 2, 3, \cdots N)/N)$. ψ_N approximates ψ either in some least-squares sense or interpolatively, by which we mean;

$$a_n(t) = \psi(\boldsymbol{x}_n, t), \phi_n(\boldsymbol{x}_m) = \delta_{nm}, (n, m = 1, 2, \cdots N). \tag{2.3}$$

The burden of solution is now shifted to the collection $\{a_n(t)\}, \{n = 1, 2, \cdots\}$. Equations for a_n are obtained through examining the set of N equations:

$$\int_{\mathcal{D}} d\boldsymbol{x} \chi_n(\boldsymbol{x}) \mathcal{F}_N\{\cdot\} = 0, (n = 1, 2, \cdots N) , \tag{2.4}$$

where, $\mathcal{F}_N$ is that $\mathcal{F}$ obtained by using (2.2) to approximate ψ, $\mathcal{D}$ is the domain of the flow, and $\chi_n(\boldsymbol{x})$ are an arbitrary set of test functions. We shall not discuss the question of the optimal χ_n for a given $\{\phi_n\}$; suffice it to say that for certain problems (L, N both linear operators) (2.4) leads back to the Rayleigh-Ritz problem, for which $\chi_n = \phi_n$ is generally best. If $\chi_n = \phi_n$, the method is Galerkin, and if we take $\chi_n = \delta(\boldsymbol{x} - \boldsymbol{x}_n)$, it is colocation (pseudo-spectral). Galerkin has the advantage of preserving existing quadratic, inviscid constant of motion of (2.2) for *arbitrary N*, whereas the colocation requires large N to accurately estimate such constraints. Early on, this fact was thought to favor the Galerkin; however examples of conservative schemes that gave poor overall results diminished the enthusiasm for Galerkin. Current folklore is that for strongly dissipative (turbulent) systems, the difference between colocation and Galerkin for a given problem measures the error of either, although counter examples favoring either method may be offered. On the otherhand, the Galerkin procedure is more difficult in practice, especially if an interpolative scheme in (2.3) is used in the colocation method. Generally, the finite element methods described above require considerably more effort than grid-point methods, because of the work required to evaluate the nonlinear terms (convolution sums). However, as compared to grid-point methods, these techniques have a much more rapid convergence, so that in general, fewer degrees of freedom are required for a given accuracy. Generally, the representation (2.2) converges exponentially (for those ϕ_n of sufficient analyticity), whereas grid point methods, algebraically. We do not need to be reminded of the enormous advantages that accrue by using fast Fourier transforms for convolution sums, in connection with the trigonometric (or Tschebychev) series for (2.2).

The methods of two-point closure, for which the basic predictive variable is $\langle u_i(\boldsymbol{x})u_j(\boldsymbol{x})\rangle$, ($\langle\cdot\rangle \equiv$ ensemble mean), should also be brought under the purview of our discussion. Such techniques offer an *in principle* economy, because of the smooth nature of ensemble means, as opposed to the chaotic fluctuations of the amplitude $\boldsymbol{u}$, for individual realizations. However, as yet, satisfactory evolution equations for $\langle\cdot\rangle$ have not been found; indeed, the number of flow situations which offer existing closures a severe challenge is growing. The problem is that intense structures are difficult to represent *via* an unconditioned ensemble mean variable. A typical closure for isotropic turbulence predicts $U(k, t)$, where

$\langle u_i(\boldsymbol{k})u_j(-\boldsymbol{k})\rangle = (1/2)(\delta_{ij} - k_i k_j/k^2)U(k,t)$:

$$(\partial_t + 2\nu k^2)U(k) = 2\int_\Delta dpdq\Theta(k,p,q)B(k,p,q)U(q)\{U(p) - U(k)\} \qquad (2.5)$$

Here, $B(k,p,q)$ is a geometrical factor stemming from the nonlinearity in (NS). In three-dimensions, $B(k,p,q) = \pi((p^2 - q^2)(k^2 - q^2) + k^2 p^2)\sin^2((p,q))(pq/k^3)$; in two-dimensions, $B(k,p,q) = (2/k^2)(p^2 - q^2)(k^2 - q^2)\sin((p,q))$. $\int_\Delta$ means the $(dpdq)$ integration is only over those $\boldsymbol{p}, q \ni \boldsymbol{k} = p + q$. $\Theta(k,p,q)$ is a relaxation time for triple moments. Various methods differ as how to specify these; we refer to Herring *et al.* [8] for an early summary, and to Kaneda [9], for more recent results on this point. (The form of Θ should be deduced from first principles, but early work admitted some empericism not found in the methods reviewed by Kaneda.) Evolution equations for a passive scalar spectrum $\langle \theta(\boldsymbol{k})\theta(-\boldsymbol{k})\rangle \equiv U_\theta$ are (2.5) with $B(k,p,q) \rightarrow B_\theta(k,p,q) \equiv 2\pi((p^3 q)/k \sin^2((q,k)))$, and with the $U's$ within $\{\}$ in (2.5) replaced by U_θ's. For a comprehensive account of the application of such methods to turbulent flows, see Lesieur [10].

We note that (2.5) preserves the basic, bilinear structure, except for the Θ factor, for which experience shows only a weak $k-$dependence (indeed, it has been claimed that qualitatively, valid information is obtained by taking $\Theta = ct$ (Frisch *et al.* [11]).

We mentioned above the smoothness of $U(k,t)$, and this is generally used to facilitate numerical methods to solve (2.5). At large Reynolds number R_λ, $T(k)$, $(T(k) \ni (\partial_t + 2\nu k^2)U(k) = T(k)/(2\pi k^2)$ (2.5)) consists of a compact, negative lobe (at small k, in the energy containing range $k \sim k_e$), with a broad, positive lobe extending from scales near the Taylor microscale range to the dissipation wavenumber, $k_s = (\epsilon/\nu^3)^{1/4}$. Since the range $(k_e < k < k_s)$ is large but smooth, it is sensible to choose the interpolative knots with increasing sparseness as $k \rightarrow \infty$. Such a choice would seem to preclude the possible use of interpolatory functions ϕ_n, for which $FFTs$ may be usefully employed to speed up the computations. However, Domarardzki and Orszag [12] have explored this technique for non-equidistant k_n. Early numerical methods to solve closure (Kraichnan [13]; Leith [14]) employed the Galerkin method, with simple non-overlapping square pulses (used interpolatively) for the $\phi_n(k)$ (2.2), with $x \rightarrow k$, and an exponentially increasing spacing of the knots, $(k_n$, in (2.3)). This choice of ϕ_n preserved energy conservation ($\int_0^\infty T(k)dk = 0$) at arbitrarly small N. Later work (Herring and Kraichnan [15], Herring [16], Tatsumi *et al.* [17]) utilized more analytic interpolatory functions (*viz;* B-splines) for (2.2) colocatively with a spacing of the k_n exponentially stretched Gaussian points. The latter choice does not automatically preserve inviscid quadratic constraints, but must rely on sufficiently accurate representation of $U(k)$ to enforce $\int_0^\infty T(k)dk \rightarrow 0$. It is not absolutely clear which is the best numerics, but my own numerical experimentation with Galerkin *vs* colocation, suggests the colocation method is actually superior for this problem. It is certainly much less complicated, especially for anisotropic flows, as, for example, those studied by Bertoglio [18].

The time stepping of the closure equations (2.5) is frequently done *via* a standard second order technique, such as an Adams-Bashforth or an Euler scheme (with perhaps a Crank-Nicholson treatment of viscous and conductive terms). A much better approach is to quasi-linearize the right-hand side of (2.5), and use the *eigen* modes of the linear terms to step (2.5) forward in time [16]. The *eigen*-mode calculations are expensive (of the same order as computing the convolution integral), but an order of magnitude is usually saved by using larger time steps.

The wisdom of investing resources in computing at very high R_λ, for which many degrees of freedom, (k_n), are devoted to accurately determining the dissipation range might be questioned, since we have known for some time that because of intermittency effects, the dissipation range, computed, is wrong. Perhaps a better strategy would be to compute on a more modest computational domain, and use hyperviscosity ($\nu_4 k^4$, say) to artificially shorten the dissipation range. Numerical experiments (in both two and three dimensions) have shown that such shortening has little effect on the retained degrees of freedom at lower k. Such an application of closure might be useful in exploring anisotropic and inhomogeneous effects, or the issue of how even simpler models (along the lines investigated by Yoshizawa [19]) may be constructed.

3. TWO-DIMENSIONAL TURBULENCE

Since the early work of Fornberg [20], Basdevant *et al* [2], and McWilliams [1] the suspicion has grown that an initially random, homogeneous flow field, satisfying Navier-Stokes,

$$(\partial_t + \boldsymbol{u} \cdot \nabla)\boldsymbol{u} = \nu\nabla^2\boldsymbol{u} - \nabla p \tag{3.1}$$

$$\nabla \cdot \boldsymbol{u} = 0 \tag{3.2}$$

may, under certain circumstances, develop a vorticity field, which is concentrated into isolated, well separated regions, and that these isolated vortices might eventually dominate the small-scale flow field. Figure 1 shows a contour map of Ω at late time in the simulation for such a flow field [1]. The surprising element here is the extreme intensity and compactness of the vortices, and the quiescent surrounding regions. This $\Omega(\boldsymbol{x})$ field is inconsistent with the idea

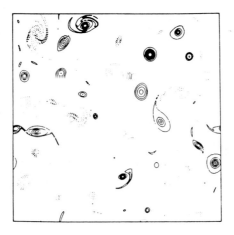

Fig. 1 Contours of a vorticity field for late-time evolution of two-dimensional turbulence. Resolution is 256^2 [1].

that the underlying statistics of $(\Omega(\boldsymbol{k}, t)$ $(\Omega(\boldsymbol{x}, t) \equiv \sum \exp(-\boldsymbol{x} \cdot \boldsymbol{k}) \Omega(\boldsymbol{k}))$ are in any sense close to multi-variate Gaussian. To see this, we need only to recall that if $\Omega(\boldsymbol{k})$ were Gaussian $K \equiv \langle \Omega^4 \rangle / \langle \Omega^2 \rangle^2$ would be 3. For the flow in Fig. 1, $K(\Omega) \sim 60$.

We may argue, *via* a form of rapid-distortion theory, that equation (3.1) is inconsistent with a near-Gaussian behavior. The argument will also give a clue as to how vortex consolidation happens, and perhaps how to incorporate it eventually in a theory [21], [22]. To this end, consider an ensemble of flows, whose initial values consist of a "large-scale" mean, $\langle \boldsymbol{u} \rangle$, plus initial small-scale Gaussian fluctuations, $\boldsymbol{u}'$, $(\boldsymbol{u} = \langle \boldsymbol{u} \rangle + \boldsymbol{u}')$. The short-time behavior for the Reynolds stresses, $R_{ij} \equiv \langle \boldsymbol{uu} \rangle_{ij}$ is implied in rapid distortion theory ($\equiv$ third-order cumulant discard) by:

$$(D/Dt + 2\nu(k))U_0(k) = S^*(1 + (k/4)\partial/\partial k)U_2(k) \tag{3.3}$$

$$(D/Dt + 2\nu(k))U_2(k) = i\langle \Omega \rangle U_2(k) + S(1/2 + (1/4)k\partial/\partial k)U_0(k), \tag{3.4}$$

$$S = -\partial_1 \langle u_1 \rangle + \partial_2 \langle u_2 \rangle + i(\partial_2 \langle u_1 \rangle + \partial_1 \langle u_2 \rangle), \tag{3.5}$$

$$\Omega = -\partial_2 u_1 + \partial_1 u_2. \tag{3.6}$$

In this complex notation, the Reynolds stress spectrum, $R_{ij}(\boldsymbol{k}, t)$ derives from $(U_0(k), U_2(k))$ $(U(\boldsymbol{k}, t) = U_0(k, t) + U_2(k, t) \exp(2i \tan^{-1}(k_2/k_1) + cc + \cdots)$, the *modal* energy spectrum)

$$\langle |u_1 + u_2|^2, (u_1 + iu_2)^2 \rangle \equiv \{U_0(k), U_2(k)\} . \tag{3.7}$$

In (3.4) $D/Dt \equiv \partial_t + \langle \boldsymbol{u} \rangle \cdot \nabla$. Omitted from (3.4) and (3.5) are contributions to R_{ij} from the triple cumulant $\langle \boldsymbol{uuu} \rangle_C$, which are initially zero and (usually assumed) slowly evolving, as compared to the terms recorded. We have also assumed that the length scales associated with two-point quantities are smaller than length scales of $\langle \nabla \boldsymbol{u} \rangle$ and $\langle \Omega \rangle$. Thus in (3.3) and (3.4), $(U_0(k), U_2(k), \cdots)$ depend parametrically on $\boldsymbol{R} \equiv (\boldsymbol{r}_1 + \boldsymbol{r}_2)/2)$, where $(\boldsymbol{r}_1, \boldsymbol{r}_2)$ are the coordinates of $\langle u_i(\boldsymbol{r}_1)u_j(\boldsymbol{r}_2) \rangle$. Similarly, $\langle \Omega \rangle$, and S are functions of $\boldsymbol{R}$. Equations (3.3) and (3.4) leave out the return to isotropy (Rotta, [23]), and omit the diffusion tendencies, the primary effects of $\langle \boldsymbol{uuu} \rangle_C$. If we include a simple parameterization of the return to isotropy as a term $-\mu U_2$ to be added to the right-hand side of (3.4), and assume the time scale more rapid for U_2 than for U_1, there results;

$$\{D/Dt + 2\nu(k)\}U_0(k) = |S|^2 \mu/(\mu^2 + \langle \Omega \rangle^2)\{1/2 + (1/4)k\partial_k\} \cdot$$

$$\{1 + (1/2)k\partial_k\}U_0(k) + \cdots . \tag{3.8}$$

For our crude estimate (3.8), we may recover the usual $E(k) \sim k^{-3}$ of two-dimensional turbulence at small scales. There is more here, however, since we have included some features that are smoothed out in the usual treatment of homogeneous turbulence. Note that for spatial regions where $|S|$ is initially large that $U_0(k)$ changes quickly, while if $|\langle \Omega^2 \rangle^{1/2} \gg S$, nothing happens. Equation (3.8) illustrates a difficulty for homogeneous closures, in that if we take an ensemble average over the large scales involving $\langle \Omega^2 \rangle$, and $\langle |S|^2 \rangle$ (a step needed to construct equations for completely homogeneous flows), we face the difficulty of evaluating a quantity that is nonlinear in $|\Omega|^2$, another closure problem. If we assume that only $\langle |\Omega|^2 \rangle$ and $\langle |S|^2 \rangle$ are needed, then these are equal for homogeneous flows, so that we would lose the stable regions. This is effectively what closures do.

Alternatively, we could develop a sub-grid procedure, based on (3.3)-(3.6) [21]. Such would lead to the depletion of turbulence in those regions where there is an excess of strain over vorticity, and (*vice versa*), a criterion considerably sharpened by the analysis of Weiss [24], who used a Lagrangian approach, with a scale-separation hypothesis. Weiss's analysis has recently been extended, and verified, numerically by Brachet *et al* [25]

4. Non-Gaussian Aspects of Three-Dimensional DNS

The degree of non-Gaussian behavior displayed by the two-dimensional DNS shown in the example found in the last section, may be an extreme, since in three dimensions, vortex structures are unstable if the vortices have any asymmetry (Pierrehumbert [26]). However, experimental data (Wyngaard and Pao [27]; Castaing *et al* [28]) suggest otherwise. In Fig. 2A, we show atmospheric measurements of Wyngaard and Pao [27] for the distribution (density) function $\mathcal{P}(x)$ for $x = \partial u / \partial t$, and in Fig. 2B the distribution function for $x = T_1 - T_0$ ($T_1 - T_0$, the temperature difference across the convective chamber) in the thermal convection experiments of Castaing *et al* [28] in helium gas. The atmospheric observations have the kurtosis of $\partial u / \partial t \sim 60$. Castaing *et al.* distinguish two types of thermal turbulence:

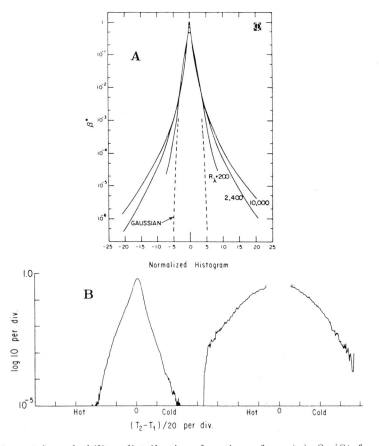

Fig. 2 Experimental probability distribution functions for: (a) $\partial u / \partial t$ for atmospheric measurements [27]; and (b) recent helium gas experiments of Castaing *et al* [28]. For Fig. 2b, the curve to the left is designated by the authors as hard turbulence, and that to the right, soft-turbulence. Ordinate is the temperature difference across the cell [28].

1) soft (for which Fig. 2B1 pertains) and hard (Fig. 2B2), with a transition at Rayleigh number $R_a = 2.5 \times 10^5$. Note that for the "soft" turbulence $\mathcal{P}(x)$ is nearly Gaussian, but for "hard" turbulence $\mathcal{P}(x) \sim exp(-\mid x \mid)$. An analogous finding has been reported by Metais and Herring [29] for the numerical simulation of stratified turbulence; without stratification (Fig. 3A), $\mathcal{P}(x = \partial u/\partial z) \sim \exp(-\mid x \mid)$, while for strong stratification (Fig. 3B), $\mathcal{P}$ is more nearly Gaussian. For these simulations $R_\lambda \sim 20$. This small value of R_λ suggests that the reasons for the exponential shape (which seems not as yet well understood) can be studied rather completely by DNS, and that it is not exclusively a large R_λ phenomenon.

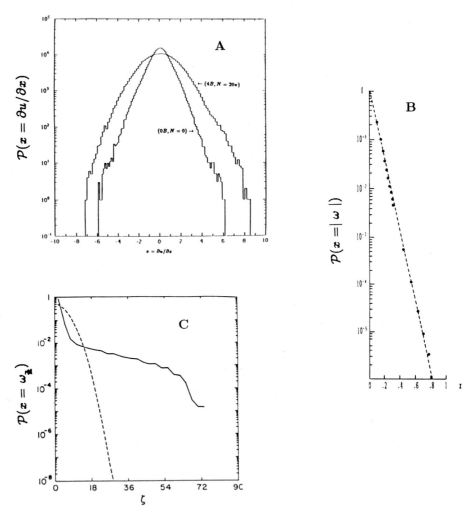

Fig. 3 Distribution functions from DNS: (a) the 64^3 resolution simulation of Metais and Herring [29], $(0B, N = 0)$ is isotropic turbulence, $(4B, N = 20\pi)$ strongly (stably) stratified flow. For both cases, $R_\lambda \sim 20$. (b) 128^3 isotropic computations of Yamamoto and Hosokawa [5]. (c) the quas-geostrophic simulations of McWilliams [37]. The dashed line is at early time, while the solid is a time after condensation into isolated vortices. Note the bi-modal signature, with high probability centered at the origin.

Figure 3C shows $\mathcal{P}(\mid \omega \mid)$, ω = vorticity, as taken from Yamamoto and Hosokawa [5], a 128^3 spectral simulation of isotropic turbulence at $R_\lambda \sim 200$ (but with an energy spectrum severly truncated at small k). The accuracy of the exponential fit here is remarkable. Figure 4 is an isosurface map of ω [5]. Note the "worm-like" concentrations of ω, perhaps reminiscent of McWilliam's isolated vorticity regions, but now with more random orientation. These authors warn, however, that regions of *very* high vorticity are more "sponge-like" than worm-like in their topography.

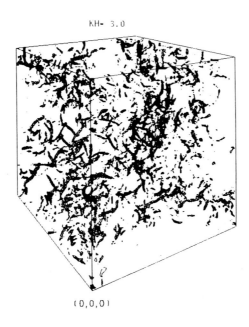

Fig. 4 Vorticity-concentration regions at $R_\lambda \sim 500$ [5].

The type of non-Gaussian behavior displayed here is different, but surely related to that encountered for the Fourier transforms of $(\boldsymbol{u}, T)$.

We may approach the issue of non-Gaussian behavior of turbulent $\boldsymbol{u}(\boldsymbol{k})$ by examining predictions of their higher moments *via* DNS, comparing results to those of closure. The issue has recently been explored by Kraichnan and Panda [30], and Chen *et al* [31], who examined DNS spectra for the nonlinearity, $\mid \boldsymbol{F}(\boldsymbol{k}) \mid^2$, in NS, where we write the NS in the form;

$$(\partial_t + \nu k^2)u_i(\boldsymbol{k}) = F_i(\boldsymbol{k}), \qquad (4.1)$$

$$\Rightarrow F_i = (\delta_{ij} - k_i k_j)/k^2 [\boldsymbol{u} \cdot \nabla \boldsymbol{u}]_j^{FT} .$$

Here FT means Fourier transform. It is thought by some that $\langle \mid \boldsymbol{F} \mid^2 \rangle$ is smaller than its Gaussian value, since the nonlinear terms organize themselves into near force-free configurations. Note that

$$\mathcal{W}(k) \equiv \langle \mid \boldsymbol{F} \mid^2 \rangle \qquad (4.2)$$

is a fourth moment of $\boldsymbol{u}(\boldsymbol{k})$. The pressure variance spectrum $\Pi(\boldsymbol{k})$ stems from that part of $\boldsymbol{F}(\boldsymbol{k})$ proportional to $k_i k_j/k^2$:

$$P(\boldsymbol{k}) = -k_i k_j/k^2 [\boldsymbol{u} \cdot \nabla \boldsymbol{u}]_j^{FT}, \qquad (4.3)$$

and is also a fourth moment in $\boldsymbol{u}$. If $u(\boldsymbol{k})$ were Gaussian, $\mathcal{W}(k)$ and $\Pi(k)$ would be

$$\mathcal{W}_G(k) = \int_\Delta B(k,p,q)U(p)U(q)dpdq \tag{4.4a}$$

$$\Pi_G(k) = \int_\Delta C(k,p,q)U(p)U(q)dpdq . \tag{4.4b}$$

Here, $B(k,p,q)$ is listed just after (2.5), and $C(k,p,q) = \pi k p q \sin^2(k,p)\sin^2(k,q)$. A formula equivalent to (4.4b) was first given by Batchelor [32]. We note that (4.4a) is very nearly the input term for (2.5). This is no accident, but stems from the idea that (2.5) is basically an "almost" Markov model of turbulence, with an eddy viscosity (the second term in (2.5)), restoring conservation in the face of a Gaussian zeroth approximation for the statistics of $\boldsymbol{F}(\boldsymbol{k})$. For the passive scalar $F_\theta = (\boldsymbol{u} \cdot \nabla \theta)^{FT}$, and the Gaussian variance spectrum $\mathcal{W}_\theta$, is (4.4a) with $U(p) \to U_\theta(p), B \to B_\theta$. The non-Gaussian parts of $\mathcal{W}$ and Π may be approximated from closure. This is best seen by recalling the perturbative derivation of closure [6], which begins by analyzing $u_i(\boldsymbol{k},t)$ into Gaussian and non-Gaussian parts;

$$u_i(\boldsymbol{k},t) = \mathcal{G}_{ij}(\boldsymbol{k},t,0)u_i^G + \int_0^t ds \mathcal{G}_{ij}(\boldsymbol{k},t,s)F_j(\boldsymbol{u}^G,s) + \cdots , \tag{4.5}$$

where, $\mathcal{G}(t,s)$ is a Response function, whose ensemble mean is related to the eddy viscosity in (2.5), and the superscript G denotes the Gaussian part. $\mathcal{G}$ is stochastic, but this happens not to be relevant to the following discussion. To obtain the non-Gaussian part of $\mathcal{W}_c \equiv \mathcal{W} - \mathcal{W}_G$ we use (4.5), in its definition, (4.2). The results are somewhat lengthy [30], and we shall not record them here. One interesting point is that the non-Gaussian part of Π is zero, in the DIA [6].

Figure 5(A,B,C,D) shows some results comparing DNS to DIA-closure for: (A,B); $(E(k), E_\theta) = 2\pi k^2(U(k), U_\theta)$; (C,D), $(2\pi k^2(\mathcal{W}, \mathcal{W}_\theta))$. Figure 6(A,B) presents the comparison of DIA to DNS for $2\pi k^2 \Pi(k)$. The initial spectra were $E(k) = E_\theta(k) = A\exp(-2(k/k_0)^2), k_0 = 4.7$, and $A \ni \int_0^\infty dk E(k,0) = 3/2$. The initial R_λ (Taylor micro-scale Reynolds number) $= 35.0$; the Prandtl number is $P_r = 1$. The DNS code is 128^3 pseudo-spectral (colocation) [33], for which truncation errors begin to be a serious problem for $k > 48$ (for Fig. 5 $k_s = 32.2$). The time at which data are shown is t=1.0 (in units of the inital eddy circulation time). At these R_λ, the DIA compares favorably with other Galilean invariant theories such as the $EDMQN$ or TFM, but the differences between theories are not very significant. We remark here that Markovian theories are ill suited as a starting point for the evaluation of $\mathcal{W}$. This may be seen by recalling that what is needed here is $\partial_t \partial_{t'} U(k,t,t')_{t' \to t}$ [34] Overall, we note a reduction in $\mathcal{W}(k)$ and $\mathcal{W}_\theta$ of about a factor of 2, for $k > k_e$, the peak wavenumber. The DIA actually over predicts this reduction at large k. On the other hand, at small k (where DNS results are more difficult to interpret), $\mathcal{W}$ and especially $\mathcal{W}_\theta$ have large, positive, non-Gaussian components. This indicates not only a strong change of the large scales, but also that such changes are organized. The magnitude of $\mathcal{W}_{\theta,c}$ at small $k < k_e$ is such as to give the practitioners of closure pause. The cross-over wavenumber where $\mathcal{W}(k) \sim 0$ is about k_e.

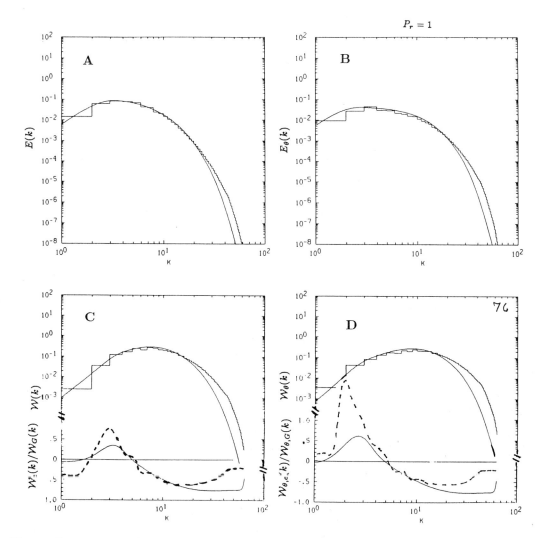

Fig. 5 Comparison of statistical turbulence theory (DIA) with DNS 128^3 after one eddy circulation time. Initial conditions are given in the text. Histogram is DNS, and solid curve is DIA. (a) $E(k,t)$; (b) $E_\theta(k,t)$; (c) $\mathcal{W}$ (see 4.2), and (d) $\mathcal{W}_\theta$. For (c) and (d), bottom curves compare non-Gaussian parts; solid is DIA, dashed is DNS.

Figure 7(A,B) compares for the DNS Gaussian to non-Gaussian spectra for $\mathcal{W}(k)$, and $\mathcal{W}_\theta$. These spectra were computed directly from the DNS code, and their Gaussian parts were computed from (4.4a)-(4.4b) using U_{DNS}.

Results for the pressure-variance spectrum in Fig. 6(A,B) show an under estimation by the DIA of about 20 percent at all k. On the other hand, Fig. 6B shows that were we to use in (4.4b) U_{DNS} instead of U_{DIA}, the reduction would be about the same indicating a small, positive departure from the Batchelor formula (4.4b).

Figure 8(A,B) shows the DNS E_θ and $\mathcal{W}_\theta$ spectra for $P_r = 8$, both total, and Gaussian parts. Figure 9(A,B) shows the same for $P_r = 0.0625$. At $P_r = 8$, $E_\theta(k)$ is quite shallow at $k > k_e$, suggesting the k^{-1} shape would be presumably obtained at $P_r \to \infty$, and at large R_λ. The truncation errors for $k > \sim 45$ are clearly visible. Figure 7(B) shows clearly the large

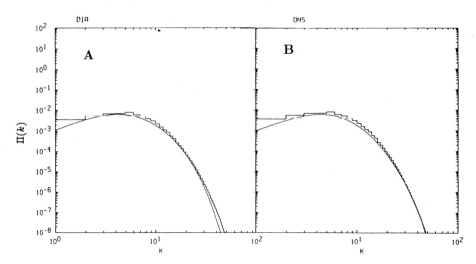

Fig. 6 Comparison of DNS pressure variance spectrum (histogram) with the predictions of (4.4b) using (a) U_{DIA}, (b) U_{DNS}.

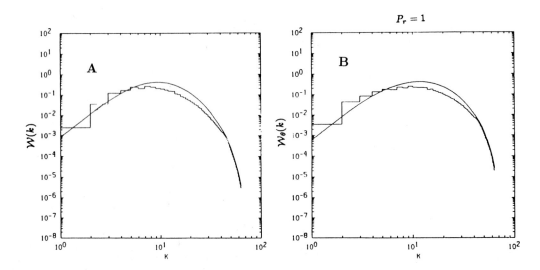

Fig. 7 Comparison of DNS $\mathcal{W}(k)$ with its Gaussian part.

negative non-Gaussian component of $\mathcal{W}_\theta$, with a corresponding positive part at $k \sim k_e$. At low P_r (8(A,B)), the $E_\theta(k)$ in much steeper, and conforms closely to the Batchelor, Howels, and Townsend [4] prescription. This is confirmed by Fig. 8B, which shows a quite small non-Gaussian part over the range $k > k_e$. However, at small $k < k_e$, the non-Gaussian part is still positive and significant.

These results suggest that the departures from Gaussianity at wind-tunnel R_λ for three-dimensional, isotropic turbulence are not overwhelming. This is despite the fact that Fig. 4 shows a strong development of vortex structures, and the fact that turbulence here would qualify as "hard" turbulence, in the sense of Libcheber (see Fig. 3C). (The "worms" of Fig. 4 are at a higher R_λ, but earlier DNS of Siggia [35] showed essentially the same structures).

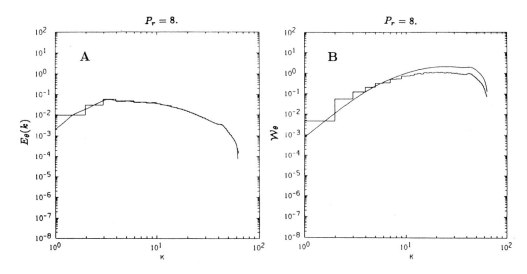

Fig. 8 $E_\theta(k)$ and $\mathcal{W}_\theta(k)$ according to DNS at $P_r = 8$. Other conditions of the run are the same as for Fig. 5. Solid curve in (a) is the smooth interpolation of the histogram used to compute the Gaussian part of $\mathcal{W}_\theta(k)$ (*via* (4.4a), with $B, U \to B_\theta, U_\theta$).

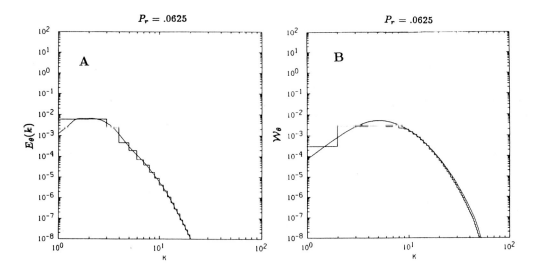

Fig.9 Same as Fig. 8, except $P_r = .0625$.

5. Concluding Comments

In this paper, we have touched upon some issues in which numerical simulations, (including some aspects of two-point closures) have revealed the physics of turbulence. We have said little about the enormous range of practical results obtained in engineering and meteorological applications. In examining DNS results here, perhaps the most interesting results were those in two- or semi-two-dimensions. Here the late-time condensation of the vortices into compact structures prohibits the naïve application of the statistical moment theory. Although certain simple applications (Sec. 2) of the statistical theory suggest what is wrong, we have as yet no clear way of incorporating structures into the theory, although some form of two-component theory is indicated. In this connection, we should recall the results of Benzi $et\ al$ [36]: if consolidated vortices are removed from the flow, the remainder seems to accord with the statistical theory (at least the $E(k) \sim k^{-3}$ is recovered).

We have also examined certain fourth moments of the flow and passive scalar fields, and compared the DNS predictions to closure (DIA). Generally, the self-organizing aspect of the nonlinearity implies a reduction of the nonlinear terms of about a factor of two. This comment applies to scales smaller than the energy peak, k_e. For larger scales, the correlated part of the force exceeds the Gaussian component by about 50 percent for the velocity field, and by a larger factor for the scalar. For the scalar, we see the same magnitude of reduction of $\mathcal{W}_\theta, k > k_e$ at high P_r, but for low P_r, the magnitude of $\mathcal{W}_{\theta,c}$ is smaller. A curious aspect of these DNS calculations is that $\mathcal{W}_c$ tends to become smaller at very high k, beyond the dissipation wavenumber, k_s. This is possibly a truncation error; we cannot believe anything beyond $k = 48$ for 128^3 pseudospectral. But the trend is there for smaller k.

References

[1] McWilliams, J. C., The emergence of isolated vortices in turbulent flows, J. Fluid Mch., 146, 21 (1984).

[2] Basedevant, C., Legras, B., Sadourny, R., and Beland, B., A study of barotropic model flows: intermittency, waves, and predictability, J. Atmos. Sci., 38, 2305 (1981).

[3] Kerr, R. M., Cascade spectra in numerical turbulence, J. Fluid Mech., in press.

[4] Batchelor, G. K., Howells, I. D., and Townsend, A., Small-scale variations of convected quantities like temperature in turbulent fluid. Part 2. The case of large conductivity, J. Fluid Mech., 5, 134 (1959).

[5] Yamomoto, K., and Hosokawa, I., A decaying isotropic turbulence pursued by the spectral method, J. Phys. Soc., Japan, 57, 1532 (1988).

[6] Kraichnan, R. H., The structure of isotropic turbulence at very high Reynolds numbers, J. Fluid Mech., 5, 497 (1959).

[7] Hussain, F., Coherent structures and turbulence, J. Fluid Mech., 173, 303 (1986).

[8] Herring, J. R., Schertzer, D., Lesieur, M., Newman, G. R., Chollet, J.-P., and Larcheveque, M., A comparative assessment of spectral closures as applied to passive scalar diffusion, J. Fluid Mech., 124, 411 (1982).

[9] Kaneda, "Attempts at statistical theories of turbulence" in $Recent\ Studies\ on\ Turbulent\ Phenomena$, T. Tatsumi, H. Maruo, and H. Takami, Eds. Association for Science Documents Information, Tokyo (1985).

[10] Lesieur, M., $Turbulent\ Fluids$, Nijhoff Publishers, Dordrecht, Boston, Lancaster, 286 pp (1987).

[11] Frisch, U., Lesieur, M., and Brisaud, A., A Markovian random coupling model for turbulence, J. Fluid Mech., 65, 145 (1974).

[12] Domarardzki, J, and Orszag, S. A., Preprint (1984).

[13] Kraichnan, R. H., Decay of isotropic turbulence in the direct interaction approximation, J. Fluid Mech., 7, 1030 (1964).

[14] Leith, C.E., Atmospheric Predictability and Two-Dimensional Turbulence. J. Atmos. Sci. 28, 145 (1971).

[15] Herring, J.R. and Kraichnan, R.H., Comparison of some approximations for isotropic turbulence. *Statistical Models and Turbulence.* p.148. Berlin: Springer-Verlag (1972).

[16] Herring, J. R., Some contributions of two-point closure to turbulence. *Frontiers in Fluid Mechanics*, S. H. Davis and J. L. Lumley, Eds., Springer-Verlag, 68 (1984).

[17] Tatsumi, T., Kida, S., and Mizushima, J., The multiple-scale cumulant expansion for isotropic turbulence, J. Fluid Mech., 85, 97 (1978).

[18] Bertoglio, J. P., A model for three-dimensional transfer in non-isotropic homogeneous turbulence. Third International Symposium on Turbulent Shear Flow. Springer-Verlag (1982).

[19] Yoshizawa, A., Statistical theory for Boussinesq turbulence. J. Phys. Soc. Japan, 46, 647 (1980).

[20] Fornberg, B., A numerical study of 2-D turbulence, J. Comp. Phys., 25, 1 (1977).

[21] Herring, J. R., Theory of Two-Dimensional Anisotropic Turbulence. J. Atmos. Sci., 32, 2254 (1976).

[22] Herring, J. R., An introduction and overview of various theoretical approaches to turbulence. Applied Mathematical Sciences no. 59: Theoretical Approaches to Turbulence, D. L. Dwoyer, M.Y.Hussaini, R.G. Voigt, Springer-Verlag, New York, Berlin, Heidelberg, Tokyo, 73 (1984).

[23] Rotta, J., Statishe Theorie Nichtshomogner Turbulenz, Z. Phys., 129, 547 (1951).

[24] Weiss, J., The dynamics of enstrophy transfer in two-dimensional hydrodynamics, La Jolla Inst. La Jolla, CA., LJI-TN-81-121, (1981).

[25] Brachet, M. E., Meneguzzi, M., Politano, H., and P.L. Sulem, The Dynamics of Freely Decaying Two-dimensional Turbulence. Preprint (1988).

[26] Pierrehumbert, R. T., Universal short-wave instability of two-dimensional eddies in an inviscid flow, Phys. Rev. Lett., 57, 3506 (1986).

[27] Wyngaard, J. C., and Pao, Y. H., Some measurements of the fine structure of large Reynolds number turbulence. *Statistical Models and Turbulence*, Springer-Verlag, Berlin Heidelberg, New York. M. Rosenblatt and C. Van Atta, Eds., 384 (1972).

[28] Castaing, B., Gunaratne, G., Heslot, F., Kadanoff, L., Libchaber, A., Thomae, S., Wu, X.-Z., Zaleski, S., and Zanetti, G., Scaling of hard thermal turbulence in Rayleigh Bernard convection, submitted to J. Fluid Mech. (1988).

[29] Metais, O., and Herring, J. R., Numerical simulations of freely evolving turbulence in stably stratified fluids. Preprint, to appear in J. Fluid Mech. (1988).

[30] Kraichnan, R. H., and Panda, B., Depression of nonlinearity in decaying isotropic turbulence. Preprint (1988).

[31] Chen, H. D., Herring, J. R., Kerr, R. M., and Kraichnan, R. H., Fourth Order Moments in Isotropic Turbulence. Preprint (1988).

[32] Batchelor, G. K., *The Theory of Homogeneous Turbulence*, (2nd edition), Cambridge University Press (1971).

[33] Kerr, R. M., Higher order derivatives correlations and the alignment of small-scale structures in isotropic numerical turbulence, J. Fluid Mech., 153, 31 (1985).

[34] Herring, J. R., A Note on Owen's Mesoscale Eddy Simulation, J. Phys. Ocean., 10, 804 (1980).

[35] Siggia, E. D., Numerical study of small scale intermittency in three dimensional turbulence, J. Fluid Mech., 107, 375 (1981).

[36] Benzi, R., Pattatnello, S., Santangelo, P., Europhys. Lett, 3, 811 (1987).

[37] McWilliams, J. C., Statistical properties of decaying geostrophic turbulence, J. Fluid Mech., in press (1988).

Theoretical and Applied Mechanics
P. Germain, M. Piau and D. Caillerie (Editors)
Elsevier Science Publishers B.V. (North-Holland)
© IUTAM, 1989

TURBULENCE AND VORTICES IN ROTATING FLUIDS

Emil J. HOPFINGER

Institut de Mécanique
Université de Grenoble, INPG and CNRS
Grenoble, France

The most striking effect of rotation on fluid motion is the tendency for the motion to become two-dimensional. Unique to rotating fluids is also the manifestation of boundary effects through Ekman pumping and inherent in these two processes is the property of rotating fluids to sustain wave motions. Considerable progress has been made recently on turbulence in rotating fluids. The mechanism of energy and enstrophy cascades is better understood and new phenomena like the transition to a rotationally dominated state and the possibility of vorticity concentration have been demonstrated. This has opened up new question concerning vortex dynamics (geostrophic and ageostrophic vortices), vortex waves, breakdown, and merging conditions in homogeneous and rotating, stratified fluid. After giving some examples of rotation effects in geophysical and industrial flows and introducing the basic concepts of rotating fluids, the paper emphasizes rotational effects on turbulence and the related problems of vortex formation and dynamics.

1. INTRODUCTION

The stimulus for research on rotating fluids stems mainly from geophysical applications. It is in these situations where the effects of rotation on fluid motions are of primary importance and are most convincingly demonstrated by pictures taken from satellites of the ocean surface or of atmospheric cloud patterns and by space probes of planets, all of which show the existence of quasi two-dimensional eddy motions. Jupiter's Great Red Spot is the most spectacular example of rotation effects. The mechanisms leading to a phenomenon like the GRS have captivated a number of scientists and various explanations have been given for instance by Maxworthy, Redekopp and Weidman [1], Marcus [2] and Sommeria, Meyers and Swinney [3].

Rotation obviously plays a key role in rotating machines where it affects the boundary layer structure on the runner blades of turbines and can give rise to intense vortex generation. The best known example of intense vortices which can cause severe damage to a turbine due to cavitation and large pressure fluctuations is the so called draft tube vortex (see Escudier [4]). The boundary layers on the runner blades may be stabilized or destabilised by rotation, affecting the shear stress. Pioneering work on this important problem is due to Johnston, Halleen and Lezius [5]. However, follow-up research has been scarce. Also, turbulence closure models for rotating turbulent flows have only been considered recently [6]. Very widespread engineering applications of rotating flows are encountered in the context of cyclone separators. To improve their efficiency, in particular the efficiency of hydrocyclones where density differences are small, it is necessary to prevent unstable flow regions and to produce optimum radial pressure gradients. Theoretical studies of flow separation in such systems have been performed recently by Greenspan and Ungarish [7]. There is also renewed interest in flows near rotating discs in confined regions. For one thing this is an interesting test case for numerical simulations of rotating fluids (Lugt and Abboud [8]) and when the depth to diameter ratio is very small it is a model of computer disc drives [9].

The three main effects of rotation, mentioned for example in Greenspan [10], are: the tendency of fluid motions to become two-dimensional in planes perpendicular to the rotation axis, the rapid spin-up of fluid columns due to Ekman pumping and the possibility of sustained wave motions. Taylor [11], first demonstrated that geostrophic equilibrium between the Coriolis force and the pressure gradient, the ingredient of two-dimensional motions, can indeed be closely approximated by experiments. Geostrophic turbulence is a higher order approximation above geostrophic equilibrium and it has been shown that it is well represented by the phenomenology of two-dimensional turbulence (Rhines [12], Colin de Verdiere [13], Griffiths and Hopfinger [14]). There are now also clear indications that three-dimensional turbulence when subjected to rotation tends toward a quasi-two-dimensional state when the Rossby number based on the turbulent velocity and length scales is sufficiently small (Hopfinger, Browand and Gagne [15], Hopfinger, Griffiths and Mory [16], Jacquin, Leuchter and Geffroy [17]). Inertial waves are believed to be intimately related with the two-dimensionalization process by transporting energy rapidly in the direction parallel to the rotation axis. A more intriguing observation is that this two-dimensionalization process may be accompanied by the generation of intense vortices [15, 18], an observation of importance for tornado genesis [19] and also for rotating engineering flows.

When the fluid layer is stratified in addition to rotation, as is usually the case in geophysical situations, phenomena unique to this situation occur. The most important of these is baroclinic instability (see Pedlosky [20]), which is the primary cause for mesoscale turbulence production. Baroclinic vortices are another interesting manifestation (Hogg and Stommel [21], Griffiths and Hopfinger [22, 23]), of the combined effects of stratification and rotation. The interactions of non-stratified or stratified rotating flows with topographies throw up a whole class of new problems which are at the forefront of research.

In Section 2 the theoretical concepts are presented which the reader should comprehend for a full understanding of the following sections. Geostrophic turbulence properties are discussed in Section 3 and the present knowledge concerning two-dimensionalization of three-dimensional turbulence is presented in section 4. Shear flows are treated in Section 5 and in Section 6 vorticity concentration into intense vortices and in particular new aspects of geostrophic vortex interactions are discussed.

2.   THEORETICAL CONCEPTS

2.1. The characteristic parameters

The physical significance of the different parameters related with rotation is best illustrated by writing down the governing equations. For an incompressible fluid in a rotating frame of reference we have

(1) $$\frac{\partial \underline{u}}{\partial t} + \underline{u} \cdot \nabla \underline{u} + 2 \Omega \underline{k} \times \underline{u} = -\frac{\nabla p}{\rho_o} + \underline{k} \, g \, \rho'/\rho_o + \nu \nabla^2 \underline{u}$$

(2) $$\nabla \cdot \underline{u} = 0 \quad .$$

The Coriolis force $-2 \Omega \underline{k} \times \underline{u}$ which appears explicitly is a deflecting force and the centrifugal force $\Omega \underline{k} \times (\Omega \underline{k} \times \underline{r})$ is, as is usual, included in the pressure. Variations in density $\rho = \rho_o (1 + \rho'/\rho_o)$ are considered to be small ($\rho'/\rho_o \ll 1$). When the variables are non-dimensionalized by horizontal and vertical length scales, L and H respectively, by corresponding velocity scales U and W, by time scale T, a density scale $\Delta\rho$ H/L, and the Coriolis parameter $2 \Omega \equiv f$ by f_0, we get

(3) $\gamma \dfrac{\partial \underline{u}}{\partial t} + \text{Ro } \underline{u} . \nabla\underline{u} + \underline{k} f \times \underline{u} = - \Delta p + (\text{Ro Ri}) \underline{k} \rho' + E \nabla^2 \underline{u}$

(4) $\nabla \cdot \underline{u} = 0$.

The parameters appearing in equation (3) are: $\gamma = 1/Tf_0$ the reduced frequency, $\text{Ro} = U/f_0 L$ the Rossby number, $E = \nu/f_0 H^2$ the Ekman number and $\text{Ri} = g \Delta \rho H/\rho_0 U^2$ a Richardson number. The combined parameter Ro Ri = 1 is a dividing boundary for rotation dominated (Ro Ri < 1) or stratification dominated (Ro Ri > 1) flows (see Maxworthy and Browand, [23]). When U = (g' H)$^{1/2}$, where g' = g$\Delta\rho/\rho_0$, the condition Ro Ri = 1 is equivalent to the internal deformation radius.

(5) $\Lambda = (g' H)^{1/2} / f_0$

Scales of motions $L > \Lambda$ are then rotation dominated and stratification is dominant when $L < \Lambda$.

2.2. Geostrophic equilibrium

When the dimensionless parameters γ, Ro, E and (Ro Ri) are much less than unity such that terms in these parameters can be neglected, the Coriolis force is closely balanced by the horizontal pressure gradient. The equation of geostrophic equilibrium is thus simply

(6) $f \underline{k} \times \underline{u} = - \nabla p/\rho_0$

This equation leads to the condition

(7) $f (\underline{k} . \nabla) \underline{u} = 0$,

known as Taylor-Proudman theorem, which states that fluid motions are invariant along a direction parallel to the rotation axis. Taylor's [11] experiments, in which a short cylindrical obstacle was displaced steadily in a radial direction at the bottom of a rotating tank, indicated the existence of a stagnant fluid column (Taylor column) above the obstacle moving with it. Hide and Ibbetson [25] investigated in some detail how the obstacle height, inertia (the Rossby number) and viscous effects (Ekman number) affect the structure of Taylor columns.

2.3. Potential vorticity

The vorticity equation is easily obtained by taking the curl of (3) giving:

(8) $\gamma \dfrac{\partial \xi}{\partial t} + (\underline{u} . \nabla) (\text{Ro } \xi + \underline{k} f) = [\text{Ro } \xi + \underline{k} f) . \nabla)] \underline{u} - \text{Ro Ri} (\underline{k} \times \nabla\rho') + E \nabla^2 \xi$.

Projection of this equation on the rotation axis z leads to:

(9) $\gamma \dfrac{\partial \omega}{\partial t} + (\underline{u} . \nabla) (\text{Ro } \omega + f) = \dfrac{WL}{HU} (f + \text{Ro } \omega) \dfrac{\partial w}{\partial z} + \dfrac{WL}{HU} \text{Ro} (\xi_H . \nabla_H) w + E \nabla^2 \omega$.

Equation (9) indicates that the velocity parallel to the rotation axis z is W ~ U H(Ro, γ)/L. This means that in the laboratory geostrophic flow conditions can be achieved even in deep fluid layers.

The dimensional vorticity in the inertial frame is $\omega_a = \omega + f$. Writing the vorticity equation (9) in dimensional form for the absolute vorticity ω_a and integrating it over the layer depth H (neglecting terms in ξ_H and in E), we obtain the barotropic potential vorticity equation

$$(10) \qquad \frac{D}{Dt}\left(\frac{\omega + f}{H}\right) = 0 \quad .$$

Introducing $u = - \partial\psi/\partial y$ and $v = \partial\psi/\partial x$ in equation (10) gives

$$(11) \qquad \frac{\partial \nabla^2 \psi}{\partial t} + J\left(\psi, \nabla^2\psi\right) + \beta\frac{\partial\psi}{\partial x} - \frac{f}{H} J\left(\psi, H\right) = 0 \quad ,$$

where $\beta = \partial f/\partial y$ and J is the Jacobian operator. A balance between the first and second terms governs 2D turbulence, between the first and third terms Rossby waves and between the first and fourth terms topographic waves. Integration over the layer depth has eliminated inertial waves.

2.4. Inertial waves

Inertial waves are of importance in deep fluid layers and are believed to play an essential part in two-dimensionalization of 3D turbulence. Their role is analogous to internal waves in stratified fluids. The wave equation is obtained by taking twice the curl of the linearized, barotropic, inviscid momentum equation (1) (Lighthill, [26]).

$$(12) \qquad -\frac{\partial^2 \nabla^2 \underline{u}}{\partial t^2} = f^2 \frac{\partial^2 \underline{u}}{\partial z^2} \quad .$$

According to the dispersion relation

$$(13) \qquad \sigma^2 = f^2 k_z^2 / (k_x^2 + k_y^2 + k_z^2) \quad ,$$

where k_x, k_y, k_z are respectively the wave number components in the directions perpendicular and parallel to the rotation axis, waves exist when the frequency $\sigma < f$. Propagation of energy at the group velocity $Cg = f k^{-1} \cos \theta$ is in the direction of the phase planes inclined at an angle θ with respect to the rotation axis. The phase velocity is $C_p = f k^{-1} \sin \theta$. The angle θ is given by $\theta = \sin^{-1}(\sigma/f)$. In the limit of very low forcing frequency, when $\sigma/f \to 0$, the group velocity is $Cg \cong f \lambda/2\pi$, where λ is the wave length of the perturbations. The time it takes for perturbations to travel throughout the depth of the fluid column is $\tau \sim H 2\pi/f \lambda$. Taking the smallest wave length perturbation equal to the Ekman layer thickness $\delta \sim (\nu/\Omega)^{1/2}$ we find

$$(14) \qquad \tau \sim (H^2 / f\nu)^{1/2}$$

which is recognized as the Ekman spin-up time [10].

3. GEOSTROPHIC TURBULENCE

Geostrophic turbulence is the chaotic nonlinear state of fluid motion where geostrophic equilibrium is nearly satisfied and the pressure is nearly hydrostatic [12]. In a barotropic fluid layer on a so called f-plane (f = constant), geostrophic turbulence reduces to two-dimensional (2D) turbulence governed by the first two terms of equation (11) to which internal viscous dissipation may be added, in the form $v\nabla^4\psi$, and Ekman dissipation - R $\nabla^2\psi$ with R = $(v/f\ H^2)^{1/2}$. Since vortex stretching is not possible in 2D turbulence the direct energy cascade is inhibited and energy cascades in fact to larger scales as was suggested by Fjørtoft [26]. The enstrophy (vorticity square) is transported to smaller and smaller scales as vorticity contours are drawn out. Simple dimensional arguments, which assume that the energy density depends only on the local wave number k and the energy flux ε, in the energy inertial range, and on k and the enstrophy flux η, in the enstrophy inertial range, give the spectral laws.

$$E(k) = C_1\ \varepsilon^{2/3}\ k^{-5/3} \qquad \text{for the inverse energy cascade range}$$

and

$$E(k) = C_2\ \eta^{2/3}\ k^{-3} \qquad \text{for the enstrophy cascade.}$$

Kraichnan [28] substantiated these spectral laws by formal theoretical arguments. Balloon dispersion experiments in the atmosphere analyzed by Morel and Larchevêque [29] seem to give observational support of a k^{-3} enstrophy cascade but in the light of numerical simulations, laboratory studies and other observations there is some doubt about the general validity of these conclusions.

3.1. Numerical simulations of 2D turbulence

Direct numerical simulations of 2D turbulence generally result in spectral slopes of - 4 or more [30]. Recent, high resolution simulations of 2D turbulence by Brachet, Meneguzzi and Sulem [31] indicate an evolution in time from k^{-4} to k^{-3} as vorticity sheets in physical space fold up. It is likely that the spectral slope would have increased again if calculations would have been continued for much longer times and isolated vortices are left behind [32].

3.2. Laboratory studies of geostrophic turbulence

In the laboratory geostrophic turbulence can be generated in a rotating fluid layer either by low frequency boundary forcing of a barotropic fluid or by baroclinic instability in a two-layer stratified fluid. Low frequency forcing was used by Colin de Verdiere [13]. Streakline photographs of particles suspended in the fluid showed the existence of two-dimensional eddy motions with the largest eddies having a scale larger than the forcing scale. The largest eddy scale being determined by Ekman dissipation $L_E \sim u/R$. The transfer of energy to scales larger than the forcing scale is a property of 2D turbulence. The freely evolving state was also considered by Colin de Verdiere. Batchelor's [33] similarity solution, modified for Ekman dissipation [13], gives a rate of increase of the length scale characteristic of the energetically dominant eddies, associated with the wave number k_1, in the form

$$(15) \qquad \frac{d\,k_1^{-1}}{dt} = T\,u\,(t_o)\,e^{-R(t-t_o)}$$

In the experiments the value of the coefficient T ranged between $(1 - 4) \times 10^{-2}$ comparable with the value of 3×10^{-2} obtained by numerical simulations [30].

Griffiths and Hopfinger [14] conducted experiments in a two-layer stratified fluid with each layer of depth H.Turbulence was produced by setting up a baroclinically unstable circular density front. The baroclinically unstable waves of wave length 7 $(g' H)^{1/2}$ / f first grew and then broke down into random eddy motions. Figure 1 shows a view of the fully developed turbulent eddy field visualized by illuminated suspended particles. The particle streaks show the instantaneous velocity field.

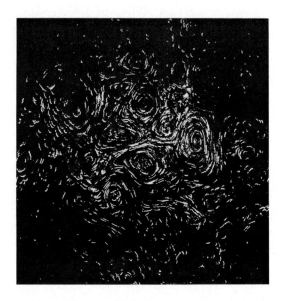

FIGURE 1
Velocity field of baroclinic turbulence. The particle streaks represents instantaneous velocities. The photograph was taken about 25 rotations periods after onset of baroclinic instability (taken from Griffiths and Hopfinger [14]).

In figure 1, the rotation is counterclockwise and the direction of the velocity is indicated by the dot at the end of the streaks; at long times most of the remaining eddies are anticyclonic. Kinetic energy decayes exponentially with a time scale depending on the critical Froude number F_c for onset of baroclinic instability [34]. An interesting observation was the blockage in eddy scale at a value $l_e \approx 2\, F_c^{1/2}\, (g' H)^{1/2}$ / f. In situations where f is not constant, this scale limitation due to baroclinic instability must be compared with the eddy Rhines radius $l_ß \sim 2\pi\, (u / ß)^{1/2}$ at which the turbulent eddies change into Rossby waves and zonal flow [30].

By extending the concept of particle pair dispersion used by Morel and Larchevêque [29] to the group of unmarked particles as shown in figure 1, Griffiths and Hopfinger [14] were able to determine a spectral law from the dispersion rate of these neutrally buoyant particles. A power law of the form

(16) $E(k) \sim k^{-2.5}$

was shown to exist over about one decade in k. Although this spectral slope differs from a k^{-3} enstrophy inertial range, the spectral fall-off is sufficiently steep to push the peak of energy dissipation to large scales as in 2D turbulence. The reasons leading to - 2.5 rather than - 3 or more are not known. It might be possible that it is due to a viscous effect (the Reynolds number of the eddies is about 500) or that energy is injected at small scales by some shear instability. A more likely explanation is that the relation between the spectral power α for a single particle, expressed by $E(k) \sim k^{-\alpha}$, and the power spectrum for relative velocity of particles separated by distance D in

$$(17) \qquad < \left(\frac{dD}{dt}\right)^2 > \sim D^n$$

is not simply $n = \alpha - 1$ as was proposed previously [14, 29, 35]. No simple relation seems to exist when $\alpha > 3$ [35] and there are now some doubts even when $\alpha \leq 3$ [37]. Consistent results are however obtained when the dispersion method is applied to 3D turbulence where $\alpha = 5/3$ [35].

4.   TWO-DIMENSIONALIZATION OF 3D TURBULENCE

According to the Taylor-Proudman theorem, any fluid motion will tend toward a two-dimensional state when the Rossby number Ro, the reduced frequency γ and the Ekman number E tend toward zero. Since in a turbulent field the turbulent Rossby number and the reduced frequency due to the turbulent forcing are identical, we need to consider only one of these two parameters. For turbulence with high rotation effects (low Rossby number) to survive, the Ekman spin-down time $f^{-1} E^{-1/2}$ must be considerably larger than the turn-over time of the energy containing eddies l/u, giving the requirement $E^{1/2} \ll Ro$.

4.1. Homogeneous turbulence

The experiments with isotropic turbulence of Ibbetson and Tritton [38] did not satisfy the conditions $E^{1/2} \ll Ro$ and therefore, with rotation, dissipation was enhanced. In the experiments of Wigeland and Nagib [39] in which grid turbulence was produced in a rotating wind tunnel, Ekman friction was unimportant but the Rossby number could not be reduced much below 1. Rotation effects on the turbulent normal stresses remained nearly negligible but had a noticeable effect on the length scale parallel to the rotation axis. In similar experiments Jacquin et al [17] were able to reduce the turbulent Rossby number $Ro_u = u/f\, L_u$ to about 0.1, where u is the r.m.s. turbulent velocity parallel to the rotation axis and L_u the corresponding longitudinal integral scale. The results showed the existence of two breakpoints: one at $Ro_u \simeq 1$, where a change in the decay rate of the turbulent normal stress perpendicular to the rotation axis occurred and also an increase of the lateral integral scale related with the normal velocity components, the other at $Ro_u \simeq 0.2$, where a change in decay of the normal stress component parallel to the rotation axis and of the corresponding longitudinal integral scale occurred.

The existence of two breakpoints is similar to what is observed in stratified turbulence. Stratification begins to affect the turbulence structure when the turbulent Froude number is about 1 or when the turbulent integral scale is about equal to the Ozmidov scale, and 3D turbulence collapse is nearly fully accomplished when the integral scale is a multiple of the Ozmidov scale or the turbulent Froude number is about 0.2 [40]. Mory and Hopfinger [41] extended these scale arguments, used in stratified turbulence, to rotating turbulence. The idea is that when the turbulent Rossby number is of order 0.2 turbulent energy radiation by inertial waves is more efficient than the transfer to small scales. A value of 0.2 implies also a turbulent forcing frequency u/L_u less than f and in this case energy radiation is nearly parallel to the rotation axis.

4.2   Oscillating grid turbulence

Oscillating grid turbulence is homogeneous in a plane parallel to the grid plane and without rotation the r.m.s. turbulent velocity decays with distance z from the grid like z^{-1} and the integral scale is proportional to z (Hopfinger and Toly [42]). Oscillating grid turbulence has the advantage that there is no mean flow which facilitates quantitative flow visualization. Hopfinger et al [15] subjected this oscillating grid turbulence to solid body rotation with the axis of rotation perpendicular to the grid plane. Particle streakline photographs taken in a plane perpendicular to the rotation axis showed a drastic change in flow structure with the appearance of coherent vortices when the fluid was rotating. A fairly large number of vortices can be identified in the streakline photograph shown in figure 2a. The vortex core of a typical vortex, nearly aligned with the rotation axis, is shown in figure 2b.

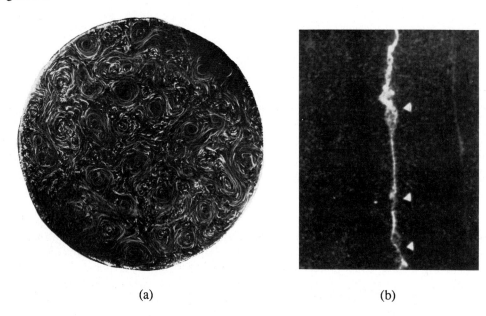

(a)   (b)

FIGURE 2
Photographs of the turbulent flow field in a rotating fluid stirred by an oscillating grid with mesh 5 cm, stroke 4 cm, oscillation frequency 6.6 Hz, Coriolis frequency $f = 4\pi$ rad s^{-1}. (a) streakline photograph in a plane perpendicular to the rotation axis, (b) view in a plane parallel to rotation axis with vortices made visual by air bubbles. The symbol ◀ indicates waves or breakdown regions.

More detailed visual observations and measurements of the mean azimuthal velocity as a function of distance from the grid permitted to locate the distance from the grid z_T where the change in flow structure occurred. The turbulence Rossby number associated with this change was 0.2 [15]. Measurements of the horizontal component (rotation axis was vertical) of the r.m.s. turbulent velocity showed a drastic reduction in spatial decay when $z > z_T$ (figure 3).

In figure 3 three zones can be distinguished: $z < z_T$ where the turbulence is practically unaffected by rotation, $z_T < z < z_R$, where turbulence is rotationally dominated but u is still weakly decaying and $z > z_R$, where u is invariant with z. The weak decay in energy with z in region 2 was explained by a reminder of small scale 3D turbulence which continues to lose energy by a direct cascade [16]. The energy balance in the whole region $z > z_T$ is a supply of energy by an inertial wave mechanism from the 3D turbulent zone and dissipation by Ekman friction. It may be noted that the vertical velocity component is not small and is of the same order as u and continues to decrease with z throughout the fluid layer. This observation is consistent with the geostrophic approximation to first order. Equation (9) indicates that to first order w ~ u Ro H/L and $\partial w/\partial z$ = constant, where L is the eddy size and H the fluid layer depth. In the experiments H/L ≈ 10 and Ro ≈ 0.1. If rotating, oscillating grid turbulence in a deep fluid layer satisfies the geostrophic approximation it does not satisfy the hydrostatic approximation. However, the ratio of vertical to horizontal pressure gradients is of order L/H, hence small.

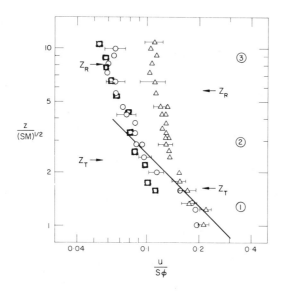

FIGURE 3

Variation with distance from the grid of the azimuthal, u, and vertical, v, components of the r.m.s. turbulent velocities. ϕ is the grid oscillation frequency, S is the stroke and M the mesh size of the grid, f = 2π rad s^{-1}; Δ, ϕ= 3.3 Hz ; o, 6.6 Hz ;   , vertical component for ϕ = 6.6 Hz. At z_T the Rossby number u/f L $\cong$ 0.2.

An indication to which degree the dynamics of this turbulence is two-dimensional is given by the spectra. These were determined, as in the experiments by Griffiths and Hopfinger [14], from the particle dispersion laws $<(dD/dt)^2>$. A spectral slope of E(k) ~ $k^{-2.5}$ was again found [35]. The inertial energy transfer is therefore predominantly toward the large scales and the dynamics is similar to the baroclinic turbulence discussed above.

When the forcing was stopped and the turbulence was allowed to evolve freely, an increase in eddy size and eddy spacing by successive vortex pairings was observed [41]. Doubling in eddy size required 30 initial turnover timescales in agreement with Colin de Verdiere's experiments [13] and numerical simulations [30].

5.   ROTATING SHEAR FLOWS

Rotation effects on shear flows depend on the orientation of the shear vorticity vector with respect to the rotation axis. Three basic configurations can be distinguished depending on whether the rotation vector is

i)   perpendicular to the shear vorticity vector and also to the shear plane,

ii)   perpendicular to the shear vorticity vector but lies in the shear plane, and

iii)   parallel to the shear vorticity vector and the shear plane.

Configuration i) is the Ekman layer situation which is of primary importance in rotating fluids and has therefore received considerable attention in the past. The two later configurations are less well understood and less well known and it is worthwhile to spend some time on them in this paper. These configurations are mainly encountered in industrial problems related with rotating machines, cyclones and other vortex flows.

5.1. Rotating channel flow

Two-dimensional fully developed turbulent channel flow with the rotation vector being spanwise (configuration iii) was studied by Johnston et al [5]. By writing down the transport equations for the turbulent Reynolds and normal stresses, Johnston et al concluded that rotation effects in the wall layers depend on the parameter

(18) $S = - f / (\partial \overline{u} / \partial y)$

where $\overline{u}$ is the mean velocity and by convention Ω or f is positive when rotating counterclockwise. When $S < 0$ rotation enhances the turbulence level and when $S > 0$ it decreases it. Bradshaw [43] introduced an equivalent Richardson number

(19) $Ri = \dfrac{-f\,(\partial \overline{u}/\partial y) - f)}{\left(\partial \overline{u}/\partial y\right)^2} = S(S+1)$

By analogy with stratified flows $Ri > 0$ indicates a stabilizing effect of rotation on the flow and $Ri < 0$ a distabilization. Near the wall, where $\partial \overline{u} / \partial y \gg f$, $Ri \approx S$. The magnitude of the parameter S is representative of the inverse of a Rossby number ($\partial \overline{u} / \partial y \sim U/\delta$)

(20) $\overline{Ro} = U / f\delta$

where δ is the boundary layer thickness. In the experiments of Johnston et al the Rossby number defined with $U = u_m$, the average velocity in the channel and $\delta = b$, the half width of the channel was always > 5. For the sake of comparison with turbulence without mean shear discussed in section 4 it is of interest to calculate the corresponding turbulent Rossby number. Taking the integral scale about equal to b and the r.m.s. turbulent velocity $u/u_m \approx 0.08$, the value is > 0.4. The results obtained by Johnston et al concerning the stabilizing influence of rotation on the side wall where $S > 0$, reproduced in figure 4, show that complete relaminarisation is possible. As is seen in figure 4, the transition Rossby number boundary depends strongly on the mean flow Reynolds number $Re = u_m\,2b/\nu$ for values of Re up to about ten times the critical Reynolds number without rotation. The extrapolation of the transition boundary (solid lines in figure 4) beyond $\overline{Ro}^{-1} \approx 0.2$ would suggest that the critical value of $\overline{Ro}^{-1}$ would grow nearly linearly with Re. It is more likely that the transition boundary asymptotes at a maximum value of $\overline{Ro}^{-1}$ which would correspond to a turbulent Rossby number of about 0.2. The decrease in turbulence level and relaminarisation is naturally accompanied by a drastic decrease in wall shear stress on the stable side. On the unstable side the wall shear stress is increased by rotation due to Taylor-Görtler vortices. Johnston et al measured an increase of about 15 % in shear velocity.

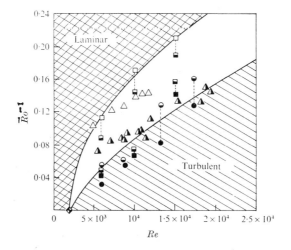

FIGURE 4

Flow regimes on the stable wall in a rotating channel [5]. Open symbols give lower limit of laminar flow, filled symbols indicate turbulent flow and half filled transitional flow (by courtesy of Prof. Johnston).

Stabilization and destabilization are easily understood in terms of the Rayleigh criterion which requires that the centrifugal force be respectively greater or smaller than the radial pressure gradient due to rotation. On the destabilized side the flow is Rayleigh unstable when the center velocity in the channel $\overline{U}_c$ is larger than order $b\Omega$. This agrees with equation (19) with $\partial \overline{u} / \partial y \sim \overline{U}_c / b$.

5.2   Free shear flows

The Rayleigh criterion also applies to free shear flows. Depending on the sign of

(21) $\overline{S} = - f \delta / (U_1 - U_2)$

a free shear layer might be destabilized ($\overline{S} < 0$) or stabilized ($\overline{S} > 0$). In (21) δ is the shear layer thickness and $(U_1 - U_2)$ is the velocity difference across the shear layer. According to the Rayleigh criterion destabilization requires in addition to $\overline{S} < 0$ that $(U_1 - U_2) > \Omega \delta$. Tritton [44] reported experiments with a free shear layer which show very strikingly the effect of rotation. Figure 5a illustrates the stabilizing effect in which case the 2D vortex structure is inforced but vortex pairing 'seems' to be inhibited; there is however no apparent reason why rotation should prevent merging of vortices which is a purely 2D process. When conditions are destabilizing the 2D structures are destroyed (figure 5b) and are probably replaced by disorganized motions with possibly streamwise vortices as long as rotation is not too strong.

An interesting shear flow is the wake where according to the Rayleigh criterion one side is stabilized (the cyclonic vortex side) and the other is destabilized. This tendency is demonstrated in figure 6 which was taken at the Coriolis table of the Institut de Mécanique by G. Chabert d'Hières. In this experiment the Rossby number U/f D ≈ 0.8, where D is here the diameter of the cylinder, which satisfies the conditions of destabilization $\Delta U > \Omega \delta$, where ΔU is the maximum velocity defect (of order U) and δ is the half width of the wake (of order D).

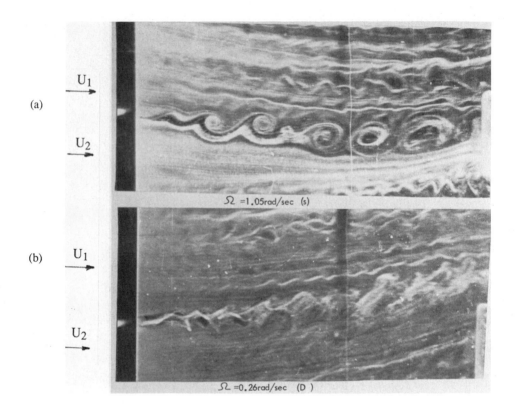

FIGURE 5

Photographs showing rotation effects on shear layer studied by Tritton [44]. In both cases $U_1 > U_2$ and in (a) rotation is clockwise, giving $\bar{S} > 0$ (stabilizing); (b) anticlockwise rotation ($f > 0$), giving $\bar{S} < 0$ (by courtesy of Prof. Tritton).

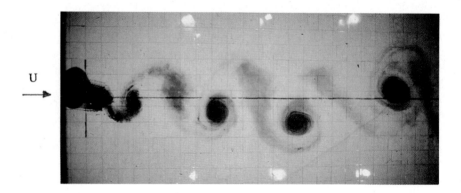

FIGURE 6

Cylindrical obstacle wake with rotation, showing that the anti-cyclonic vortex side is destroyed. Rotation is anticlockwise.

5.3. Rotating pipe flow

Intuitively, rotation should have a stabilizing effect on circular pipe flow (configuration ii) but observations show the contrary. The critical Reynolds number for instability due to finite amplitude perturbations decreases from 2500 for zero rotation to about 900 when the swirl number $M = \Omega r_0 / \overline{U}_m$ has a value of about 4 [45], where r_0 is the pipe radius and $\overline{U}_m$ the average velocity. The linear inviscid stability of rotating Poiseuille flow has been considered theoretically by Pedley [46], Maslowe [47] and others. The flow was found to be stable for $M = 0$ and unstable to asymmetric disturbances for finite values of M, with maximum amplification when $M \sim 1$. A physical explanation for this instability has been given by Pedley [45] in terms of the Rayleigh criterion related with the perturbations. The general picture is however likely to be more subtle [47].

6.   VORTEX FORMATION AND DYNAMICS

Vortex dynamics is central to turbulent flows in general and it stands to reason that in rotating flows it is of primary importance because of the presence of background vorticity. Forcing of fluid layers with background vorticity causes stretching or contraction of vortex lines. The forcing can have different origins: convective instability, boundary forcing by turbulent boundary layers, topographic forcing or barotropic and baroclinic instabilities. Weak forcing gives rise to quasi-geostrophic vortices and strong forcing can give rise to very intense vortices like hurricanes and tornadoes.

6.1. Intense vortices

An example of intense vortex formation by convective forcing of a rotating fluid layer is shown in Chandrasekhar [48] and by Boubnov and Golitsyn [49]. Figure 2 shows a striking example of vorticity concentration by turbulent boundary forcing. Hopfinger et al [15] were able to show that the cyclonic vortices generated in the rotating oscillating grid turbulence configurations have a core vorticity which is one to two orders of magnitude higher than the background vorticity. The number of vortices per unit area is proportional to the rotation rate for fixed forcing and the vortex spacing l_s varies like the turbulent Ekman layer thickness ($l_s \sim f^{-1/2}$). Boubnov and Golitsyn [49] also found a linear dependency of the vortex density on rotation rate in the highly non-linear regime and it seems that this linear relation is a general property of strongly forced rotating fluids.

An interesting phenomenon are the kink waves described by Hopfinger et al [15] which travel along the vortices (figure 2b) and help to extract energy from the region of three-dimensional turbulence. Some of these waves have been shown to be solitary waves (Hopfinger and Browand [50]) governed by the non-linear Schrödinger equation to which solutions were obtained by Hasimoto [31] and by Leibovich, Brown and Patel [32]. It was also thought that these non-axisymmetric waves were responsible for vortex breakdown and at least three different modes of breakdown have been identified [15]. These breakdown zones are regions of local production of small scale turbulence and hence regions of energy dissipation. The dynamic importance of energy transport by the vortex waves and dissipation in the breakdown regions is however difficult to evaluate and remains an open question. Also, the relation of kink waves with vortex breakdown remains speculation. An attempt by Maxworthy, Hopfinger and Redekopp [53] to establish a connection between the two remained unsuccessful.

A physical model of vorticity concentration in a rotating fluid layer forced by three-dimensional turbulence has been proposed by Maxworthy et al [53]. At the edge of the turbulent layer (an Ekman layer) the local Rossby number is about 0.2 which means that turbulent forcing is at a frequency well below f. Energy transport by inertial waves is then nearly parallel to the rotation axis. In the transient experiment fluid columns of conical shape, forming the envelope of inertial waves, are seen to move axially away from the turbulent region at a velocity close to the group velocity of inertial waves [54]. These cones cause a convergence of the flow between them as is indicated schematically in figure 7. It is plausible that this process is maintained in the steady state. Other explanations have also been proposed. Lundgreen [55] suggested the existence of persistent suction sites within the turbulent layer. Mory and Caperan [56] considered the linear marginal stability of a rotating turbulent field forced at one boundary. Although this stability formulation cannot be compared directly with the experiments by Hopfinger et al [15] a stability theory could explain for instance the selection of the vortex spacing and the focalization of energy at this mode. There have been speculations that this mechanism of intense vortex formation by random forcing is relevant to tornado genesis [19]. Boundary currents could also act as forcing zones for generation of vortices in the layer below [57] and in general a turbulent Ekman layer is a potential forcing region.

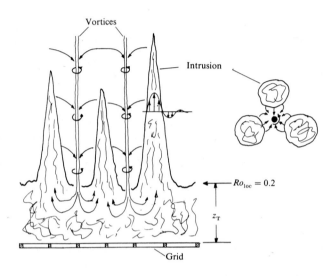

FIGURE 7

Schematic diagram of an intense vortex generation mechanism above a turbulent boundary layer.

FIGURE 8

Example of a draft tube vortex in a Francis turbine made visible by cavitation in the vortex core (by courtesy of Sté Neyrpic).

It is of interest to show here an example of the draft tube vortex in a turbine because it is a beautiful illustration of concentrated vortices in rotating fluids. Such a vortex behind a Francis turbine wheel, illustrated in figure 8, which can cause large pressure fluctuations and hence vibrations of the turbine and damage due to cavitation, is formed when the turbine is operated at partial load or at over load and there is residual swirl in the draft tube. If the formation mechanism of the draft tube vortex is understood, the reasons for the spiraling are less clear. Explanations range from flow separation on the turbine runner blades to vortex breakdown phenomena, the latter being more probable [4].

6.2. Geostrophic vortices

Geostrophic vortices (relative vorticity small compared with f) in homogeneous fluid are, except for Ekman layers and associated weak Ekman pumping, essentially two-dimensional vortices. Such eddies are characterized by a core of radius R of high vorticity with nearly potential flow outside. Long-range interactions (distance between vortices d >> R) are well accounted for by point vortex interactions (Batchelor [58]). For a vortex pair of opposite signs the point vortex interaction laws remain a good approximation even when $d \sim 2R$ and such a vortex pair translates with speed $V = \Gamma / 2\pi d$, where Γ is the circulation. The core structure is however of importance in the formation of coherent vortex couples, characteristic of modons, which have a translation motion well above that of an ordinary vortex pair. Their formation seems to require the existence of a screen of vorticity of opposite sign around each core [59] but this raises the question of vortex stability.

Two vortices of like sign undergo an orbital motion with angular velocity $\zeta = \Gamma / \pi d^2$ when d >> R. Strong interaction of like sign vortices causes deformations of the vortices due to mutual straining, followed by rapid merging when the separation is less than a certain critical value. A critical separation distance was determined by numerical integration of the two-dimensional Euler equations for vortices with constant or piecewise constant core vorticity with circular boundaries of radius R or a V-state configuration [60]. For vortices of the same strength (same Γ) these calculations give a critical distance $d/R = 3.2$. For separations smaller than critical, merging takes place in a time well below or of the order of the orbital time $2\pi \zeta^{-1}$. The first experiments with two-dimensional vortex pairs of like sign, generated in a shallow fluid layer, were performed by Caperan and Maxworthy [61]. The results show that the vortices attract one another and eventually amalgamate regardless of their initial separation. The time required can be a multiple of the orbital time. This absence of a critical distance was attributed to an alternating shedding and diffusion of vorticity outside a typical core region. Griffiths and Hopfinger [23] studied merging conditions in a rotating tank in homogeneous and stratified fluid. In a homogeneous fluid layer, the background rotation has no effect on the dynamics of vortex interaction but Ekman pumping makes cyclones and anticyclones evolve differently. Cyclones are comparable with the two-dimensional vortices of Caperan and Maxworthy because in both the vortices spin down due to Ekman pumping. For cyclonic pairs no stability boundary was found, a result which agrees with that of Caperan and Maxworthy. However, anticyclonic vortex pairs indicated the existence of a critical distance $d/R \simeq 3.3$, where R is defined as $R = \Gamma / 2\pi v_m$, with v_m being the maximum tangential velocity. This critical distance is indicated in figure 10 in the limit when the Rossby radius $\Lambda = 0$.

Eddies in oceans and atmospheres are generally strongly influenced by vertical density stratification. The interaction of baroclinic geostrophic vortices is therefore of interest both in studies of the dynamics of individual eddy interactions and of baroclinic geostrophic turbulence. Using a baroclinic point vortex model, Hogg and Stommel [21] showed that the potential energy increases when two point vortices of the same sign approach one another. This energy increase would suggest that coalescing of baroclinic vortices is less likely than it is for their barotropic counterpart. Vortex interactions of finite core vortices in a two-layer stratified rotating fluid have been experimentally investigated by Griffiths and Hopfinger [22, 23] for values of Λ / R ranging from 0 to 4. Dipole interaction studies [22] were motivated by the eddy heat transport configuration called "heton", introduced by Hogg and Stommel. During these dipole interactions vortices of the same sign in the same layer were occasionally seen to merge and this lead to a more systematic study of vortex merging conditions [23]. An example of the coalescence of two anticyclones in the top layer of a two layer stratified fluid is shown in figure 9 for conditions $d/R = 4.5$, and $\Lambda/R = 2.5$. The vortices merge in a fraction of the orbital time which is 44 T_Ω with T_Ω, the rotation period, being indicated on the counter seen in figure 9. The scenario of merging displayed in this figure is qualitatively similar to that of barotropic vortices. The shedding of vorticity into spiral arms is characteristic of vortex pairing and seems to be necessary for axisymmetrization [62] of the combined core which is at first elliptical in shape.

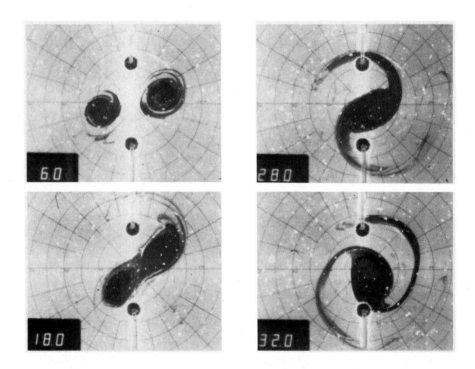

FIGURE 9

Photographs of merging of two anticyclones in the top layer of a two-layer stratified fluid [23]. $d = 18$ cm, $\Lambda = 10$ cm, $R = 4$ cm. The time elapsed in rotation periods, $T_\Omega = 4 \pi/f = 6.28$ s, is indicated on the counter. The vortex Reynolds number $\Gamma/\nu \approx 2.5 \cdot 10^3$ and the Ekman spin-down timescale is 300s.

The critical vortex separation d/R for the whole range of Λ/R studied by Griffiths and Hopfinger is reproduced in figure 10. The numbers next to the data points give the time until merging, non-dimensionalized by $2\pi \zeta^{-1}$. This time correspond to the state where the two vortices just begin to overlap (illustrated in the second frame in figure 9). When the distance between vortices is just less than critical the merging time is a fraction of the orbital time except when $0 < \Lambda /R < 0.5$, where baroclinic instability is possible and competes with merging. The behaviour for $\Lambda/R > 4$ is speculative; it is thought that the critical distance would further increase and then decrease again to approach the barotropic value when $\Lambda/R \to \infty$ where the two layers are decoupled. Recent contour dynamics calculations of two-layer geostrophic V-states by Polvani, Zabusky and Fliert [63] disagree wholly with experiments.

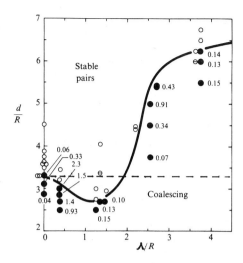

FIGURE 10

Critical vortex separation normalized by vortex core radius, d/R, as a function of Rossby radius, Λ/R, for anticyclones [23]. Filled symbols indicate coalescing pairs and open symbols stable pairs. - - - - -, contour dynamics calculations [63].

In the attempt to explain the change of the critical distance with Λ Griffiths and Hopfinger [23] determined the change in vortex structure with stratification. They solved the inviscid geostrophic potential vorticity equation for a baroclinic Rankine vortex and obtained an expression for the tangential velocity around a vortex in the two layers in the form:

$$
\left.
\begin{aligned}
\frac{v_a}{R\omega_o} &= \frac{1}{4}\frac{r}{R} + \frac{1}{2}K_1\left(\frac{R}{\Lambda}\right)I_1\left(\frac{r}{\Lambda}\right) \\[2ex]
\frac{v_b}{R\omega_o} &= \frac{1}{4}\frac{r}{R} - \frac{1}{2}K_1\left(\frac{R}{\Lambda}\right)I_1\left(\frac{r}{\Lambda}\right)
\end{aligned}
\right\} \quad r < R
$$

(22)

$$
\left.
\begin{aligned}
\frac{v_a}{R\omega_o} &= \frac{1}{4}\frac{R}{r} + \frac{1}{2}I_1\left(\frac{R}{\Lambda}\right)K_1\left(\frac{r}{\Lambda}\right) \\[2ex]
\frac{v_b}{R\omega_o} &= \frac{1}{4}\frac{R}{r} - \frac{1}{2}I_1\left(\frac{R}{\Lambda}\right)K_1\left(\frac{r}{\Lambda}\right)
\end{aligned}
\right\} \quad r > R
$$

where subscript 'a' refers to the layer where the vortex of core vorticity ω_o is initiated and 'b' to the opposite layer. K_1 and I_1 are the modified Bessel functions. In the limit $R/\Lambda \to 0$ the result for the point vortex is recovered from the expression for $r > R$ and is [21].

(23) $$v = \frac{1}{2} \frac{s}{r} \left[1 \pm K_1 \left(\frac{r}{\Lambda} \right) \right]$$

The constant $s = R^2 \omega_0 / 2$ is the vortex intensity and $2 \pi s$ is the vortex strength. The circulation Γ is a function of r even when $r > R$ except in the barotropic limit where $\Gamma = \pi s$. In this limit ($\Lambda/R \to 0$) the velocities are

(24) $v_a = v_b =$
$$\begin{cases} \frac{1}{2} \frac{s}{R^2} r & , r < R \\[2ex] \frac{1}{2} \frac{s}{r} & , r > R \end{cases}$$

For intermediate values of Λ/R, in particular $\Lambda/R \sim 1$, the top-layer velocity outside the core decreases more rapidly than r^{-1}. The induced far field velocity is thus a smaller fraction of the maximum velocity and this may explain why baroclinic vortex pairs are more stable when $\Lambda/R \sim 1$. When $\Lambda/R \gg 1$ the maximum velocity at $r = R$ is $v_a = \frac{1}{2} R \omega_0 = s/R$ which is twice that of $\Lambda/R \to 0$, equation (24). Outside the core $v_a \sim r^{-1}$ over a distance comparable with R, then decreases more rapidly than in a barotropic vortex and asymptotes at the barotropic velocity limit when $r \gg \Lambda$. Considering this behaviour the vortex structure gives no explanation of the increased critical distance when Λ/R is large. Ocean eddies, like the Gulf Stream warm core rings or others, generally lie in the range $\Lambda/R < 1$ and coalescence of such eddies is therefore a rare phenomenon. Hurricanes on the other hand have $\Lambda/R > 10$ and such tropical cyclones are known to attract each other when d is of order 10 R.

Vortex merging in the presence of a mean shear has been extensively studied by Marcus [2] in a number of fascinating numerical experiments. He considered an azimuthal single layer and also two-layer flow in an annulus with and without a beta-plane effect (varying Coriolis frequency with radius or equivalent latitude). The azimuthal flow with anticyclonic shear had a uniform potential vorticity and was neutrally stable. Viscous effects were neglected. On this flow Marcus superposed spots of excess or defect vorticity $\pm \omega_e$. Calculations show that a spot with vorticity opposite to the shear (cyclonic vortices) is drawn out into filaments and fragmented by Kelvin-Helmholtz instability. But, anticyclonic vortices keep their coherency in an anticyclonic shear even in chaotic ambiant motion due to shear instability of drawn-out vorticity bands. Two or more anticyclones located at different azimuthal positions eventually pair up into one stable vortex because small differences in radial positions make them collide. The required condition for the formation of a stable vorticity spot is that ω_e is of the same order as the mean shear and of the same sign, as mentioned above. A beta-effect is not essential but helps to determine the zonal shear. A finite value of the internal deformation radius Λ affects the vorticity distribution in the vortex but does not alter the basic processes.

Sommeria, Meyers and Swinney [3] conducted experiments in a fast rotating annulus with a beta-plane ($\beta = 2 \Omega s/H$, where s is the bottom slope). Fluid was pumped into the tank through a ring of holes at some inner radius and was withdrawn at some larger radius. This gave rise to a counter-rotating azimuthal jet with a cyclonic shear. This jet was shown to become unstable, forming cyclonic vortices, the number of which decreased with increasing flow rate. At high flow rate only one stable vortex remained in an ambient turbulent flow. These experiments are the cyclonic counterpart of the numerical simulations of Marcus for anticyclonic shear[*]. Both the numerical

---

[*]   In the experiments anticyclonic vortices in an anticyclonic shear were, according to the authors, unstable. It is of
     interest to recall that this was also the case in the shear flows discussed in section 5, where Rayleigh instability
     was given as an explanation. In the experiments by Sommeria et al. the Rossby number is too low for Rayleigh
     instability to occur and there is no reason why the equivalent of the numerical simulations could not be achieved.

model and the laboratory experiments are relevant to large vortex formation phenomena in nature of which Jupiter's Great Red Spot is the most striking example. A view of the GRS is shown in figure 11. It exists in chaotic (turbulent) zonal shear flow (figure 11) and smaller vortices advected along the latitude collide and amalgamate with it. The models suggest that turbulent mixing leads to a state of uniform potential vorticity and any non-uniformities are swept into the isolated vorticity patch. Because of this energy radiation by Rossby waves is inhibited.

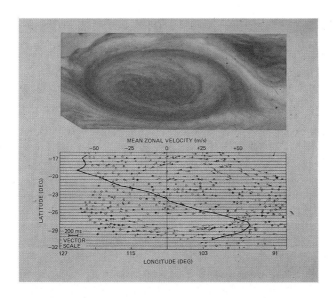

FIGURE 11

View of Jupiter's Great Spot with an indication of the mean zonal shear averaged along a latitude (by courtesy of Prof. Maxworthy).

7. CONCLUDING REMARKS

Each section in this paper would clearly have required a separate lecture. However, since the paper addresses both non-specialist and specialists, it was necessary to give a broader coverage of the topic specified in the title and this at the risk of seeming occasionally superficial. The aim has been to transmit an unified view of geostrophic turbulence and two-dimensionalization of turbulence and of how these are related with present day research on vortex interactions and dynamics. It was also important to show how rotating shear flows which have applications in rotating machines fit into these general concepts.

Progress in the fundamental understanding of rotation effects on fluid motions has come primarily from research motivated by geophysical problems and this, I think, is also clear from the present paper. Nevertheless, theoretical development is almost exclusively based on the geostrophic approximations. Ageostrophic phenomena like the two-dimensionalization process of turbulence discussed in section 4 and the related intense vortex formation (section 6) are, however, frequent. Tornadoes and hurricanes are atmospheric examples of ageostrophic flows and sub-mesoscale coherent vortices in the ocean [64] have ageostrophic features. Vortex flows in flow machinery also fall outside the framework of quasi-geostrophic theory. Research is to be encouraged along these lines of ageostrophic phenomena.

Important problems related with the general topic of this paper have not been treated. These include in particular fronts and baroclinic instability. There are new developments on instability which are of general interest [65]. Topographic effects and aspects of flow separation in rotating fluids are also in the mainstream of related research. More engineering oriented problems like flows driven by rotating discs in enclosures, initially studied in the context of gas turbines, attract renewed attention because of occasional vibrations of pick-up heads in computer disc drives [9]. Rotating flows with two phases, in particular in hydrocyclones, open up a whole range of new problems [7] and would have deserved some discussion.

REFERENCES

[1]   Maxworthy, T., Redekopp, L.G. and Weidman P., On the production and interaction of planetary solitary waves: Application to the Jovian Atmosphere, Icarus 33 (1978) 388.

[2]   Marcus, P.S., Numerical simulation of Jupiter's Great Red Spot, Nature 331 (1988) 693.

[3]   Sommeria, J., Meyers, S.D. and Swinney, H.L., Laboratory simulation of Jupiter's Great Red Spot, Nature 331 (1988) 1.

[4]   Escudier, M., Confined vortices in flow machinery, Ann. Rev. Fluid Mech. 19 (1987) 27.

[5]   Johnston, J.P., Halleen, R.M. and Lezius, D.L., Effects of spanwise rotation on the structure of two-dimensional fully developed turbulent channel flow, J. Fluid Mech. 56 (1972) 533.

[6]   Bardina J., Fertziger, J.H. and Rogallo, R.S., Effect of rotation on isotropic turbulence: computation and modelling, J. Fluid Mech. 154 (1985) 321.

[7]   Greenspan, H.P. and Ungarish, M., On the enhancement of centrifugal separation, J. Fluid Mech. 157 (1985) 359.

[8]   Lugt, H.J. and Abboud, M., Axisymmetric vortex breakdown with and without temperature effects in a container with a rotating lid, J. Fluid Mech. 179 (1987) 179.

[9]   Abrahamson, S.D., Koga, D.J. and Eaton, J.K., Flow visualization and spectral measurements in a simulated rigid disk drive, to be published.

[10]   Greenspan, H.P., the Theory of Rotating Fluids (Cambridge University Press, 1968).

[11]   Taylor, G.I., Experiments on the motion of solid bodies in rotating fluids, Proc. Roy. Soc. (London) A104 (1923), 213.

[12]   Rhines, P.B., Geostrophic Turbulence, Ann. Rev. Fluid Mech. 11 (1979) 401.

[13]   Colin de Verdiere, A., Quasi-geostrophic turbulence in a rotating homogeneous fluid, Geophys. Astrophys. Fluid Dynamics, 15 (1980) 213.

[14]   Griffiths, R.W. and Hopfinger, E.J., The structure of mesoscale turbulence and horizontal spreading at ocean fronts, Deep Sea Res. 31 (1984) 245.

[15]   Hopfinger, E.J., Browand, F.K. and Gagne Y., Turbulence and waves in a rotating tank, J. Fluid Mech. 125 (1982) 505.

[16]   Hopfinger, E.J., Griffiths, R.W. and Mory, M., The structure of turbulence in homogeneous and stratified rotating fluids, J. Mec. Théor. Appl., numéro spécial (1983) 21.

[17]   Jacquin, L., Leuchter, O. and Geffroy, P., Experimental study of homogeneous turbulence in the presence of rotation, to appear.

[18]   McEwan, A.D. Angular momentum diffusion and the initiation of cyclones, Nature 260 (1976) 126.

[19]   Hopfinger, E.J. and Browand, F.K., Intense vortex generation by turbulence in a rotating fluid, in Bingtsson, L. and Lighthill, J. (eds.), Intense Atmospheric Vortices (Springer Verlag, 1982) 285.

[20]   Pedlosky, J., Geophysical Fluid Dynamics (Springer Verlag, 1979).

[21]   Hogg, N.G. and Stommel, H.M., the heton, an elementary interaction between discrete baroclinic geostrophic vortices and its implications concerning eddy heat-flow, Proc. Roy. soc. (London) A 397 (1975) 1.

[22]   Griffiths, R.W. and Hopfinger, E.J., Experiments with baroclinic vortex pairs in a rotating

fluid, J. Fluid Mech. 173 (1986) 501.

[23] Griffiths, R.W. and Hopfinger, E.J., Coalescing of geostrophic vortices, J. Fluid Mech. 178 (1987) 73.

[24] Maxworthy, T. and Browand, F.K., Experiments in rotating and stratified flows: oceanographic applications, Ann. Rev. Fluid Mech. 7 (1975) 273.

[25] Hide, R. and Ibbetson, A., An experimental study of Taylor columns, Icarus, 5 (1966) 279.

[26] Lighthill, J., Waves in Fluids (Cambridge University Press, 1978).

[27] Fjørtoft, R., On the changes in the spectral distribution for kinetic energy for two-dimensional nondivergent flow, Tellus 5 (1953) 225.

[28] Kraichnan, R., Inertial ranges in two-dimensional turbulence, Phys. Fluids 10 (1967) 1417.

[29] Morel, P. and Larchevêque, M., Relative dispersion of constant level balloons in the 200 mb general circulation, J. Atmos. Sci. 31 (1974) 2189.

[30] Rhines, P. Waves and turbulence on a beta-plane, J. Fluid Mech. 69 (1975) 417.

[31] Brachet, M.E., Meneguzzi, M. and Sulem, P.L., Small-scale dynamics of high-Reynolds number two-dimensional turbulence, Phys. Rev. Lett. 57 (1986) 683.

[32] McWilliams, J.C., Emergence of isolated coherent vortices in turbulent flow, J. Fluid Mech. 146 (1984) 21.

[33] Batchelor, G.K. Computation of the energy spectrum in homogeneous two-dimensional turbulence, Phys. Fluids 12 Suppl. (1969) 233.

[34] Griffiths, R.W. and Linden, P.F., The stability of vortices in a rotating stratified fluid, J. Fluid. Mech., 195 (1981) 283.

[35] Mory, M. and Hopfinger, E.J. Structure functions in a rotationally dominated turbulent flow, Phys. Fluids 29 (7) (1986) 2140.

[36] Babiano, A., Basdevant, C. and Sadourny, R., Structure functions and dispersion laws in two-dimensional turbulence, J. Atmos. Sci. 42 (1985) 941.

[37] Maxworthy, T., Caperan, Ph. and Spedding, G.R., An experimental study of decaying stratified 2D turbulence (in prep.).

[38] Ibbetson, A. and Tritton, D.J., Experiments on turbulence in a rotating fluid, J. Fluid Mech. 68 (1975) 639.

[39] Wigeland, R.A. and Nagib, H.M., Grid generated turbulence with and without rotation about the streamwise direction, IIT Fluids and Heat Transfer Report R78-1 (1978).

[40] Hopfinger, E.J., Turbulence in stratified fluids: A review, J. Geophys. Res. 92 n° C5 (1987) 5287.

[41] Mory, M. and Hopfinger, E.J., Rotating turbulence evolving freely from an initial quasi 2D state, in: Frisch, U., Keller, J.B., Papanicolau, G. and Pironneau, O. (eds), Macroscopic Modelling of Turbulent Flows (Springer Verlag, 1985).

[42] Hopfinger, E.J. and Toly, A.-J., Spatially decaying turbulence and its relation to mixing across density interfaces, J. Fluid Mech. 78 (1976) 155.

[43] Bradshaw, P., The analogy between streamline curvature and buoyancy in turbulent shear flow, J. Fluid Mech. 36 (1969) 177.

[44] Tritton, D.J., Experiments on turbulence in geophysical fluid dynamic. I. - Turbulence in rotating fluids, in: Soc. Italiana di Fiscia, Bologna (ed.), Turbulence and Predictability in Geophysical Fluid Dynamics and Climat Dynamics (1985).

[45] Nagib, H.M., Lavan, Z. and Fejer, A.A., Stability of pipe flow with superposed solid body rotation, Phys. Fluids, 14 (1971) 766.

[46] Pedley, T.J., On the stability of viscous flow in a rapidly rotating pipe, J. Fluid Mech. 35 (1969) 97.

[47] Maslowe, S.A., Instability of rigidly rotating flows to non-axisymmetric disturbances, J. Fluid Mech. 64 (1974) 307.

[48] Chandrasekhar, S. Hydrodynamic and Hydromagnetic stability (Clarendon Press, 1961).

[49] Boubnov, B.M. and Golitsyn, G.S., Experimental study of convective structures in rotating fluids, J. Fluid Mech. 167 (1986) 503.

[50] Hopfinger, E.J. and Browand, F.K., Vortex solitary waves in a rotating turbulent flow, Nature 295 (1982) 1.

[51] Hasimoto, H., A soliton on a vortex filament, J. Fluid Mech. 51 (1972) 477.

[52] Leibovich, S., Brown, S.N. and Patel, Y., Bending waves on inviscid columnar vortices, J. Fluid Mech. 173 (1986) 595.

[53] Maxworthy, T., Hopfinger, E.J. and Redekopp, L.G., Wave motions on vortex cores, J. Fluid Mech. 151 (1985) 141.

[54] Dickinson, S.C. and Long, R.R., Oscillating grid turbulence including effects of rotation, J. Fluid Mech. 126 (1983) 315.

[55] Lundgreen, T.S., The vortical flow above the drain-hole in a rotating vessel, J. Fluid Mech. 155 (1985) 381.

[56] Mory, M. and Caperan, Ph., On the genesis of quasi-steady vortices in a rotating turbulent fluid, J. Fluid Mech. 185 (1987) 121.

[57] Griffiths, R.W. and Hopfinger, E.J., Gravity currents moving along a lateral boundary in a rotating fluid, J. Fluid Mech. 134 (1983) 357.

[58] Batchelor, G.K., An Introduction to Fluid Dynamics (Cambridge University Press, 1967).

[59] Couder, Y. and Basdevant, C., Experimental and numerical study of vortex couples in two-dimensional flows, J. Fluid Mech. 173 (1986) 225.

[60] Overman, E.A. and Zabusky, N.J., Evolution and merging of isolated vortex structures, Phys. Fluids 25 (1982) 1297.

[61] Caperan, Ph. and Maxworthy, T., An experimental investigation of the coalescence of two-dimensional finite-core vortices, Phys. Fluids (submitted).

[62] Melander, M.V., McWilliams, J.C. and Zabusky, N.J., Axisymmetrization and vorticity gradient intensification of isolated 2D-vortices, J. Fluid Mech. (submitted).

[63] Polvani, Zabuski, N.J. and Fliert, G., Two-layer geostrophic V-states and merger. I. Constant potential vorticity lower layer, J. Fluid Mech. (submitted).

[64] McWilliams, J.C. Sub-mesoscale, coherent vortices in the ocean, Reviews of Geophysics 23 (1985) 165.

[65] Klein, P. and Pedlosky, J., A numerical study of baroclinic instability at large supercriticality, J. Atmos. Sciences 43 (1986) 1243.

Theoretical and Applied Mechanics
P. Germain, M. Piau and D. Caillerie (Editors)
Elsevier Science Publishers B.V. (North-Holland)
© IUTAM, 1989

MECHANISMS OF TOUGHENING IN CERAMICS

John W. HUTCHINSON

Division of Applied Sciences
Harvard University
Cambridge, MA 02138 USA

1. INTRODUCTION

Considerable effort has been expended to produce structural ceramics with enhanced toughness with some success, although toughening mechanisms which remain effective at high working temperatures remain elusive. Parallel efforts are underway for concrete and polymeric materials. A micro-mechanics of toughening of brittle solids is beginning to emerge. For ceramics, the mechanisms include transformation toughening, toughening by a metallic particulate phase, and fiber reinforcement. Brittle polycrystalline or multi-phase solids also have "inherent" toughening mechanisms connected with their heterogeneity which are not well understood. An overview of some of these mechanisms is given in this talk with the aim of highlighting the approaches and issues from a mechanics perspective. Transformation toughening and toughening by a metallic phase will be discussed briefly first. The bulk of the talk is concerned with the status of inherent toughening mechanisms in single phase polycrystalline ceramics.

2. TRANSFORMATION TOUGHENING

A theory of transformation toughening began to emerge just over five years ago after it had been discovered experimentally that a second phase constituent which undergoes a stress-induced martensitic phase transformation could be used to toughen structural ceramics. In the intervening years theory and experiment have reinforced one another, and there now exists a reasonable understanding of many aspects of the phenomenon [1]. Materials have been produced recently with toughnesses higher than ever before recorded for ceramics. One of the first clear successes of the mechanics modeling was the prediction of crack growth resistance for transformation toughened materials (i.e. R-curve behavior in fracture terminology). Stable crack growth was then observed and within the last few years R-curves have been measured for a number of zirconia toughened materials.

The mechanics approach involves the formulation of a phenomenological continuum model of the stress-strain behavior of the two-phase material and then the use of this constitutive model to determine the effect of transformation on behavior at a macroscopic crack tip. The approach parallels that taken to understand the role of plasticity in the fracture of metals. The two most important ingredients to the constitutive theory are the multi-axial stress condition for nucleation of transformation and the characterization of the transformation strain at the continuum level. For zirconia particles transforming from the tetragonal to the monoclinic phase, a 4% volume expansion occurs. A shear transformation accompanies the dilatation but the amount which occurs at the continuum level is uncertain because a particle usually transforms into multiple layers with alternating shearing. The simplest constitutive models are based on a critical mean stress condition for transformation and a purely dilatational transformation strain, although these are undoubtedly too simple for a fully quantitative theory.

The role of transformation in toughening is examined within the context of the continuum theory by the extent to which the intensity of the near-tip stress field is altered, or shielded, by the zone of transformation induced at the tip. According to calculations based on the simplest constitutive model, the transformation zone has no net effect on the near-tip intensity for a stationary crack. Only when the crack advances leaving behind a wake of transformed material does significant crack tip shielding take place. Toughening is inherently tied to stable crack growth resistance.

3.   CERMETS ~ TOUGHENING BY METAL PARTICLES

Tungsten carbide/cobalt used for some time as a material for cutting tools exploits its metal phase to toughen an otherwise brittle matrix material. Any macroscopic crack which occurs in the material leaves behind metal particles, or ligaments, which bridge the crack surface just behind the crack tip restraining its opening. For this mechanism to be effective it is essential that the particles be well-bonded to the matrix and that the advancing crack tip be drawn to particles so that bridges are left behind. Basic mechanics models [2] relate the near-tip stress intensity to the remote, or applied stress intensity, as altered by the bridging particles. Imposition of the condition that the near-tip stress intensity be at the level needed to crack the matrix (times a factor proportional to the area fraction of the matrix) leads to the toughening enhancement due to bridging. This enhancement depends on the ratio of the work of fracture of the particles to the work of fracture of the unreinforced matrix. The modeling of the toughening enhancement stands at a somewhat perplexing state. To bring the models in line with experimental data for several systems (Al_2O_3/Al and WC/Co) requires that the bridging stress developed in the particles be more than twenty times their uniaxial flow stress. Constraint by the surrounding matrix does raise the bridging stress to levels as large as 6 or 7 times the uniaxial flow stress, as model experiments have revealed [3]. But a level of twenty times the flow stress seems unrealistically large.

A tentative modification of the bridging model which might bring it into line with experiments involves the idea that the stress intensity at the tip may have to be well above the level needed just to crack the matrix if the crack front is trapped by the particles. The trapping of a bowed dislocation line by arrays of obstacles is an analogous phenomenon. Preliminary calculations [4] suggest that stress intensity levels between two and three times the critical matrix toughness level may be required to advance the crack front through a line of particles with an area fraction around 20%. Thus the critical crack-tip toughness for a bridged crack should (perhaps) be the critical intensity required to advance the trapped crack front rather than that of the pure matrix. Clearly, there are important aspects of the phenomenon that remain obscure.

4.   TOUGHNESS OF SINGLE PHASE POLYCRYSTALLINE CERAMICS

It can fairly be said that the mechanics of the toughness of single phase polycrystalline ceramics is not even qualitatively understood. Consider alumina, Al_2O_3, at room temperature as an example. The toughness for intrinsic grain boundary cracking (measured in terms of a critical energy release rate) is generally agreed to be about $2Jm^{-2}$; the toughness associated with the most cleavable planes of a single crystal is roughly $8Jm^{-2}$; and toughness measured in the macroscopic cracking of high quality polycrystalline alumina usually lies between 20 and 50 Jm^{-2}, depending on the grain size. The mystery is why the polycrystalline toughness is so high given the observational fact that the macroscopic crack does advance by predominantly grain boundary cracking. Obviously, the notion of a "path of least resistance" is not at play here. Various inherent mechanisms of toughening have been suggested to explain the source of the polycrystalline toughness [5] and these will be reviewed here from the mechanics perspective. It is our opinion that the mechanism (or mechanisms) has not been identified with any convincing certainty, and this is borne out by the divergence of views of workers in the field. A quantitative understanding of the inherent mechanisms of polycrystalline toughness of brittle solids is likely to impact our understanding of toughening mechanisms in general.

Central to each of the mechanisms which have been considered is the heterogeneity at the micro-scale of the polycrystal. Grains and grain boundaries have random orientation. Most single crystals are elastically anisotropic resulting in stress concentrations (singularities) at grain boundary junctions and vertices. Non-cubic crystals have a certain degree of anisotropy in thermal expansion properties; and, because polycrystalline materials are usually formed at sintering temperatures, at room or operating temperatures there are usually substantial residual stresses varying from crystal to crystal. These residual stresses increase with increasing grain size, and above a critical grain size most non-cubic polycrystalline ceramics undergo spontaneous micro-cracking along grain boundaries upon cooling from the fabrication temperatures to room temperature. For Al_2O_3, this critical grain size is on the order of 200 to 400 μm. Below the critical grain size grain boundary cracks are stable until they finally coalesce with the macro-crack. The micro-cracks are thought to be nucleated at grain boundary junctions or vertices where local stress concentrations occur [6]. They typically run over one grain boundary facet until they arrest at another grain boundary junction [7].

Thus, a macroscopic crack making its way through a brittle polycrystalline ceramic must contend with having to deflect to accommodate the varying orientations of the grain boundaries, with residual stresses varying from grain to grain, with the possibility of a zone of stress induced micro-cracks at its tip, and with any uncracked ligaments left behind the tip bridging its opening.

4.1. Crack Deflection Toughening

Macroscopic cracks in many polycrystalline ceramics do advance along grain boundaries since the grain boundary toughness is substantially less than the lowest cleavage toughness of the crystals. The variation in the orientation of the grain boundary facets and in the residual stress force the crack to advance in a tortuous manner which is not readily analyzed. The simplest approach of adding the additional surface energy associated with the extra area of grain boundary surface generated by the tortuosity is not valid because part of the energy supplied to create the crack surface comes from the residual stress. Most workers in the field seem to accept the outcome of the highly approximate analysis given in [8] where it is concluded that deflection effects in materials with more or less equiaxial grains may account for about a factor of 2 increase in toughness (measured in energy units) above the grain boundary value. The modeling does not specifically account for residual stress effects, and it is conceivable that there is a contribution to toughening from crack front trapping by residual stress.

4.2. Micro-crack Toughening

The idea behind the mechanism of micro-crack toughening is that stable grain boundary micro-cracks are nucleated by the high stresses in the vicinity of the macroscopic crack tip. These micro-cracks then lower the stress experienced by the tip. This shielding effect has been studied from two vantage points: one involving a smearing out of the effects of the micro-cracks appropriate to a zone of profuse micro-cracking and the other treating the interaction of the macro tip with discrete micro-cracks. No clear-cut conclusion from these theoretical studies has yet emerged, but an attempt will be made below to summarize the current status.

Direct experimental evidence of profuse micro-cracking surrounding a macro-crack is limited, partly because of the difficulty of making such observations. Detailed documentation of a micro-crack zone in a two phase alumina/zirconia material has recently been published [7], including information on the spatial distribution of the micro-cracks and on their orientation. In this particular material there appeared to be no preferred orientation of the micro-cracks relative to the stresses at the macro-crack tip, as might be expected given the large and essentially random variation of the residual stresses in the grains. A similar study of a relatively large grain, single phase alumina [9] revealed no micro-crack zone but did provide evidence of substantial crack bridging by uncracked grains, as will be discussed further in the next subsection. In any case, direct experimental evidence for the micro-crack toughening mechanism in single phase polycrystalline ceramics does not yet exist.

The theoretical approach to analyzing the shielding effect of a zone of profuse micro-cracks surrounding the macro-crack tip has close parallels to the approach taken to transformation toughening [10-12]. There are two separate contributions to shielding -- one arising from the reduced effective moduli in the zone and the other due to the partial release of residual stress which is formally equivalent to a transformation strain. When the micro-cracks are randomly orientated the reduced moduli remain isotropic and the equivalent transformation strain is a pure dilatation. Just as in the case of transformation toughening, the effect of the release of residual stress on shielding is proportional to the square root of the zone size and the shielding increases as the crack advances leaving micro-cracks in its wake. By contrast, the shielding contribution due to the reduced moduli is independent of the zone size, is relatively insensitive to the shape of the zone, and increases, but only slightly, with crack advance. Thus, any R-curve effect is mainly due to the release of residual stress. An attempt was made in [7] to make a detailed comparison between the theoretical micro-cracking models and experimental data for the alumina/zirconia material using observed distributions of micro-cracks and measured information on the release of residual stress. In this particular case, the two contributions to the toughness were roughly equal resulting in a total toughness increase of about a factor of four (measured in energy units). This level of increase appears to be representative; that is, significantly higher toughening should probably not be expected from profuse micro-cracking.

The second theoretical approach to micro-crack toughening employs two-dimensional, plane strain solutions for a macro-crack interacting with discrete micro-cracks in the vicinity of its tip, as represented by the studies in [13-17]. These investigations focus on how the stress intensity at the macro-crack tip is altered -- lowered or elevated -- by the configuration of micro-cracks. They have exclusively considered the weakening of the solid and have not included the contribution from the release of residual stress. A recent unpublished study in [17] gives the maximum mutual shielding possible in the interaction between a semi-infinite macro-crack loaded in mode I and a single micro-crack. That is, the configuration of the micro-crack is varied to minimize the energy release rate of the macro-crack subject to the constraint that the energy release rate at either tip of the micro-crack itself not exceed that of the macro-crack. The resulting minimum energy release rate of the macro-crack is reduced to 46% of the "applied" level and one of the tips of the micro-crack attains this same level. In other words, the maximum toughening which can be achieved by a single micro-crack is roughly a factor of 2 in energy terms, assuming no contribution from the release of residual stress is taken into account. For reasons now discussed, we believe this is likely to be representative of the maximum toughening increment expected from micro-crack shielding.

As already mentioned, shielding is independent of the size of the micro-crack zone. Moreover, calculations in [10] reveal that an annular zone of randomly oriented profusely distributed micro-cracks with a central core of uncracked material centered at the macro-crack tip results in very little change in the stress intensity at the macro-crack. Thus, one concludes that it is mainly the micro-cracks nearest to the macro-crack tip which determine the net shielding, or anti-shielding, of the macro-tip. This conclusion appears to be borne out by recent calculations for large numbers of discrete micro-cracks. Since the maximum mutual shielding from a single micro-crack is about .5, we expect this is more or less representative for any array. The mechanics of what determines a typical array has thus far not been addressed.

4.3.   Macro-crack Bridging

Several mechanisms involving bridging of a macro-crack tip have been identified as playing an important role in toughening: metallic particles in brittle matrices discussed in Section 3 and ceramic fibers in ceramic matrices. Clear evidence of bridging by uncracked grains has been observed in large grain, single phase alumina [9]. This material also displayed significant R-curve behavior with the crack resistance still increasing after several millimeters of crack advance. The bridging grains were observed to survive as far back as one hundred grain diameters behind the tip.

The role the bridges play in reducing the stress intensity at the crack front (i.e. shielding the lead crack tip) is readily understood and analyzed [18]. The puzzling aspect of the phenomenon is how the bridging grains are able to survive far back from the lead tip since the stress intensity levels where the crack surface impinges on these grains are exceptionally high according to any relatively straightforward estimate. One plausible suggestion [19] invokes local residual stresses at the bridging grains for their "protection". Suppose, for example, that prior to any cracking a grain destined to be a bridge supports a residual tensile stress normal to the potential crack plane. Surrounding this grain on the same plane will be a compressive stress which is largest at the equator of the grain and falls off away from it. This compressive region will tend to protect the grain from any crack engulfing it on the plane in question. Statistically, only a relatively small fraction of the grains will be favorably oriented and positioned to be so protected, and thus the area fraction of surviving bridges is not expected to be large. A first quantitative model of the protective role of residual stress [17] shows some promise in rationalizing the survivability of the bridges but not for bridges as far back as those observed in [9]. At this writing, the issue remains a puzzle.

4.4.   Concluding Remarks on Toughening Mechanisms in Polycrystalline Ceramics

We have tried to make the case that a micro-mechanics model of the inherent toughening of single phase polycrystalline ceramics does not yet appear to exist. Specifically, with alumina as an example, an accepted explanation does not exist for how the toughness of the polycrystalline material (in energy units) can be on the order of 10 to 20 times the intrinsic grain boundary toughness, even though the cracking process is largely grain boundary cracking. It may be that the several mechanisms discussed above act in concert adding up to give the inherent toughness of the polycrystalline material. It seems equally likely that there may be other mechanisms or other features to the mechanisms mentioned above which remain to be uncovered. Further background, insight, and access to the literature can be found in the recent review articles [1, 5, 20].

ACKNOWLEDGEMENTS

This work was supported in part by the National Science Foundation under Grant NSF-MSM-88-12779, by the Materials Research Laboratory under Grant NSF-DMR-86-14003, and by the Division of Applied Sciences, Harvard University.

REFERENCES

[1]　Evans, A. G. and Cannon, R. M., Acta metall. 34 (1986) 761.

[2]　Budiansky, B., Amazigo, J. C. and Evans, A. G., J. Mech. Phys. Solids, 36 (1988) 167.

[3]　Eberhardt, J. and Ashby, M. F., "Flow Characteristics of Highly Constrained Metal Wires", Report, Engineering Department, Cambridge University, Cambridge, England (December 1987).

[4]　Fares, N. and Rice, J. R., work in progress, Division of Applied Sciences, Harvard University.

[5]　Freiman, S. W., Ceramic Bulletin, **67** (1988) 392.

[6]　Tvergaard, V. and Hutchinson, J. W., J. Am. Ceram. Soc., **71** (1988) 157.

[7]　Rühle, M., Evans, A. G., McMeeking, R. M., Charalambides, P. G., and Hutchinson, J. W., Acta metall. **35** (1987) 2701.

[8]　Faber, K. T. and Evans, A. G., Acta metall. **31** (1983) 565.

[9]　Swanson, P. L., Fairbanks, C. J., Lawn, B. R., Mai, Y-W., and Hockey, B. J., J. Am. Ceram. Soc. **70** (1987) 279.

[10]　Hutchinson, J. W., Acta metall. **35** (1987) 1605.

[11]　Ortiz, M., J. Appl. Mech. **54** (1987) 54.

[12]　Charalambides, P. G. and McMeeking, R. M., Mechanics of Materials **6** (1987) 71.

[13]　Kachanov, M., Int. J. Solids Structures **23** (1987) 23.

[14]　Rubinstein, A. A., Int. J. Fracture **27** (1985) 113.

[15]　Horii, M. and Nemat-Nasser, S., J. Mech. Phys. Solids **35** (1987) 601.

[16]　Rose, F., Int. J. Fracture **31** (1986) 233.

[17]　Shum, D. and Hutchinson, J. W., work in progress, Division of Applied Sciences, Harvard University.

[18]　Mai, Y.-W. and Lawn, B. R., J. Am. Ceram. Soc. **70** (1987) 289.

[19]　Swain, M. V., J. Mater. Sci. Lett. **5** (1986) 1313.

[20]　Clarke, D. R. and Faber, K. T., J. Phys. Chem. Solids **48** (1987) 1115.

Theoretical and Applied Mechanics
P. Germain, M. Piau and D. Caillerie (Editors)
Elsevier Science Publishers B.V. (North-Holland)
IUTAM, 1989

MULTIPHASE FLOW IN EXPLOSIVE VOLCANIC AND GEOTHERMAL ERUPTIONS

Susan Werner KIEFFER

United States Geological Survey
Flagstaff, Arizona, U.S.A. 86001

ABSTRACT: Explosive volcanic or geothermal eruptions are those eruptions (or parts of eruptions) in which the continuous phase of the erupting fluid is, or becomes, gas and the dispersed phase is liquid or solid. In explosive eruptions, fluids can flow at velocities exceeding 100 meters/second (m/s); vent velocities up to 500 and 600 m/s have been inferred from field evidence. These eruptions discharge fluids that range in composition from nearly pure H_2O to heavily mass-laden mixtures of vapor, rock, molten droplets of magma, and ash. The high velocity in explosive eruptions is gained by conversion to kinetic energy of enthalpy of hot rock and gas in reservoirs that are under lithostatic or hydrostatic pressure. Shallow eruptions in which reservoir pressures are only of the order 1-10 MPa (10-100 bars) can eject rocks up to 20 km laterally; deeper eruptions in which the lithostatic pressure is on the order of 100-1000 MPa (1-10 kbar) can eject plumes that rise into the stratosphere. A characteristic of the complex multicomponent, multiphase fluids in volcanic and geothermal environments is their low sound velocity. When gas is the continuous phase, sonic velocities are typically on the order of 100 m/s; in liquid with a small fraction of gas bubbles, the sonic velocity can be as low as 1-10 m/s. In explosive eruptions or flow from geothermal wells, fluid velocities can be greater than the sonic velocity. Therefore, both subsonic and supersonic flow can occur in geologic settings. The importance of the low characteristic velocities of geologic fluids has not been widely recognized and, as a result, the importance of supersonic flow in geologic processes has generally been underestimated.

1. INTRODUCTION

The study of volcanoes extends back to the Greek philosophers, but perhaps the earliest definition appropriate for an overview of volcanism is that of Daly (1911, p. 48): "Volcanic action is the working of the extrusive mechanism which brings to the earth's surface rock matter or free gas, initially at the temperature of incandescence." Since Daly's time, the term volcanic eruption has been used to describe "the emission or ejection of volcanic materials at the earth's surface from a crater or pipe or fissure" (American Geologic Institute Dictionary of Geologic Terms, 1974). The words "volcanic materials" include gas, ash, rock, and magma (melted rock that contains volatile gases and is typically at temperatures between 700 and 1200°C). For this IUTAM symposium on two-phase flow, the focus of this paper is on the so-called "explosive" eruptions.

Volcanic eruptions are classified into two end members depending on the nature of the ejecta: magmatic eruptions in which fresh magma is erupted, and other eruptions (which carry a variety of names described below) in which no fresh magma is erupted. The term geothermal eruption was used in the title of this paper as a very informal term that implies an eruption of hot water, often containing ash or rock fragments, from a vent in ground heated at depth--examples include hydrothermal eruptions, hydrothermal explosions, geyser eruptions, fumarole emissions, phreatic eruptions, and the blow-out of drilled geothermal wells. A hydrothermal eruption is an emission of water with or without entrained rocks from a geothermal system (Figure 1); the term includes both hydrothermal explosions and geyser eruptions (Muffler and others, 1971). Geyser eruptions are intermittent (often nearly periodic) discharges of hot water from a vent. A hydrothermal explosion is the violent eruption of water confined in near-surface rock, causing such a large scale disruption that a large proportion of solid debris is expelled along with water and steam.

The nonmagmatic eruptions often, but not uniquely, occur in geothermal areas that lie several kilometers above magma bodies that provide a heat source. For example, at the geothermal areas of Yellowstone, U.S.A., or of the North Island of New Zealand the magma may reside at depths of a few to perhaps 10 km. Heat from the magma drives large convection cells of water. At the surface of the earth, the water emerges in the geothermal areas (Figure 1).

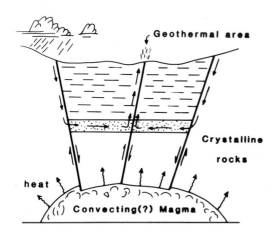

FIGURE 1. Schematic diagram of the main features of a geothermal system (after White, 1967). Hot magma at depth is a source of both heat and heated fluids with dissolved material (for example, H_2O, NaCl, SO_2, SiO_2, etc.). A convective column of hot water develops over the magma; the water is provided from the surface or near-surface (ground-water) environments. It descends on the periphery of the magma body because it is cold, is heated as it nears the magma body, and rises back to the surface to form steaming ground, hot springs, geysers, fumaroles, mud volcanoes, and hydrothermal explosions.

When magma is closer to the surface, it also provides heat to ground-water and can cause so-called phreatic eruptions in which the ground water carries rock fragments (but not magma). Phreatomagmatic eruptions result from direct magma-water interactions, and magma appears as a product in the ejecta (see Colgate and Sigurgeirsson, 1973 for a description of lava-sea water interactions; see summary of larger-scale magma-water interactions in Lorenz, 1987). It is obvious that, unless one knows where magma resides within a volcanic system, it may be difficult to decide whether to call a given eruption "hydrothermal" or "phreatic", and it may be difficult whether to chose between the words "eruption" and "explosion".

Magmatic eruptions occur at a great many scales and have many different surface manifestations--ranging from the effusion of liquid flows (with dispersed gas bubbles and crystals) to gassy "explosive" eruptions in which gases dissolved from the magma have become the continuum phase and carry dispersed droplets of silicate melt, ash, and/or crystals. At depth (order of tens of kilometers) magmas are generally undersaturated in volatiles, and their ascent toward the surface is driven by buoyancy (for summary, see Marsh, 1984). The magma that emerged at Mount St. Helens is one of the most thoroughly studied in terms of its relations to the fluid dynamics of the eruption column that it formed (Rutherford and others, 1985; Carey and Sigurdsson, 1985). As pressure on a magma decreases during an eruption [for example, as it rises in the conduit or if a sudden event (like a landslide) changes the pressure conditions in the whole mountain], bubbles of gas exsolve (Figure 2). When the relative volume of the bubbles reaches about 75% of the total volume (Sparks, 1978) the magma fragments and gas becomes the continuous phase, i.e., the eruption becomes explosive. The references cited above suggest that the disruption level for the magma that formed the tall plinian column at Mount St. Helens was at a depth of 580 m. To date, for fluid-mechanical modeling, when magma is in the dispersed explosive regime, it is treated by dusty-gas thermodynamic models; thus, although there is a significant difference in geochemistry and petrology between hydrothermal, phreatic, phreatomagmatic, and magmatic eruptions, these differences are blurred in fluid-mechanical models of eruptions.

Just as there is a continuum of terminology used to describe these eruptions and the fluids that erupt, there is a continuum of fluid-dynamical behavior between the flow of hot springs, geysers, phreatic eruptions, and the so-called "explosive" class of volcanoes. This continuum will be illustrated in this paper, and common elements of thermodynamics and fluid-dynamics will be described.

The eruptions categorized as "explosive" are so varied that it is difficult to generalize about the processes in detail, but I here propose a working definition that the term "explosive volcanism" be used as a descriptive term for those eruptions, or parts of eruptions, in which the continuous phase of the erupting fluid is, or becomes, gas, and the dispersed phase is liquid or solid. It should be noted that this definition is descriptive, not mechanistic, and purposely avoids constraints on scale or duration.

Volcanologists face a challenging inversion problem in their science, illustrated schematically in Figure 2a. Processes that are occurring inside a volcano, or within flow fields erupting from a volcano, must be inferred from remote observations, often very incomplete. Traditionally, the study of unobserved volcanic eruptions has been based on reconstruction of events from the record of deposits laid down in the waning stages of eruptions and, in a few cases, erosion surfaces created during the waxing stages of the eruptions. Even when volcanoes are erupting, they are often remote and the most common data are "far-field" data: barograph records, seismic records, knowledge of how far the volcano has spread deposits, or, recently, satellite data. In the last few decades, as the population of humans and geologists has increased, more "near-field" data--such as eyewitness accounts and actual photographs--of eruptions have become available. Although the detail and amount of data now available are far greater than ever before, the problem is still an inverse problem, not a direct analysis or laboratory experiment.

In the 1950's and 1960's, it was common practice to take far-field barograph or seismic data from volcanic eruptions and to use formulas derived from nuclear and chemical tests to express the energy released by an eruption (see Williams and McBirney, 1972, Chapter 4 for a summary of such analyses). In fact, it became an accepted practice in volcanology to express the energy of an eruption in "megatons" because equations from the engineering literature on bombs were used. Fluid dynamicists should be aware that although there is at least an implicit analogy between volcanic eruptions and chemical and nuclear explosions in the use of equations and terminology, volcanologists generally do not believe that either chemical or nuclear detonation sources in the sense of bombs are involved in the eruption process.

A problem with inversion of far-field data is that it cannot be extrapolated to near-source or internal conditions if the processes generating the far-field data are different and cannot be mathematically scaled. This fact is being recognized by modern analyses of far-field data: Older techniques relied on extrapolating far-field data linearly backwards to a hypothetical source region of small dimensions and instantaneous energy release. It is now more common for researchers to generate physically plausible synthetic source conditions and time-histories and to try, thereby, to reproduce far-field data. For example, Banister (1984), and Reed (1987) consider the pressure-time history of the May 18, 1980 avalanche, lateral blast, and Plinian column at Mount St. Helens in analysis of barograph records. Kanamori and Given (1982) and Kanamori and others (1984) generated synthetic source force-time models to analyze seismic data from the same event. Numerical modeling of steady fluid flow in volcanic conduits and plumes has been very successful in explaining near-field observations of plumes and both near-field and far-field volcanic deposits (Wilson, and others, 1980 and earlier work; Wohletz and others, 1984; Sparks and others, 1986; Sparks, 1987; for review, see Wilson and others, 1987). Furthermore, laboratory work on simulating some aspects of eruption is providing experimental constraints on processes (e.g., Wohletz and McQueen, 1984; Kieffer and Sturtevant, 1984).

In spite of the advances in analyzing and modeling internal volcanic processes, their flow fields, and their relation to near-field and far-field data, controversies about the nature of source processes still exist because of the concepts brought from "explosion" literature into volcanic models. A brief review of the evolution of ideas about "pressure" in volcanoes will illustrate the problem.

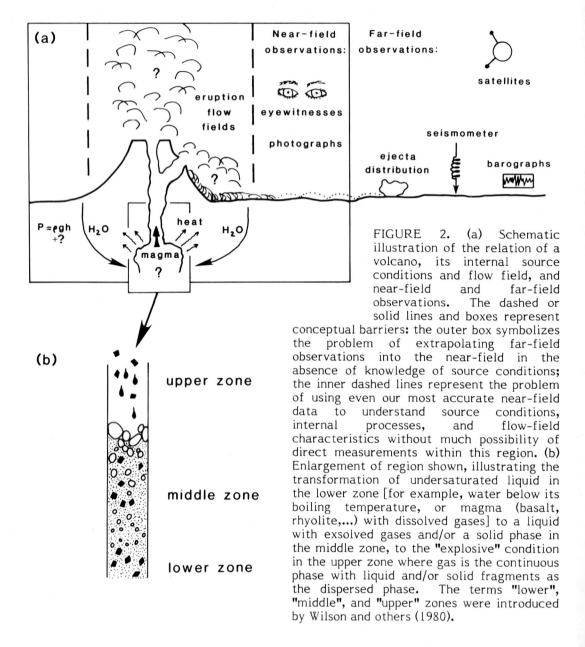

FIGURE 2. (a) Schematic illustration of the relation of a volcano, its internal source conditions and flow field, and near-field and far-field observations. The dashed or solid lines and boxes represent conceptual barriers: the outer box symbolizes the problem of extrapolating far-field observations into the near-field in the absence of knowledge of source conditions; the inner dashed lines represent the problem of using even our most accurate near-field data to understand source conditions, internal processes, and flow-field characteristics without much possibility of direct measurements within this region. (b) Enlargement of region shown, illustrating the transformation of undersaturated liquid in the lower zone [for example, water below its boiling temperature, or magma (basalt, rhyolite,...) with dissolved gases] to a liquid with exsolved gases and/or a solid phase in the middle zone, to the "explosive" condition in the upper zone where gas is the continuous phase with liquid and/or solid fragments as the dispersed phase. The terms "lower", "middle", and "upper" zones were introduced by Wilson and others (1980).

There can be at least two separate "plumbing" systems within a volcano. As indicated in Figure 2, under equilibrium conditions in a volcano constructed only of solid material, the pressure at any depth, h, is primarily the "lithostatic" pressure, $P_l = \rho_r g z$, where ρ_r is the average rock density and g is the acceleration of gravity. If the volcanic edifice is highly fractured and contains fluid that has contact with the atmosphere (e.g., water, or water + gases), there may be a separate hydrothermal system under "hydrostatic" pressure, $P_h = \rho_f g z$, where ρ_f is the density of the fluid. Volcanic eruptions commonly involve the depressurization of both a magmatic and a hydrothermal system. A magma or hydrotheramal system might sustain a relatively modest added pressure (due, for example, to pressure transmission by "squeezing" on the chamber by the surrounding rocks (Whitney and Stormer, 1986). The excess pressure (indicated in Figure 2a by a question mark) is limited by the failure criteria for the volcano, usually the tensile strength. As a rule of thumb, 1 km of (fractured) rock at density 2000 kg/m^3 gives a lithostatic pressure of 20 MPa (200 bars), and 1 km of hot liquid water gives a hydrostatic pressure of about 9 MPa

(90 bars). Therefore, typical pressures in volcanic edifices that stand a few km in height above their surroundings are a few hundred bars, and the tensile strength is negligible. At greater depths, it has been proposed that the equilibrium pressure on a magma could be as much as 100 bars per 1 g/cm^3 of density differential for every 1 km thickness of the magma body: for example, if a magma body lies at 10 km depth, and if the average density of the rocks overlying it is 3000 kg/m^3, then the lithostatic pressure on the top of the magma body is roughly 3 kbars. If the magma body has a density of 2000 kg/m^3 (2.0 g/cm^3) and is 1 km thick, the excess pressure is 100 bars. This is a small fraction of the lithostatic pressure, but is a function of depth and size of the magma body, relative densities of the magma and surrounding rock, and strength of the volcano.

After several decades of struggle with the problem of the relation between measured or inferred vent velocities and source pressure (Matuzawa, 1933a, b, c; Gorshkov, 1959, 1960; Decker and Hadikusumo, 1961; Richards, 1965; Melson and Saenz, 1973, Fudali and Melson, 1972; Wilson, 1972), the issue of the relation between source pressure and initial vent velocity was clarified by Self and others (1979), who demonstrated by use of the compressible Bernoulli equation that pressures of only a few to a few tens of MPa (10-100 bars) are required to produce vent velocities of hundreds of meters per second. The required pressures are proportional to the gas content of the erupting fluid. Reservoir pressures of a few hundred bars are entirely within the realm of steady-state conditions within volcanoes (Figure 2a).

The fluid-dynamics and pressure-velocity relations of magmas originating much deeper than those discussed above have been investigated throughout the late 1970's and early 1980's. Numerous papers have demonstrated that flow driven simply by lithostatic pressures (or lithostatic pressure plus a modest confining pressure allowed by the rocks of the edifice) in volcanic reservoirs can produce the observed volcanic phenomena. These models typically assume that reservoirs of volatile-rich magma are under hydrostatic pressure at 7-10 km depth and that an eruption is initiated by fracture propagation to the earth's surface (see Shoemaker and others, 1962; Shaw, 1980). For rheology, the models consider transformation of the magma from an undersaturated liquid to a liquid with entrained bubbles to an "explosive" gas containing entrained ash or molten silicate fragments. For an equation of state, simple pseudogas theory is used (see Section 4), and the equations of conservation of momentum and continuity are solved for a vertical conduit and a turbulent plume. All of the models show that the enthalpy in the magma would be sufficient to overcome internal friction and potential energy, and would drive a jet from the vent at the Earth's surface to velocities of several hundred m/s. The process of transformation of the magma (believed to be undersaturated in the reservoir) into a gassy high-velocity jet is illustrated schematically in Figure 2(b). Depending on the buoyancy of the jets and the amount of atmospheric entrainment, such jets could form plumes that rise into the stratosphere or could fall back to the ground surface to form a ground-hugging flow. These models were related to observed rising or falling columns at volcanic eruptions (Wilson, 1976; Wilson and others, 1980; Carey and Sigurdsson, 1985; Sparks, 1987; see also review by Wilson and others, 1987), and many useful inferences about mass flux and ash distribution were obtained. Furthermore, the models were consistent with experimental constraints on magma composition and volatile content, and with petrologic observations of the pressures and temperatures that had occurred in the history of minerals and rock fragments entrained in the magma.

The propulsion of ejecta for many kilometers by shallow, laterally directed blasts has also been shown to be plausibly explained by decompression of a volatile-rich multicomponent fluid by available lithostatic pressures of only ~10 MPa (~100 bars) (Kieffer, 1981a,b; Eichelberger and Hayes, 1982). While the relative roles of gas decompression and gravity have not been worked out in detail (because the solution of the equations of motion with all of these force terms is too complex), the analyses demonstrate that there is sufficient enthalpy available in the hot rocks and compressed gases of even very shallow volcanic reservoirs under normal lithostatic or hydrostatic conditions to produce the observed or inferred ejecta velocities.

This paper provides some examples of "explosive volcanism", summarizes the thermodynamic and fluid dynamic phenomena believed to occur, and, especially, stresses the point that observed eruption phenomena appear to be produced simply from the conversion of enthalpy stored in water and hot rocks under normal lithostatic or hydrostatic conditions to kinetic energy.

2. EXPLOSIVE VOLCANISM

A simple natural "explosion" is the eruption of a geyser--particularly, a fountain geyser. A fountain geyser has a pool of water (usually cooled by surface evaporation) overlying a conduit of hot water. Eruptions consist of the emergence of large bubbles of steam through the water in the pool (Figure 3). The entire eruption may consist of the emergence of only one bubble. The breaking of the bubble at the surface of the pool is rapid, and the event is thus perceived as "explosive" in resembling the surface detonation of a bomb, as well as being explosive in the working sense defined in Section 1.2 of this paper. Each eruption of a fountain geyser may consist of only one to as many as a dozen discrete bubble-bursts, or may consist of such a rapid succession of bubble bursts that the cooled water in the pool over the conduit is entirely discharged and a jet-like, continuous eruption lasting for several minutes may occur. In this latter case, the sense of an "explosion" is lost and Perret's (1924) sense of "emission" is obtained because of the steady nature of the flow at the vent and the steady structure of the eruption column. However, under the working definition proposed, this whole continuum of fountain geyser eruption styles is explosive because the events are gas-driven. The problems of brevity versus longevity, and non-steady versus steady flow, are ones which those who grapple with defining "explosive volcanism" must contend.

FIGURE 3. Strökkur Geyser, Iceland, in eruption. The steam bubble that has just emerged and broken through the surface pool of water is approximately 2 m in diameter at the base; each eruption consists of one or two such "explosions". Note the wave moving out into the cool water of the fountain pool as the steam bubble bursts.

Temperatures have been measured in fountain geysers (for example, Rinehart, 1974), and are on the order of 10°C higher at the bottom of the geyser than the surface (atmospheric) boiling temperature. In Narcissus Geyser, Yellowstone National Park, the boiling temperature at the atmospheric pressure of 0.075 MPa (0.75 bars) is 93°C. The basin of Narcissus is shaped like an hour glass and is 6.5 m deep. Over a time scale of several hours, the basin fills with water and the water at 6.5 m depth (the base of the pool) slowly heats toward the local saturation temperature of 105°C. Occasional small steam bubbles form and rise upward. If the individual bubbles are too small compared to characteristic dimensions for heat transfer, they collapse in the cooler water toward the top of the pool; conversion of the latent heat in the steam heats water, and convective overturn is a common result of heat transfer upward in geyser conduits prior to eruptions. If, at Narcissus Geyser, the amount of water heated to 105°C is large enough, the bubble grows, reaches the surface, and bursts through in an eruption that throws steam and water of the pool upward and outward (as shown in Figure 3 for Strökkur Geyser, Iceland). The enthalpy available from the isentropic decompression of liquid water at 105°C to a mixture of water and steam at 93°C is 0.79 kj/kg. If this amount of enthalpy is entirely converted to kinetic energy, the mixture could attain a maximum velocity of 40 m/s, and--in theory--a height of 40 m. However, in fountain geyser eruptions, much of the energy is expended in setting water in the pool into motion as the bubble grows and rises (see waves in Figure 3) and the typical eruption height is more on the order of 10 m. An additional reason that the theoretically attainable height is not attained is that the bubble grows and rises into cooler water, and therefore, some of the enthalpy that is theoretically available in a strictly isentropic expansion is lost as steam condenses onto the bubble walls as cooler water is encountered in the fountain. Nevertheless, this simple example demonstrates that "a little enthalpy goes a long way" when converted to kinetic energy.

The "explosive" breaking of a single bubble characteristic of Strŏkkur geyser is relatively rare. Typically, even fountain geysers produce a steady jet for some prolonged (minutes) time. The best examples of steady jets are produced by columnar geysers, geysers that do not have surface pools of water, but have, instead, vents topped by cones. Old Faithful Geyser has not been the world's tallest Geyser (Steamboat Geyser in Yellowstone was measured at 160 m height during eruptions in the 1960's), but it is an excellent and well-known example of a columnar geyser (Figure 4). Old Faithful's eruptions begin with short ejections of water, probably mimicking those of Strŏkkur, although the activity is initiated 5 m below the ground surface and is difficult to document in detail (see Kieffer, 1984a for details). This phase of the eruption is known as preplay (Figure 5). As an eruption develops over an interval of 20 to 40 seconds, the fluid column rises in a serious of bursts to a height between 30 and 50 m; I have called this interval unsteady flow (Figure 5). The number of bursts varies from 1 to 8--4 bursts are shown in Figure 5, and frame (d) in Figure 4 probably represents a burst, and frame (e) probably represents one of the brief pauses between bursts. Steady flow follows the unsteady bursting for about 30 s during which the column stays at maximum height and during this time, surging occurs at a frequency of 1-2 Hz (Figure 5). After this sequence, the eruption declines over an interval of up to 3 minutes; during this time, the height of the column steadily drops to about 10 m height (Figure 5).

A final example of two-phase flow in volcanic and geothermal environments can be seen in flow from geothermal wells. These bores are such excellent large-scale "laboratory" experiments for geologists when data from the drilling, testing, and operation are available (for example, White and others, 1975) that their flow should be considered as an example of an "eruption", even though it is artificially produced. The bores are typically 0.15-0.30 m diameter and--when high-enthalpy wells are turned on and off for testing--they display a spectrum of flow conditions ranging from subsonic to supersonic at the exit plane (Figure 6). When shut down, geothermal wells must be carefully managed because accumulation of noncondensible gases at the well-head has caused "explosions". In extreme cases, runaway flow through the bores has caused ground failure and uncontrolled flow through fissures and adjacent geothermal ground, resulting in surface features not unlike "hydrothermal explosion features" (e.g., the Rogue bore, New Zealand).

FIGURE 4 (next page). Series of photographs of the early stages of eruption of Old Faithful. (a) Photo taken about 1 second after fluid emerges from the surface to form an "eruption". The erupting fluid is emerging as the low white mass in the center of the photo. The cloud of condensed steam to the near left is the residual fluid from a burst of preplay just prior to this photograph (prehaps at about t = -5 sec). At the far left, another faint cloud of steam can be seen in front of the dark hillside, a remnant of fluid from an earlier burst of preplay. (b) t~4 sec. (c) t~7 sec. (d) t~10 sec. (e) t~13 s. (f) t~6 sec. (g) t~19 sec--nearly steady state conditions. Note that by t = 13 sec (e), there is clearly fluid descending down the right side of the jet, and that during the interval between (f) and (g) the descending "sheath" of fluid has dramatically increased the apparent width of the geyser (use background tree and snow patterns for this comparison). For scale, the geyser is about 40 m maximum height in (g).

FIGURE 4: Caption on previous page

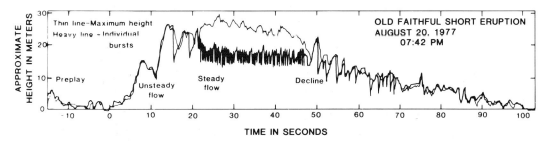

FIGURE 5. Graph of height vs. time of a typical eruption of Old Faithful geyser, showing the four stages of eruption discussed in the text. Light line is the height of the top of the water-steam column; heavier line traces individual pulses of water, not visible in Figure 4 because they occur at later times during the steady flow. They are easier to see on movie films than in still photos. The heavy line was made by tracing the path of individual pulses of water as they became visible on a movie in order to document the frequency of the surging, 1-2 bursts per second. By convention, time t = 0 is taken as the time after which water remains above the surface of the ground. The end of an eruption is the time at which the last liquid water is ejected. An eruption can continue for more than 4 minutes, with the last two minutes typically resembling the period between 60 and 80 seconds on this graph.

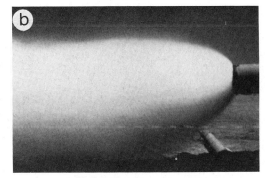

FIGURE 6. A geothermal bore in the Imperial Valley, California, being tested; bore diameter about 10" (25 cm). (a) Low flow rate; barely choked (sonic) flow. (b) Higher flow rate, possibly higher enthalpy fluid. Note the lateral expansion of the fluid at the exit plane of the bore. This is Prandtl-Meyer expansion of a fluid whose pressure is higher than atmospheric at the exit plane, and is characteristic of a supersonic plume. The asymmetry of the plume on the top and bottom sides is presumeably due to differences in abundance of entrainable air between the top (exposed to the atmosphere) and the bottom (close to the ground).

The fluid-dynamic continuum between "explosive" and "emmisive" geyser eruptions is replicated in volcanic eruptions. The events of March, April, and May, 1980, at Mount St. Helens, Washington, U.S.A. showed this complete continuum of eruption styles.   The volcano first erupted on March 27, 1980, after a repose of about one century; the initial eruptions were phreatic (no juvenile magma was ejected) and created a small crater in the summit area (Figure 7). Over a period of several weeks, repeated eruptions enlarged this crater, and fall-back of ejecta and slumping of ice and rocks from the steep walls created a rubble filled crater (Figure 8). Water was sometimes seen in the bottom of the crater by observers flying overhead in aircraft, and thus the crater can be thought of--in a very general way--as a fountain geyser in which the fountain is continually filled and refilled with both water and debris. The crater became several hundred meters deep, and phreatic eruptions emanated from the bottom of the crater (Figure 9).

FIGURE 7. A view of the summit crater on the afternoon of March 27, 1980, several hours after an eruption formed this crater. Note the fractures that run across the icefield that covers the top of the mountain.   Estimated diameter of the summit crater is 100 m. Photo by David Frank.

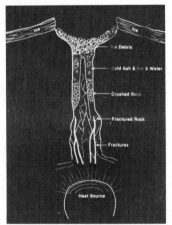

FIGURE 8. A schematic cross-Section through the summit of Mount St. Helens during the eruptions of late March and early April, 1980. The crater was on the order of 200 m deep; the depth and detailed shape of the conduit shown are not known, nor is the exact location and nature of the heat source.

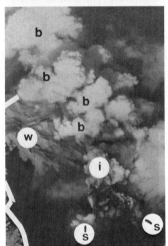

FIGURE 9. View from a helicopter nearly directly over the crater (inadvertently) as an eruption begins. The rim of the crater is traced by the white line. The rocks of the inner crater wall are indicated by (w); bursts of condensed steam from "preplay" eruptions (presumeably from hot water ponded in the base of the crater) are indicated by (b) (compare with steam bursts visible in Figure 4 (a) and (b)). The plume of the initial phase of the eruption deep within the crater is indicated by (i), and a basal zone of steam formation is indicatd by (s)--this presumeably arises from boiling of ground water in the debris in the bottom of the crater.

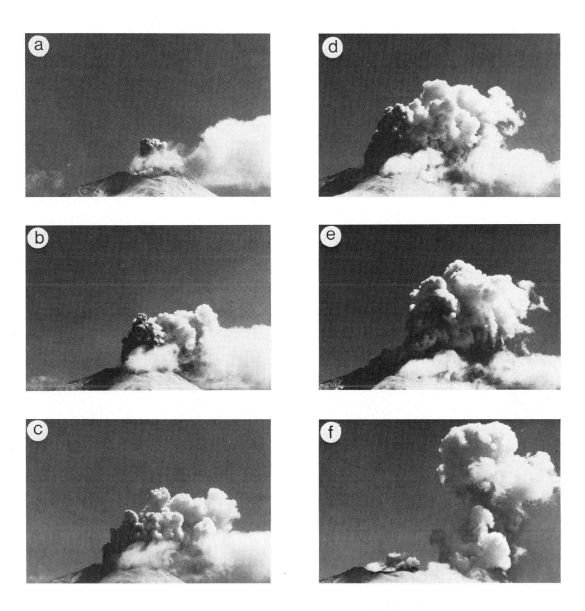

FIGURE 10. Sequence of eruptions from Mount St. Helens in March, 1980. The eruptions originate in the floor of a crater that is probably on the order of 100 m below the visible rim, and the maximum height of dark ejecta (in (e)) is probably on the order of 100 m. (a) First available photo, perhaps 10 seconds after the eruption was initiated in the bottom of the crater (refer back to Figure 9). Note the cloud of steam to the right, indicating preplay activity. (b) t~70 sec. Note flow segregation between dark ash (generally on the left) and the light steam being carried to the right by the wind in this and all remaining photos. (c) t~100 sec; (d) t~130 sec. (e) t~160 sec--eruption is probably over; (f) t~220 sec.

FIGURE 11. Photographs of the lateral blast. (a) May 18, prior to 8:32 a.m. onset of eruption. (b) ~10 sec. (c) ~50 sec. (d) ~65 sec. (e) ~70 sec. (f) ~2 min. (g) ~2.5 minutes. Photographs copyrighted by Keith Ronnholm, Seattle, Washington, U.S.A. Reprinted with permission.

As emphasized above, the March and April eruptions at Mount St. Helens were "phreatic"--mixtures of hot ground water and crushed rock. A sequence of photos of a typical (but short) eruption is shown in Figure 10; compare with the sequence of Old Faithful pictures in Figure 4. "Preplay" has clearly occurred, as indicated by the remnant of the steam to the right, but any "preplay" or "unsteady initiation" of the eruption was difficult to document from the viewers location for the sequence of Figure 10. However, the eruption probably resembled that of Figure 9 before it emerged into view above the summit of the crater in the sequence of Figure 10, with "preplay" occurring before the eruption became visible over the summit. Unsteady flow appears to be occurring in Figures 10(a) and (b); steadier conditions in (c) and (d); and the eruption is waning or over by (e). This short "volcanic" eruption is certainly of the class that volcanologists call "explosive", and yet it has a well-defined steady flow stage. The whole eruption is within an order of magnitude in height, duration of preplay, unsteady and steady flow stages, and duration to an eruption of Old Faithful--differing only in the nature of the erupting fluid. The entrained rock fragments add mass to the flow, and problems of two-phase flow become immediately apparent; for example, in Figure 10 note the segregation of the heavy (dark) material that rises and then falls back into the crater, and the light (white) steam that drifts buoyantly downwind.

On May 18, 1980, "explosive volcanism" of a different fluid-dynamic and thermodynamic nature developed. The phreatic eruptions changed to magmatic eruptions when a large landslide from the flank of the volcano exposed a shallow hydrothermal system and a deeper magma reservoir to a sudden pressure unloading. This eruption was well-documented by seismographs, barographs, satellites, and eyewitnesses. The eruption consisted of two parts: (1) a laterally-directed "blast" in which a heavily particle-laden fluid devastated, within a few minutes, an area of about 600 km^2 north of the mountain (Figure 11), and (2) a vertically-directed column (plinian column) that was more-or-less steady for at least 9 hours (Figure 12) and was accompanied by intermittent pyroclastic flows. Note that both types of eruptions would be included in the proposed use of the term "explosive volcanism".

Eruptions much larger than those discussed here have occurred in the geologic past, and their general characteristics can be reconstructed from studies of deposits and structures [e.g., see Fenner (1923) and Hildreth (1987) for discussion of the eruption in 1912 that formed the Valley-of-Ten-Thousand Smokes in Alaska; Christiansen (1984) for a discussion of the history of the large structures in Yellowstone; and Smith (1979) for a discussion of ash flows characteristics and processes]. Reconstruction of deep-seated eruption dynamics at maars in the Hopi Buttes motivated Shoemaker and others (1962) to propose the first shock-tube analogy for volcanic processes, an idea developed quantitatively by McGetchin (1968) in a study of another deep, old structure called the Moses Rock diatreme in Utah, U.S.A. Space precludes examination of these larger structures in this paper.

FIGURE 12. The vertical plinian column that erupted from Mount St. Helens, May 18, 1980. Photo by Austin Post, U.S. Geological Survey. The summit diameter of the mountain from which the plume emerges is approximately 1 km diameter; the vent feeding the plume is probably much smaller (~100 m) as inferred from the size of the initial magma domes that emerged from the vent in June through August, 1980, and from calculations of mass flux by Carey and Sigurdsson (1985).

In summary, these examples illustrate that volcanologists and geologists who study hydrothermal eruptions or flow through geothermal wells use the adjective "explosive" quite broadly in descriptions of eruptions of varying duration, steadiness, and scale. In the next Section, the nature of volcanic and geothermal fluids will be examined, and the dynamics of their eruption from a variety of reservoirs will be summarized in more quantitative terms. It will be shown that no additional energy (such as chemical or nuclear detonations, or excess pressure-volume decompression) beyond that available from conversion of available enthalpy is required to explain observed and inferred properties of volcanic flow fields.

3. THE NATURE OF VOLCANIC AND GEOTHERMAL FLUIDS

3.1. Water--Thermodynamic properties of a ubiquitous volcanic fluid.

The ubiquitous fluid of volcanic and geothermal relevance is water. It occurs in nearly its purest natural form (i.e., not containing large amounts of dissolved gases or solids) in the Upper Basin of Yellowstone National Park. Therefore, analysis of Old Faithful Geyser provides a basis for an understanding the dynamics of eruption of simple fluids. However, the complexities that result from changes of phase as an eruption cycle proceeds (see Figure 13) make quantitative modeling of even the simplest geysers difficult.

In analysis of complicated geologic processes, many simplifications must be made. An important consideration is whether the fluid-dynamic history can be considered to be adiabatic, isobaric, isothermal, isenthalpic, or isentropic. Because of the rapid ascent of material in volcanic vents (for example, fluid probably traverses the 20 m long conduit of Old Faithful in a few tenths of a second), flow is often approximated as adiabatic. For the same reason, if there are no shock waves, and if friction is negligible, the flow is also often approximated as isentropic to simplify the analysis. Computer models allow relaxation of these assumptions (Wilson and others, 1980; Wohletz and others, 1984). However, a simple temperature-entropy (TS) diagram is very informative for illustration of the range of phase changes that can occur in H_2O under volcanic conditions.

Two styles of eruption can be defined, depending where the thermodynamic path lies relative to the critical point of H_2O: (a) "low-entropy" eruptions, in which the initial entropy is less than the critical point entropy. In such eruptions, fluids that start as a liquid or supercritical fluid and follow paths with $dS \geq 0$ produce a vapor phase (such a path is shown for Old Faithful in Figure 13), and (b) "high-entropy" eruptions, in which a fluid that starts as a vapor or supercritical fluid and follows a path with $dS \geq 0$ can produce liquid. In both cases, a significant entropy increase of the fluid at high enough temperature and pressure conditions--as might be caused by heat transfer from conduit walls or from entrained fragments--can cause the two-phase field to be by-passed. An example of the thermodynamic path of H_2O decompressing from 1000°C along an isentropic path (vertical), and along other paths determined by heat transfer from entrained ash particles (discussed later) is shown in Figure 13.

3.2. An example of the influence of the sound speed of water and its different phases on conduit resonances.

To a fluid dynamicist, sound speed is an important parameter--perhaps the most important single parameter--in analyses of "explosive" volcanism, and it is one of the most poorly known (and poorly defined) quantities for the fluids with which volcanologists must work. Sound speed is such an important parameter because the relation of fluid velocity to sound speed determines the flow regime, the stability of standing waves, and the change of the first-order thermodynamic parameters along the path of a conduit (for example, Thompson (1972, pp. 278) concisely contrasts variations of velocity and pressure in converging and diverging channels under subsonic and supersonic flow conditions).

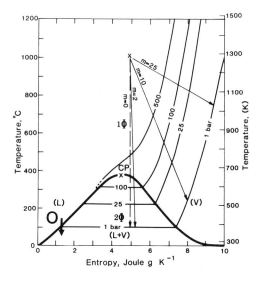

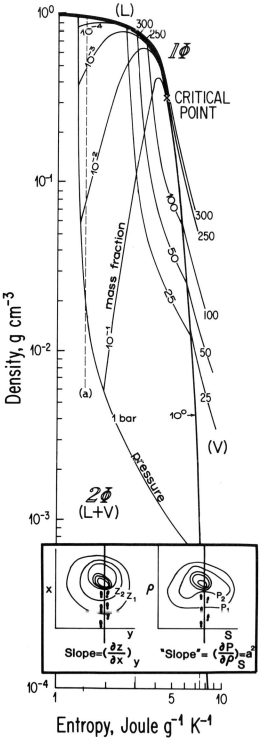

FIGURE 13 (above). Temperature-entropy diagram for H_2O, showing various decompression paths discussed in the text. The one-phase and two-phase fields are indicated by 1Φ and 2Φ, respectively, with (L) indicating liquid and (V) indicating vapor. On the left, decompression of liquid water originating at the base of Old Faithful's conduit at 116°C (2.3 bars) to a two-phase mixture at 93°C, 0.8 bar is shown by the path O. On the right are a series of lines that show the thermodynamic paths of the vapor phase when mixtures containing different mass ratios (m) of solids are decompressed isentropically. The initial condition of T = 1000° is chosen to represent a typical hot magma. Transfer of heat from the entrained mass causes the vapor entropy to increase as shown.

FIGURE 14 (left). Entropy-density relations for H_2O to illustrate variations of sound speed across the phase diagram. Entropy is relative to a triple-point entropy, S (triple-point)=0. Data are from Keenan and others (1969) steam tables. The single-phase field is shown as 1Φ (with L for liquid, V for vapor), and the two-phase (liquid+vapor, L+V) field as 2Φ. Contours of constant pressure (isobars) are shown in 25-bar increments. In the two-phase field, contours of constant mass fraction vapor (isopleths) are shown. Simplified from Kieffer and Delany (1979).

The variation of sound speed (c) for pure H_2O over the range of pressure (P), temperature (T), and compositions with which volcanologists must work is shown graphically on density-entropy (ρS) diagram in Figure 14. On this diagram, sound speed, $c^2=(\partial P/\partial \rho)_s^{1/2}$, is the directional derivative along vertical lines of constant entropy. Thus, the sound speed is inversely proportional to the vertical gradient of the isobaric contours and, as illustrated by the inset, by viewing these diagrams as "topographic maps", the reader can visualize the changes of sound speed with pressure, temperature, and mass-fraction vapor. The change of isobar spacing graphically shows that the sound speed of water in its different phases and mixtures can vary by three orders of magnitude [see Kieffer (1977) for detailed calculations; and Kieffer and Delany, 1979 for details of ρS diagrams].

As an example of the influence of fluid sound speed on properties of interest to geologists, consider the problem of the resonant longitudinal frequencies of a conduit of length L filled with a fluid, as illustrated in Figure 15. From seismic records around geothermal plumbing systems or volcanoes, we can often measure a resonant frequency, and would like to be able to invert the data to determine source dimensions. In Figure 15, three possible states of the fluid are depicted on the left (reproduced from Figure 2b). If a conduit containing a fluid is perturbed, it will--to a first approximateion-- resonate at a characteristic frequency, f_{long} given by simple open-ended or closed-ended organ pipe formulas shown at the top of Figure 15. The resonance frequency changes as the sound speed changes, and, because the sound speed can change by two or three orders of magnitude as an undersaturated fluid changes to a two-phase mixture, the resonant frequencies can change by the same factor. In volcanic situations, we can measure resonant frequencies by seismic techniques, but rarely have any direct measurement of the conduit dimensions (L) or the physical state of the fluid. Therefore, the large variation of fluid sound speeds that can occur leaves an ambiguity in interpretation of the measured frequencies--sound speed changes and conduit dimension changes cannout be decoupled easily. A detailed application of this theory to interpretation of geothermal noise at Old Faithful can be found in Kieffer (1984a), and to volcanic tremor in Leet (1988).

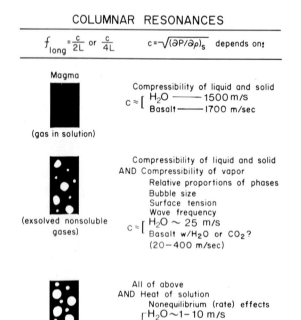

COLUMNAR RESONANCES

$f_{long} = \frac{c}{2L}$ or $\frac{c}{4L}$ $c = \sqrt{(\partial P/\partial \rho)_s}$ depends on:

Magma

(gas in solution)

Compressibility of liquid and solid
$$c \approx \left[\begin{array}{l} H_2O \longrightarrow 1500 \text{ m/s} \\ Basalt \longrightarrow 1700 \text{ m/sec} \end{array} \right.$$

(exsolved nonsoluble gases)

Compressibility of liquid and solid
AND Compressibility of vapor
   Relative proportions of phases
   Bubble size
   Surface tension
   Wave frequency
$$c \approx \left[\begin{array}{l} H_2O \sim 25 \text{ m/s} \\ Basalt \text{ w/}H_2O \text{ or } CO_2? \end{array} \right.$$
(20–400 m/sec)

(exsolved soluble gases)

All of above
AND Heat of solution
   Nonequilibrium (rate) effects
$$c \approx \left[\begin{array}{l} H_2O \sim 1-10 \text{ m/s} \\ Basalt \text{ w/}H_2O? \\ Basalt \text{ w/}CO_2? \end{array} \right.$$

FIGURE 15. Schematic illustration of a column containing a fluid (left), its resonant longitudinal frequencies, f, (top), and the effect of change of sound speed with state of the fluid in the column (right).

3.3. Dynamics of eruption of Old Faithful.

After an eruption, the conduit of Old Faithful is empty of fluid, and recharges slowly with water and heat over an interval of very approximately 60 minutes. Preplay begins about 30 minutes before an eruption. During recharge, the water in the conduit becomes, on the average, hotter. However, most of the water in the conduit is below the reference boiling curve for pure H_2O during the recharge cycle and even as the eruption is initiated, as shown in Figure 16a (Birch and Kennedy, 1972). Only the top few meters of water are at reference boiling curve conditions (Figure 16a, the shaded area). Each jet of preplay probably is erupted from this top few meters of the conduit fluid. The fluid-flow might be rather similar to that shown in Figure 3 for Strŏkkur Geyser, and the thermodynamics similar to that discussed in Section 2 for Narcissus Geyser. This part of the eruption simply cannot be documented in detail for Old Faithful because the water surface lies out of sight about 5 m below the ground surface.

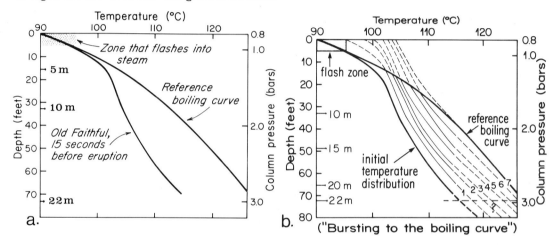

FIGURE 16. (a) Illustration of pressure-depth-temperature relations in Old Faithful Geyser. The initial temperature distribution is indicated; note that below a few meters depth the temperature is everywhere below the reference boiling curve for pure liquid water. (b) Curves (labelled 1-7) represent pressure-temperature distributions after successive bursts have unloaded water from the conduit during unsteady flow--the choice of 7 bursts is arbitrary (from Kieffer, 1984a).

Explanation of processes following preplay becomes difficult because of the complex relation of the fluid P-T conditions to the reference boiling curve conditions, Figure 16a. The fluid is volatile, but everywhere lies in a different relation to its local saturation curve. The problem is that of unsteady two-phase flow from a vertical shock tube of complex dimensions in a gravitational field; a semi-quantitative model for the unsteady flow is given in Kieffer (1984a) and cannot be repeated here because of length constraints.

During steady flow, a two-phase fluid flows from the conduit at a rate estimated at more than 6.8 cubic meters/sec (1800 gallons per minute) (E. Robertson, pers. commun., 1977). During this part of the eruption, the hottest water at 116°C decompresses to 93°C. If this occurs isentropically, 4 weight percent of the liquid is converted to vapor. An enthalpy decrease of 3.89 kj/kg is available for kinetic energy, theoretically sufficient to accelerate the fluid to a velocity of 88 m/s. Inversion of the observed height of the geyser during the steady phase of the eruption (50 m maximum height) to a vent velocity using the theory of negatively buoyant plumes (Turner, 1966) gives a vent velocity of 78 m/s, a value in surprisingly good agreement with the velocity predicted simply from the enthalpy change of the fluid.

There are many theories and data sets regarding the problem of two-phase flow under relatively well-controlled laboratory conditions, particularly regarding the problem of choking of the flow. Several of these arguments and empirical formulas derived from studies suggest that the flow from Old Faithful is choked during steady flow (Kieffer, 1988, in press). The simplest argument to summarize is the following: It is likely that choking takes place at a constriction about 5 m below the exit plane (Kieffer, Hutchinson, and Westphal, unpublished measurements of diameter). Assume that the fluid is at about 0.1 MPa at this point. The fluid would be 3 weight per cent vapor, and would have a density of 19.7 kg/m^3. The estimated area of the choke plane is 0.15 m^2. The equilibrium sound speed for a mixture of the calculated composition is 45 m/s. Using these numbers, the calculated mass flux is 132 kg/s (2100 gal/min), in satisfactorily close agreement with the estimated discharge of 114 kg/s (1800 gal/min) to suggest that the flow is choked.

At the exit plane, if the liquid and vapor were in equilibrium, the mixture would be 4 weight percent vapor, which has an equilibrium sound speed of about 54 m/s. With an exit velocity of approximately 78 m/s, the implied Mach number at the exit plane is ~1.5, barely supersonic given the uncertainties of measurement, modeling, and definition of sound speed in two-phase flow. Thus, Old Faithful is probably a weakly supersonic, two-phase, pressure-balanced jet (compare with geothermal bore photos, Figure 6). Most geysers and volcanoes on the Earth will not be highly underexpanded and supersonic (like the geothermal well in Figure 6) because rocks are so weak that any overpressure and lateral expansion in a plume will erode divergent Sections into the top of conduits (Wilson and others, 1980; Kieffer, 1984(b)). An exception to this generality occurs when sectors of a mountain collapse and rapidly reduce the pressure on volatile material within the mountain, as will be discussed in Section 4.3. In preparation for this discussion, I now turn away from the "relatively simple" problem of two-phase liquid-vapor flow and turn to the problem of dusty gas flow in nature.

4. DUSTY GASES AND WORSE

4.1. Pseudogases.

Most volcanoes, of course, do not erupt a fluid as simple as pure H$_2$O. In the early phases of the Mount St. Helens eruptions of 1980, when the eruptions were phreatic and very geyser-like (Figures 4 and 10), the vapor phase was H$_2$O, but it was (relatively) lightly laden with rock and ash fragments. When the eruption changed into the lateral blast, (Figure 11), the fluid was a mixture of gas, volcanic rock, and magma in the explosive state. When it changed to a purely magmatic phase (the plinian eruption shown in Figure 12), the mixture was inferred to be 4.6 weight percent water (Rutherford and others, 1985; Carey and Sigurdsson, 1985). As discussed in Section 1, regardless of the origin of the volatiles, fluid dynamic models currently specify only mass ratios (m) of solid (or molten droplets, ash,...) to vapor components and treat these fluids as pseudogases (homogeneous fluids with heat transfer accounted for by modification of the isentropic exponent γ in the perfect gas law, and fluid density accounted for by modification of the gas constant R (see top of Figure 17). For relatively dilute mixtures (say, with mass ratios of solid to vapor phases (m) of order 1), a thermodynamic model of a volcanic equation-of-state probably does not badly err by using conventional dusty-gas models for the equation of state and sound speeds (Figure 17).

However, in using pseudogas or even more complicated dusty gas theory for analysis of volcanic and geothermal flow problems, we are typically stretching the assumptions of pseudogas theory--for example, the value of m for the juvenile magma at Mount St. Helens was about 20. As another example, Figure 18 shows some of the "particulate" material trapped by a vehicle engulfed by the lateral blast of May 18, 1980: representation of whatever multicomponent fluid carried this debris as a homogeneous pseudogas is clearly an oversimplification! The problems of multicomponent equations of state and descriptions of rheology are ubiquitous in continuum descriptions of the properties of geologic fluids [for example, magma rheology changes dramatically if the magma contains enough bubbles and crystals that they interact (Shaw and others, 1968)]. Descriptions of densely-laden fluids, such as the mixture that may have produced the deposit shown in Figure 18, and of complex multicomponent, multiphase non-dilute fluids are needed for many geologic problems.

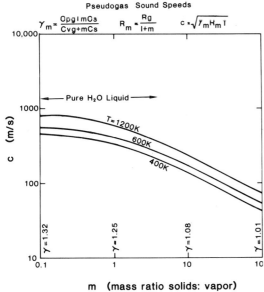

Pseudogas Sound Speeds

$$\gamma_m = \frac{Cpg\,|m Cs}{Cvg + mCs} \qquad R_m = \frac{Rg}{1+m} \qquad c = \sqrt{\gamma_m H_m T}$$

FIGURE 17. Dependence of sound speed on mass loading for typical volcanic fluids. Symbols in the top lines are: γ_m, the isentropic exponent of the mixture; c_{pg}, c_{vg}, and c_s, the heat capacities respectively at constant pressure and constant volume of the gas phase, and of the entrained solid phase; R_m and R_g are the gas constants of the mixture and of the gas phase respectively; T is temperature in Kelvin.

FIGURE 18. Tractor with sample of debris trapped from the lateral blast. Note that trapped debris ranges in size from large logs to fine dust. Initial temperatures of this debris may have ranged from about 20°C for the trees to about 700°C for the volcanic component of the ash and rocks. This makes it very difficult to apply pseudogas theory to materials like this with any confidence!

The difference in thermodynamic history of the vapor phase between an eruption of pure water and water containing hot entrained ash is shown on the T-S diagram of Figure 13. The heat transferred to the steam phase increases its entropy, and causes the eruption to approach more nearly isothermal conditions than if H_2O alone is the driving volatile. It is this heat transfer that allows volcanic eruptions containing entrained hot materials to attain such high velocities. The pseudogas theory with quasi-one-dimensional flow equations (or two-dimensional inviscid flow for lateral blasts) has been the foundation of volcanic fluid dynamics for the last decade (Wilson, and others, 1987).

4.2. Applications of pseudogas theory in steady-state flow.

As an example of the use of pseudo-gas and multiphase flow theory as currently used in volcanology, consider numerical models of the phases of the May 18, 1980 lateral blast at Mount St. Helens. The steady flow of the plinian column and its relation to the dacitic magma properties were examined by Carey and Sigurdsson (1985) and a brief comparison of their methods and treatment with the above summary of the steady flow of Old Faithful will demonstrate the analogy between flow of a volatile liquid and magma at shallow levels. One substantial difference between the two analyses is that the fluid-mechanical model proposed for Old Faithful allows the flow to become supersonic, whereas many current models for steady flow in volcanic systems constrain the flow to be sonic or subsonic at the exit plane (further discussion of this point can be found in Kieffer, 1984b).

As discussed above, the fluid in Old Faithful is driven to velocities of the order of 70 m/s by decompression from 116°C (2.3 bars pressure) to 93°C (0.7 bars pressure). This decompression (if in equilibrium) produces 4 weight per cent vapor; this vapor drives the eruption. Rutherford and others (1985) analyzed the dacite from the Plinian eruption of Mount St. Helens and found 4.6 weight percent dissolved volatiles, so, to first order the volatile content driving an eruption of Old Faithful and the Plinian column at Mount St. Helens were identical. The major differences in the two eruptions are that the conduit pressure at which the magma of Mount St. Helens became "explosive" was calculated to be 160 bars (Carey and Sigurdsson, 1985), the temperature of the fluid was greater (about 900°), and the vent was larger (order of 100 m). With these differences accounted for, the calculated steady-state fluid velocity was 200-300 m/s, and the calculated mass flux was about 2×10^7 kg/s. Thus, although there are differences in the equations of motion and boundary conditions used in modeling Old Faithful and volcanic Plinian columns, this comparison shows that conceptually there is an analogue between steady geyser flow of a simple volatile liquid and magma when it is its gassy phase.

4.3. The lateral blast at Mount St. Helens.

Kieffer (1981a,b; 1982; 1984b) formulated a supersonic flow model to explain the pattern of tree blow-down (Figure 19) within the devastated area of Mount St. Helens by using two-dimensional supersonic flow theory and a pseudo-gas equation-of-state (Figure 17).

The devastated area extended amazingly far east and west considering the fact that the landslide exposed a scarp of hot rock and fluids that faced almost due north. One challenging problem was to explain the shape of the devastated area (Figure 19), and at least the generalities of tree blowdown pattern indicated in the figure. In order to model the blast, the fluid flow was assumed to have reached quasi-steady flow conditions. In order to compare the model calculations with observed field evidence (shape and size of the devastated area, tree blowdown pattern), it is assumed that these characteristics were imprinted on the landscape during this time. The lateral blast actually consisted of a waxing phase in which the existing landscape was severely eroded (see Kieffer and Sturtevant, in press, 1988) and a waning phase in which a complex set of deposits were laid across the devastated area (Lipman and Mullineaux, 1981). The geologic complexities were greatly simplified for a fluid flow analysis; none-the-less, the dimensions, shape, and tree-blowdown patterns are mimicked rather remarkably by simple models of steady-state nozzle dynamics.

The simplified model and assumptions are illustrated in Figure 20. It was assumed that the mountain contained a complex mass of magma, hydrothermal rock, steam and water on the morning of May 18, just before the lateral blast began, and that the pressure in the mountain was lithostatic at an average value of 125 bars (12.5 MPa)-- appropriate to an average depth of 650 m with overlying (fractured) rock and ice of average density 2000 kg/m^3. The eruption was triggered by the failure of the north flank, resulting in three closely spaced landslides (Figure 20). These landslides exposed the hot rock and volatile material suddenly to atmospheric pressure (0.87 bars, 0.087 MPa). The initial average temperature was assumed to have been 600 K, and the initial density to have been 1000 kg/m^3. The vent was assumed to be 1 km in east-west dimensions and 0.25 km in vertical dimension, and 0.5 km deep (to give the volume of ejecta measured as having been erupted).

The two-dimensional equations of axisymmetric flow from the vent were solved by two methods: (a) hand calculations using the method of characteristics with empirical relations from JANNAF (1975) to define the flow boundary, results shown in Figure 21, and (b) a numerical calculation of the inviscid core of the flow only (this figure can be found in Kieffer, 1984b).

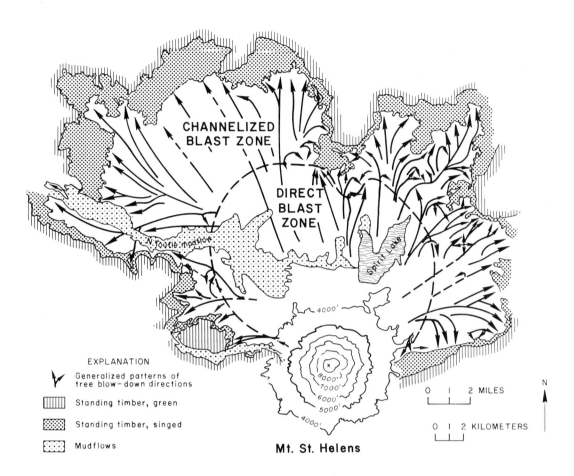

FIGURE 19.   A simplified map of the devastated area and flow streamlines (arrowed vectors) as they were indicated by the alignment of fallen trees. For simplicity, the devastated area has been divided into three regions: (1) an internal <u>direct blast zone</u> in which the direction of blowdown of the trees is relatively independent of large-scale topographic features; (2) an outer <u>channelized blast zone</u> where the direction of tree blowdown is strongly influenced by topography; and (3) the <u>singed zone</u> where trees were still standing, but were singed.

The resulting model (Figure 21) suggests that the inner "direct blast zone" was the result of exposure to supersonic flow conditions, and that the outer "channelized blast zone" was beyond the Mach disk shock was the result of exposure to subsonic flow. Although gravity was not included in the compressible flow equations used for the model, field evidence suggests that gravity became the dominant force driving the flow in the channelized blast zone.   At the edge of the channelized zone, the erupting fluid had decompressed sufficiently, and had dropped enough of its suspended particulate material, that it became buoyant and lifted from the land surface (Kieffer, 1981b; Sparks and others, 1986).

Because the reservoir was at such high pressure compared with ambient pressure, the flow would have choked at the vent at a pressure of about 7.5 MPa (75 bars), with a choke (sonic) velocity of 104 m/s--the observed initial velocity of the blast was measured at 100 m/s, so the theory is in good agreement with near-source observations. Upon leaving the vent at sonic velocity, the material, driven by the 75 bars vent pressure, would have expanded supersonically into the lower pressure atmosphere, with a Prandtl-Meyer expansion angle calculated to be 96°. This Prandtl-Meyer expansion from the north toward both the west and east caused the devastation of the northern sector of the volcano as the flow diverged. Note the good match between the model expansion and the boundary of the devastated area close to the volcano.

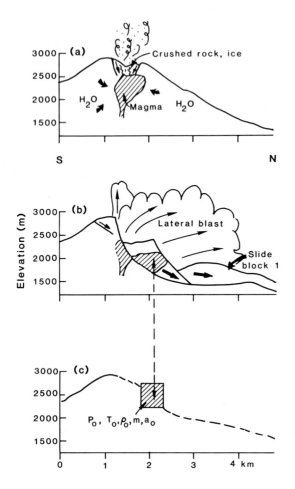

FIGURE 20. Initial conditions assumed for the model of the Mount St. Helens lateral blast. (a) Schematic cross-Section of the mountain April-May 18, 1980, prior to the lateral blast. (b) Schematic diagram of the initiation of the blast by the failure of the north flank and the formation of three landslide blocks. (c) Simplified model of the source of the blast material. This material is assumed to have been confined in a reservoir and to have accelerated to choked (Mach 1) conditions at the vent. For computational purposes, the exit Mach number of the flow was taken as 1.02.

The pressure decreased rapidly away from the vent, dropping below ambient atmospheric pressure outside a zone about 6 vent diameters from the source (the zone of sub-atmospheric pressure is shown as the stippled pattern in Figure 21). Beyond this, at a distance of about 11 km, a Mach disk shock brought the flow back from supersonic to subsonic conditions: no calculations have been done beyond this zone, and it is likely that the flow attained nearly atmospheric pressure conditions after passage through the Mach disk shock.

The coincidence of the calculated supersonic flow field with the direct blast area led the author to speculate that the direct blast zone near the volcano had been subjected to supersonic flow. It is worth noting that the flow was internally supersonic--the flow front moved at a nearly constant velocity of 100 m/s over a distance of 20 km, whereas calculated internal flow velocities are as high as 300 m/s within the supersonic zone. The distinction between flow front velocities and internal flow velocities in supersonic flow is very important for volcanologists to remember, because photographs or satellite observations reveal only flow-front velocities, whereas the record of erosion and sedimentation on the ground is an accumulated record of waxing and waning stages. Kieffer and Sturtevant (1984) demonstrated that flow front velocities depend on the relative densities of the jet and the surrounding atmosphere. A dense jet entering a relatively low-density atmosphere, e.g., the lateral blast, will maintain nearly sonic velocity for large distances because the effect of entrainment is small. The high particle velocities within the supersonic flow zone are reduced to the flow-head velocity by the Mach disk and lateral intercepting shocks, and by further expansion and entrainment. Other relations between the model and observations are given in detail in Kieffer (1981a,b).

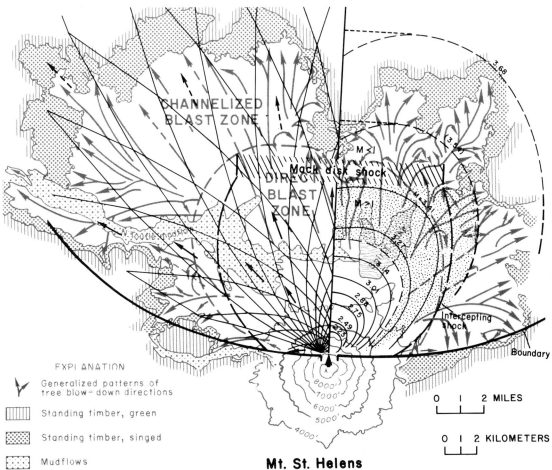

Mt. St. Helens

FIGURE 21. Results of model calculations superposed on map of the devastated area. Results of hand-calculations discussed in text with JANNAF criteria for shape of boundary. The differences between this model and the computer model discussed in Kieffer (1984b) are not significant at the scale of the geologic uncertainties and model simplifications. All length dimensions, x and y, are normalized to vent diameter, d. Contours of constant thermodynamic properties are shown, and their exact values can be found in Kieffer (1981b and 1984b) for the two models respectively. Small arrows indicate calculated local flow directions. The flow diverges through rarefaction waves (not shown, see Kieffer, 1981b for details) that reflect from the flow boundaries to form the intercepting shocks shown. Across these shocks the flow velocities decrease, the pressure increases, and the flow changes direction. The intercepting shocks coalesce across the flow to form the strong Mach disk shock. In most of the flow, the pressure is subatmospheric (stippled zone). It rises back toward atmospheric pressure across the Mach disk and intercepting shocks. The location of the Mach disk shock is not predicted by the numerical procedures; it is estimated from the JANNAF (1975) equations. Note the approximate coincidence of the supersonic part of the supersonic part of the jet and the direct blast zone.

It is difficult to convey the magnitude of the lateral blast (which, by scale of the geologically conceivable explosive eruptions mentioned in Section 2 was a very small volcanic eruption), so a comparison with one of the more impressive displays of power created by humans may be helpful.

Imagine that the rockets of a Saturn 5 are pointed at you, as shown in Figure 22. These rockets, which carried Apollo astronauts to the Moon, consisted of five F-1 liquid-oxygen/kerosene motors. The mass/flux area at the exit of an F-1 motor is about 25 $g/s/cm^2$; that of the lateral blast at the Mount St. Helens vent was 240 times as great. The power per unit area of the F-1 motors was approximately 0.8 $Mwatt/cm^2$; that of the lateral blast was three times greater. The saturn 5 power was delivered over five rockets covering roughly 50 m^2; the power at Mount St. Helens flowed out of a vent more than 2,000 times this area. (Thus, the reader might now try to imagine 2000 Saturn V engines firing simultneously over 180° of arc!)

FIGURE 22. Five F-1 engines on a Saturn 5 rocket.

The total power of the five Saturn 5 motors was about 4×10^5 megawatts; that of the blast was nearly 16,000 times as great. The thrust of the Saturn 5 was 7.5 million pounds (3.3×10^7 N) that of the blast was 10^5 greater. According to this model, which presumeably represents the peak flow conditions during the blast (rather than a truly steady flow condition), the mass flux was about 10^4 grams per second per square centimeter ($g/s/cm^2$). The thermal flux or power per unit area was 2.5 megawatts per square centimeter ($Mwatts/cm^2$). The total energy was 24 Mt, of which 7 Mt was dissipated during the blast itself, and the remaining 17 Mt was dissipated during the almost simultaneous condensation of the steam in the blast and subsequent cooling of steam and rock to ambient temperature in the weeks following May 18.

5. CONCLUSIONS REGARDING THE TERM "EXPLOSIVE VOLCANISM"

The intent of this paper has been to summarize our current understanding of "explosive volcanism" in a way that permits an appreciation of the fundamental principles of flow involving gassy multiphase mixtures in which enthalpy of a fluid under relatively modest pressure is converted into kinetic energy of an expanding flow field upon initiation of an eruption. When there are hot solid or liquid fragments suspended in the vapor phase, the enthalpy available is enormous. Given that geologic constraints cannot uniquely define the source parameters and internal flow conditions, there will always be uncertainties in our knowledge of the details of the thermodynamics and flow field properties. In addition to this, there is always the frustration that geologists can rarely, if ever, do a full-scale laboratory experiment in order to resolve those controversies that inevitably arise when scientific theories must be based on incomplete data. Nevertheless, models of the types discussed here have worked remarkably well in the reconstruction of likely fluid flow processes at Mount St. Helens and in geysers, and if such models can be made available for predicting the potential consequences of eruptions at other active volcanoes of the world, strategies of hazards mitigation might achieve a more quantitative basis. In the future development of these models, it is important that the fluid dynamics community help us refine our equations-of-state, our models of gas-particle interaction, our problems with erodible boundary layers, and with the equations of momentum which must be solved for the problem of compressible flow in a gravitational field. There are many important and interesting volcanologic problems awaiting both laboratory and theoretical work by fluid dynamicists.

ACKNOWLEDGMENTS

I thank E.M. Shoemaker for many discussions empahsizing the need to consider carefully the use of the word "explosive" in conjunction with volcanic processes, H. R. Shaw for calling my attention to the early work of Daly and the context in which it should be read and both H.R. Shaw and H. Sigurdsson for generous and thoughtful reviews of the original manuscript.

REFERENCES

------, American Geological Institute, Dictionary of Geological Terms, (Anchor Books,Garden City, 1974), 545pp.

Banister, J.R., Jour. of Geophys. Res., **89**, 4895-4904 (1984).

Birch, F., and Kennedy, G.C., Geophys. Monographs, **16**, 329-336 (1972).

Carey, Steven, and Sigurdsson, Haraldur, Jour. of Geophys. Res., **90**, 2948-2958, (1985).

Christiansen, R.L., in Explosive Volcanism: Inception, Evolution and Hazards, (National Academy Press, Washington, D.C., 1984), pp. 84-95.

Colgate, S.A., and Sigurgeirsson, T., Nature, **244**, 552-555 (1973).

Daly, R.A., Proc. of the Amer. Acad. of Arts and Sci., **XLVII(3)**, 47-122 & 5 plates (1911).

Decker, R.W., and Hadikusumo, D., Jour. of Geophys. Res., **66**, 3497-3511 (1961).

Eichelberger, J.C., and Hayes, D.B., Jour. of Geophys. Res., **87**, 7727-7738 (1982).

Fenner, C.N., National Geogr. Soc. Contributed Tech. Papers, Katmai Series, 1 (1923).

Fudali, R.F., and Melson, W., Bull. Volcanologique, **35**, 383-401 (1972).

Gorshkov, G.S., Bull. Volcanologique, **20**, 77-112 (1959).

Gorshkov, G.S., Bull. Volcanologique, **23**, 141-144 (1960).

Hildreth, Wes, Bull. of Volcanology, **49**, 680-693 (1987).

JANNAF [Joint Army, Navy, NASA, Air Force], JANNAF Handbook of Rocket Exhaust Plume Technology (Chem. Prop. Inform. Agency Publ. 263, Chapter 2, 1975), 237 pp.

Kanamori, H., and Given, J.W., Jour. of Geophys. Res., **87**, 5422-5432 (1982).

Kanamori, Hiroo, Given, J.W., and Lay, Thorne, Jour. Geophys. Res., **89**, 1856-1866 (1984).

Keenan, J.H., Keyes, F.G., Hill, P.G., and Moore, J.G., Steam Tables, (John Wiley, New York, 1969), 196 pp.

Kieffer, S.W., Jour. of Geophys. Res., **82**, 2895-2904 (1977).

Kieffer, S.W., Nature, **291**, 568-570 (1981a).

Kieffer, S.W., in: The 1980 Eruptions of Mount St. Helens, Washington, Lipman, P.W. and Mullineaux, D.R. (eds.), U.S. Geological Survey Professional Paper 1250 (U.S.Government Printing Office, Washington, 1981b), pp. 379-400.

Kieffer, S.W., in: The Satellites of Jupiter, D. Morrison, Ed. (Univer. of Arizona Press Tucson, 1982), pp. 647-723.

Kieffer, S.W., Jour. of Volcanology and Geothermal Res., **22**, 59-95 (1984a).

Kieffer, S.W., in Explosive Volcanism: Inception, Evolution and Hazards, (National Academy Press, Washington, D.C., 1984b), pp.143-157.

Kieffer, S.W., Geologic Nozzles, in press Spec. Lectures in Phys., Springer-Verlag, 1988.

Kieffer, S.W. and Delany, J.M., Jour. of Geophys. Res., **84**, 1611-1620 (1979).

Kieffer, S.W., and Sturtevant, Bradford, Jour. Geophys. Res., **89**, 8253-8268 (1984).

Kieffer, S.W., and Sturtevant, Bradford, Jour. of Geophys. Res., in press (1988).

Leet, R., Journal of Geophys. Res., **93**, 4835-4849 (1988)

Lipman, P.W., and Mullineaux, D.W., Eds. The 1980 Eruptions of Mount St. Helens, Washington, U.S. Geological Survey Professional Paper 1250 (U.S. Government Printing Press, Washington, D.C., 1981), 844 pages + 1 map.

Lorenz, V., Chemical Geology, **62**, 149-156 (1987).

Marsh, Bruce D., in Explosive Volcanism: Inception, Evolution, and Hazards, (Geophysics Study Committee, National Academy Press, Washington, D.C., 1984), pp. 67-83.

Matuzawa, T., Bull. Earthquake Res. Inst. Tokyo, 11, (a,b,c) on pp. 329, 347, 732 (1933).

McGetchin, T.R., Ph.D. thesis, Calif. Instit. of Technol., Pasadena (1968).

Melson, W.G., and Saenz, R., Bull. Volc., **37**, 416-437 (1973).

Muffler, L.J.P., White, D.E., and Truesdell, A.H., Geol. Soc. Amer. Bull., **82**, 723-740 (1971).

Perret, F.A., Carnegie Institution of Washington Publication 339, 151 pp. (1924).

Reed, J.W., Jour. of Geophys. Res., **92**, 11,979-11,992 (1987).

Richards, A.F., Nature, **207**, 1382-1383 (1965).

Rinehart, J.S., EOS (Trans. of the American Geophys. Union), **52**, 1052-1062 (1974).

Rutherford, M.J., Sigurdsson, H., Carey, S., and Davis, A., Jour. Geophys. Res., **90**, 2929-2947 (1985).

Self, Stephen, Wilson, Lionel, and Nairn, I.A., Nature, **277**, 440-443 (1979).

Shaw, H.R., in Physics of Magmatic Processes, R.B. Hargraves (ed.) (Princeton Univ. Press, Princeton, N.J., 1980), pp. 201-264.

Shaw, H.R., Wright, T.L., Peck, D.L., and Okamura, R., Amer. Jour. Sci., **266**, 255-264 (1968).

Shoemaker, E.M., Roach, C.H., and Byers, F.M., Jr., in Petrologic Studies, A Volume to Honor A.F. Buddington, (Geological Society of America, Boulder, Colorado, 1962), pp. 327-355.

Smith, R.L., Geological Society of America Special Paper 180, pp. 5-27 (1979).

Sparks, R.S.J., J. Volcanol. Geotherm. Res., **3**, 1-37 (1978)

Sparks, R.S.J., Bull. of Volcanology, **48**, 3 16 (1987).

Sparks, R.S.J., Moore, J.G., and Rice, C.J., Jour. Volc. Geoth. Res., **28**, 257-274 (1986).

Sparks, R.S.J., and Wilson, L., Geophys. Jour. Royal Astron. Soc., **69**, 551-570 (1982).

Thompson, P.A., Compressible-Fluid Dynamics (McGraw Hill, New York, 1972), 665 p.

Turner, J.S., Jour. of Fluid Mechanics, **26**, 779-792 (1966).

White, D.E., American Jour. of Science, **265**, 641-684 (1967).

White, D.E., Fournier, R.O., Muffler, L.J.P., and Truesdell, A.H., U.S. Geological Sur vey Professional Paper 892, (United States Gov. Printing Office, Washington, D.C., 1975), 70 pp.

Whitney, J.A., and Stormer, J.C., Science, **231**, 483-485 (1986).

Williams, Howell, and McBirney, A.R., Volcanology (Freeman, Cooper and Co., San Francisco, 1979), 391 pp.

Wilson, Lionel, Geophys. Jour. of the Royal Astronomical Society, **30**, 381-392 (1972).

Wilson, Lionel, Geophys. Jour. Royal Astronomical Society, **45**, 543-556 (1976).

Wilson, Lionel, Sparks, R.S.J., and Walker, G.P.L., Geophys. Jour. of the Ro yal Astronomical Society, **63**, 117-148 (1980).

Wilson, Lionel, Pinkerton, H., and Macdonald, R., Ann. Rev. Earth and Planet. Sci.. **15**, 73- 95 (1987).

Wohletz, K.H., McGetchin, T.R., Sandford, M.T.,II, and Jones, E.M., Jour. Geoph. Res., **89**, 8269-8286 (1984).

Wohletz, K.H., and McQueen, R.G., in Explosive Volcanism: Inception, Evolution and Hazards, (National Academy Press, Washington, D.C., 1984), pp. 158-169.

Theoretical and Applied Mechanics
P. Germain, M. Piau and D. Caillerie (Editors)
Elsevier Science Publishers B.V. (North-Holland)
© IUTAM, 1989

SEPARATION BUBBLES

M. KIYA

Department of Mechanical Engineering
Hokkaido University
Sapporo, 060, Japan

A separation bubble is a recirculating region of flow bounded by a separated shear layer and a solid surface, being characterized by rolled-up vortices in the shear layer and their interaction with the surface. This paper reviews the properties of those turbulent separation bubbles in low-subsonic regime which are nominally two-dimensional or axisymmetrical. Discussions are made mainly based on experimental results obtained for the separation bubble at the square leading edge of a long blunt plate. The properties of this particular flow are likely to be basically similar to those of a wide range of the separation bubbles.

1. INTRODUCTION

Separation bubbles are a region of recirculating flow which is bounded by a separated shear layer and a solid surface. At high Reynolds numbers they are turbulent, and characterized by strong rolled-up vortices in the shear layer and their interaction with the surface. Turbulent separation bubbles play an important role in a number of practical flow configurations including wings, turbomachinery impellers, diffusers, combustors, heat exchangers, buildings. These separation bubbles usually include complicated boundary conditions, so that basic studies on separation bubbles have been performed for idealized configurations with simple boundary conditions which are illustrated in figure 1. They are (i) a backward-facing step (Bradshaw & Wong [1], Eaton & Johnston [2],Chandrsuda & Bradshaw [3], Troutt et al. [4], Bhattacharjee et al. [5], Driver et al. [6]), (ii) a blunt plate (Ota & Narita [7], Kiya et al. [8], Kiya and Sasaki [9][10], Cherry et al. [11], Welch et al. [12]), (iii) a blunt circular cylinder (Ota & Motegi [13], Kiya & Nozawa [14]), and (iv) a normal plate with a long splitter plate (Ruderich & Fernholz [15], Castro & Haque [16]).

Separation bubbles fall in the category of complex turbulent flows which were reviewed by Bradshaw [17] in his Sectional Lecture at the XIVth ICTAM, 1976, Delft. Among the four configurations, the second and the third are the simplest configurations in the sense that Reynolds number is the only parameter required to define the flow. The first configuration has at least two governing parameters, viz. Reynolds number and the ratio of the boundary-layer thickness at separation to the step height. Nevertheless this flow has most widely been studied. Bradshaw & Wong's [1] paper includes review of papers before 1972 while Eaton & Johnston [18] reviewed papers published between 1972 and 1980. A review of the blunt-plate separation bubbles is presented by Kiya [19][20]. Simpson's [21] review includes the configurations (i), (ii) and (iv) in addition to detached flows on streamlined surfaces, an extensive list of papers being also given.

The reviews mentioned above, together with individual papers, reveal that properties of the separation bubbles (i)-(iv) are basically similar except for those which are peculiar to each configuration. The purpose of this paper is

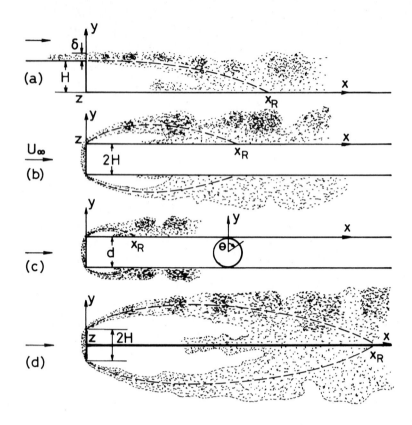

FIGURE 1

Fundamental configurations of separation bubbles and definition of main symbols: (a) backward-facing step, (b) blunt plate, (c) blunt circular cylinder, (d) normal plate with splitter plate. δ is boundary-layer thickness at separation. Length of bubble is approximately to scale.

to discuss those general properties of the separation bubbles which have been revealed by recent studies, rather than to give an extensive review of separation-bubble flows in general. Most of the discussions will be made on the basis of results obtained for the leading-edge separation bubble (ii) [8]–[11]. Results for the other configurations will be presented where appropriate. A special emphasis is placed on properties of large-scale vortices and flow unsteadiness in the reattachment region of the shear layer. Finally, consideration is confined to the flows in a low-subsonic regime and at Reynolds numbers Re of the order of 10^3–10^5. Reynolds number is defined in terms of the main-flow velocity U_∞ and the representative length, which is defined in figure 1, of the configurations.

2. GENERAL PROPERTIES OF SEPARATION BUBBLES

Symbols are defined in figure 1. The x-axis is taken in the longitudinal direction, the y-axis normal to the reattachment surface and the z-axis normal to the x- and y-axes. In (iii) the azimuthal angle θ replaces the z-axis. The origin of the coordinate system is at the concave corner for (i) and (iv) and at the separation edge for (ii) and (iii). The boundary layer along the front face of the plate (ii) separates at the edge to form a separated shear layer. At Reynolds numbers of the order of 10^3–10^5, the boundary layer is laminar. This is also the case for other configurations (iii) and (iv) owing to large

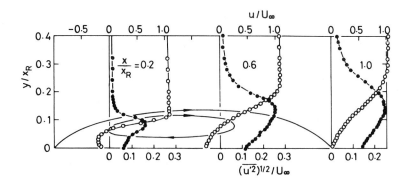

FIGURE 2

Time-mean streamlines, time-mean longitudinal velocity U and r.m.s. value of longitudinal velocity fluctuation u' in separation bubble at leading edge of blunt plate [9]. ——, streamlines; o, U; •, $(\overline{u'^2})^{1/2}$.

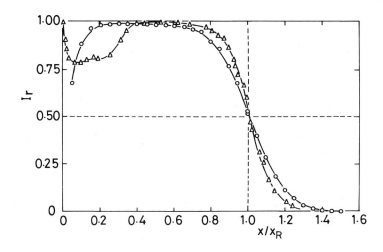

FIGURE 3

Variations of reverse-flow intermittency I_r near surface [9][14]. o, blunt plate; △, blunt circular cylinder.

acceleration of flow at the edge of the boundary layer. The plate is long enough for interaction between the shear layers on both sides to be neglected.

In terms of the time-mean flows, the shear layer reattaches onto the surface at a position whose longitudinal coordinate is denoted by x_R. A streamline starting from the separation edge and terminating at the reattachment position is a dividing streamline. A region enclosed by the dividing streamline and the surface is a recirculating region. The time-mean streamlines in the recirculating region are shown in figure 2, together with the time-mean longitudinal velocity U and the r.m.s. value of its fluctuating component u' [9].

The separated shear layer is subject to the Kelvin–Helmholtz instability, thus rolling up to give rise to concentrated vortices. These vortices coalesce to yield larger and larger vortices downstream until the reattachment position is reached. Thus the flow in the separation bubble is unsteady, especially near

the reattachment position where the vortices have largest length scales. An
aspect of the flow unsteadiness can be represented by the reverse-flow intermittency I_r which is the fraction of time during which the flow at a fixed
position is in the upstream direction. Results are shown in figure 3 for the
configurations (ii) and (iii) [14]. It is worth noting that I_r never attains
to unity even in the middle of the separation bubbles. The reattachment position has been defined as a longitudinal position where I_r measured near the
surface takes the value of 0.5. This definition yields the reattachment position which is the same as that defined as a position where the time-mean shear
stress at the surface is zero [15].

The reattachment length, which is the longitudinal distance between the separation edge and the reattachment position, strongly depends on the configurations; that is, for the backward-facing steps x_R is approximately 6 step
heights, provided that the boundary layer at separation is laminar and its
thickness is less than approximately 0.1 step heights [1][3]; for the blunt
plates x_R is approximately 5 half thicknesses [9][11]; for the blunt circular
cylinders x_R is 1.6 diameters [14][22]; for the normal plates with the splitter
plate x_R is approximately 19 half heights of the plate [15][16]. It will be
demonstrated in what follows that the reattachment length is a characteristic
length scale for vortical motions in the separation bubbles. This was first
shown by Mabey [23] for central frequencies of the surface-pressure fluctuations at the reattachment position of a wide range of separation bubbles. As
a measure of the longitudinal extent of highly unsteady flow near the reattachment position, we define the reattachment region rather arbitrarily as the
longitudinal range where the reverse-flow intermittency near the surface is
between 0.1 and 0.9.

3. VISUALIZED FLOW PATTERNS

Real flows in the separation bubbles are much more complicated than what one
might imagine from the time-mean pattern of figure 2. Figure 4 shows instantaneous flow patterns obtained by visualization, which was made by tracers
introduced into the flow through a thin slit immediately downstream of the
separation edge and entrained into the shear layer from the low-velocity side.
The tracers do not show an actual region of vorticity but a region where
vorticity must have been concentrating in an appropriate amount.

Figure 4(a) demonstrates the formation of larger and larger spanwise vortices
by coalescence with increasing longitudinal distance from the separation edge.
Vorticity seems to be shed from the bubble as a series of more-or-less discrete
turbulent structures, which will hereinafter be called large-scale vortices.
Characteristic streamwise spacings between the large-scale vortices are of the
order of $(0.6-0.8)x_R$ [9][11].

The spanwise vortices are not actually spanwise or two-dimensional but have
spanwise structures as shown in figure 4(b). The vortices are fairly two-
dimensional only within a short distance downstream of the separation edge.
Beyond this distance which is a function of Reynolds number, the vortices burst
into three-dimensional structures following fairly periodic spanwise deformation. Wavelengths of the deformation seem to increase with increasing longitudinal distance. Even after the bursting, we observe more-or-less discrete
turbulent structures aligned pseudo-periodically to the spanwise direction.
These structures have approximately the same convection velocity, so that they
look like a large spanwise vortex as a whole. Some aspects of the separation
bubbles can be described in terms of such spanwise vortices [8][24]. The
pseudo-periodic spanwise arrangement of the turbulent structures is likely to
become more and more random with increasing Reynolds number.

An end view at the reattachment position, figure 4(c), shows a number of

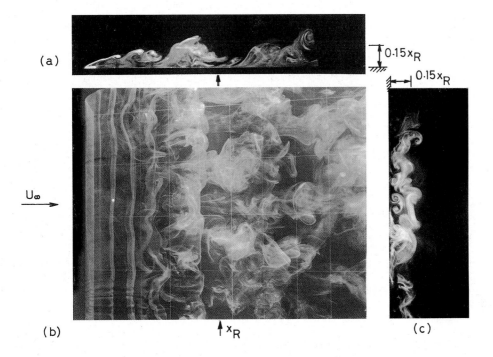

FIGURE 4
Flow patterns in blunt-plate separation bubble at Re = 1200. (a) Pattern in
midspan (x, y)-plane, (b) pattern in (x, z)-plane, and (c) secondary-flow
pattern in (y, z)-plane at $x/x_R = 1.0$. Patterns (b) and (c) are taken at same
instant but pattern (a) is not.

mushroom-like structures. They are presumably legs of a pair of counter-
rotating longitudinal vortices which are inclined at an angle to the main-flow
direction. Details of the end view depend on phases of the longitudinal motion
of the spanwise vortices, the mushroom-like structures being most evident in
the middle of two consecutive spanwise vortices. The longitudinal vortices,
which are produced in the early part of the shear layer, grow to have a large
scale in the reattachment region, where their interaction with the surface
yields the complicated system of longitudinal vortices with a wide range of
length scales, as seen in figure 4(c). End views obtained by another visualization technique are shown in figure 5, where time advances downwards [25].
They are visualized by hydrogen bubbles generated from seven wires arranged in
the yz-plane which is located at $x/x_R = 0.95$ close to the reattachment position. Since a normal plane immediately downstream or upstream of the plane of
the wires was illuminated by a thin sheet of light, we observe a local secondary-flow pattern of the forward or reverse flow, respectively.

Mushroom-like structures in the secondary-flow patterns of the forward flow,
figure 5a, move towards the surface with increasing time, while those of the
reverse flow, figure 5b, are lifted up from the surface. This again demonstrates the existence of the longitudinal vortices. It is likely that there
are characteristic spanwise spacings between the structures, values of which
are of the order of $(0.2-0.4)x_R$ [25]. The lift up of the structures is most
frequent in the reattachment region. Since the lift up *pumps up* fluid near
the surface, it might possibly be a mechanism producing a maximum heat transfer
in the reattachment region. This is consistent with the view that the heat
transfer is controlled by the turbulence level near the surface [26].

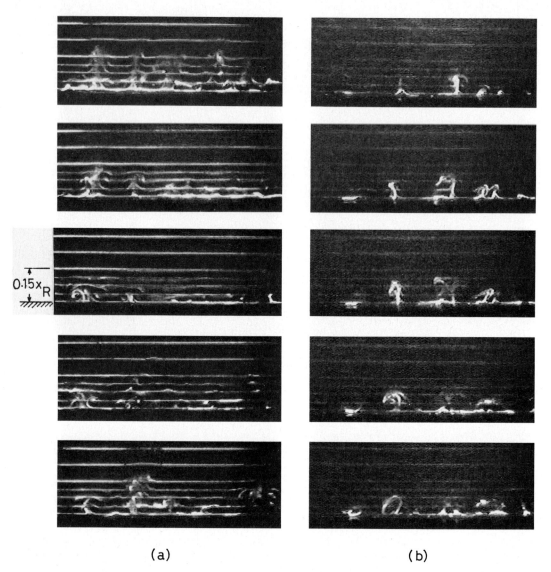

(a)   (b)

FIGURE 5

Secondary-flow patterns of (a) forward flow and (b) reverse flow near reattachment position $x/x_R = 0.95$ in blunt-plate separation bubble at Re = 1200 [25]. Time advances downwards. Time step Δt between two consecutive photographs is $U_\infty \Delta t/x_R = 0.17$ for (a) and 0.11 for (b).

Another mechanism for the maximum heat transfer is suggested by Thompson et al. [24] on the basis of a numerical simulation using two-dimensional discrete-vortex method. They found that a maximum instantaneous local heat transfer occurs on the front side of large-scale vortices existing in the reattachment region because this side of the large-scale vortices draw most effectively cooler outer fluid towards the heated surface. Inside the separation bubble the surface is covered with recirculating fluid with relatively low velocity, which leads to lower values of the instantaneous and time-mean heat transfer. On the other hand, the large-scale vortices downstream of the reattachment region are implicitly assumed to be so diffused that they cannot draw outer fluid towards the surface as effectively as those in the reattachment region. Both this mechanism and the foregoing one are proposed based on insufficient evidence, exact mechanisms remaining to be disclosed.

4. TURBULENCE PROPERTIES OF SEPARATED SHEAR LAYER

4.1. Reynolds Stresses

Properties of the separated shear layer are different from those of a classical plane mixing layer between a uniform flow and a fluid at rest. The shear layer has a stabilizing curvature all the way from the separation edge to the reattachment position. It is also affected by turbulent fluid transported from the reattachment region by the reverse flow, the turbulent fluid being entrained into the low-velocity side of the shear layer. Castro & Haque [16] argue that effects of the entrained turbulent fluid **are much** greater than those of the curvature.

An effect of the entrained turbulence can be demonstrated by a structural parameter $\overline{u'v'}/q'^2$, where $q'^2 = u'^2 + v'^2 + w'^2$, v' and w' being the fluctuating velocity components in the y- and z-directions. This is shown in figure 6 plotted against the non-dimensional vertical coordinate $(y - y_0)/\Lambda$, where Λ is the vorticity thickness normalized by the actual velocity difference across the shear layer ΔU and y_0 denotes the y-position where $U = 0.67U_\infty$ [16]. The low-velocity side is characterized by a much lower value of $\overline{u'v'}/q'^2$ than that in the plane mixing layer. Although the entrained fluid presumably has significant shear stress, this is perhaps uncorrelated with the shear stress in the local turbulent fluid of the shear layer. Consequently, whilst turbulence energy rises substantially, the shear stress does not [16].

The levels of the normal and shear stresses normalized by $(\Delta U)^2$ rise all the way from the separation edge to the reattachment position, and thereafter fall rapidly. Figure 7 shows the variations of the maximum values of the normal stresses as the flow develops [16]. The rapid decrease of the stresses downstream of the reattachment position is among the most poorly understood fea-

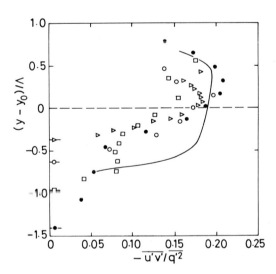

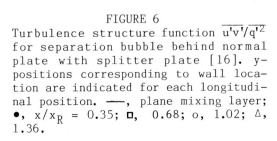

FIGURE 6

Turbulence structure function $\overline{u'v'}/q'^2$ for separation bubble behind normal plate with splitter plate [16]. y-positions corresponding to wall location are indicated for each longitudinal position. ——, plane mixing layer; ●, $x/x_R = 0.35$; □, 0.68; o, 1.02; △, 1.36.

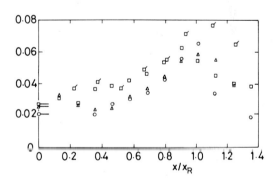

FIGURE 7

Maximum normal Reynolds stresses normalized by actual velocity difference squared $(\Delta U)^2$ in separated shear layer behind normal plate with splitter plate [16]. □, ◻, $\overline{u'^2}$; △, $\overline{v'^2}$; o, $\overline{w'^2}$. Plane mixing-layer values are shown at left.

tures of the reattaching shear layers [18]. Bradshaw & Wong [1] suggest that the zero normal-velocity condition at the surface and the strong longitudinal pressure gradient causes the large-scale vortices to be roughly torn in two in the reattachment region. The resulting rapid reduction in length would be responsible for the decrease in the Reynolds-stress levels. However, later studies [4][9][10][11][15] throw doubt on the tearing of the large-scale vortices.

Another mechanism is proposed by Troutt et al. [4]. They note that the reattachment region coincides with an apparent halt in the vortex-coalescence activity due to the close proximity of the surface. Since the vortex coalescence is an important Reynolds-stress generation mechanism in plane mixing layers [27], they argue that the inhibited vortex coalescence can cause the rapid decrease in the Reynolds stresses. It seems to the author, however, that this theory does not explain the continued decrease downstream of the reattachment region. The stabilizing curvature of the time-mean streamlines in the reattachment region cannot account for the continued decrease either [18].

Figure 7 also shows that, in the early part of the flow, the axial normal stress $\overline{u'^2}$ is larger than the transverse stress $\overline{v'^2}$ by rather more than in the plane mixing layer. Together with the substantially smaller vorticity thickness in this region, this is perhaps evidence for *flapping* of the shear layer, which will be discussed in detail in Section 5.2. For the blunt-plate separation bubbles, the amplitude of the flapping is approximately $0.005x_R$ at $x/x_R = 0.2$ and tends to generate a longitudinal velocity fluctuation u' of the order of $0.1U_\infty$ [10]. Assuming that this is uncorrelated with the local longitudinal velocity fluctuation of the shear layer, the measured normal stress $\overline{u'^2}$ is estimated to be increased by 36% by the flapping. This increase is consistent with the experimental result of figure 7.

4.2. Integral Scales and Motion of Vortices

The variations of the integral time scales and length scales describe how vortices in the shear layer grow as the flow develops. Figure 8 shows the time scale of the velocity fluctuation u', which is denoted by $T_{u'}$ [10][11]. The

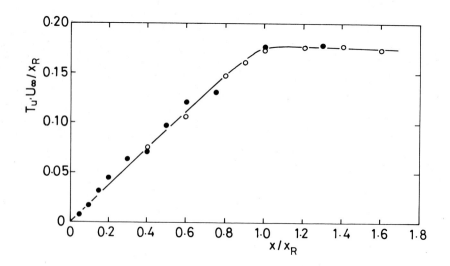

FIGURE 8
Variation of integral time scale $T_{u'}$ of longitudinal velocity fluctuation at outer edge of shear layer of blunt-plate separation bubble [9]. o, measured by X-wire probe; •, measured by single-wire probe.

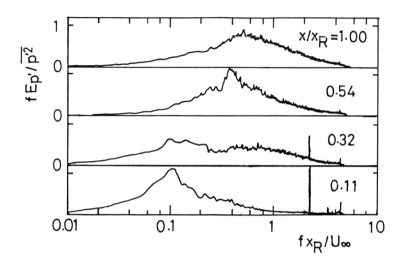

FIGURE 9

Power spectra of surface-pressure fluctuation $E_{p'}$ at several positions in blunt circular-cylinder separation bubble [28].

growth of $T_{u'}$ is seen to be approximately linear up to the reattachment position. Since the convection velocity of the vortices U_c are fairly constant, being $(0.5-0.6)U_\infty$, all the way from the separation edge to a position well downstream of the reattachment position [14], the longitudinal length scale $L_{xu'}$, which is the time scale multiplied by the convection velocity, grows in the same way. The spanwise length scale $L_{zu'}$ also increases approximately linearly in the downstream direction [9].

An important feature of figure 8 is that the longitudinal length scale $L_{xu'}$ is almost constant downstream of the reattachment position. This is also the case for the spanwise length scale of the surface-pressure fluctuations $L_{zp'}$. These results, together with the visualized flow patterns of figure 4(a,b), indicate that the vortex coalescence is inhibited downstream of the reattachment position owing to the effect of image vortices inside the plate.

The representative length scales of the vortices in the reattachment region appear to be given by those of the surface-pressure fluctuations, being $L_{xp'}$ = $0.11x_R$ and $L_{zp'}$ = $0.21x_R$ [9][11]. Hence the aspect ratio $L_{zp'}/L_{xp'}$ is approximately 2. For the blunt circular cylinders of figure 1, on the other hand, the corresponding values are $L_{xp'}$ = $0.10x_R$ and $L_{\theta p'}$ = $0.11x_R$, where $L_{\theta p'}$ is the circumferential length scale, and hence the aspect ratio is approximately unity. The value of $L_{\theta p'}$ is consistent with the experimental result that the flow in the reattachment region of the blunt circular cylinder has a cellular structure with nine cells [28].

The large-scale vortices are pseudo-periodically shed from the bubble with a frequency f_v of the order of $(0.6-0.7)U_\infty/x_R$ [9]. The existence of such vortex shedding is demonstrated by a broad peak of the power spectra of the surface-pressure fluctuations in the reattachment region. The spectra are shown in figure 9 for the blunt circular cylinder [28]. Similar results are also obtained for the blunt plate [9][11]. Values of the frequency f_v are almost the same for the four configurations of figure 1 [6][11][28]. In the early part of the bubble, the pressure spectra have another peak at a lower frequency $f_1 \doteq 0.1U_\infty/x_R$. This peak is further evidence for the flapping motion of the shear layer, being observed for all of the four configurations in figure 1 [9][11][16][28]. Values of the frequency f_1 are approximately the same for the configurations.

4.3. Structure of Large-scale Vortices in Reattachment Region

Structure of large-scale vortices in the reattachment region is important in practical devices because they are responsible for large r.m.s. values of the surface pressure and the shear stress at the surface [15][23][29]-[31] and for high heat- and mass-transfer rates [24][26]. Moreover the pseudo-periodic nature of the motion of these vortices can produce cyclic structural and thermal loading. In this section, we discuss those large-scale vortices which are associated with larger surface-pressure fluctuations than a threshold level. Results are obtained by a phase-averaging technique in which the surface-pressure waveform at the reattachment position is employed as a conditioning signal [10]. This technique has been suggested by a two-dimensional numerical

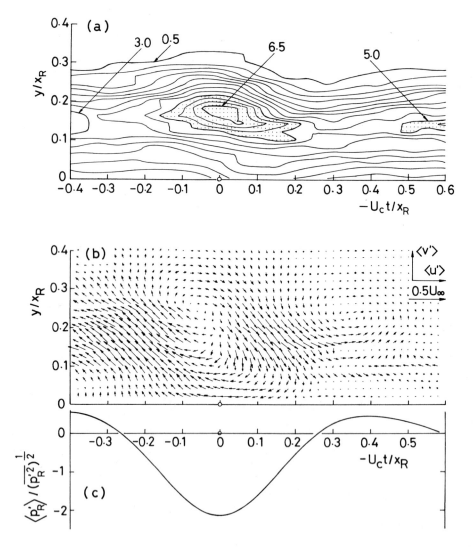

FIGURE 10

Space-time (y, t) distributions of (a) contour lines of high-frequency turbulent energy $\langle u_T'^2 \rangle$ normalized by $U_\infty^2/10^4$ (contour interval 0.5), shaded area having values higher than 0.5, (b) fluctuating velocity vector $(\langle u' \rangle, \langle v' \rangle)$ and (c) surface-pressure fluctuation at reattachment position $\langle p_R' \rangle$ [10]. Time t is synchronized with negative peak values (valleys) of p_R' lower than -1.8 units of its r.m.s. value. Detection frequency of valleys is approximately one-third of vortex-shedding frequency.

simulation [8] which shows that the surface pressure is negative beneath a large-scale vortex, attaining a definite minimum right below its centre.

Figures 10 and 11 show phase-averaged distributions of the fluctuating velocity components ($\langle u' \rangle$, $\langle v' \rangle$) and ($\langle u' \rangle$, $\langle w' \rangle$) and high-frequency turbulent energy $\langle u_T'^2 \rangle$ in the space-time (y, t) and (z, t) domains. Here u_T' is components of the longitudinal velocity fluctuation with frequencies of the order of the Kolmogoroff frequency. A region where $\langle u_T'^2 \rangle$ attains a maximum is assumed to be a central part of a vortex. This assumption is supported by the observation that turbulent energy of high-frequency components attains a definite maximum at the centre of a vortex in the wake of a normal plate [32].

Figure 10(a,b) shows a large system of flow rotating in the clockwise sense, at the centre of which $\langle u_T'^2 \rangle$ attains a maximum. Hence this rotating system can be interpreted as a real vortex. Figure 11(a,b) indicates that a centre rotating in the clockwise sense at $(-z, -U_c t)/x_R \doteqdot (0.28, -0.1)$ is a real vortex. From symmetry there should be another centre rotating in the counter-clockwise direction at the image point $(-0.28, -0.1)$. Hence these centres are probably legs of a pair of counter-rotating vortices. The most economical interpretation of these space-time distributions, although it may not be the only one, is

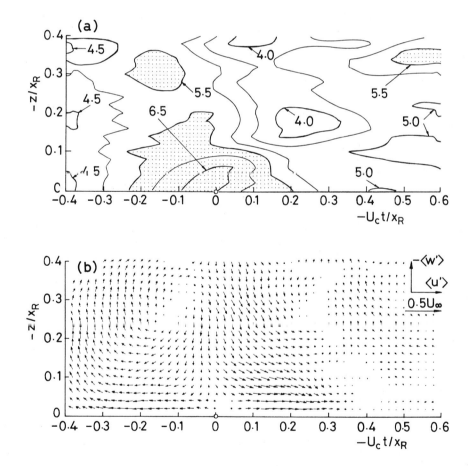

FIGURE 11

Space-time (z, t) distributions of (a) contour lines of high-frequency energy $\langle u_T'^2 \rangle$ near centre of shear layer in normalized form (contour interval 0.5), shaded areas having values greater than 5.5, and (b) fluctuating velocity vector ($\langle u' \rangle$, $\langle w' \rangle$) near centre of shear layer [10].

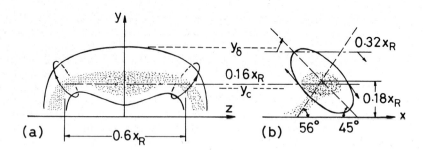

FIGURE 12

Illustration of large-scale vortices in reattachment region: (a) structure in
(y, z)-plane viewed from upstream side; (b) section in (x, y)-plane [10].

that the large-scale vortices are U-shape vortices as illustrated in figure 12.
The large-scale vortices produce a large part of Reynolds stresses. This is
demonstrated in figure 13 in terms of $-\langle u'\rangle\langle v'\rangle$, which gives the extent to
which the velocity fluctuations associated with the vortices contribute to the
overall Reynolds shear stress. The shear stress is likely to be produced near
the front and back edges of the vortices.

It is difficult to recognize the U-shape vortices in visualized flow patterns
such as figures 4 and 5. Although a number of pairs of counter-rotating

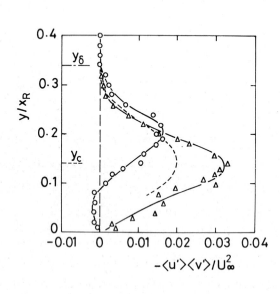

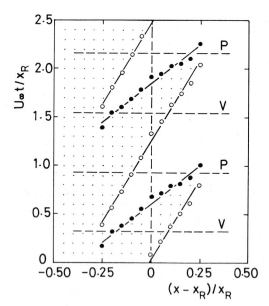

FIGURE 13

Contribution of large-scale structures
to Reynolds shearing stress in reattachment region in terms of velocity
product $\langle u'\rangle\langle v'\rangle$ at time of peaks (o)
and valleys (Δ) of dp_R'/dt [10]. Peaks
and valleys appear approximately near
back edge and front edge of large-
scale vortices, respectively. ---,
global Reynolds shear stress. y_δ and
y_c are edge and centre of shear layer.

FIGURE 14

Motion of reverse-flow regions (shaded
area) in space-time (x - x_R, t) domain
in terms of motion of zeros A and B of
longitudinal velocity near surface.
o, A; ●, B; origin of t is arbitrary;
horizontal broken lines indicate times
when phase-averaged pressure fluctua-
tion $\langle p_R'\rangle$ attains peaks P and valleys
V.

vortices are observed in these patterns, most of them are much smaller than the U-vortices. This is somewhat surprising in view of the fact that the valleys of the pressure waveform lower than the employed threshold level are detected at a frequency as much as one-third of the vortex-shedding frequency [10]. In other words, the U-vortices should have been observed rather frequently. It is possible that the U-vortices are distorted by small-scale structures to such an extent that the former is unable to be visually identified. It is also poss-ible that vorticity has so diffused beyond the mushroom-like structures that they have no definite relation with vorticity distribution.

5. FLOW UNSTEADINESS IN SEPARATION BUBBLES

There are two kinds of pseudo-periodic unsteadiness in the separation bubbles. One is associated with the motion of the large-scale vortices in the reattach-ment region. The other is the flapping motion of the shear layer, which is accompanied by contraction and enlargement of the bubbles as will be shown in Section 5.2. The pseudo-periodic nature of the two unsteady flows is well manifested in the broad peaks of the power spectra of the surface-pressure fluctuations presented in figure 9. Such peaks also appear in the power spec-tra of the velocity fluctuation u' near the edge of the shear layer [11].

5.1. Unsteadiness Due to Motion of Large-scale Vortices

First we discuss the unsteady flow due to the motion of the large-scale vor-tices. An aspect of the unsteadiness can be described by the motion of re-verse-flow regions [10]. A reverse-flow region is defined as a region where the instantaneous longitudinal velocity near the surface is negative. In [10], the velocity at a distance $0.005x_R$ from the surface is chosen rather arbitrari-ly. This velocity is short-time averaged with an averaging time of approxi-mately one-tenth of the vortex-shedding interval $1/f_v$; the short-time averaged velocity will be called the surface velocity u_s for convenience.

A boundary between a reverse-flow region and a forward-flow region is defined by either $u_s = 0$ and $du_s/dt < 0$ or $u_s = 0$ and $du_s/dt > 0$ when u_s is measured as a function of time at a fixed position x; the former is denoted by zero A and the latter by zero B. If we could measure the spatial distribution of u_s at a fixed time, zeros A and zeros B would be identified as positions where $u_s = 0$, $du_s/dx > 0$ and $u_s = 0$, $du_s/dx < 0$, respectively. Positions of zeros A and zeros B are perhaps not far from positions where the instantaneous shear stress at the surface is zero.

The key technique for determining the motion of the zeros is the pressure-conditioned sampling of the velocity u_s. At a fixed longitudinal position in the reattachment region, the surface velocity is sampled to find zero A and zero B. The time at which zero A or B is detected is then employed as the synchronization time to phase-average the surface-pressure fluctuation at the reattachment position p_R'. A peak P and a valley V emerge in each phase-averaged pressure at time t_P and t_V, respectively. On the assumption that the peak and the valley correspond to particular phases of the motion of the large-scale vortices, the time t_P or t_V is taken as an origin to measure the time when zero A or zero B appears. This gives the motion of zeros A and B in the space-time ($x - x_R$, t) domain.

The result is shown in figure 14. This demonstrates how the longitudinal extent and the number of reverse-flow regions vary with time, depending on phases of the motion of the large-scale vortices. Zeros B move with a speed of approximately $0.7U_\infty$, which is much greater than the speed $0.3U_\infty$ of zeros A. The average speed $0.5U_\infty$ is almost equal to the convection velocity of the vortices. It is worth mentioning that peak P appears when zero B has caught up with zero A, that is, when the reverse-flow region in the downstream side

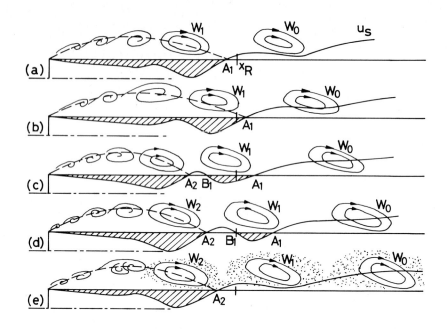

FIGURE 15

Relation between motion of large-scale vortices and that of zeros A and B of longitudinal velocity near surface; shaded areas are reverse-flow regions; broken lines are dividing streamlines [10].

disappears. On the other hand, valley V occurs when the second reverse-flow region has just emerged.

Based on these results we can construct a relation between the motion of the reverse-flow regions and that of the large-scale vortices, as is sketched in figure 15 for one cycle of shedding of the vortices. At phase (a) a vortex W_0 has been shed from the bubble, and the next vortex W_1 is approaching to the reattachment region. There is no reverse flow beneath the vortex W_0 partly owing to its acceleration downstream of the reattachment region [14] and partly owing to a decay of its strength caused by turbulent diffusion of vorticity. The vortex W_1 and the zero A_1 have moved a little downstream at phase (b). At phase (c) a forward-flow region A_2B_1 appears on the back side of the vortex W_1, giving rise to a detached reverse-flow region beneath this vortex. The pressure fluctuation p_R' attains a valley V approximately at this phase (c). The zero B_1 moves faster than the zeros A_1 and A_2, hence at phase (d) this difference in speeds having extended the forward-flow region A_2B_1 and contracted the reverse-flow region A_1B_1. Finally, just before phase (e), the zero B_1 has caught up with the zero A_1, thus the reverse-flow region vanishing at phase (e). Then the next cycle of the vortex shedding begins. The above interpretation is obviously made on the basis of the large *spanwise* vortices which are discussed in Section 3. Since actual large-scale vortices are three-dimensional as shown in figures 4, 5 and 12, instantaneous spanwise distributions of zeros A and B are not uniform accordingly. It is expected that data of how the motion of the zeros are correlated in the spanwise direction will give us good information on structure of the large-scale vortices. Such data have not been obtained yet.

5.2. Low-frequency Flapping Motion

The second unsteadiness has central frequencies f_1 of approximately $0.1U_\infty/x_R$, which is approximately one-sixth of the vortex-shedding frequency f_v. This value of the frequency is obvious in the power spectra shown in figure 9. The existence of the low-frequency motion is further supported by the following features: (i) a long tail appears in the auto correlation of the velocity fluctuation u' near the time-mean centre of the shear layer at x/x_R = 0.2 [9]; (ii) values of the longitudinal normal stress $\overline{u'^2}$ are much larger than those of the transverse normal stress $\overline{v'^2}$ in the early part of the bubble [16]; (iii) the cross correlation of low-pass filtered surface-pressure fluctuations, the cut-off frequency being $2f_1$, has high values for two positions largely separated in the longitudinal direction x/x_R = 0.2 and 1.0 [10].

Features (i) and (ii) can be interpreted by a low-frequency transverse oscillation of the shear layer. Moreover, feature (iii) suggests that the low-frequency motion is of large scale in the sense that its effect is felt throughout the separation bubble. This phenomenon is called the flapping or the flapping motion.

The flapping motion is confirmed by longitudinal velocity profiles which are obtained by the use of a pressure-conditioned sampling in the early part of the bubble [10]. First the sampling of the velocity is made only when the low-pass filtered surface pressure at a position, x/x_R = 0.2, in the early part of the separation bubble attains maxima greater than a threshold level. Second the sampling is made for minima of the pressure lower than another threshold level of the same value but of the opposite sign. This procedure is based on the fact that the low-pass filtered surface-pressure and longitudinal-velocity fluctuations are well correlated in the first half of the bubble. Thereby the flapping motion is found to have an amplitude of $0.005x_R$, but this rather small amplitude is sufficient to cause u' fluctuation of the order of $0.1U_\infty$ in the central part of the shear layer [10]. The significance of this apparent fluctuation in measurements of the longitudinal normal stress has been discussed in Section 4.1.

The flapping motion is accompanied by contraction and enlargement of the separation bubble [6][9][10][33]. This can be demonstrated by the motion of zeros of the low-pass filtered surface velocity u_{s1} [10]. A position where u_{s1} is zero is interpreted as an instantaneous reattachment position of short-time averaged flows. Two kinds of zeros exist in the waveform of u_{s1} measured at a fixed position. One is characterized by u_{s1} = 0 and du_{s1}/dt < 0 (denoted by zero A_1) and the other by u_{s1} = 0 and du_{s1}/dt > 0 (denoted by zero B_1). The motion of zeros A_1 and B_1 is obtained in the same way as that of zeros A and B mentioned in Section 5.1. At a fixed position x near the time-mean reattachment position, zeros A_1 are searched for and the time at which A_1 is detected is then employed as the synchronization time against which the low-pass filtered pressure fluctuation at the reattachment position p'_{R1} is phase-averaged. The same procedure is also followed for zeros B_1. The phase-averaged pressure $\langle p'_{R1} \rangle$ is found to attain a peak P_1 and a valley V_1.

Assuming that the peaks and the valleys of $\langle p'_{R1} \rangle$ correspond to particular phases of the contraction and enlargement of the separation bubble, the times when zeros A_1 and B_1 appear at the fixed position are measured relative to the times when the peaks P_1 or the valleys V_1 are detected. By repeating the same measurement at several longitudinal positions, the motion of zeros A_1 and B_1 is obtained in the space-time $(x - x_R, t)$ domain. The result is presented in figure 16. Zeros A_1 move upstream while zeros B_1 move downstream. The speeds of A_1 and B_1 are roughly $0.1U_\infty$ and $0.2U_\infty$, respectively, at the middle of the enlargement and contraction process; that is, the contraction is much swifter than the enlargement as suggested by Eaton & Johnston [33]. This is also obtained in numerical simulations in two-dimensions by Thompson et al. [24] and Kondoh & Nagano [34].

The downward motion of the shear layer induces the contraction of the bubble while the upward motion induces the enlargement. The flapping motion has much

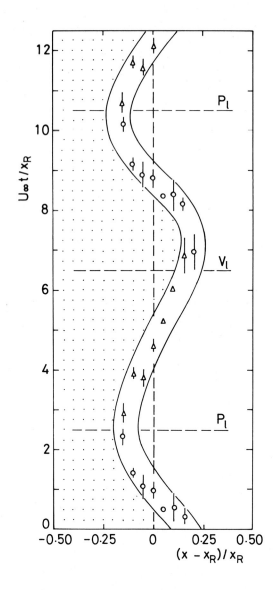

FIGURE 16

Low-frequency contraction and enlargement of separation bubble in space-time (x −x_R, t) domain in terms of motion of zeros A_1 and B_1 of low-pass filtered longitudinal velocity near surface [10]. Δ, A_1; o, B_1; flags added to symbols indicate experimental uncertainty; horizontal broken lines indicate times when phase-averaged pressure $<p'_{R1}>$ attains peaks P_1 and valleys V_1; shaded area is reverse-flow region; origin of time t is arbitrary.

larger spanwise length scale than that associated with the large-scale vortices [10]. This is consistent with Eaton & Johnston's [33] conjecture that the flapping motion is roughly two-dimensional. Moreover there is evidence that the flapping motion is accompanied by a change in strength of the large-scale vortices in the reattachment region, stronger vortices being shed during the contraction process than the enlargement process [10]. As mentioned in Section 4.2, the flapping motion is common to all the configurations in figure 1. It is observed even when the main flow contains turbulence of the order of 10% [11]. Hence it is likely that the flapping motion is an inherent feature of nominally two-dimensional or axisymmetrical separation bubbles.

A exact mechanism of the flapping motion is not understood yet. Eaton & Johnston [33] suggest that this is caused by an instantaneous imbalance between the entrainment from the recirculation zone and the reinjection of fluid from the reattachment region. They conjecture that an unusual event causes a short-time breakdown of the spanwise vortices in the shear layer which would temporarily reduce the entrainment rate and thereby cause an increase in the volume of recirculating fluid. This increase will move the shear layer away from the surface and increase the short-time averaged reattachment length, which is the enlargement process. The mechanisms proposed by Cherry et al. [11] and Driver et al. [6] are more-or-less similar to the above one if the words 'a short-time breakdown of the spanwise vortices' in [33] are replaced by 'a temporary interruption to shear-layer growth/coalescence process' in [11] and by 'disorder of roll-up and pairing process' in [6]. However, the assumption that a breakdown of the spanwise vortices causes the upward motion of the shear layer and the enlargement of the bubble seems to be inconsistent with the well-established fact that the free-stream turbulence which tends to reduce the spanwise length scale causes reduction of the time-mean bubble length. Although Eaton & Johnston [33] assume an *unusual* event for the cause of the breakdown of the spanwise vortices, the pseudo-periodic nature of the flapping motion suggests that it is induced by feedback of disturbances from the reattachment region to the separation edge. The disturbances are possibly generated by the change in strength of the large-scale vortices in the reattachment region, propagating upstream as pressure wave to cause the low-frequency modulation of the shear layer. This is the mechanism responsible for low-frequency motions in cavity flows, shear layers impinging on corners, jets impinging on wedges, etc. (Rockwell [35]).

The contraction and enlargement of the separation bubble have not been identified on visualized flow patterns within the author's knowledge. Cherry et al.'s [11] flow visualization by smoke might demonstrate such flow patterns. They show that shedding of vortices from the bubble varies through a range of phases. There is a temporal cessation in the pseudo-periodic shedding of the large-scale vortices from the bubble and an necking of the shear layer downstream. On the other hand, there also exist vigorous shedding phases, in their terminology, which give the most significant upward shifts of the shear layer from the surface. The former might correspond to a phase of the contraction process while the latter might indicate a phase of the enlargement process. There is, however, no direct evidence that this correspondence is true. Well-controlled experiments are needed to reveal detailed properties and exact mechanisms of the flapping motion.

One may suspect, from the pseudo-periodic nature of the flapping motion, that it is stimulated by acoustics as in the case of Welsh & Gibson [36]. This possibility is excluded by that the flapping motion is common to all the configurations in figure 1 which are investigated in wind tunnels with different dimensions and in different main-flow velocities, and that the values of the non-dimensional frequency $f_1 x_R / U_\infty$ are approximately the same.

6. CONCLUSION

Recent developments in the measurement of highly turbulent, unsteady flows have considerably increased our understanding of turbulent separation bubbles. The properties of the separation bubble discussed in this review are likely to be similar to a wide range of nominally two-dimensional or axisymmetrical separation bubbles. Some of the future problems to be studied are (i) understanding a feedback mechanism sustaining the flapping motion, (ii) obtaining in more detail structure of three-dimensional large-scale vortices by multi-point, conditional-sampling measurements, (iii) obtaining instantaneous spanwise structure of forward- and reverse-flow regions near the reattachment position, (iv) revealing mechanisms responsible for high heat- and mass-transfer rates in the reattachment region, (v) understanding a mechanism of the rapid **decay** of Reynolds stresses downstream of the reattachment position, and (vi) manipulating structure of flow in separation bubbles.

A full direct or LES simulation has not yet been performed for turbulent separation bubbles. Previous numerical simulations use turbulence models based on averaged versions of the Navier-Stokes equations to obtain tolerable agreement with measurements. According to Perry [37], these models should be regarded as sophisticated interpolation schemes which cannot be applied with confidence to flow situations which are outside of a certain range of observation; also these models contain complex quantities which are difficult to interpret physically. Moreover the turbulence models have been developed on the basis of information on simple turbulent shear flows. This state of the turbulence models, together with the unsteady nature of flow in the separation bubbles associated with the flapping motion and the shedding of the large-scale vortices, indicates that agreement between calculated results and measurements might be fortuitous. On the other hand, full direct or LES simulations can make invaluable contribution to solving some of the problems mentioned above and to understanding detailed structure in general of turbulent separation bubbles. It is hoped that such simulations will be realized within several years.

REFERENCES

[1] Bradshaw, P. and Wong, F.T.Y., J. Fluid Mech. 52 (1972) 113.
[2] Eaton, J.K. and Johnston, J.P., Turbulent Flow Reattachment: an Expeimental Study of the Flow and Structure Behind a Backward-facing Step, Report MD-39, Thermosciences Div., Dept. of Mechanical Engineering, Stanford University, 1980.
[3] Chandrsuda, C. and Bradshaw, P., J. Fluid Mech. 110 (1981) 171.
[4] Troutt, T.R., Scheelke, B. and Norman, T.R., J. Fluid Mech. 143 (1984) 413.
[5] Bhattacharjee, S., Scheelke, B. and Troutt, T.R., AIAA J. 24 (1986) 623.
[6] Driver, D.M., Seegmiller, H. and Marvin, J.G., AIAA J. 25 (1987) 914.
[7] Ota, T. and Narita, M., Trans. ASME, J. Fluids Engng 100 (1978) 224.
[8] Kiya, M., Sasaki, K. and Arie, M., J. Fluid Mech. 120 (1982) 219.
[9] Kiya, M. and Sasaki, K., J. Fluid Mech. 137 (1983) 83.
[10] Kiya, M. and Sasaki, K., J. Fluid Mech. 154 (1985) 463.
[11] Cherry, N.J., Hillier, R. and Latour, M.E.M.P., J. Fluid Mech. 144 (1984) 13.
[12] Welch, L.W., Hourigan, K., Flood, G.J. and Welsh, M.J., Three Dimensional Separated Flow Over the Surface of a Bluff Body, in: Proc. 9th Australasian Fluid Mech. Conf., Auckland, New Zealand, 8-12 December 1986 (University of Auckland, School of Engineering, 1986) pp. 496-499.
[13] Ota, T. and Motegi, H., Trans. ASME, J. Appl. Mech. 47 (1980) 1.
[14] Kiya, M. and Nozawa, T., Trans. JSME, Ser. B, 53 (1987) 1183 (in Japanese).
[15] Ruderich, R. and Fernholz, H.H., J. Fluid Mech. 163 (1985) 283.
[16] Castro, I.P. and Haque, A., J. Fluid Mech. 170 (1987) 430.
[17] Bradshaw, P., Complex Turbulent Flows, in: Koiter, W.T., (ed.), Theoretical and Applied Mechanics (North-Holland, Amsterdam, 1976) pp. 103-113.
[18] Eaton, J.K. and Johnston, J.P., AIAA J. 19 (1981) 1093.
[19] Kiya, M., Vortices and Unsteady Flow in Turbulent Separation Bubbles, in: Proc. 9th Australasian Fluid Mech. Conf., 8-12 December 1986, Auckland, New Zealand (The University of Auckland, School of Engineering, 1988) pp. 1-6.
[20] Kiya, M., Structure of Flow in Leading-edge Separation Bubbles, in: Smith, F.T. and Brown, S.N., (eds.), Boundary-layer Separation (Springer-Verlag, Berlin, 1987) pp. 57-71.

[21] Simpson, R.L., Two-dimensional Turbulent Separated Flow: Vol. 1, Rep. no. AGARD-AG 287-VOL-1, 1986.

[22] Ota, T., Trans. ASME, J. Appl. Mech. 42 (1975) 311.

[23] Mabey, D., J. Aircraft 9 (1972) 642.

[24] Thompson, M.C., Hourigan, K. and Welsh, M.C., Int. Communications in Heat and Mass Transfer 13 (1986) 665.

[25] Kiya, M., Tamura, H. and Ishikawa, R., J. of the Flow Visualization Society of Japan 9 (1987) 171 (in Japanese).

[26] Vogel, J.C. and Eaton, J.K., Trans. ASME, J. Heat Transfer, 107 (1985) 922.

[27] Browand, F.K. and Weidman, P.D., J. Fluid Mech. 76 (1976) 127.

[28] Kiya, M., Mochizuki, O., Tamura, H., Nozawa, T., Ishikawa, R. and Kushioka, K., Turbulence Structure in the Leading-edge Separation Zone of a Blunt Circular Cylinder (to be presented at 4th Asian Conf. of Fluid Mech., 19-23 August 1989, Hong Kong).

[29] Katsura, J., On Wind Pressure Fluctuations on the Side Surface of Model With Long Rectangular Sections, in: Proc. 2nd National Conf. on Wind Engineering Research, June 1975, Colorado State University, pp. IV-20-1-IV-20-3.

[30] Sasaki, K. and Kiya, M., Bulletin of JSME 28 (1985) 610.

[31] Hillier, R. and Dulai, B.S., Pressure Fluctuations in Turbulent Separated Flow, in: Proc. 5th Int. Symp. on Turbulent Shear Flows, 7-9 August 1985, Cornell University, Ithaca, New York, pp. 5.13-5.18.

[32] Kiya, M. and Matsumura, M., J. Fluid Mech. 190 (1988) 343.

[33] Eaton, J.K. and Johnston, J.P., Low-frequency Unsteadiness of a Re-attaching Turbulent Shear Flow, in: Bradbury, L.J.S., Durst, F., Launder, F.W., Schmidt, F.W. and Whitelaw, J.H., (eds.), Turbulent Shear Flows 3 (Springer-Verlag, Berlin, 1982) pp. 162-170.

[34] Kondoh, T. and Nagano, Y., Computational Study of Separating and Re-attaching Flows Behind a Backward-Facing Step, in: Proc. of 1st CFD Symp., 22-24 December 1987, Tokyo, pp. 421-424 (in Japanese).

[35] Rockwell, D., AIAA J. 21 (1983) 645.

[36] Welsh, M.C. and Gibson, D.C., J. Sound Vib. 67 (1979) 501.

[37] Perry, A.E., A description of Eddying Motions and Turbulence, in: Proc. 9th Australasian Fluid Mech. Conf., 8-12 December 1986, Auckland, New Zealand (The University of Auckland, School of Engineering, 1986) pp. 7-12.

Theoretical and Applied Mechanics
P. Germain, M. Piau and D. Caillerie (Editors)
Elsevier Science Publishers B.V. (North-Holland)
© IUTAM, 1989

THE PENDULUM FROM HUYGENS' *HOROLOGIUM* TO SYMMETRY BREAKING
AND CHAOS

John W. MILES

Institute of Geophysics and Planetary Physics
University of California, San Diego
La Jolla, California 92093

The history of the pendulum, beginning with the seventeenth-century work of Galileo and
Huygens, is reviewed. Numerical experiments that illustrate symmetry breaking, period
doubling and the transition to chaotic motion for swinging oscillations of harmonically
driven, planar and spherical pendulums are described. Analytical approximations for these
phenomena, for breaking of the resonance curve (a precursor of symmetry breaking), and
for inverted oscillations of the pendulum are developed. Some analytical and numerical
results for rotational oscillations and for parametrically excited oscillations also are
described.

1. INTRODUCTION

The pendulum is both paradigm and prototype of the simple oscillator and is isomorphic to such modern
nonlinear oscillators as the Josephson junction. Its scientific study goes back to Galileo (c. 1632), and
almost everything that we knew about it prior to the present decade was discovered by Huygens in 1659.
Its very name implies an ineluctable regularity, as in Edgar Allan Poe's "The Pit and the Pendulum", but
we now know that it is capable of bizarre and chaotic behaviour that would have been inconceivable during the first three-hundred years following Huygens' discoveries.

1.1. Galileo

Galileo is reported to have noticed, at the age of seventeen, that each swing of a lamp, suspended by a
long chain from the roof of the cathedral in Pisa, seemed to take the same time, which he measured by his
own pulse [Robertson (1931), p.76; Wolf (1950), p.28]. This report may be apocryphal, but Galileo manifestly believed, on the basis of his experiments, that the period of the pendulum is independent not only
of the weight of the bob but also the amplitude of oscillation. Speaking through Salviati (*Two New Sciences*, First Day [129]), he says "the vibrations [of a pendulum with a cork bob] whether large or small,
are all performed in time-intervals which are not only equal among themselves, but also equal to the
period of a pendulum of the same length with a lead bob." Galileo also determined that the period of a
simple pendulum is proportional to the square root of its length (*ibid.* [139]), but it was left to Huygens to
prove this result and to determine the constant of proportionality.

Galileo proposed a pendulum clock, but neither he nor his followers appear to have constructed a working
model, which also was left to Huygens.† He did, however, construct a pendulum "pulsimeter" for the
comparative measurement of pulse rates (Wolf 1950, p.111).

*This work was supported in part by the Physical Oceanography, Applied Mathematics and Fluid
Dynamics/Hydraulics programs of the National Science Foundation, NSF Grant OCE-81-17539,
by the Office of Naval Research, Contract N00014-84-K-0137, 4322318 (430), and by the DAR-
PA Univ. Res. Init. under Appl. and Comp. Math. Program Contract N00014-86-K-0758 admin-
istered by the Office of Naval Research.
†See Favaro (1891) and Robertson (1931, pp.75-111) for reviews of the evidence in support of the
contrary claims of some of Galileo's followers.

1.2. Huygens

Christiaan Huygens published his *Horologium Oscillatorium* in 1673. In it, he describes the design of the pendulum clock that he had invented in 1656 and constructed in 1657, calculates the period of small oscillations of a simple pendulum and the equivalent length of a compound (physical) pendulum, determines the cycloid as the pendulum-bob trajectory for which period is independent of amplitude, and proves that the suspending thread of such a pendulum must osculate a congruent cycloid. These last four results, as well as that for the centripetal acceleration of a particle moving on a circle (in which Huygens anticipated Newton), all date from the year 1659, which, for intensity of scientific creativity, may be compared with 1665-6 for Newton or 1905 for Einstein.

In Part II of the *Horologium*, Huygens invokes Galileo's law for freely falling bodies and assumes small displacement to obtain (although not in algebraic form)

$$T = 2\pi(l/g)^{1/2} \equiv T_0 \tag{1.1}$$

for the period of a simple pendulum of length l. He recognizes that, contrary to Galileo's belief, the period of the pendulum increases with amplitude and asserts (in Part I of the *Horologium*) that "the ratio of the times corresponding to any arcs can be defined by numbers based on certain calculations and can be as exact as one could wish." For example, he states that the ratio of the period for oscillations of amplitude $\frac{1}{2}\pi$ to T_0 is $34/29 = 1.172$, which he subsequently improved to $2623/2222 = 1.1805$. His method of calculation, which appears to have been published only with the 1934 edition of his complete works (*Oeuvres* **18**, 374-387), is essentially geometrical. The analytical result

$$T/T_0 = (2/\pi)K(k) , \quad k = \sin\frac{1}{2}\alpha , \tag{1.2}$$

where K is a complete elliptic integral of the first kind and modulus k and α is the amplitude, was given by Euler (1776) in the form of both an integral and a power series in k. In particular, the period for $\alpha = \frac{1}{2}\pi$ is given by

$$T/T_0 = 1 + \sum_{n=1}^{\infty} \frac{1^2 \cdot 3^2 \cdots (2n-1)^2}{2^2 \cdot 4^2 \cdots (2n)^2 2^n} \quad (\alpha = \frac{1}{2}\pi) , \tag{1.3}$$

from which Euler calculated $T/T_0 = 1.1805$ (subsequently corrected by his editor to 1.1803; the apparent agreement with Huygens appears to be coincidental).

The series (1.3) converges rather slowly (like 2^{-n}). Euler did obtain more rapidly converging series, but evidently did not apply them to this last example. It is worth noting that

$$T/T_0 = 1 + 2 \sum_{n=1}^{N} \text{sech}(n\pi) \quad (\alpha = \frac{1}{2}\pi) , \tag{1.4}$$

which follows from the expansion of K in the Jacobi nome, coincides to three decimals for truncations at $N = 1$ and 2, respectively, with Huygens' fractional approximations $34/29$ and $2623/2222$.

In Part IV of the *Horologium*, Huygens addresses the problem of determining the equivalent length of a (planar) pendulum of arbitrary shape. The problem had been posed by Mersenne soon after Galileo's research and attacked unsuccessfully by Descartes, who had introduced the notion of a "center of agitation", but it was Huygens who gave precise meaning to this term through his result (in modern form)

$$l = I/md \tag{1.5}$$

for the length of a simple pendulum with the same period as a rigid body with its center of mass at a distance d from the axis of rotation and a moment of inertia I. Huygens gives the result in the form

$$l = \sum_n m_n d_n^2 / \sum_n m_n d_n \qquad (1.6)$$

for any number of point masses m_n at distances d_n from the axis of rotation and uses this result to obtain l for various cylinders, a cone, and a sphere. As Millikan, Roller & Watson (1937) remark, "This was the second *great* problem to be solved in kinetics, the first (that of falling bodies) having been solved by Galileo .... It was the first successful attempt to deal with the kinetics of a rigid body; in fact, it was in this connection that the concept of ... the moment of inertia came to light."

Huygens develops Parts II-IV of the *Horologium* (Part I describes the construction and testing of his clock) in great detail, almost all in geometrical language with many diagrams and no algebraic equations. [We are reminded that Lagrange, in his epochal formulation of analytical mechanics (1788), boasted that "one would not find a single diagram in his work."] In contrast, in Part V in only seven pages he describes the conical pendulum and the construction of a clock using this pendulum and states, without proof, thirteen theorems on centrifugal force. Most of the latter work antedates the *Horologium* and is reported in *De vi centrifuga*, which appears to have been written in 1659, but was published posthumously in 1703. Some of these theorems, including Huygens' result for centripetal acceleration, were communicated as anagrams to Henry Oldenburg, Secretary of the Royal Society, in a letter dated 4 September, 1669.

1.3. The Pendulum as a Scientific Instrument

The pendulum was recognized already in Newton's time as a superb instrument for the differential measurement of gravity, and modern pendulum gravimeters and seismometers are accurate to one part in ten million.

The earliest such measurement appears to be that of Jean Richer, who, having been sent to Cayenne in 1672 by the French Academy to observe an opposition of Mars, found that, to beat seconds, a pendulum must be made shorter at Cayenne than at Paris, "a discovery that marked the beginning of speculations as to the exact shape of the Earth." (Wolf 1950, p.67). Newton refers to Richer's observations and to similar observations by Halley and others in support of his calculation of the oblateness of the Earth in consequence of the centripetal variation of gravity (*Principia*, Book III, Proposition XX, Problem IV).

1.4. Influence of the Pendulum on Fluid Mechanics

Huygens' results apply strictly only to a pendulum *in vacuo*, and it was immediately recognized that the surrounding medium is responsible not only for dissipation but also for a dimunition of the period. It was at first thought that the latter effect is due solely to buoyancy, for which a correction was inferred from hydrostatic principles, but in 1828 Bessel pointed out that the virtual inertia of the medium could be significant. Bessel represented this inertia as the product of the displaced mass and an empirical constant k, which he and others determined experimentally. Poisson subsequently calculated $k = \frac{1}{2}$ for a sphere on the assumption of incompressible, irrotational flow. The calculation of both inertial and dissipative effects was made by Stokes (1850), who used his results to calculate the terminal velocity of a small water droplet in air, remarking that "The pendulum thus, in addition to its other uses, affords us some interesting information relating to the department of meteorology."

2. SYMMETRY BREAKING AND CHAOTIC MODULATION FOR THE SPHERICAL PENDULUM

Symmetry breaking in mechanics dates from Euler's (1744) determination of the buckling load of a column under a vertical load. It may occur in many ways for symmetrical forcing of a pendulum, but perhaps the most easily understood is the nonplanar response of a spherical pendulum to a small, horizontal, planar oscillation of its pivot (Miles 1962, 1984).

The spherical pendulum is perhaps the simplest example of a two-degree-of-freedom oscillator with identical natural frequencies and serves as a counterexample to the remarkable claims of both Lagrange and Laplace that such equality implies a linear (in t) growth of the solution of the corresponding differential equation.† It also serves as a prototype of internal resonance, in which energy may be transferred between the two degrees of freedom (which are uncoupled in the linear approximation) through nonlinear coupling, with the result that nonplanar motion may be excited by a planar displacement of the pivot if the driving frequency approximates the natural frequency.

Let l be the length of the pendulum, $\omega_0 = (g/l)^{\frac{1}{2}}$ its natural frequency, and δ its damping ratio ($\delta \lesssim 1$ for damping $\lesssim$ critical). We suppose that its pivot is subjected to the horizontal displacement

$$x_0 = \varepsilon l \cos \omega t \quad (0 < \varepsilon \ll 1) \tag{2.1}$$

and seek the horizontal displacement (x, y) of the pendulum bob for t sufficiently large to ensure the decay of transients associated with the initial conditions. The equations of motion have the form

$$\left[\frac{d^2}{dt^2} + 2\delta\omega_0 \frac{d}{dt} + \omega_0^2 \right](x,y) + \mathbf{N} = \varepsilon\omega_0^2 l (\cos\omega t, 0), \tag{2.2}$$

where $\mathbf{N}$ comprises the nonlinear terms and is cubic in the present approximation. If $\mathbf{N}$ is neglected (2.2) implies $y = 0$, but if δ is sufficiently small and ω approximates ω_0 (*resonance*) the nonlinear terms are manifestly significant and may transfer energy between the two degrees of freedom, in consequence of which the symmetric solution(s), for which $y = 0$, although still admissible, may not be stable.

We begin our analysis of (2.2) by remarking that, in the first approximation, $(d^2/dt^2)(x,y) \doteq -\omega^2(x,y)$ and cancels the term $\omega_0^2(x,y)$ at resonance ($\omega = \omega_0$); accordingly, if δ is sufficiently small the forcing term of amplitude ε must be balanced by the cubic $\mathbf{N}$, from which it follows that x and y must be $O(\varepsilon^{1/3})$. This last conclusion continues to hold if $\omega^2 - \omega_0^2$ and δ are $O(\varepsilon^{2/3})$, which suggests the introduction of the damping and tuning parameters

$$\mu \equiv \frac{2\delta}{\varepsilon^{2/3}}, \quad \nu \equiv \frac{\omega^2 - \omega_0^2}{\varepsilon^{2/3}\omega^2} \tag{2.3a,b}$$

and the assumption of a T-periodic ($T \equiv 2\pi/\omega$) solution of the form

$$(x,y) = \varepsilon^{1/3} l [(p_1, p_2)\cos\omega t + (q_1, q_2)\sin\omega t] + O(\varepsilon l), \tag{2.4}$$

in which $O(\varepsilon l)$ represents the third and higher harmonics generated by the cubic nonlinearity. Solving for $p_1, \cdots$, calculating the dimensionless energy

$$E \equiv \lim_{\varepsilon \to 0} \left\{ \frac{\frac{1}{2}m[\dot{x}^2 + \dot{y}^2 + \omega_0^2(x^2 + y^2)]}{\frac{1}{2}mgl\varepsilon^{2/3}} \right\} = p_1^2 + q_1^2 + p_2^2 + q_2^2, \tag{2.5}$$

and choosing $\mu = 1/4$ we obtain the resonance curve(s) plotted in Fig. 1 (in which $\sqrt{E}$ may be interpreted as a dimensionless, r.m.s. amplitude).

---

†Cf. Thomson & Tait (1912), who remark that Lagrange's error in the first (1788) edition of *Mécanique Analytique* is carried through to both the second (1811) and posthumous (1853) editions. Lamb, in a footnote (p.382) to the 1912 edition of Thomson & Tait, states that Lagrange's error was noticed and corrected by Weierstrass in 1858.

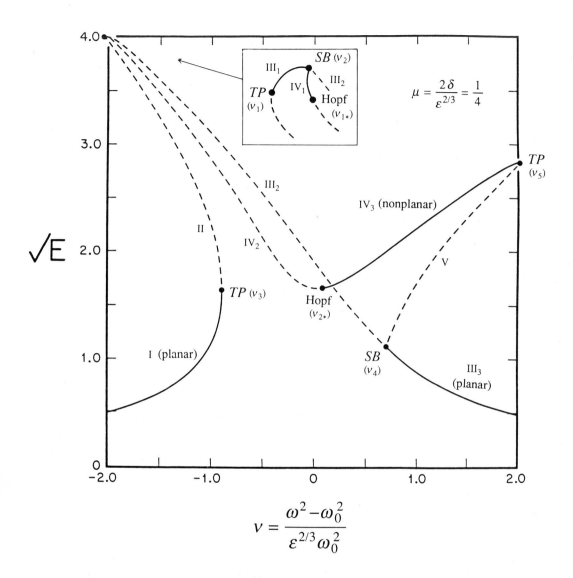

Fig. 1. Resonance curves for the spherical pendulum for $\mu = 1/4$. Branches I-III correspond to planar motion and branches IV and V to nonplanar motion. The solid/dashed segments correspond to stable/unstable oscillations. See text for details.

The solid/dashed lines in Fig. 1 comprise stable/unstable states. Branches I (which extends to $v = -\infty$) and III_3 (which extends to $v = \infty$) represent stable planar solutions; in addition, there is a very small branch III_1, at the top of the resonance curve between branches II and III_2, that is stable but difficult to realize. Branches II and III_2 represent unstable, planar solutions. Branch IV_3 represents stable, non-planar solutions; in addition, there is a small branch IV_1, between IV_2 and the junction of branches III_1 and III_2 that is stable but, like III_1, difficult to realize. Finally, branches IV_2 and V represent unstable, non-planar solutions. The junctions between branches I and II, II and III_1, and IV_3 and V, v_3, v_1 and v_5, respectively, are turning-point (saddle-node) bifurcations; the triple junctions of III_1, III_2 and IV_1 and of III_2, III_3 and V, v_2 and v_4, respectively, are symmetry-breaking (pitchfork) bifurcations; the junctions between IV_1 and IV_2 and between IV_2 and IV_3, v_{1*} and v_{2*}, respectively, are Hopf bifurcations. All seven of these bifurcations ($v_1 < v_2 \cdots < v_5, v_{1*} < v_{2*}$) exist for $0 < \mu < 0.433$, for which the configuration of Fig. 1 is representative. The number of bifurcations decreases from seven to zero as the damping parameter μ increases from 0.433 to 0.625; there are no non-planar fixed points, and the resonance curve represents only stable planar solutions, for $\mu > 0.625$. Numerical experiments imply that if a single, stable harmonic oscillation exists for a particular value of v the solution ultimately will tend to that point; if either two or three such states exist (it is impossible to have more than three in the present problem) for a particular v, which state is attained depends on the initial conditions. There is some range of v in which there are no stable harmonic oscillations if $\mu < 0.441$; if $\mu < 0.433$ this range is $v_3 < v < v_{2*}$, where v_3 is the turning-point bifurcation between the planar branches I and II, and v_{2*} is the Hopf bifurcation that separates the non-planar branches IV$_2$ and IV$_3$.

If $v > v_5$ the solution following the decay of an initial transient is asymptotic to a T-periodic, planar oscillation on branch III$_3$. If v is slowly decreased the solution will follow III$_3$ to the symmetry-breaking bifurcation v_4 (although a solution started from arbitrary initial conditions may terminate in a T-periodic, non-planar oscillation on branch IV$_3$). As v is decreased through v_4 the solution jumps to a T-periodic, non-planar oscillation on branch IV$_3$ and remains there until v reaches the Hopf bifurcation at v_{2*}.

Stable, T-periodic solutions are impossible for $v_3 < v < v_{2*}$, but the solution remains approximately sinusoidal, and (2.4) may be generalized by allowing (p_1, q_1, p_2, q_2) to depend on the slow time

$$\tau = \tfrac{1}{2}\varepsilon^{2/3}\omega t . \tag{2.6}$$

The substitution of (2.4) into (2.2) then yields the evolution equations

$$\left[\frac{d}{d\tau} + \mu\right](p_1, q_1, p_2, q_2) + (P_1, Q_1, P_2, Q_2) = (0, 1, 0, 0) , \tag{2.7}$$

where $P_1, \cdots$ are cubic functions of $p_1, \cdots$. The divergence of (P_1, Q_1, P_2, Q_2) vanishes; accordingly, the divergence of the system (2.7) in the four-dimensional phase space is

$$\frac{\partial \dot{p}_1}{\partial p_1} + \frac{\partial \dot{q}_1}{\partial q_1} + \frac{\partial \dot{p}_2}{\partial p_2} + \frac{\partial \dot{q}_2}{\partial q_2} = -4\mu \quad (\dot{p}_1 \equiv dp_1/d\tau , \cdots) , \tag{2.8}$$

from which it follows that an element of volume in the phase space contracts like $\exp(-4\mu\tau)$, and any trajectory ultimately must be confined to a limiting subspace of dimension smaller than four [cf. Lichtenberg & Lieberman (1983)].

Numerical experiments for $\mu = \frac{1}{4}$ and $v_3 < v < v_{2*}$ yield the p_1–p_2 projections and the power spectra shown in Fig. 2 (Miles 1984). [Note that $(x, y) = \varepsilon^{1/2}l(p_1, p_2)$ at $\omega t = 0, 2\pi, \cdots$, so that (p_1, p_2) may be regarded as a Poincaré projection.] The asymptotic $(\tau \to \infty)$ trajectories for $v_{2*} = 0.072 > v > -0.04$, e.g. $v = 0.05$ in Fig. 2, are periodic limit cycles that project as ovals in the p_1–p_2 plane; the corresponding spectra of E (or, equivalently, of the envelope of the sinusoidal carrier) comprise the fundamental frequency F_1 and its harmonics, with F_1 decreasing from $F_* = 0.160$, the (analytically) calculated frequency of the nascent limit cycle at the Hopf-bifurcation point, with decreasing v. Very weak period doubling is observed at $v = -0.04$, and the half-order subharmonics increase in strength as v is decreased from -0.04 to -0.144; see $v = -0.10$ in Fig. 2. Period quadrupling (i.e. a second period doubling) first occurs, with decreasing v, in the interval $(-0.13, -0.14)$, and the quarter-order subharmonics increase in strength down to -0.144; see $v = -0.142$ in Fig. 2. Period octupling (a third period doubling) is observed at $v = -0.144$; see Fig. 2. The metamorphosis of the trajectories from almost elliptical ovals through increasing convolutions at each period doubling is illustrated by the results for $v = 0.05, -0.10, -0.142$, -0.144 and -0.145 in Fig. 2 after allowing for the fact that the initial condiitons for $v = -0.144$ were altered to obtain the complement of the trajectory that had been obtained for initial conditions identical with those for $v = 0.05, -0.10$ and -0.142. [The phase-space equations (2.8) are invariant under the reflection $(p_2, q_2) \to -(p_2, q_2)$ and therefore admit asymptotic solutions in complementary pairs; which member of the pair is attained depends on the initial conditions.]

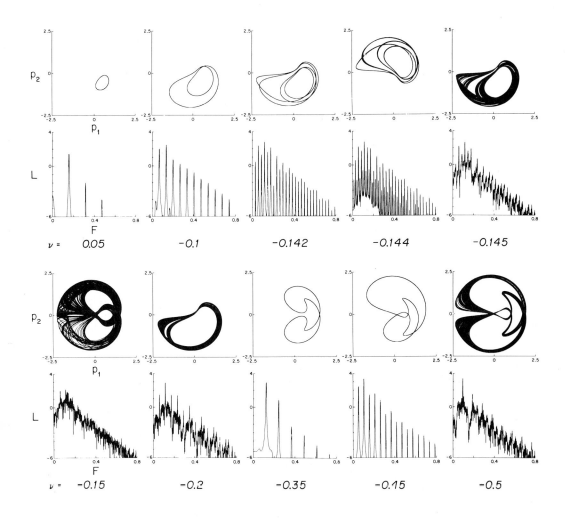

Fig. 2. Phase-plane projections and power spectra for spherical pendulum for $\mu = 1/4$ ($v_3 = -0.92$, $v_{2*} = 0.07$); p_1 and p_2 are defined by (2.4), $L = \log_{10} E$, where E is given by (2.5), and $F = $ (spectral frequency)$/\omega_0$ (so that $F = 1/(2\pi)$ is the fundamental frequency).

Chaotic trajectories occur throughout the v interval $[-0.145, -0.20]$. The typical $p_1 - p_2$ trajectory within this interval, e.g. $v = -0.15$ (see also $v = -0.5$) in Fig. 2, appears to be symmetric with respect to $p_2 = 0$ after a sufficiently long time, although it may spend substantial intervals in either the top or bottom half of this pattern, transferring from one to the other half at aperiodically spaced times that are quite sensitive to the initial conditions. The available evidence (broad-banded spectra, sensitivity to initial conditions, and intermittent transfer between two complementary domains of the phase space) strongly suggests that these trajectories lie on a strange attractor of the type discovered by Lorenz (1963) and Ruelle & Takens (1971). The trajectories for $v = -0.145$ and -0.20 (see Fig. 2) are exceptional in being confined to one of the two (upper and lower) domains of the putative attractor. The complement of the trajectory shown in Fig. 2 for $v = -0.145$ was obtained by altering the initial conditions from $(p_1, q_1, p_2, q_2) = (0,0,0,1)$ to $(0,0,1,0)$; on the other hand, essentially the same trajectory as that shown for $v = -0.20$ was obtained with each of the three sets of initial conditions $(0,0,0,1)$, $(0,0,1,0)$ and $(0,\frac{1}{2},\frac{1}{2},\frac{1}{2})$.

The trajectories for $v = -0.21$ and -0.22 are limit cycles, with period doubling as v increases from -0.22 to -0.21. Those for $v = -0.23$ and -0.24 are chaotic and appear to lie on strange attractors. Those for $-v = 0.25 \ (.01) \ 0.30 \ (.05) \ 0.45$ are periodic, with period doubling at -0.25 as v increases to -0.24 and at -0.40 and -0.45 as v decreases to -0.50, where a strange attractor again occurs; see $v = -0.35, -0.45$ and -0.50 in Fig. 2. It seems likely that period quadrupling and octupling could be observed through finer sampling over the spectral intervals that separate those of chaotic trajectories, just as in the v interval $(-0.13, 0.145)$. Moreover, the appearance of strange attractors within the narrow interval $(-0.22, -0.25)$ suggests that there could be additional such subintervals within the interval $(-0.30, -0.50)$. The v interval $(-0.50, -0.92)$ also contains subintervals of both periodic and chaotic trajectories.

3. SYMMETRY BREAKING AND CHAOS FOR THE DIRECTLY FORCED PLANAR PENDULUM

Symmetry breaking of the forced oscillations of a planar pendulum has been observed by D'Humieres *et al.* (1982) in their numerical solution of

$$\ddot{\theta} + 2\delta\omega_0\dot{\theta} + \omega_0^2\sin\theta = \varepsilon\omega_0^2\sin\omega t , \tag{3.1}$$

where ε now is the ratio of the maximum external torque to the maximum gravitational torque acting on the pendulum, and by Blackburn *et al.* (1987) in an experimental study. If ε is sufficiently small the solution of (3.1) is T-periodic and symmetric, in the senses that

$$\theta(t+T) = \theta(t) , \quad \theta(t+\tfrac{1}{2}T) = -\theta(t) \quad (T \equiv 2\pi/\omega) , \tag{3.2a,b}$$

and may be expanded in a Fourier series that contains only the odd harmonics of ω. But if ε exceeds a certain critical value, ε_s (the subscript s signifies symmetry breaking), the solution in some frequency interval may be T-periodic but asymmetric in that it satisfies (3.2a) but violates (3.2b) and comprises both odd and even harmonics. This asymmetric, T-periodic solution is stable in only a small subdomain of the $\delta, \varepsilon, \omega$-parameter space and becomes unstable through a period-doubling bifurcation that typically leads to chaos through a period-doubling cascade; see Fig. 3.

We proceed on the assumption that $\delta \ll 1$ and the hypothesis that the truncation†

$$\theta = \theta_0 + \alpha\sin(\omega t - \phi) \tag{3.3}$$

provides an at least qualitatively valid description of the solution of (3.1). Perhaps the most satisfactory way of determining θ_0, α and ϕ is to regard them as generalized coordinates in an average-Lagrangian formulation (which procedure admits generalization to higher-order truncations of the Fourier series); however, it is more direct, and yields the same results, to substitute (3.3) into (3.1) and take moments with respect to 1, $\cos\omega t$ and $\sin\omega t$ (the method of harmonic balance) to obtain

$$J_0(\alpha)\sin\theta_0 = 0 , \quad 2\delta\alpha(\omega/\omega_0) = \varepsilon\sin\phi , \quad 2J_1(\alpha)\cos\theta_0 - \alpha(\omega/\omega_0)^2 = \varepsilon\cos\phi , \tag{3.4a,b,c}$$

where $J_n \equiv J_n(\alpha)$ is a Bessel function of the first kind.

It follows from (3.4a) that there are three distinct solutions of the form (3.3): $\theta_0 = 0$, which corresponds to symmetric oscillations; $\alpha = j_{0n}$ $(j_{01} = 2.40, \ j_{02} = 5.52, \cdots$ are the zeros of $J_0)$ and $0 < |\theta_0| < \pi$, which corresponds to asymmetric oscillations; $\theta_0 = \pi$, which corresponds to symmetric oscillations of the inverted pendulum. We provisionally assume $\alpha \leq j_{01}$ on the hypothesis that $\alpha > j_{01}$ lies outside of the domain of the truncation (3.3), but see §§3.5-3.7.

---

†The development of (3.4)-(3.14) is from Miles (1988a).

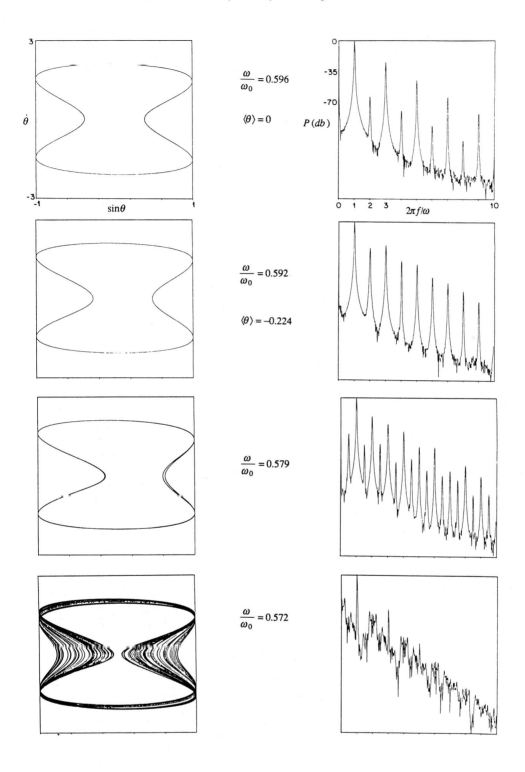

Fig. 3. Phase-plane trajectories and power spectra for the planar pendulum described by (3.1) for $\delta = 1/8$ and $\varepsilon = 1/2$, showing the progression from symmetric oscillations ($\omega/\omega_0 = 0.596$) through symmetry breaking ($\omega/\omega_0 = 0.592$) and period doubling ($\omega/\omega_0 = 0.579$) to chaos ($\omega/\omega_0 = 0.572$). The scales for the lower plots are identical with those in the top plots.

3.1. Symmetric swinging oscillations

Setting $\theta_0 = 0$ and eliminating the phase angle ϕ between (3.4b,c), we obtain the resonance curve(s)

$$\alpha\{[(\omega/\omega_0)^2 - \Omega]^2 + 4\delta^2(\omega/\omega_0)^2\}^{1/2} = \varepsilon \quad (\Omega \equiv 2J_1/\alpha) ; \tag{3.5}$$

see Fig. 4.

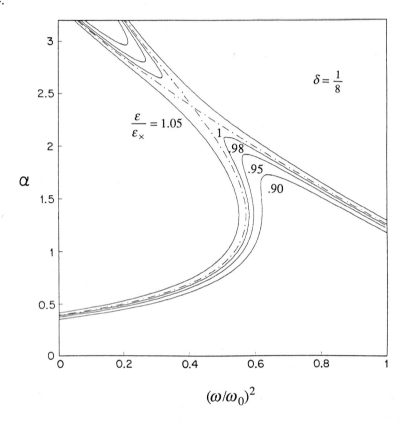

Fig. 4. Resonance curves for planar pendulum as determined from (3.5) for $\delta = 1/8$ and $\varepsilon/\varepsilon_\times = 0.9, 0.95, 0.98, 1 \ (-\cdot-)$ and 1.05. All points above $\alpha = 2.40$, between the turning points for $\varepsilon < \varepsilon_\times$, or above the turning point on the left branch for $\varepsilon > \varepsilon_\times$ correspond to unstable states.

Resonance, as defined by $\phi = \frac{1}{2}\pi$ (so that the external moment is in phase with the damping moment), occurs at $(\omega/\omega_0)^2 = \Omega$ (the *spine* of the resonance curve), and is possible if and only if $\varepsilon < \varepsilon_*$, where, in the present approximation

$$\varepsilon_* = 2\delta(2\alpha J_1)^{1/2}_{max} = 3.16\delta , \quad \alpha_* = \alpha_0 = 2.40 , \quad \omega_*/\omega_0 = .657 . \tag{3.6a,b,c}$$

The resonance curve has a maximum (which is on the spine only in the limit $\delta \downarrow 0$) if and only if $\varepsilon < \varepsilon_\times$, where

$$\varepsilon_\times = \varepsilon_*[1 - 1.16\delta^2 + O(\delta^4)] . \tag{3.7a}$$

The resonance curve for $\varepsilon = \varepsilon_\times$ crosses itself at $\alpha = \alpha_\times$ and $\omega = \omega_\times$, where

$$\alpha_\times = \alpha_*[1 - 0.80\delta^2 + O(\delta^4)] , \quad \omega_\times = \omega_*[1 - 1.5\delta^2 + O(\delta^4)] . \tag{3.7b,c}$$

The resonance curves in the neighborhood of this crossing approximate hyperbolae (see Fig. 4), but the branches above $\alpha = \alpha_\times$ for $\varepsilon < \varepsilon_\times$ comprise stable states only in an $O(\delta^2)$ subdomain of $\alpha_\times < \alpha < \alpha_*$.

If $\varepsilon < \varepsilon_\times$ (and $\delta^2 \ll 1$) the resonance curves for $\alpha < \alpha_\times$ have two turning points, at which $d\omega/d\alpha = 0$, and the segment connecting these turning points comprises only unstable states. If $\varepsilon > \varepsilon_\times$ the resonance curve consists of two separate branches. The right branch is monotonic ($d\alpha/d\omega < 0$), while the left branch has a single turning point and intersects $\omega = 0$ at the two points determined by $2J_1(\alpha) = \varepsilon$ if $\varepsilon < 1.16$ or disappears if $\varepsilon > 1.16$. That segment of the left branch above the turning point comprises only unstable states.

The present approximations are in error by at most a few percent for $\varepsilon < \varepsilon_*$ and $\alpha < \alpha_*$, but the contributions of higher harmonics may be (but see §3.5) much larger if $\varepsilon > \varepsilon_*$, and it then is preferable to plot the root-mean energy

$$\langle E \rangle^{\frac{1}{2}} = \langle \tfrac{1}{2}\dot\theta^2 + 1 - \cos\theta \rangle^{\frac{1}{2}} = [\tfrac{1}{4}\alpha^2 (\omega/\omega_0)^2 + 1 - J_0(\alpha)]^{\frac{1}{2}} \tag{3.8}$$

rather than α (note that $\sqrt{\langle E \rangle} \to \alpha/\sqrt{2}$ as $\alpha \to 0$). This result is graphically indistinguishable from the result of Bryant's (1988) numerical analysis (in which the number of terms included in the Fourier expansion of θ is sufficient to reduce the error in $\langle E \rangle$ to 10^{-3}) for $\varepsilon \leq 0.3$ and only marginally distinguishable for $\varepsilon = 0.4$, but differs significantly therefrom for $\varepsilon = 0.5$; see Fig. 5.

3.2. Symmetry breaking and asymmetric oscillations

It follows from (3.4a) that symmetry is broken at $\alpha = \alpha_s = 2.40$ in the present approximation and from $\alpha_s > \alpha_\times$ that the symmetry-breaking bifurcations, at which symmetric oscillations lose stability to asymmetric oscillations, occur on the right branches of the separated resonance curves. (The present approximation also predicts symmetry-breaking bifurcations above the turning points on the unstable segments of the left branches of the separated resonance curves, but these bifurcations are not significant.)

The asymmetric oscillations are stable only in a rather small parametric domain and become unstable through a period-doubling bifurcation, which is followed, for increasing ε or decreasing ω, by a period-doubling cascade (Fig. 3). Bryant (1988) has calculated the first three bifurcations (period doubling, quadrupling and octupling) in this cascade for $\delta = 1/8$ and $\varepsilon = 1/2$; see Fig. 6, in which the locus of stable asymmetric oscillations is $P_1 P_2$.

If $\delta \ll 1$ the symmetry-breaking bifurcation, $\omega = \omega_s$, may lie to the left of the right-hand turning-point bifurcation, $\omega = \omega_+$, in which case the solution of (3.1) may be asymptotic (as $\omega t \uparrow \infty$) to either a symmetric oscillation on the lower branch of the resonance curve (3.5) or an asymmetric oscillation, and which of these two states is realized for $\omega_s < \omega < \omega_+$ depends on the initial conditions.

3.3. Resonance in the limit $\delta, \varepsilon \downarrow 0$

The T-periodic solution of (3.1) in the limit $\delta \downarrow 0$ with $\varepsilon = O(\delta)$ is given by (Miles 1988a)

$$\theta(t) = 2\sin^{-1}(k\,\mathrm{sn}\,\tau) + \sum_1^N \delta^n \theta_n(\tau) + O(\delta^{N+1}), \quad \tau \equiv (2K/\pi)(\omega t - \phi), \quad \frac{\omega}{\omega_0} = \frac{\pi}{2K}, \tag{3.9a,b,c}$$

where sn is a Jacobi elliptic sinθ, K is a complete elliptic integral of modulus k, and k is determined by (3.9c). The requirement that the work done by the external moment equal the dissipation implies

$$\varepsilon \omega_0^2 \langle \dot\theta(t) \sin\omega t \rangle = 2\delta\omega_0 \langle \dot\theta^2(t) \rangle. \tag{3.10}$$

Substituting (3.9a) into (3.10) and letting $\delta \downarrow 0$, we obtain

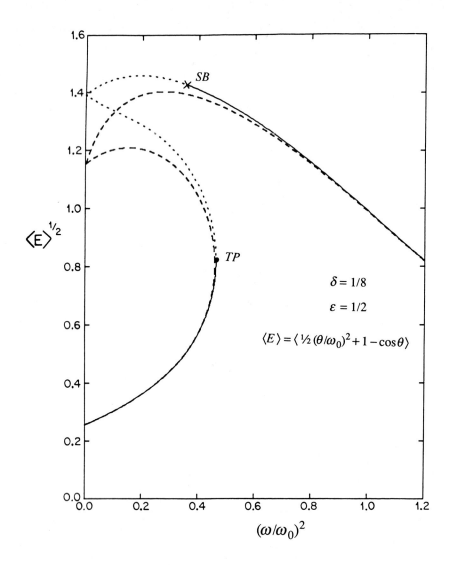

Fig. 5. Resonance curves for planar pendulum for $\delta = 1/8$ and $\varepsilon = 1/2$ as determined from (3.5) and (3.8) (– – –) and by Bryant (1988) through numerical collocation for stable/unstable (——/···) states.

$$(\varepsilon/2\delta)\sin\phi = R\,(\omega/\omega_0)\,, \qquad (3.11)$$

where R is plotted in Fig. 7. It follows that resonance, as defined here, is impossible for $\varepsilon > 2\delta R_* \equiv \varepsilon_*$, where $R_* = 1.642$ is the (only) maximum of R. This compares with $\varepsilon_*/2\delta = 1.58$ from (3.6a), which therefore is in error by 3.7% in the limit $\delta \downarrow 0$.

The higher-order functions θ_n satisfy inhomogeneous Lamé equations of the term

$$\theta_n'' + (1 - 2k^2\mathrm{sn}^2\tau)\theta_n = f_n(\tau) \qquad (3.12)$$

and may (at least formally) be expanded in Lamé functions. The resulting prediction of the symmetry-breaking bifurcation for $N = 1$ yields $\omega_s = 0.613$ for $\delta = 1/8$ and $\varepsilon = \frac{1}{2}$, in reasonable agreement with $\omega_s/\omega_0 = 0.595$ from Bryant's (private communication) numerical simulation.

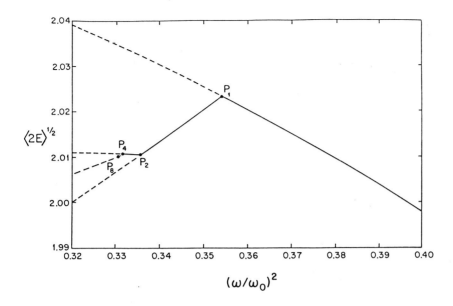

Fig. 6. Bifurcating resonance curves for symmetric, asymmetric, period-2, and period-4 oscillations (——/−−− for stable/unstable states), as determined by Bryant (1988).

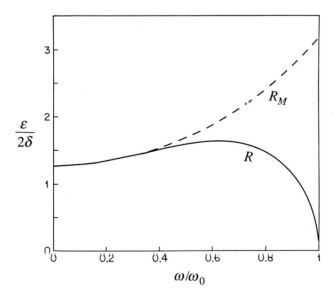

Fig. 7. The ratio $\varepsilon/2\delta$ at resonance ($\phi = \frac{1}{2}\pi$) for the planar pendulum in the limit $\delta \downarrow 0$ (3.11) (——) compared with Melnikov's criterion (3.14) (−−−)

3.4. Melnikov's criterion

Melnikov's criterion for the transition to chaos through bifurcation from the homoclinic orbit

$$\sin \tfrac{1}{2}\theta_h = \tanh \omega_0 t , \quad \theta_h = 2\,\mathrm{sech}\,\omega_0 t \tag{3.13a,b}$$

is given by (Guckenheimer & Holmes 1983, §4.5)

$$\frac{\varepsilon_M}{2\delta} = \frac{\displaystyle\int_0^\infty \dot\theta_h^2(t)\,dt}{\displaystyle\omega_0\int_0^\infty \dot\theta_h(t)\cos\omega t\,dt} = \frac{4}{\pi}\cosh\left[\tfrac{1}{2}\pi\frac{\omega}{\omega_0}\right] \equiv R_M\left[\frac{\omega}{\omega_0}\right], \tag{3.14}$$

which exceeds R everywhere in $0 < \omega < \omega_0$ but is quite close thereto for $\omega < 0.4\omega_0$ (see Fig. 7). Letting $\delta = 1/8$ and $\varepsilon = 1/2$, we obtain $\omega = 0.651$ from (2.14), for the transition to chaos, which compares with $\omega = 0.575$ from numerical simulation (Fig. 6). It should be emphasized that Melnikov's criterion effectively assumes that the transition to chaos occurs for $\phi = \tfrac{1}{2}\pi$ (on the spine of the resonance curve), which is not possible for $\varepsilon > \varepsilon_*$ except in the limit $\omega \downarrow 0$. Moreover, Melnikov's criterion is associated with the subharmonic-bifurcation cascade nT ($n = 1, 2, 3, \cdots$) (l.c.a.), which contrasts with the period-doubling cascade $2^n T$ ($n = 0, 1, 2, \cdots$) that follows symmetry breaking.

3.5. Resonance curves for $\varepsilon \gg \delta$

As ε is increased above that value at which the left branch of the resonance curve disappears ($\varepsilon = 1.16$ in the approximation of §3.1), loops begin to form in the remaining branch (see Fig. 8a). Points to the left of the first symmetry-breaking bifurcation ($\omega/\omega_0 = 0.870/1.057$ in Fig. 8a/b) are unstable except for short segments that connect symmetry-breaking bifurcations on the left to turning points on the right: e.g. $\omega/\omega_0 = (0.390, 0.403), (0.264, 0.268), (0.203, 0.205)$ in Fig. 8a.

In the stable interval $\omega/\omega_0 > 0.870$ of Fig. 8a the amplitude α ranges from 0 to 0.82π as ω/ω_0 decreases from ∞ to the bifurcation point. The amplitude ranges for the remaining three, stable intervals (see above) are $(2.61\pi, 2.80\pi)$, $(4.60\pi, 4.84\pi)$ and $(6.66\pi, 6.76\pi)$, respectively, which imply that: symmetric oscillations in the first stable interval do not reach the upward vertical; those in the second stable interval pass through the upward vertical ($\theta = \pi$) once in each direction during each period and continue through the downward vertical ($\theta = 2\pi$) before reversing; those in the third stable interval pass through the upward vertical twice ($\theta = \pi, 3\pi$) and continue through the downward vertical ($\theta = 2\pi, 4\pi$) twice consecutively during each period; etc.

The resonance curve in Fig. 8b is predicted by (3.5) within an error that is barely discernible at the scale of the drawing (the maximum discrepancy in amplitude is roughly 2%), and the symmetry-breaking bifurcations above the first are predicted by the higher odd zeros of $J_0(\alpha)$, $\alpha = j_{03} = 8.65$, $j_{05} = 14.93$, $\cdots$ (the higher even zeros, j_{02}, j_{04}, $\cdots$ lie on unstable segments of the resonance curve). Moreover, the amplitudes in the stable intervals differ asymptotically by increments of 2π, as predicted by the asymptotic approximation to $J_0(\alpha)$. This agreement is surprising in view of the Fourier truncation implicit in the approximation $\theta = \alpha\sin(\omega t - \phi)$;[†] however, Bryant's numerical solutions reveal that the higher harmonics are indeed small for $\delta = 1/8$ and $\varepsilon = 2$ — e.g., α_3/α_1 (ratio of third harmonic to fundamental) $= 0.017/0.020/0.018$ at $\omega/\omega_0 = 1.057/0.356/0.235$.

3.6. Inverted oscillations

The counterpart of the resonance curve (3.5) for $\theta_0 = \pi$ is

$$\alpha\{[(\omega/\omega_0)^2 + \Omega] + 4\delta^2(\omega/\omega_0)^2\}^{1/2} = \varepsilon . \tag{3.15}$$

This curve (Fig. 9) has a single turning point ($d\omega/d\alpha = 0$) and opens to the right if $\varepsilon > 1.16$; it has separate, upper and lower branches if $\varepsilon < 1.16$. The upper branch comprises only unstable states, but a

---

[†] One is reminded of Gilbarg & Paolucci's (1953) remark that 'Equations have often been successful beyond the limits of their original derivation, and indeed this ... is one of the hallmarks of a great theory.'

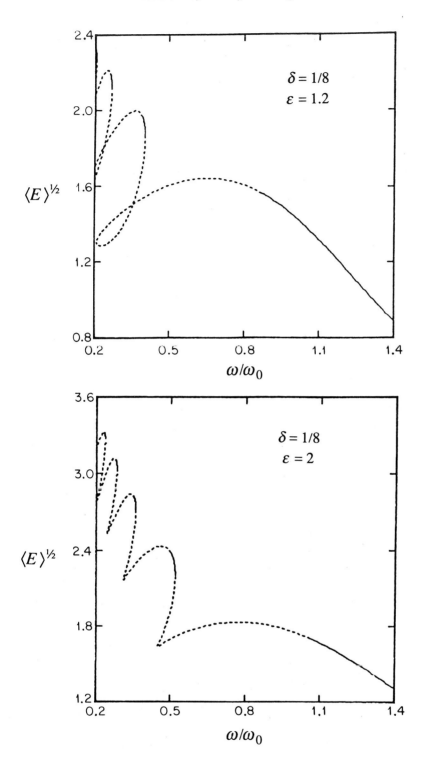

Fig. 8a,b. Resonance curves for planar pendulum for $\delta = 1/8$ and $\varepsilon =$ (a) 1.2 and (b) 2, as determined by Bryant (private communication) through numerical collocation for stable/unstable (———/– – –) states. The curve for $\varepsilon = 2$ is well approximated by (3.5) and (3.8).

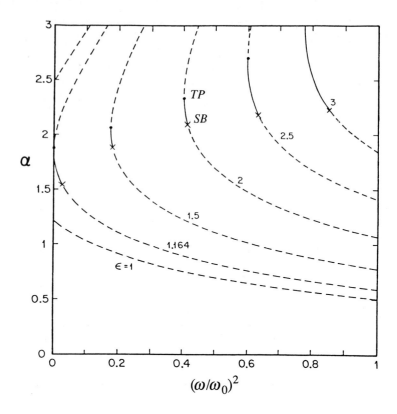

Fig. 9. Resonance curves for symmetric oscillations of inverted pendulum for $\delta = 1/8$ and $\varepsilon = 1$, 1.164, 1.5, 2 and 2.5 (——/–––– for stable/unstable states), as determined from (3.15). The dots/crosses are turning points/symmetry-breaking bifurcations.

small segment of the lower branch, just below the turning point and above a symmetry-breaking bifurcation, comprises stable states (Miles 1988b).

Numerical integration of (3.1) for $\delta = 1/8$ and $\varepsilon = 2.5$, for which stable inverted oscillations are predicted in $0.77 < \omega/\omega_0 < 0.79$, yielded oscillations with $\theta_0 = \pi \mod 2\pi \pm 0.01$ for $0.725 \le \omega/\omega_0 \le 0.792$, symmetry breaking at $\omega/\omega_0 = 0.794$, period doubling at $\omega/\omega_0 = 0.796$, and chaos at $\omega/\omega_0 = 0.800$; see Figs. 10 and 11.

Bryant (private communication) finds, by comparison with his numerical results, that (3.15) also accurately predicts the higher branches of the resonance curve, which resembles those of Fig. 8, if ε is not too small (the maximum error in $\langle E \rangle^{1/4}$ vs ω for $\delta = 1/8$ and $\varepsilon = 2$ is less than 2%). He also finds that the minimum value of ε for stable inverted oscillations for $\delta = 1/8$ is $\varepsilon = 0.48$, but (3.15) is not accurate for ε this small.

3.7. Rotational oscillations

Free rotational oscillations of a pendulum are described by

$$\theta = 2\mathrm{am}(\omega_0 t/k) = \omega t + 4 \sum_{n=1}^{\infty} \frac{q^n \sin(n\omega t)}{n(1+q^{2n})}, \quad \frac{\omega}{\omega_0} = \frac{\pi}{kK} \quad (\delta = \varepsilon = 0), \qquad (3.16a,b)$$

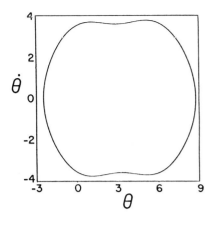 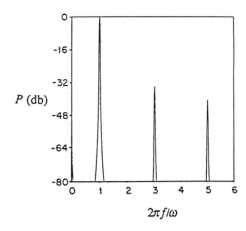

Fig. 10. Inverted oscillations just above the turning-point frequency, $\omega/\omega_0 = 0.725$, for $\delta = 1/8$ and $\varepsilon = 2.5$ after the decay of the initial transient. The mean value of θ, which has been subtracted out in the calculation of the power spectrum, is $\langle\theta\rangle = 3.155 \bmod 2\pi$. The three peaks in the power spectrum represent the fundamental and the third and fifth harmonics.

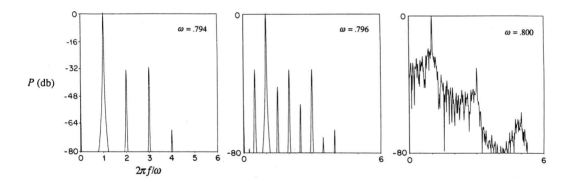

Fig. 11. The power spectra (after subtracting out the mean) for inverted oscillations at symmetry breaking ($\omega/\omega_0 = 0.794$), period doubling ($\omega/\omega_0 = 0.796$), and the initiation of chaotic modulation ($\omega/\omega_0 = 0.800$) for $\delta = 1/8$ and $\varepsilon = 2.5$. Note the appearance of the second harmonic (cf. Fig. 10), which implies symmetry breaking, for $\omega/\omega_0 = 0.794$, and of the subharmonic and its multiples, which imply period doubling, for $\omega = 0.796$. [$\omega \equiv \omega/\omega_0$ in above figures.]

where am is Jacobi's amplitude of modulus k, K is a complete elliptic integral, and q is the Jacobi nome. The angular velocity $\dot{\theta}$ is periodic and positive-definite and has the mean value ω. Perturbation solutions for $\delta \ll 1$ and $\varepsilon = O(\delta)$ may be constructed as in §3.3 and also have a mean angular velocity of ω, but the oscillation about this mean is stronger, and $\dot{\theta}$ typically is not positive-definite. These solutions are stable in a finite subdomain of the $\delta, \varepsilon, \omega$ space, but are excited only for a rather narrow range of initial conditions. .

Rotational oscillations for $\varepsilon \geq 1$ exhibit a much richer variety of features than those for $\varepsilon \ll 1$. In particular, the mean angular velocity may be $p\omega/q$, where p and q are integers, and the period is qT:

$$\theta = \frac{p}{q}\omega t + \sum_{n=0}^{\infty} \alpha_n \cos\left[\frac{n\omega t}{q} - \phi_n\right]. \tag{3.17}$$

The amplitude $\theta - (p/q)\omega t$ may exceed 2π, in consequence of which the pendulum periodically executes one or several complete revolutions in the direction opposite to the mean rotation; see, e.g., Fig. 12. Only a few analytical results are available for these more complicated rotational oscillations, and the available numerical results are not yet adequate to provide an even qualitatively valid description of all of the possible motions.

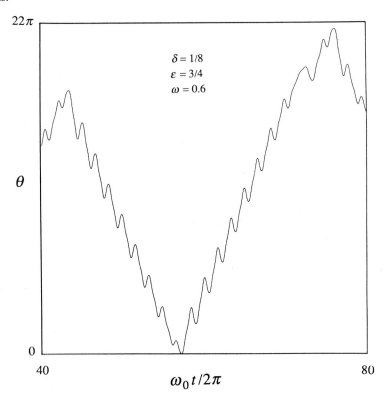

Fig. 12. Reversing rotational oscillations for $\delta = 1/8$, $\varepsilon = 3/4$ and $\omega/\omega_0 = 0.6$. [$\omega \equiv \omega/\omega_0$ in figure.]

4. PLANAR PENDULUM: HORIZONTAL OSCILLATION OF PIVOT

Bryant & Miles (1988a) have considered symmetry breaking and chaotic motion of a planar pendulum that is driven by a prescribed, horizontal displacement $\varepsilon(g/\omega^2)\sin\omega t$ (an acceleration of $-\varepsilon g \sin\omega t$) of its pivot. The equation of motion is

$$\ddot{\theta} + 2\delta\omega_0\dot{\theta} + \omega_0^2\sin\theta = \varepsilon\omega_0^2\cos\theta\sin\omega t , \qquad (4.1)$$

where θ, δ and ω_0 are defined as in (3.1). The solutions of (4.1) are evidently equivalent to those of (3.1) if the constraint $|\theta| \ll 1$ is imposed, but there are significant differences in the nonlinear regime.

The assumption of sinusoidal motion of the form (3.3) leads to a set of moment equations similar to (3.4). The resulting resonance curve for symmetric ($\theta_0 = 0$) oscillations crosses itself if $\varepsilon > 4.63\,\delta = \varepsilon_\times$. This crossing differs significantly from that associated with breaking of the resonance curve (cf. Fig. 4), which also is predicted in the present case by, but lies outside of the domain of, the approximation of sinusoidal motion. The analytical approximation is indistinguishable (on the scale of the present plots) from the numerical calculation for $\delta = 1/8$ and $\varepsilon < \varepsilon_\times$ and is quite good below the crossing for $\varepsilon = 0.8$ (Fig. 13a). The resonance curve is triple-valued in $\langle E \rangle^{1/2}$ between the turning points at $\omega/\omega_0 = 0.707$ and 0.800, and the symmetric oscillations are stable except on the segment connecting these points.

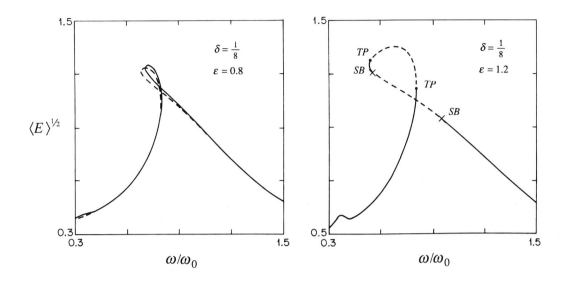

Fig. 13. Resonance curves for horizontal oscillation of pivot of planar pendulum for: (a) $\delta = 1/8$ and $\varepsilon = 0.8$, as determined through a sinusoidal approximation $(---)$ and numerical integration of (4.1); (b) $\delta = 1/8$ and $\varepsilon = 1.2$, as determined through numerical integration of (4.1). The dots/crosses are turning points/symmetry-breaking bifurcations ($\underline{\quad\quad}/---$ for stable/unstable states).

The resonance curve for symmetric oscillations with $\delta = 1/8$ and $\varepsilon = 1.2$ is plotted in Fig. 13b. As in Fig. 13a, symmetric oscillations are stable to the lower turning point at $\omega = \omega_+ = 0.803\omega_0$ and then unstable to the upper turning point at $\omega = \omega_- = 0.533\omega_0$ (the unstable section is dashed in the figure). However, in contrast to Fig. 13a, the symmetric oscillations in Fig. 13b are stable only for a small interval below the upper ($\omega = \omega_-$) turning point, are unstable for a much larger (dashed) section, and are again stable as ω increases further. Asymptotic solutions of (4.1) on the unstable sides of the latter two stability boundaries show that both are symmetry-breaking bifurcations, with the symmetric oscillations losing stability to asymmetric oscillations. As ε is increased from 1.2 to 2.0 the resonance curves for symmetric oscillations retain the same form as in Fig. 13b, with the short stable section just below the upper turning point becoming almost vanishingly small, and the unstable section following it increasing in length at both ends.

The asymmetric states that emerge from the symmetry-breaking bifurcations have only a small range of stability (cf. §3.2) and lose stability through a period-doubling bifurcation that leads to chaos through a period-doubling cascade.

The numerical integration of (4.1) also yields: $3T$-periodic, symmetric solutions, which (with increasing ε) lose stability through a symmetry-breaking bifurcation and progress to chaos through a period-doubling cascade; inverted, $3T$-periodic solutions of the form

$$\theta = \pi + \sum_{n=0}^{N} \alpha_{2n+1} \cos\left[\left[\frac{2n+1}{3}\right]\omega t - \phi_n\right] ; \qquad (4.2)$$

rotational oscillations of the form (3.17).

5. PARAMETRICALLY EXCITED OSCILLATIONS

Now suppose that the pendulum is driven by the prescribed, vertical acceleration $\varepsilon g \sin \omega t$ of its pivot (cf. §4). The equation of motion is

$$\ddot{\theta} + 2\delta\omega_0\dot{\theta} + \omega_0^2(1+\varepsilon\sin\omega t)\sin\theta = 0,\qquad(5.1)$$

where θ, δ and ω_0 are defined as in (3.1). It is evident that (5.1) admits the solution $\theta = 0$, but if ε is sufficiently large a wide variety of oscillations is possible. The excitation then is described as parametric, in reference to the periodic variation of the effective gravitational acceleration. I do not know who first observed oscillations of this type, but their scientific study appears to date from Faraday's (1831) discovery of oscillations of frequency $\tfrac{1}{2}\omega$ on the surface of a liquid in a container subjected to a vertical oscillation of frequency ω. The first analytical treatment of the problem appears to be due to Rayleigh (1883), who showed that the linear approximation (so that $\sin\theta \approx \theta$) to (5.1) admits solutions of the form

$$\theta = \alpha \sin \tfrac{1}{2}(\omega t - \phi)\qquad(5.2)$$

if (for $\delta^2 \ll 1$)

$$\varepsilon < 4\delta, \quad (\tfrac{1}{2}\omega/\omega_0)^2 < 1 + \tfrac{1}{2}(\varepsilon^2 - 16\delta^2)^{\frac{1}{2}}.\qquad(5.3a,b)$$

The assumption of the sinusoidal motion (5.2) without the restriction $|\theta| \ll 1$ leads to a set of moment equations similar to (3.4). If $\varepsilon_0 < \varepsilon < \varepsilon_*$, where [cf. (5.3a)]

$$\varepsilon_0/4\delta = (1-\delta^2)^{\frac{1}{2}}, \quad \varepsilon_*/4\delta = 1.110[1 - 0.11\delta^2 + O(\delta^4)],\qquad(5.4a,b)$$

these moment equations yield a closed resonance curve such as that shown in Fig. 14 for $\delta = 1/8$ and $\varepsilon/\varepsilon_0 = 1.05$. If $\varepsilon > \varepsilon_*$ the two branches remain separate, as shown in Fig. 14 for $\varepsilon/\varepsilon_0 = 1.2$. The intersections of these resonance curves with $\alpha = 0$ are given by $(\tfrac{1}{2}\omega/\omega_0)^2 = \beta_\pm$, where [cf. (5.3b)]

$$\beta_\pm = 1 - 2\delta^2 \pm \tfrac{1}{2}(\varepsilon^2 - \varepsilon_0^2)^{\frac{1}{2}},\qquad(5.5)$$

and $\alpha = 0$ is a stable state for $(\tfrac{1}{2}\omega/\omega_0)^2 > \beta_+$ or $(\tfrac{1}{2}\omega/\omega_0)^2 < \beta_-$.

Symmetry breaking occurs on the right branches of the separated resonance curves (there is a small parametric domain in which symmetry breaking is predicted on the right branches of the closed resonance curves, but this has not been confirmed by numerical integration). The $2T$-periodic, symmetric oscillations lose stability to $2T$-periodic, asymmetric oscillations as ω is decreased through the symmetry-breaking bifurcation. These asymmetric oscillations are stable in only a rather small range of ω and lose stability to $4T$-periodic, asymmetric oscillations through a period-doubling bifurcation, which typically leads (with decreasing ω) to chaos through a period-doubling cascade.

Many other stable states — e.g. T-periodic, asymmetric oscillations — interspersed with spectral windows in which $\theta = 0$ is stable, are possible if ε is not small, but these have not yet been fully investigated.

It has long been known that the statically unstable state $\theta = \pi$ of the parametrically excited pendulum is stable for sufficiently large ε. The effects of damping on the stability of this state can be shown to be uniformly $O(\delta^2)$ in (5.1), so that the perturbation $\theta_1 \equiv \theta - \pi$ is governed by

$$\ddot{\theta}_1 - \omega_0^2(1+\varepsilon\sin\omega t)\theta_1 = 0 \quad (\delta, |\theta_1| \ll 1).\qquad(5.6)$$

This is Mathieu's equation, for which the solutions are known to be stable if ε lies within an interval of $\beta \equiv (\tfrac{1}{2}\omega/\omega_0)^2$ that may be approximated by

$$(2\beta)^{\frac{1}{2}} \le \tfrac{1}{2}\varepsilon \le \beta + 1 \quad (\beta \ge \tfrac{1}{2}).\qquad(5.7)$$

$2T$-periodic oscillations about $\theta = \pi$ also may be stable, but only for fairly large amplitudes, and their calculation requires the inclusion of higher harmonics.

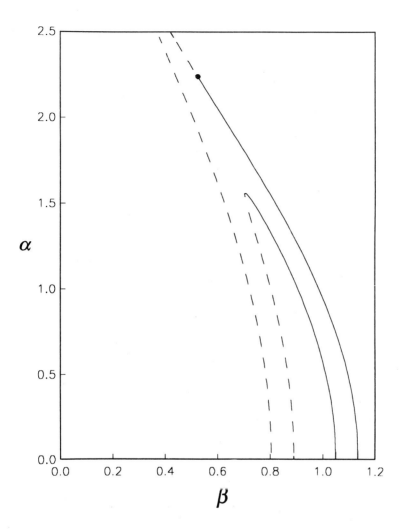

Fig. 14. The resonance curves for symmetric, subharmonic oscillations of the parametrically excited pendulum described by (5.1), as determined from the moment equations implied by the approximation (5.2), for $\delta = 1/8$ (for which $\varepsilon_*/\varepsilon_0 = 1.12$) and $\varepsilon/\varepsilon_0 = 1.05$ (lower curve) and 1.2. The solid/dashed segments comprise stable/unstable states, and the dot denotes the symmetry-breaking bifurcation. Note that $\alpha = 0$ is a stable state for $\beta \equiv (\tfrac{1}{2}\omega/\omega_0)^2 > \beta_+$ or $< \beta_-$, where $\beta_\pm$ is given by (5.5).

CONCLUSION

The pendulum has played a role in the evolution of mechanics that is matched only by that of the planets. From the regularity observed by Galileo in the cathedral in Pisa to the chaos of Lorenz, it has been our touchstone, and so it will continue to be.

ACKNOWLEDGEMENTS

I am indebted to Janet Becker, Peter Bryant and Diane Henderson for aid with the numerical work.

REFERENCES

Blackburn, J.A., Yang, Z-J., Vik, S., Smith, H.J.T. & Nerenberg, M.A.H. 1987 Experimental study of chaos in a driven pendulum. *Physica* **26D**, 385-395.

Bryant, Peter J. 1988 Appendix B to Miles (1988a).

Bryant, Peter & Miles, John 1988a On horizontal forcing of a weakly damped pendulum. *Physica D* (*sub judice*).

Bryant, Peter & Miles, John 1988b On parametrically excited oscillations of a weakly damped pendulum (in preparation).

D'Humieres, D., Beasley, M.R., Huberman, B.A. & Libchaber, A. 1982 Chaotic states and routes to chaos in the forced pendulum. *Phys. Rev. A* **26**, 3483-3496.

Euler, L. 1744 "De curvis elasticis" in *Methodus inveniendi lineas curvas maximi minimive proprietate gaudentes* (Lausanne); cited by S. Timoshenko, *Theory of Elastic Stability*, McGraw-Hill, New York, 1936, p.64.

Euler, L. 1776 De motu oscillatorio penduli cuiuscunque dum arcus datae amplitudinis absolvit. *Acta acad. sci. Petropolitanae* **1777**: II, 159-182; *Opera Omnia* (II) **7**, 39-60 (E503).

Faraday, M. 1831 On a peculiar class of acoustical figures, and on certain forms assumed by groups of particles upon vibrating elastic surfaces. *Phil. Trans. R. Soc. Lond.* **121**, 299-340.

Favaro, Antonio 1891 Galileo Galelei e Cristano Huygens. Nuovi documenti sull'applicazione del pendolo all'orologio. *Mem. del Reale Istituto Veneto* **24**, 389-418.

Galilei, Galileo 1914 *Two New Sciences* (Transl. by H. Crew and A. deSalvio), MacMillan, New York.

Gilbarg, D. & Paolucci, D. 1953 The structure of shock waves in the continuum theory of fluids. *J. Rat'l Mech. Anal.* **2**, 617-642.

Guckenheimer, J. & Holmes, P. 1983 *Nonlinear Oscillations, Dynamical Systems and Bifurcation of Vector Fields*. Springer-Verlag, New York.

Huygens, C. 1673 *Horologium Oscillatorium*. Vol. 18 of *Oeuvres Complètes* (1934); translated as *The Pendulum Clock or Geometrical Demonstrations Concerning the Motion of Pendula as Applied to Clocks* by R.J. Blackwell, Iowa St. U. Press (1986). The *Horologium*, published in 1658, gives the design of Huygens' clock but does not contain the dynamical discoveries of 1659.

Huygens, C. 1934 *Oeuvres Complètes*. Martinus Nijhoff, La Haye.

Lagrange, J.L. 1788 *Mécanique Analytique*. Chez la Veuve Desaint, Paris.

Lichtenberg, A.J. & Lieberman, M.A. 1983 *Regular and Stochastic Motion*. Springer-Verlag, New York, p. 382.

Lorenz, E.N. 1963 Deterministic nonperiodic flow. *J. Atmos. Sci.* **20**, 130-141.

Miles, J.W. 1962 Stability of forced oscillations of a spherical pendulum. *Quart. Appl. Math.* **20**, 21-32.

Miles, J.W. 1984 Resonant motion of a spherical pendulum. *Physica* **11D**, 309-323.

Miles, J.W. 1988a Resonance and symmetry breaking for the pendulum. *Physica* **D** (in press).

Miles, J.W. 1988b Directly forced oscillations of an inverted pendulum. *Phys. Lett. A*.

Millikan, R.A., Roller, D. & Watson, E.C. 1937 *Mechanics, Molecular Physics, Heat, and Sound*. Ginn, New York, p.342.

Newton, Isaac 1946 *Newton's Principia* (Translated by P. Motte (F. Cajori). University of California Press.

Rayleigh, Lord 1883 On maintained vibrations. *Phil. Mag.* **15**, 222-235; *Scientific Papers* **2**, 188-193.

Robertson, J.D. 1931 *The Evolution of Clockwork.* S.R. Publishers, East Ardsley, Wakefield, Yorkshire, U.K. (1972).

Ruelle, D. & Takens, F. 1971 On the nature of turbulence. *Comm. Math. Phys.* **20**, 167-192; **23**, 343-344.

Stokes, G.G. 1850 On the effect of the internal friction of fluids on the motion of the pendulum. *Trans. Camb. Phil. Soc.* **9** (part 2), 8-106; *Math. and Phys. Papers* **3**, 1-141.

Thomson, W. & Tait, P.G. 1912 *Principles of Mechanics and Dynamics.* Cambridge University Press, Vol. 1, §343m.

Wolf, A. 1950 *A History of Science, Technology, and Philosophy in the 16th and 17th Centuries.* G. Allen & Unwin Ltd.

Theoretical and Applied Mechanics
P. Germain, M. Piau and D. Caillerie (Editors)
Elsevier Science Publishers B.V. (North-Holland)
© IUTAM, 1989

COMPUTATIONAL MICROMECHANICS

Alan NEEDLEMAN

Division of Engineering
Brown University
Providence RI 02912

Topics in the computational micromechanics of large plastic flow and damage evolution in ductile solids are discussed. Numerical challenges presented by such problems include the modelling of plastic flow localization and the creation of new free surface due to progressive micro-rupture. The prediction of macroscopic ductility and toughness from a continuum analysis of the micro-scale failure mechanism is illustrated. Calculations of micro-scale deformation processes and failure modes are also reviewed.

1. INTRODUCTION

A major challenge for computational mechanics is the direct calculation of macroscopic ductility and toughness in terms of parameters characterizing measurable (and controllable) features of the material's microstructure. For 'engineered' materials, i.e. modern alloys and composites, the ability to predict the effects of microstructural variations on ductility and toughness would have a major impact on material design. Developing this capability is as much a modelling issue as a computational one and, in fact, advances on these two fronts are closely coupled.

Computational studies are reviewed that aim at developing a quantitative micromechanics description of damage evolution in ductile solids. Attention is focussed on circumstances where crystallographic slip is the mechanism of plastic flow and where damage arises from micro-scale void nucleation, growth and coalescence. Two classes of calculations are discussed. One class involves the use of microstructurally based constitutive relations to analyze problems involving large plastic flow and damage evolution. The other class concerns the use of computational continuum mechanics to analyze micro-scale deformation and failure processes, in order to develop more accurate microstructurally based constitutive relations.

Strain localization issues are a main focus of the discussion, because strain localization is a characteristic feature of large strain inelastic response and because calculations of strain localization phenomena present significant numerical challenges. Another focus is the development and use of constitutive relations that allow for a complete loss of stress carrying capacity with the associated creation of new free surface, without any additional failure criterion being employed, so that fracture arises as a natural outcome of the deformation process.

2. GENERAL FORMULATION

2.1 Field Equations

The purpose here is to establish notation and to define field quantities; a more complete development is given by Tvergaard in this volume [1]. A Lagrangian formulation is employed where each material particle is labelled by its position, $\mathbf{x}$, in a conveniently chosen reference configuration and these labels, together with time, serve as the set of independent variables.

In the current configuration the material point initially at $\mathbf{x}$ is at $\bar{\mathbf{x}}$. The displacement vector $\mathbf{u}$ and the deformation gradient $\mathbf{F}$ are defined by

$$\mathbf{u} = \bar{\mathbf{x}} - \mathbf{x} \qquad \mathbf{F} = \frac{\partial \bar{\mathbf{x}}}{\partial \mathbf{x}} \tag{2.1}$$

The rate of deformation tensor is defined by

$$\mathbf{D} = \frac{1}{2}(\dot{\mathbf{F}} \cdot \mathbf{F}^{-1} + \mathbf{F}^{-T} \cdot \dot{\mathbf{F}}^T) \tag{2.2}$$

where $(\ )^{-T}$ is the inverse transpose, $\mathbf{a} \cdot \mathbf{b} = a^i b_i$ and $(\dot{\ })$ is $\partial(\ )/\partial t$.

The principle of virtual work is expressed in terms of the nonsymmetric nominal stress tensor $\mathbf{s}$, which is related to the force, $\mathbf{df}$, transmitted across a material element by

$$\mathbf{df} = \mathbf{n} \cdot \mathbf{s} \, \mathrm{d}S \tag{2.3}$$

where $\mathrm{d}S$ and $\mathbf{n}$ give the area and orientation of a material element in the reference configuration.

For a 'standard' boundary value problem, with tractions prescribed over part of the surface and displacements on the remainder, the principle of virtual work is written as

$$\int_V \mathbf{s} : \delta\mathbf{F} \, dV = \int_S \mathbf{T} \cdot \delta\mathbf{u} \, dS - \int_V \rho \frac{\partial^2 \mathbf{u}}{\partial t^2} \cdot \delta\mathbf{u} \, dV \tag{2.4}$$

where V, S and ρ are the volume, surface and mass density, respectively, of the body in the reference configuration. In quasi-static analyses the last integral on the right hand side of (2.4) is taken to vanish identically. Also, $\mathbf{A} : \mathbf{B} = A^{ij} B_{ji}$ and

$$\mathbf{T} = \mathbf{n} \cdot \mathbf{s} \tag{2.5}$$

Inelastic constitutive relations are more conveniently expressed in terms of the symmetric Cauchy, $\boldsymbol{\sigma}$, or Kirchhoff, $\boldsymbol{\tau}$, stress measures which are related to $\mathbf{s}$ through

$$\mathbf{s} = \mathbf{F}^{-1} \cdot \boldsymbol{\tau} = \det(\mathbf{F}) \, \mathbf{F}^{-1} \boldsymbol{\sigma} \tag{2.6}$$

where $\det(\mathbf{F})$ is the ratio of the volume of a material element in the current configuration to its volume in the reference configuration.

2.2. Constitutive Relations

Inelastic material response is specified through a flow rule for the plastic part of the rate of deformation tensor, $\mathbf{D}^p$. Elastic material response takes the form of a relation between $\mathbf{D}^e = \mathbf{D} - \mathbf{D}^p$ and some suitable objective stress rate. Standard kinematic relations are then employed to express the constitutive relation in terms of field quantities entering the principle of virtual work, (2.4).

The flow rule can be expressed as

$$\mathbf{D}^p = \dot{\Lambda} \frac{\partial \Phi}{\partial \boldsymbol{\tau}} \tag{2.7}$$

where Φ is the flow potential surface.

For a rate independent solid, the relation (2.7) has two or more branches; in the simplest case there are two, one corresponding to plastic loading and the other to elastic unloading. Furthermore, with plastic isotropy and if 'normality' holds $\Lambda = (D\tau/Dt) : (\partial\Phi/\partial\tau)$, where $D\tau/Dt$ is the material time derivative of τ and $\Phi = 0$ is the yield surface.

A fundamental distinction between rate dependent material behavior and rate independent behavior, with implications for the numerical analysis of strain localization phenomena, is that for the rate dependent solid $\dot\Lambda$ does not depend on incremental quantities. For rate dependent material response, $\dot\Lambda$ depends on the current stress and deformation state and on current values of internal variables characterizing the material hardness. For example, for a rate dependent Mises solid, (2.7) becomes

$$\mathbf{D}^p = \frac{3\dot{\bar\epsilon}}{2\bar\sigma}\tau' \tag{2.8}$$

where

$$\tau' = \tau - \frac{1}{3}(\tau : \mathbf{I})\mathbf{I} \qquad \bar\sigma^2 = \frac{3}{2}\tau' : \tau' \tag{2.9}$$

With power law rate hardening,

$$\dot{\bar\epsilon} = \dot\epsilon_0 [\bar\sigma/g(\bar\epsilon)]^{1/m} \tag{2.10}$$

The function $g(\bar\epsilon)$, with $\bar\epsilon = \int \dot{\bar\epsilon}\,dt$, represents the effective stress versus effective strain response in a tensile test carried out at a strain-rate such that $\dot{\bar\epsilon} = \dot\epsilon_0$, m is the strain rate hardening exponent and (2.10) is supplemented with an evolution equation for g; a simple power law relation is specified by $g \sim (1 + \bar\epsilon/\epsilon_0)^N$, where ϵ_0 and N are material constants. For structural metals at room temperature, values of m between 0.002 and 0.02 are typical. For this range of m, the constitutive equations are a 'stiff' system and special care is required for their time integration, e.g. [8].

For a rate independent solid with a smooth flow potential surface, the stress rate-strain rate relation has the form

$$\dot{\mathbf{s}} = \begin{cases} \mathbf{K}_{tan} : \dot{\mathbf{F}} & \text{for plastic loading} \\ \mathbf{K}_{elastic} : \dot{\mathbf{F}} & \text{for elastic unloading} \end{cases} \tag{2.11}$$

In (2.11) the tensor of moduli depend on incremental quantities through the loading-unloading condition. By way of contrast, for the rate dependent solid

$$\dot{\mathbf{s}} = \mathbf{K}_{elastic} : \dot{\mathbf{F}} - \dot{\mathbf{P}} \tag{2.12}$$

The plasticity is embodied in $\dot{\mathbf{P}}$ and, furthermore, $\dot{\mathbf{P}}$ is independent of incremental field quantities as, of course, are the elastic moduli.

Regardless of whether the material is characterized as rate independent or rate dependent, accurate representations of large plastic flow and damage phenomena require a more detailed description of material behavior than is embodied in the Mises idealization.

Yield Surface Vertex Effects

Within the framework of continuum slip theory, crystal deformation is described as arising from two distinct physical mechanisms; elastic deformation due to distortion of the lattice and crystallographic slip (shearing along certain preferred lattice planes in certain preferred lattice directions) which leaves the lattice undisturbed. The combination of slip plane and slip direction is termed a slip system. For example, in face centered cubic crystals there are 12 slip systems; the slip planes are the 4 cube diagonals and within each slip plane the slip directions are the 3 face diagonals.

For rate independent material response, Schmid's law states that yielding occurs on a slip system when the resolved shear stress on that slip system attains a critical value. Let $\mathbf{m}^{*(\alpha)}$ and $\mathbf{s}^{*(\alpha)}$, respectively, denote the current orientation of the slip plane normal and the slip direction for slip system α. The yield surface is the boundary of the region $\mathbf{s}^{*(\alpha)} \cdot \boldsymbol{\tau} \cdot \mathbf{m}^{*(\alpha)} < Y_0^{(\alpha)}$, where $Y_0^{(\alpha)}$ is the critical resolved shear stress for system α. If two or more slip systems are active, yielding occurs at a vertex of the yield surface. Furthermore, Hill [2] has shown that this vertex structure is transmitted to the macroscopic constitutive description of aggregates of crystals, i.e. polycrystalline solids. The importance of this is that, at a vertex, the stiffness associated with an abrupt change in strain path is much less than the elastic stiffness that governs the response if the yield surface is smooth.

In what may be the first computational mechanics investigation, Taylor [3,4] calculated the stress-strain response in uniaxial tension and compression of an aggregate of face centered cubic crystals. In the Taylor [3,4] model, each grain of the polycrystal is treated as an isolated single crystal and assumed to undergo uniform straining, with the strain in each grain being that of the aggregate. Although more sophisticated, 'self-consistent' averaging schemes have been employed in small deformation determinations of polycrystal constitutive response, e.g. Hutchinson [7], nearly all calculations of large strain polycrystal stress strain behavior have used a Taylor-type uniform strain assumption. A fundamental difficulty that has plagued large strain calculations is that the choice of active slip systems is not unique. Therefore, neither is the predicted large strain constitutive behavior which depends strongly on the computed crystal lattice rotations.

Accounting for material rate sensitivity resolves the constitutive non-uniqueness, [9,10,11]. A finite deformation rate dependent Taylor [3,4] type model has been developed by Asaro and Needleman [11] and results based on that model are used to illustrate some key features of large strain polycrystal inelastic response. With $\dot{\gamma}^{(\alpha)}$ denoting the shear rate on slip system α

$$\mathbf{D}^p + \boldsymbol{\Omega}^p = \dot{\mathbf{F}} \cdot \mathbf{F}^{-1} - \dot{\mathbf{F}}^* \cdot \mathbf{F}^{*-1} = \sum \dot{\gamma}^{(\alpha)} \mathbf{s}^{*(\alpha)} \mathbf{m}^{*(\alpha)} \qquad (2.13)$$

Here, $\mathbf{F}^*$ includes the elastic deformation and any rigid body rotations. Note that the kinematics of crystallographic slip leads to a specification of a plastic spin, $\boldsymbol{\Omega}^p$, as well as a plastic rate of deformation. The full finite deformation framework for single crystal constitutive relations is described in [12,13,14] and an overview of the continuum slip framework and its physical background is given in [35]. Specific crystal constitutive relations are described in [15] and [10,16]. For rate dependent material behavior, each $\dot{\gamma}^{(\alpha)}$ is a function of the resolved shear stress on slip system α and current slip system hardness. On the other hand, in the rate independent theory a choice of active (plastically loading) slip systems needs to be made and this set may not be unique. The fundamental distinction between the rate dependent and rate independent descriptions is that plastic flow in the rate dependent theory depends on quantities evaluated at the current state and is independent of their rates.

Figure 1a shows constant offset effective plastic strain surfaces following plane strain tension that are computed in the following manner. The aggregate is subject to the prescribed loading history, unloaded and then reloaded along various trajectories in stress space. Each point in Fig. 1a corresponds to one probe of the constant offset surface. These surfaces are not true yield surfaces, since there is no sharp yield point for the viscoplastic material characterization, nor are they surfaces of constant flow potential. However, "yield surfaces" of this type are what are commonly measured using procedures like those simulated in the analysis.

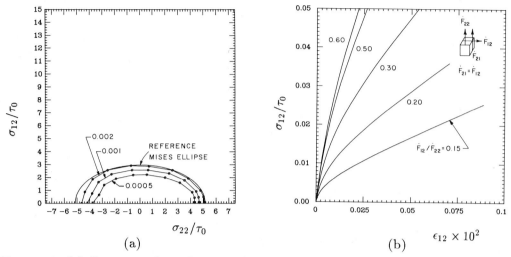

(a)   (b) $\epsilon_{12} \times 10^2$

FIGURE 1. (a) Constant offset effective plastic strain surfaces following plane strain tension to an effective strain of 0.23. (b) Shear stress-shear strain curves for various loading paths following plain strain tension to an effective strain of 0.23. From [11].

Figure 1b shows the calculated shear stress-shear strain response with the deformation gradient rate prescribed as shown in the insert. There is no unloading in this case. For rate dependent solids, the flow potential surface is necessarily smooth, which implies that the initial slope must be the elastic shear modulus, Rice [17]. The initial response in Fig. 1b is indeed elastic, but after very small shear strains, the shear stiffness drops to a small fraction of its elastic value for imposed ratios $\dot{F}_{12}/\dot{F}_{22} \lesssim 0.5$. It is the shear modulus for modest departures from proportional loading that is of relevance to plastic instability phenomena and there is a clear vertex-like reduction in shear stiffness after small, but finite shear strain increments, even though the curvature of the constant offset surfaces in Fig. 1a is not much sharper than the comparison Mises surface. With much larger departures from proportional straining, the shear stiffness increases, eventually to elastic values.

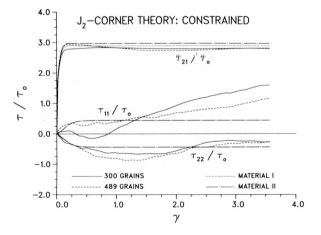

FIGURE 2. Stress response for two J_2 corner theory solids compared with two polycrystal calculations. From [18].

One use of physical plasticity theories is to calibrate phenomenological theories. Harren et al. [18] have carried out calculations of large strain simple shear response based on the polycrystal model of [11] and compared predictions of this physical plasticity theory with various phenomenological constitutive relations. Figure 2 shows this comparison for the rate independent J_2 corner theory of Christoffersen and Hutchinson [19]. The effective stress-effective strain curve input to the J_2 corner calculation came from matching a plane strain compression polycrystal simulation. The remaining corner theory parameters are taken as specified in [19] and [20]. These corner theory parameters have also been used in various numerical localization analyses, e.g. [21,22,23].

Progressive Microrupture

Experimental studies of ductile fracture have shown the central role played by microvoid nucleation and growth in metals, e.g. [24,25]. The voids nucleate mainly at second phase particles by decohesion of the particle-matrix interface or by particle fracture, and final rupture involves the growth of neighboring voids to coalescence. A constitutive framework for progressively cavitating solids developed by Gurson [26] has been used to analyze a variety of ductile failure phenomena and a version, [27], of this constitutive relation for rate dependent matrix material behavior is outlined. The voids are represented in terms of a single parameter, the void volume fraction, f, and the flow potential has the form

$$\Phi = \frac{\sigma_e^2}{\bar{\sigma}^2} + 2q_1 f^* \cosh\left(\frac{3q_2 \sigma_h}{2\bar{\sigma}}\right) - 1 - q_1^2 f^{*2} = 0 \tag{2.14}$$

Here, σ_e is the Mises effective stress, $\sigma_h = \frac{1}{3}\boldsymbol{\sigma} : \mathbf{I}$ is the hydrostatic stress, and $\bar{\sigma}$ is the strength of the matrix material. The parameters q_1 and q_2 were introduced by Tvergaard [28] to bring shear band bifurcation predictions of the Gurson [26] constitutive relation into closer agreement with corresponding results of full numerical analyses of a periodic array of voids.

The function f^* was proposed in [29] to account for the effects of rapid void coalescence at failure. Initially $f^* = f$, as originally proposed by Gurson [26], but at some critical void fraction, f_c, the dependence of f^* on f is increased in order to simulate a more rapid decrease in strength as the voids coalesce,

$$f^* = \begin{cases} f & f \le f_c \; ; \\ f_c + \dfrac{f_u^* - f_c}{f_f - f_c}(f - f_c) & f > f_c \end{cases} \tag{2.15}$$

The constant f_u^* is the value of f^* at zero stress in (2.14), i.e. $f_u^* = 1/q_1$. As $f \to f_f$, $f^* \to f_u^*$, the material loses all stress carrying capacity and new free surface is created.

The evolution of the void volume fraction results from the growth of existing voids and the nucleation of new voids. The rate of increase of void volume fraction due to the growth of existing voids is determined from plastic incompressibility of the matrix material,

$$\dot{f} = \dot{f}_{growth} + \dot{f}_{nucleation} = (1 - f)\mathbf{D}^p : \mathbf{I} + \dot{f}_{nucleation} \tag{2.16}$$

The plastic part of the rate of deformation, $\mathbf{D}^p$, is taken in a direction normal to the flow potential and is given by

$$\mathbf{D}^p = \dot{\Lambda}\frac{\partial \Phi}{\partial \boldsymbol{\sigma}} \tag{2.17}$$

Setting the plastic work rate equal to the matrix dissipation,

$$\boldsymbol{\sigma} : \mathbf{D}^p = (1 - f)\bar{\sigma}\dot{\bar{\epsilon}} = \dot{\Lambda}\boldsymbol{\sigma} : \frac{\partial \Phi}{\partial \boldsymbol{\sigma}} \tag{2.18}$$

determines the plastic flow proportionality factor, $\dot{\Lambda}$. For fully dense materials, $f \equiv 0$, and this constitutive relation reduces to (2.8)–(2.10), except that the flow rule is phrased in terms of Cauchy stress, $\boldsymbol{\sigma}$, rather than Kirchhoff stress, $\boldsymbol{\tau}$. Since the matrix material is plastically incompressible, this difference is inconsequential (large volume changes can occur for the void-matrix aggregate and, for the aggregate, the distinction between these two stress measures is important).

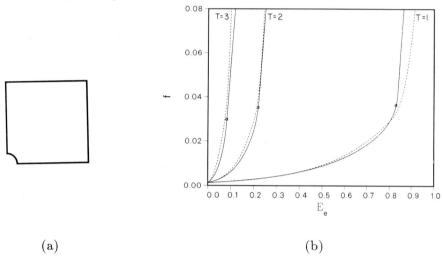

(a)   (b)

FIGURE 3. (a) Region analyzed numerically in the axisymmetric cell model used to obtain the results in (b). (b) Comparison of computed void volume fraction evolution with predictions of the modified Gurson [26] model using $q_1 = 1.25$, $q_2 = 1.0$, $f_c = 0.03$ and $f_f = 0.13$. The initial void volume fraction is $f_0 = 0.0013$. From [30].

The form of (2.14) was arrived at in [26] through an approximate rigid-plastic limit analysis of a thick walled spherical shell. Koplik and Needleman [30] have recently carried out finite element cell model analyses for an axisymmetric representation of a uniform array of spherical voids, fully accounting for void interaction effects and for void shape changes. The matrix material was represented by the rate dependent Mises relation (2.8)–(2.10). Figure 3b shows a comparison of finite element results for void volume fraction evolution with corresponding predictions of the Gurson [26] constitutive relation at three triaxiality levels. With the macroscopic effective stress denoted by Σ_e, and the macroscopic hydrostatic stress by Σ_h, the triaxiality T is given by

$$T = \frac{\Sigma_h}{\Sigma_e} \tag{2.19}$$

For a range of material parameters and triaxiality levels, rather good agreement was found in [30] using $q_1 = 1.25$ and $q_2 = 1.0$. Furthermore, the cell model calculations show a shift in strain state to a mode of uniaxial straining at which point the plastic deformation localizes to the ligament between neighboring voids. This event was associated with the accelerated void growth accompanying coalescence and was used to specify the parameters f_c and f_f in (2.15). In [30], the effect of cell aspect ratio was considered. The stress-strain response, at least for low void volume fractions, was found to be well approximated as a function of void volume fraction, independent of cell aspect ratio, but the strain at the initiation of accelerated void growth was sensitive to the cell aspect ratio.

3. STRAIN LOCALIZATION ISSUES

Localization is a common, if not inevitable, outcome of large plastic flow in a wide variety of solids and manifests itself through the concentration of deformation into one or more narrow bands, with the band direction set by the material's constitutive relation. In certain circumstances, concentration of deformation into a more or less well-defined band can be induced by the boundary constraints, but this does not constitute localization in the sense used here. The theoretical foundation for the analysis of localization in plastic solids is presented by Rice [31] and is reviewed by Tvergaard [1] in this volume. Some issues concerning the solution of localization problems using grid based numerical methods are discussed in this section.

3.1 Localization under quasi-static loading conditions

Consider an infinite planar block of height h subject to simple shearing displacement boundary conditions, Fig. 4, and neglect nonlinear kinematic effects that lead to possible normal strains or normal stresses. One solution to this boundary value problem is a state of homogeneous shear. At some stage of the deformation history, the possibility of bifurcation into a shear band mode can be considered. Equilibrium requires that the stress state remain homogeneous so that $\dot{\tau}_b = \dot{\tau}_o$, where $()_b$ and $()_o$ denote quantities inside and outside the band in Fig. 4. For a rate independent solid, (2.10) reduces to $\dot{\tau} = G_{tan}\dot{\gamma}$ for plastic loading and $\dot{\tau} = G_{elastic}\dot{\gamma}$ for elastic unloading. Earliest bifurcation becomes possible when

$$G_{tan}(\dot{\gamma}_b - \dot{\gamma}_o) = 0 \tag{3.1}$$

A non-trivial solution to (3.1) is possible when G_{tan} vanishes.

In the postbifurcation regime, the material inside the band continues to undergo plastic loading, while the material outside the band is elastically unloading (if the material outside the band were to continue plastic loading, the shear strain rates in the two regions would be identical and there would be no band). The total shear strain rate is $\dot{\gamma} = (1-\beta)\dot{\gamma}_o + \beta\dot{\gamma}_b$ where $\beta = h_b/h$.

Rate equilibrium and the constitutive relation give the overall shear stress–shear strain response as

$$\dot{\tau} = \frac{G_t\dot{\gamma}}{[\beta + (1-\beta)\frac{G_t}{G}]} \tag{3.2}$$

The value of β is undetermined by the analysis and the solution is inherently non-unique. Furthermore, the presence of an initial imperfection does not resolve this non-uniqueness, [36]. Hence, the band of localized deformation is arbitrarily narrow. As a consequence, numerical solutions to localization problems for rate independent solids exhibit an inherent mesh dependence. Furthermore, global quantities such as the overall stiffness characteristics of the body depend on the mesh size used to resolve the band of localized deformations. In calculations, if the discretization does not lead to artificial shear band broadening, the band is one element wide. If this element size is chosen appropriately for the physical situation being modelled, such finite element solutions can be physically relevant.

For the rate dependent solid, the homogeneous solution is unique and bifurcation into a shear band mode is ruled out. The one dimensional constitutive relation is $\dot{\tau} = G_{elastic}\dot{\gamma} - \dot{P}$, where $\dot{P}$ is independent of rate quantities. Bifurcation then requires,

$$G(\dot{\gamma}_b - \dot{\gamma}_o) = 0 \tag{3.3}$$

Since the elastic shear modulus, G, is positive, the unique solution to (3.3) is $\dot{\gamma}_b = \dot{\gamma}_o$. In fact, bifurcation into any mode is excluded, [36]. With a band type imperfection, localization does occur. It is worth noting that material rate dependence eliminates inherent mesh sensitivity by giving rise to a unique solution and not by smoothing out the strain discontinuity across the band. The solution is sensitive to initial conditions; for example,

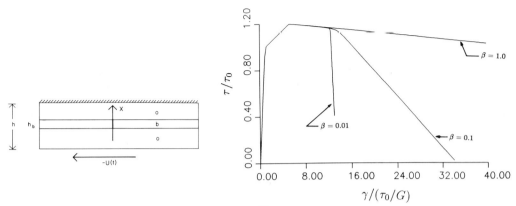

FIGURE 4. Simple shearing of a planar strip.

FIGURE 5. Rate dependent shear stress-shear strain response, $\beta = h_b/h$. From [36].

in the simple shear problem, the band width is set by the initial inhomogeneity thickness as shown in Fig. 5, where β denotes the width of the imperfection band. Although the solution is unique, numerical stability becomes an issue for nearly rate independent solids, as discussed in [36]. The simple shear problem is highly degenerate, but the conclusion holds in a more general context; the rate dependent formulation eliminates the pathological mesh dependence inherent to the rate independent formulation and provides a framework for assessing the effects of various physical mechanisms on band evolution, e.g. damage processes, [53], and heat conduction, [37].

3.2 Localization under dynamic loading conditions

Regardless of the formulation, the mesh sets the minimum band width at one element spacing. This is illustrated by some results from [38], where the dynamics of shear band development from internal material inhomogeneities under plane strain compressive loading was investigated. The material was characterized as an elastic-viscoplastic von Mises solid, (2.8)–(2.10), so that wave speeds remain real even in the softening regime. The phenomenology of shear band development under dynamic loading conditions was found to be much the same as under quasi-static loading conditions. For example, shear bands did not occur for a strain hardening solid but, with softening (i.e. the function g in (2.10) reaching a maximum and then decreasing), shear bands developed at 45° from the compression axis, as expected based on a quasi-static "material instability" analysis.

The process of band formation involves a cascade of finger-like plastic strain contours emanating from the inhomogeneity and from a small region where the band intersects the impacted surface, in a continually sharpening band. Although there is no single event that can be identified with shear band initiation, there is a reasonably well identifiable strain above which all further straining occurs within a narrowing band, as illustrated in Fig. 6. With localization, the speed of propagation, in the direction normal to the band, of plastic strain contours essentially vanishes. The initial thickness of the band reflects the size of the inhomogeneity, but, as shearing develops in the band, the region of high strain rate narrows to one element width and the discreteness of the mesh significantly affects subsequent response. Progressively narrowing bands are observed under dynamic loading conditions, [39]; in other circumstances, [40], band width appears to be more nearly constant.

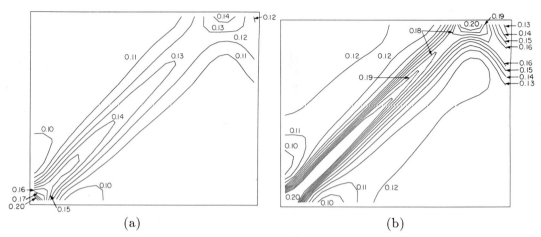

FIGURE 6. Contours of constant effective plastic strain for a softening solid at two stages of deformation. From [38].

4. COMPUTATIONAL CONSIDERATIONS

Within the context of a displacement finite element method, the sharp strain gradients that accompany localization can only be accurately resolved if the band interfaces follow element boundaries. Unless special care is used to align element boundaries along band directions, for example with appropriately oriented 'crossed triangle' quadrilateral elements, the mesh introduces a shear band broadening that can control the course of strain localization and can mask effects of physical response mechanisms on localization. For example, if quadrilateral isoparametric elements are used in the analysis, there are only two directions parallel to the mesh lines. The number of such directions can be increased by building the quadrilaterals from four crossed-triangular elements. In this type of mesh, sharp gradients can develop along the diagonals as well as the sides of the quadrilaterals. Finite element analyses of shear bands based on fine meshes of quadrilateral elements built up from four 'crossed' constant strain triangles do give sharply localized deformation modes.

However, the crossed triangle finite element formulation is specifically geared to two dimensional problems, it does require careful mesh design to account for likely directions of localization and the results are rather sensitive to the orientation of the mesh relative to the directions of localization. More recently, Ortiz, Leroy and Needleman [41] have proposed a method in which a deformation mode carrying a suitable strain discontinuity is added to the element. The effectiveness of this method has been demonstrated in a variety of localization problems, [41,42,43], and appears to permit accurate localization solutions to be obtained using rather general isoparametric finite element meshes.

Other important computational issues arise in connection with constitutive descriptions, such as the Gurson [26], model for which the stress carrying capacity vanishes so that new free surface is created in the course of a calculation. The stable relaxation of nodal forces to the stress free state can be accomplished using the 'element vanish' technique of Tvergaard [44]. When the failure condition is met in an element ($f^* = f_u^*$), the element is taken to vanish in that it no longer contributes to the virtual work (2.4) but the nodal points associated with vanished elements are not removed. In order to avoid numerical difficulties associated with nearly vanished elements, the elements are actually taken to vanish in the computations slightly before $f^* = f_u^*$, typically at $f^* = 0.9f_u^*$. The small residual nodal forces are then released gradually. Additionally, the highly non-proportional stress redistribution accompanying the creation of new free surface requires small time steps, as discussed in [66].

The actual discretization of the principle of virtual work, (2.4), by finite element methods follows the same procedures as in the small strain theory, e.g. [45]. Similarly, time integration schemes used in small deformation problems, for both quasi-static and dynamic loading histories, can be used for finite deformation analyses, e.g. [45,46].

5. LOCALIZATION IN SOLIDS WITH A YIELD SURFACE VERTEX

5.1 Single Crystals

The interaction between two material instability modes is illustrated in an analysis of Pierce et al. [10]. Within the continuum slip framework for crystal plasticity, Asaro [47] carried out a bifurcation analysis for a two dimensional model crystal oriented for symmetric double slip. Three essentially different types of bifurcation mode were identified. Two are relevant here; one is a shear band mode which involves continued double slipping in the shear band; another is a localization mode where shearing occurs parallel to the tensile axis and the shearing mode itself leads to unloading on one of the currently active slip systems. Pierce et al. [10,16] carried out finite element analyses of plane strain tension using the two dimensional crystal model of [47] and, in most of their calculations, sharp shear bands that propagate across the gage section were found. However, for a physically significant range of parameters, the latter instability mode is activated and, as illustrated in Fig. 7 and discussed in [10,16], this leads to 'patchy' slip. Material strain rate sensitivity plays an essential role for the case shown in Fig. 7, since the boundary value problem for the corresponding rate independent solid was not amenable to numerical solution and, in fact, solutions might actually not exist in the (singular) rate independent limit. The extent to which the nonuniformity of deformation associated with patchy slip depends on the discretization is not known. Patchy slip induces kinematic constraints that inhibit the propagation of a shear band across the gage. Hence, if shear bands are regarded as a failure mode, this is a case where an instability mode, the 'patchy' slip mode, delays the onset of failure and increases the crystal's ductility. Patchy slip like deformation modes have been observed within grains of polycrystals [48] and the implications of such modes for the large strain constitutive response of polycrystals remain to be quantified. An initial study of the effects of non-uniform deformation on polycrystal stress-strain response is described in [49].

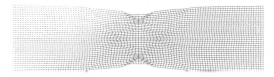

FIGURE 7. Deformed finite element mesh showing the non-uniformity of deformation associated with patchy slip. A shear band does develop, but it does not propagate across the gage. Each quadrilateral consists of four 'crossed' triangles. From [10].

5.2 Phenomenological Corner Theory

In order to explore the influence of a yield surface vertex on the course of localization, Tvergaard et al. [21] analyzed plane strain tension of a rectangular block using the J_2 corner theory of Christoffersen and Hutchinson [19] to characterize the material behavior. J_2 corner theory [19] is a rate independent constitutive relation that has a vertex and permits shear bands to emerge in strain hardening solids. Various small initial thickness inhomogeneities are prescribed in [21] to initiate diffuse necking. As seen in Fig. 8, the particular band pattern that occurs depends on the initial geometric imperfection but, regardless of the imperfection, the orientation of the band, or bands, is in good agreement with that predicted from a band bifurcation analysis. Figure 9 shows the shear band pattern in a bend specimen, [22], where the pre-localization deformation pattern is quite inhomogeneous. The extent to which the shear band broadening on the tensile side of the bend specimen is a consequence of the discretization is not known. In Figs. 8 and 9, each quadrilateral consists of four 'crossed' triangles.

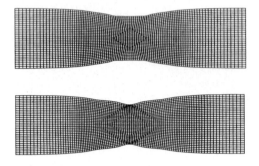

FIGURE 8. Two plane strain tension calculations from [21].

FIGURE 9. Pure bending from [22].

6. FLOW LOCALIZATION AND FAILURE IN POROUS PLASTIC SOLIDS

6.1. Failure in Plane Strain and Round Tensile Bars

Typical phenomenologies that develop in tensile specimens of structural metals are illustrated in Fig. 10 which show a characteristic feature of ductile fracture in structural metals; in the round bar tension test, Fig. 10a, diffuse necking is followed by a cup-cone type fracture, while in a plane strain tension test of *the same material*, Fig. 10b, failure occurs in a localized shearing mode with relatively little diffuse necking.

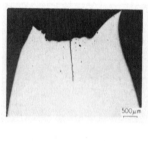

(a)

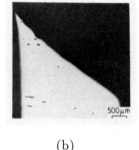

(b)

FIGURE 10. Effect of stress state on failure of steel specimens. (a) Axisymmetric tension. (b) Plane strain tension. From [50].

A finite element analysis of necking and failure in the round bar tensile test based on the modified Gurson [26] constitutive relation, with rate independent matrix material behavior, has been carried out in [29]. A crack forms in the center of the neck and propagates across the specimen leading to the configuration shown in Fig. 11a, where the shaded region corresponds to material that has undergone a complete loss of stress carrying capacity. The tendency for the crack to zig-zag is a consequence of shear localization being inhibited by the additional plastic work associated with the hoop strains that accompany

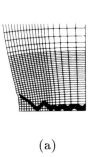

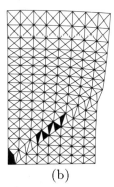

(a)   (b)

FIGURE 11. (a) From [29]. Computed failure mode in the neck of an axisymmetric tensile specimen. (b) From [51]. Computed failure mode in the neck of a plane strain tensile specimen. Triangular finite elements which have undergone a complete loss in stress carrying capacity are painted black.

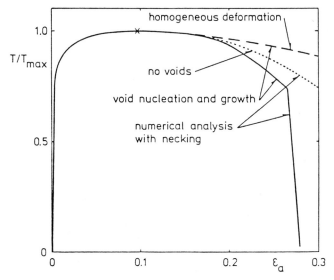

FIGURE 12. Load versus imposed strain for axisymmetric tensile specimens. From [29].

shearing in the axisymmetric geometry. As the free surface is approached this axisymmetric constraint is relaxed, permitting the cone of the cup-cone fracture to form. There is no corresponding geometrical constraint in plane strain tension so that once initiated, a shear band can propagate across the entire specimen as illustrated in Fig. 11b, from [51], which was obtained using the rate dependent version of Gurson's [16] constitutive relation. The *matrix material* in Fig. 11 is always strain hardening; softening is a consequence of increasing porosity. Also, the width of the localization band is set by the mesh, with rate independent material behavior, and by the geometric imperfection, with rate dependent material behavior (provided the mesh is sufficiently refined). It is also worth noting that predicting the development of the different failure modes in the round bar and plane strain tension tests is outside the scope of classical plasticity theory. In particular, in plane strain, a strain hardening Mises solid does not show any tendency for the deformation mode to shift to one involving localized shearing.

The development of failure has a significant effect on the overall load-displacement response as shown for the axisymmetric specimen in Fig. 12. The rather abrupt load drop is associated with the initiation of a crack in the center of the specimen. A similar sharp load

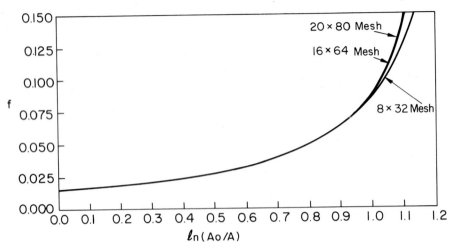

FIGURE 13. Development of porosity at the center of a necked axisymmetric tensile bar for three finite element meshes.

drop is found for the plane strain tension specimen in [51].

An important computational issue in these calculations concerns convergence as the mesh is refined. While this issue remains to be fully addressed, an axisymmetric necking calculation has been carried out, [52], based on the rate dependent version of the Gurson [16] constitutive relation and using three finite element meshes. The mesh resolutions are 8×32, 16×64 and 20×80, where the first number refers to the number of elements across the radius and the second number specifies the number of elements in the axial direction. Figure 13 shows the porosity evolution at the center of the neck for the three meshes and illustrates the convergence. The plots are terminated at the point where $f = f_c = 0.15$ in the center of the neck. As shown in Fig. 12, the load drops very rapidly once initial failure occurs so that the strain to failure will differ little from the strain at the terminal points on the curves in Fig. 13. A fine mesh is needed to analyze crack propagation and even the 20×80 mesh is not adequate to fully resolve details of the local crack tip stress field as it propagates across the specimen.

6.2. Notched Bars

The results in the previous section illustrate the ability of the Gurson [26] constitutive framework to qualitatively predict observed ductile failure behaviors. A meaningful quantitative comparison between predictions and experiment is more complex because of the path dependent and progressive nature of ductile fracture. Becker et al. [53] have compared quantitative predictions of void growth, strength and ductility with detailed measurements in round notched bars. Various notch geometries were studied in order to obtain different stress histories. The tensile specimens were machined from partially consolidated and sintered iron powder compacts in order to minimize nucleation effects and focus on void growth and failure issues.

The only experimentally determined quantities input into calculations in [53] were the uniaxial stress-strain curve for the matrix material and the initial void volume fraction. The parameters q_1, q_2, f_c and f_f entering the porous plastic constitutive description were chosen to provide a reasonable fit of both strength and void growth predictions with results of micro-mechanical models of periodic arrays of voids of the type discussed in Section 2.2. Figures 14 and 15 show results for one of the notch configurations in [53]. Eight specimens with this geometry were used and tests were terminated prior to fracture in order to measure

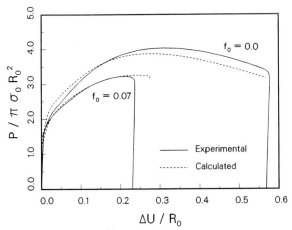

FIGURE 14. Experimental and calculated load-extension behavior of fully dense and 7% initially porous notched bars. From [53].

the evolution of porosity and deformed specimen geometry.

Figure 14 shows a comparison of the load extension response of the near full density specimen and a 7% initially porous notched bar with the predictions from the finite element calculations. The calculation for the fully dense material was terminated when the extension reached the failure extension of the specimen. The other calculation was stopped shortly after the void fraction in the element at the center of the notch reached the critical value, f_c, since previous calculations (e.g. Fig. 12) have shown that beyond this value, the load drops rapidly and little additional overall extension occurs.

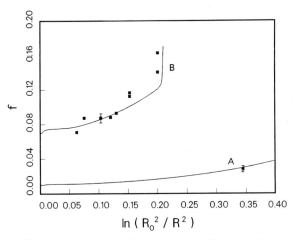

FIGURE 15. Void growth as a function of strain for 1% and 7% initially porous notched bars. From [53].

The void fractions measured at the center of the eight specimens with 7% nominal initial void fraction describe the evolution of porosity with deformation. These measurements and the void fraction measured at the center of the notched bar with a 1% initial void fraction are shown in Fig. 15 together with the predictions from the finite element calculations. The predictions are quite good at lower strains but the model underpredicts the void growth at large strains. The points at $\ln(R_0^2/R^2) = 0.153$ are from two separate specimens, which indicates that the results are reproducible. The results in [53] indicate that the constitutive framework (2.14)–(2.18) gives reasonably accurate predictions of void growth and of the effects on strength and ductility for a wide range of initial void volume fractions and levels of stress triaxiality. Becker et al. [53] conclude that key issues that need to be addressed to enhance the modelling capability are the matrix material constitutive characterization and improved quantification of void nucleation phenomena and non-uniform void distribution effects.

6.3. Cracks

In a more or less macroscopically uniform field, a continuum description of ductile rupture can, at least in principle, be based on material parameters of a kind entering constitutive descriptions, such as strength or strain hardening values, and on ratios of geometric parameters, e.g. the volume fraction of void nucleating particles. On the other hand, macroscopic measures of resistance to crack initiation and growth, e.g. the critical stress intensity factor, must depend on a characteristic length associated with the material, if only from dimensional considerations, [54,55]. For example, the characteristic length may be particle spacing or grain size. In order to avoid an inherent mesh dependence of numerical results, the boundary value problem must incorporate a characteristic length scale.

One aim of computational micromechanics is the prediction of macroscopic fracture behavior (crack length versus applied load and/or time) from an analysis of the fracture mechanism operating on the microscale. In addition to the inherent complexity of the phenomenon – fracture processes in real materials are complex and often more than one fracture mechanism is operative – the range of size scales involved makes computation difficult.

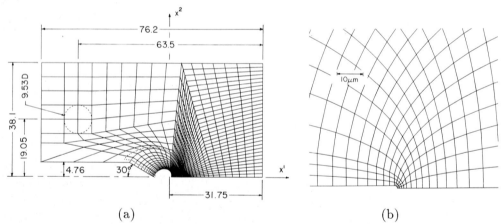

(a)   (b)

FIGURE 16. Finite element mesh modelling the compact tensile specimen geometry. (a) Outer mesh with the specimen dimensions in mm. (b) The mesh ahead of the crack. Each quadrilateral consists of four 'crossed' triangles. From [57].

Nevertheless, some initial progress has been made. In [66], a fracture mechanism involving two size scales of void nucleating particles was analyzed and crack growth was continued far enough to estimate J_{IC} and the tearing modulus, [67]. In [66] the spacing between the large particles served as a characteristic length. For a well characterized [56] model material system, an Al-Li alloy that fails by grain boundary cavitation at room temperature, Becker et al. [57] quantitatively compared measurements and predictions. A complete compact tension specimen was analyzed, with the material described by the porous plastic constitutive relation (2.14)–(2.18). The grains were modelled as roughly circular regions free of void nucleating particles and the distribution of grain boundary voids was modelled by allowing void nucleation in a band between grains. The calculations in [57] showed that the width of the porous band between grains was the key length scale for the microstructures

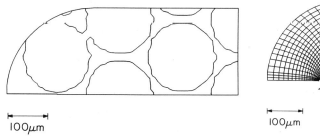

FIGURE 17. Grain distribution ahead of the crack tip with 200 μm diameter grains. From [57].

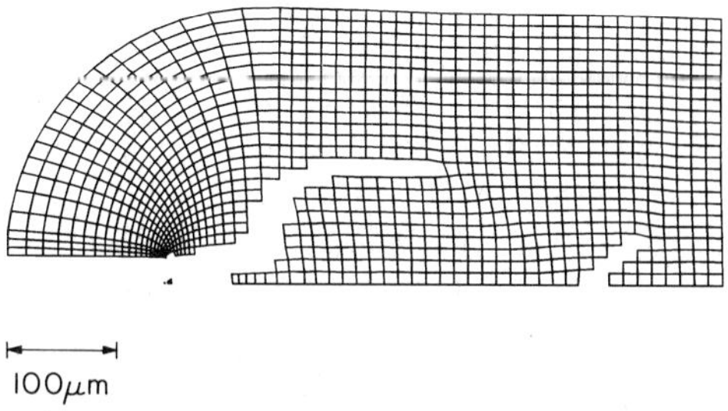

FIGURE 18. Deformed finite element mesh ahead of the crack tip showing the mode of crack growth. From [57].

analyzed. Figure 16 shows the finite element mesh and Fig. 17 shows one of the two grain sizes analyzed. In the most complete numerical calculations, the thickness of the grain boundary porous zone was taken to be 10 μm, which is two to three times the actual mean particle spacing on the grain boundary. A thick grain boundary porous zone was used because of the resource limitations imposed by storage requirements and execution time for the multistep analysis. With the width of the grain boundary porous zone and the grain size as variables in the model, the analysis in [57] quantified the fracture toughness and crack growth resistance (characterized by the tearing modulus).

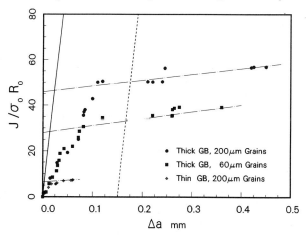

FIGURE 19. Calculated value of the J integral versus change in crack length, Δa. The chain dot lines are determined from linear regression applied to the points to the right of the dashed line. The solid line is the blunting curve. From [57].

The model is a highly idealized one and its limitations are discussed in [57]. However, there are no adjustable parameters, so a meaningful comparison with experiment can be made. All material parameters input into the analysis were independently measured with two exceptions. The void nucleation strain was inferred from uniaxial tension tests and the identification of the thickness of the grain boundary porous zone with the interparticle spacing was not a consequence of physical measurements but was suggested by finite element calculations for arrays of voids.

An outcome of the calculations was crack growth as a function of applied load. The computed crack growth characteristics were in good accord with observations. Figure 18 shows one stage of crack growth where elements that have undergone a complete loss of stress carrying capacity have been deleted from the figure. Figure 19 shows curves of Rice's J integral [58] versus change in crack length for the three calculations carried out in [57]. For the calculations with a grain boundary porous zone of 10 μm, there is sufficient crack growth to determine J_{IC} according to ASTM E-813 standards. Standard linear regression is used to fit the points beyond the dashed line and J_{IC} is determined at the intersection with the blunting line. The quantitative agreement between the model predictions and the experimentally measured values was quite good; the predicted critical stress intensity values were within a factor of two of the experimental values, while the predicted and observed values of the tearing modulus were typically within 50% of each other.

7. VOID NUCLEATION BY INTERFACIAL DECOHESION

A precise characterization of void nucleation is one of the major impediments to the quantitative description of ductile failure processes. In [59], a finite strain, cohesive zone type interface model was developed for analyzing void nucleation by interfacial decohesion. In this formulation, constitutive relations are specified independently for the material (or materials) and the interface. The constitutive relation for the interface, illustrated in Fig. 20, permits complete decohesion to occur. Since the mechanical response of the interface is specified in terms of both a critical interfacial strength and the work of separation per unit area, dimensional considerations introduce a characteristic length.

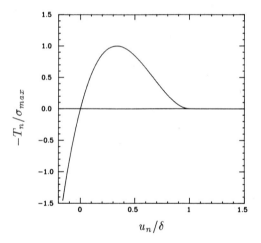

FIGURE 20. Traction versus separation distance for a purely normal mode of separation. From [59].

In [59] this model was used to analyze void nucleation from spherical rigid inclusions. The boundary value problem formulated was the axisymmetric cell model mentioned in Section 2.2, but for an array of rigid particles. The numerical results were related to the description of void nucleation within the Gurson [26] framework. Two nucleation criteria have been used in the Gurson [26] constitutive framework. One is a plastic strain criterion for which

$$\dot{f}_{nucleation} = \mathcal{D}\dot{\bar{\epsilon}} \qquad (7.1)$$

while the other is the stress based criterion

$$\dot{f}_{nucleation} = \mathcal{B}(\dot{\Sigma}_e + \dot{\Sigma}_h) \qquad (7.2)$$

In (7.1), $\mathcal{D}$ is considered a function of $\bar{\epsilon}$, while analogously in (7.2), $\mathcal{B}$ is taken to be a function of $(\Sigma_e + \Sigma_h)$ so that the quantity $(\Sigma_e + \Sigma_h)$ plays the role of a nucleation stress. The numerical results in [59] suggest an effective nucleation stress of the form $\Sigma_e + c\Sigma_h$, with $c \approx 0.35$. The best fitting value of c appears to depend on particle size; larger particles are characterized by a somewhat more strongly hydrostatic stress dependent nucleation stress. This is of particular significance since the hydrostatic stress dependence of void nucleation in (5.3) leads to a strong non-normality in the plastic flow rule, which promotes early flow localization.

This nucleation model has also been used to investigate microstructural effects on void nucleation in Al-SiC metal matrix composites, [60]. The boundary value problem analyzed in [60] is as in [59], but with cylindrical inclusions. Thus, the simulation is for a three dimensional periodic array of circular cylindrical fibers aligned end-to-end and subject to a tensile stress parallel to the fiber axes. Experimental observations were compared with patterns of void formation predicted from the calculations using appropriate values for material parameters. The calculations indicated how the qualitative features of void nucleation, such as shape, size and location, depended on specific material parameters, such as interface strength and fiber geometry. For example, the consequence of rounding the fiber corner was investigated in [60]. Figure 21a shows an observation of void nucleation at a fiber end and Fig. 21b shows the deformed finite element mesh around a fiber end in a case where the interfacial strength is high. As seen in Fig. 21, voids initiate preferentially near the fiber corner and the observed and predicted modes of void evolution are similar.

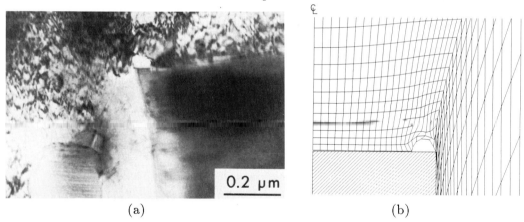

(a)	(b)

FIGURE 21. (a) Void nucleation at a fiber end in an Al-SiC composite. (b) Deformed finite element mesh in the vicinity of a fiber end. The fiber axis is aligned with the tensile axis. Each quadrilateral consists of four 'crossed' triangles. From [60].

8. DISTRIBUTION EFFECTS

Void distribution effects have been shown to play a significant role in limiting ductility, both experimentally, [61] and theoretically, [62]. Becker [62] analyzed void growth and coalescence in a small material element, with the material characterized by the modified Gurson [26] constitutive relation, but with a nonuniform initial distribution of the void volume fraction. Becker [62] found a significantly smaller fracture strain for a solid with a non-uniform porosity distribution than for a solid with a uniform porosity at the average value. There was a negligible effect of the nonuniformity prior to the localization that precedes fracture, [62]. Similarly, in [30], it was found that the stress-strain response was insensitive to the cell aspect ratio, but that the strain at the initiation of accelerated void growth was sensitive to the cell aspect ratio.

In [63], a porous plastic solid subject to plane strain tension and containing a doubly periodic array of clusters of cylindrical voids is analyzed. The void distribution within each cluster is irregular. Figure 22 shows one distribution and the predicted macroscopic stress-strain response for three distributions, along with a cell model calculation using a cell geometry as in Fig. 3a. The material is characterized as a viscoplastic solid, (2.8)–(2.10). The directionality of the distribution has its main effect on the aggregate flow strength; an increased density of voids at 45° to the tensile axis, as in Fig. 22a, gives a significant decrease in strength. The magnitude of the strength reduction decreases with decreasing void area fraction, [63]. Also note that the amount that the overall stress-strain response changes with increasing grid resolution depends on the distribution.

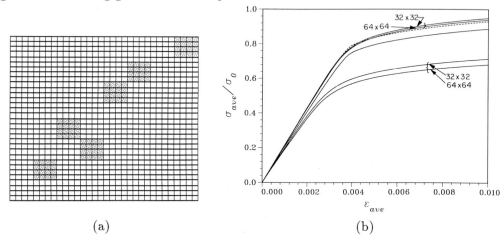

(a)   (b)

FIGURE 22. (a) Void pattern on a 32×32 quadrilateral 'crossed triangle' mesh where the voids are the shaded area. (b) Stress strain curves for various void patterns with a void area density of 0.09375. Two calculations are repeated with a 64×64 mesh. The dashed line is a unit cell calculation with a single cylindrical void and the distribution in (a) is the lowest stress-strain curve. The material is power law hardening with $N = 0.1$. From [63].

9. CREEP CRACK GROWTH

In Section 6.3, the direct simulation of fracture in terms of the microstructural failure mechanism was emphasized. Another objective of computational micromechanics studies of fracture is to relate macroscopic characterizing parameters to quantities characterizing the microscale fracture process. In [64] this aim was pursued in an analysis of transient creep crack growth under conditions of plane strain and small scale creep. A plane strain tensile elastic asymptotic crack tip field was imposed at a large distance from the initial crack tip.

The material is characterized as an elastic-power law creeping material, with an additional contribution to the rate of creep deformation arising from a given density of cavitating grain boundaries. A finite deformation phenomenological constitutive description of this process is given in [32,33] and is reviewed by Tvergaard [1] in this volume. The expression for the inelastic (creep) part of the rate of deformation is

$$\mathbf{D}^p = \dot{\epsilon}_0 \left(\frac{\bar{\sigma}}{\sigma_0}\right)^n \left[\frac{3\,\boldsymbol{\sigma}'}{2\,\bar{\sigma}} + \rho\left\{\frac{3}{2}\frac{n-1}{n+1}\frac{\boldsymbol{\sigma}'}{\bar{\sigma}}\left(\frac{s-\sigma_n}{\bar{\sigma}}\right)^2 + \frac{2}{n+1}\left(\frac{s-\sigma_n}{\bar{\sigma}}\right)\bar{\mathbf{n}} \otimes \bar{\mathbf{n}}\right\}\right] \tag{9.1}$$

where n is the creep exponent, $\mathbf{x} \otimes \mathbf{y}$ denotes the tensor product having components $x^i y^j$ and

$$\boldsymbol{\sigma}' = \boldsymbol{\sigma} - \sigma_m \mathbf{I} \qquad \sigma_m = \frac{1}{3}(\boldsymbol{\sigma} : \mathbf{I}) \qquad \bar{\sigma}^2 = \frac{3}{2}\boldsymbol{\sigma}' : \boldsymbol{\sigma}' \qquad s = \bar{\mathbf{n}} \cdot \boldsymbol{\sigma} \cdot \bar{\mathbf{n}} \tag{9.2}$$

Here, s represents the macroscopic normal stress on cavitating facets with normal $\bar{n}$ in the current configuration. The parameters σ_n and ρ are internal variables; σ_n is the average normal stress in the vicinity of the voids and ρ is a measure of the density of the cavitating facets. The evolution equation for σ_n comes from the description of grain boundary void growth by the combined processes of grain boundary diffusion and plastic dislocation creep in the adjoining grains. It is important to emphasize that this evolution equation and the flow rule (9.1) are outcomes of detailed micromechanical modelling. An alternative framework for analyzing creep rupture and creep crack growth is discussed by Chaboche [65] in this volume.

Two cases were analyzed in [64]. With the rate of crack growth faster than the rate of growth of the creep zone, it is found that Hui-Riedel [34] singular fields dominate, whereas when the rate of crack growth is slower than the rate of growth of the creep zone, HRR [5,6] type fields dominate. Within the context of small strain theory for a mathematically sharp crack, the existence of Hui-Riedel [34] singular fields at the crack tip does not depend on crack speed, but the finite strain region at the crack tip suppresses the Hui-Riedel [34] fields in the calculation for the slowly growing crack.

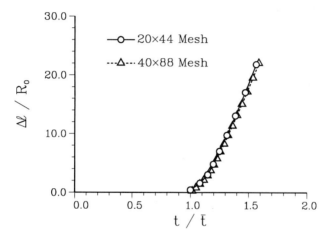

FIGURE 23. Crack growth, $\Delta\ell$, versus time for two meshes. From [64].

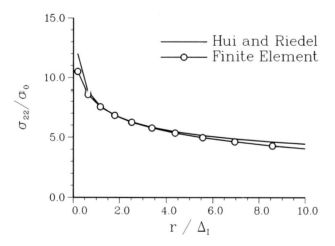

FIGURE 24. Comparison of Hui-Riedel [34] and finite element predictions. From [64].

Although the slow growing, HRR field dominated case is the situation of most significance with regard to applications, a comparison between numerically computed near tip field quantities and the Hui-Riedel [34] asymptotic fields is of interest. For a crack growing at a rate $\dot{\ell}(t)$, Hui and Riedel [34] find that, when $n > 3$, the asymptotic stress and strain fields take the following form,

$$\sigma_{ij} = A_n \sigma_0 \left[\frac{\dot{\ell}(t)}{\dot{\epsilon}_0(E/\sigma_0)r} \right]^{1/(n-1)} \tilde{\sigma}_{ij}(\theta, n) \tag{9.3}$$

In (9.3) $\dot{\ell}(t)$, the crack growth rate, is undetermined by the asymptotic analysis and, in fact, is unspecified by usual continuum analyses. It is determined by the micromechanics of the failure process and Fig. 23 shows computed the crack length history for two meshes. In these calculations, as well as in the slow growing crack calculations, crack growth occurred straight ahead along the initial crack line, [64]. Figure 24 shows a comparison between the finite element and the Hui-Riedel [34] stress state with $\dot{\ell}(t)$ calculated from Fig. 23. The failure condition used in the constitutive model corresponds to a condition for void coalescence along a cavitating grain boundary facet, [32,33]. An additional contribution to the overall strain to failure comes from the linking-up of these cracks. The results in [64] show a significant dependence of crack growth rate on the interval over which the linking-up process occurs and suggest that explicit micro-mechanical modelling of this process merits attention.

10. CONCLUDING REMARKS

There is real potential for the development of a quantitative theory of ductile fracture processes, but much remains to be done. One need is for rate dependent phenomenological plasticity theories that exhibit vertex-like plastic response. An important further step would be the incorporation of physically based models of porosity evolution into such constitutive relations. This would provide a framework for analyzing plasticity dominated localization leading to failure as well as strain localization induced by microrupture. Another need is a characterization of porosity, or more generally of second phase distributions, that goes beyond the one parameter volume fraction description.

Attention has been confined to plastic flow by crystallographic slip and damage evolution by micro-scale void nucleation, growth and coalescence. Modelling of other plastic deformation mechanisms and damage processes has not been touched on, including for example twinning, phase changes and cleavage. Additionally, only monotonic loading histories have been considered. The interaction between competing damage processes is an important topic, where only a few initial steps have been taken, e.g. [68]. It is also worth noting that all boundary value problem solutions discussed here have been for two dimensional geometries. Real microstructures are, of course, three dimensional. Effective numerical procedures for three dimensional problems involving localization and progressive failure are important for further progress and some steps have been taken [41,42,43].

Implications of a rate dependent constitutive relation have been illustrated in several contexts, e.g. crystal plasticity and strain localization. The key feature of the rate dependent formulation is that plastic flow is a function of quantities evaluated at the current state and is independent of their rates. With this constitutive formulation, even problems involving localization and failure are 'well-posed' and convergence with increasing mesh refinement can be obtained. In essence, for localization problems, the rate dependent formulation introduces a characteristic length into the problem. Of course, in any particular circumstance, there is the question as to what the appropriate characteristic length scale is. Physically appropriate and mathematically tractable ways of incorporating length scales into continuum descriptions of localization and failure merit further exploration.

ACKNOWLEDGMENTS

Support for my contributions to the work reviewed here from the solid mechanics programs of NSF and ONR, and from the Materials Research Group at Brown University, funded by NSF, is gratefully acknowledged. I also appreciate the support provided by NSF grant MSM-8618007 for preparation of this paper.

REFERENCES

[1] Tvergaard, V., this volume.
[2] Hill, R., *J. Mech. Phys. Solids* **15** (1967) 79.
[3] Taylor, G.I., *J. Inst. Metals* **62** (1938) 307.
[4] Taylor, G.I., in Stephan Timoshenko 60th Anniversary Volume, Lessels, J.M. (ed.), (Macmillan, New York, 1939) p. 218.
[5] Rice, J.R. and Rosengren, G.F., *J. Mech. Phys. Solids* **16** (1968) 1.
[6] Hutchinson, J.W., *J. Mech. Phys. Solids* **16** (1968) 13.
[7] Hutchinson, J.W., *Proc. Roy. Soc. Lond.* **A319** (1970) 247.
[8] Peirce, D., Shih, C.F. and Needleman, A., *Comp. Struct.* **18** (1984) 875.
[9] Pan, J. and Rice, J.R., *Int. J. Solids Struct.* **19** (1983) 973.
[10] Peirce, D., Asaro, R.J. and Needleman, A., *Acta Metall.* **31** (1983) 1951.
[11] Asaro, R.J. and Needleman, A., *Acta Metall.* **33** (1985) 923.
[12] Hill, R., and Rice, J.R., *J. Mech. Phys. Solids* **20** (1972) 401.
[13] Rice, J.R., *J. Mech. Phys. Solids* **19** (1971) 433.
[14] Hill, R., and Havner, K.S., *J. Mech. Phys. Solids* **30** (1982) 5.
[15] Asaro, R.J. and Rice, J.R., *J. Mech. Phys. Solids* **25** (1977) 309.
[16] Peirce, D., Asaro, R.J. and Needleman, A., *Acta Metall.* **30** (1982) 1087.
[17] Rice, J.R., *J. Appl. Mech.* **37** (1970) 728.
[18] Harren, S., Lowe, T.C., Asaro, R.J. and Needleman, A., Analysis of Large-Strain Shear in Rate-Dependent FCC Polycrystals: Correlation of Micro and Macromechanics, *Phil. Trans. Roy. Soc. Lond.*, in press.
[19] Christoffersen, J. and Hutchinson, J.W., *J. Mech. Phys. Solids* **27** (1979) 465.
[20] Hutchinson, J.W. and Tvergaard, V., *Int. J. Mech. Sci.* **22** (1980) 339.
[21] Tvergaard, V., Needleman, A. and Lo, K.K., *J. Mech. Phys. Solids* **29** (1981) 115.
[22] Triantafyllidis, N., Needleman, A. and Tvergaard, V., *Int. J. Solids Struct.* **18** (1982) 121.
[23] Needleman, A. and Tvergaard, V., in Finite Elements – Special Problems in Solid Mechanics, Oden, J.T. and Carey, G.F., (eds.), (Prentice-Hall, Englewood Cliffs, NJ, 1983) p. 94.
[24] Rogers, H.C., *Trans. TMS-AIME* **218** (1960) 498.
[25] Gurland, J. and Plateau, J., *Trans. ASM* **56** (1963) 442.
[26] Gurson, A.L., Plastic Flow and Fracture Behavior of Ductile Materials Incorporating Void Nucleation, Growth and Interaction, Ph.D. Thesis, Brown University, 1975.
[27] Pan, J., Saje, M. and Needleman, A., *Int. J. Fract.* **21** (1983) 261.
[28] Tvergaard, V., *Int. J. Fract.* **17** (1981) 389.
[29] Tvergaard, V. and Needleman, A., *Acta Metall.* **32** (1984) 157.
[30] Koplik, J. and Needleman, A., Void Growth and Coalescence in Porous Plastic Solids, *Int. J. Solids Struct.*, in press.
[31] Rice, J.R., in Proc. 14th Int. Congr. Theoret. Appl. Mech., Koiter, W.T., (ed.), (North-Holland, Amsterdam, 1977) p. 207.
[32] Tvergaard, V., *Acta Metall.* **32** (1984) 1997.
[33] Tvergaard, V., *J. Mech. Phys. Solids* **32** (1984) 373.
[34] Hui, C.Y. and Riedel, H., *Int. J. Fract.* **17** (1981) 409.

[35] Asaro, R.J., *Adv. Appl. Mech.* **23** (1983) 1.

[36] Needleman, A., *Comp. Meth. Appl. Mech. Engr.* **67** (1988) 69.

[37] LeMonds, J. and Needleman, A., *Mech. Matl.* **5** (1986) 339.

[38] Needleman, A., Dynamic Shear Band Development in Plane Strain, *J. Appl. Mech.*, in press.

[39] Marchand, A. and Duffy, J., *J. Mech. Phys. Solids* **36** (1988) 251.

[40] Vardoulakis, I., Goldscheider, M. and Gudehus, G., *Int. J. Num. Analyt. Meth. Geomech.* **2** (1978) 99.

[41] Ortiz, M., Leroy, Y., and Needleman, A., *Comp. Meth. Appl. Mech. Engr.* **61** (1987) 189.

[42] Leroy, Y., and Ortiz, M., Finite Element Analysis of Strain Localization in Frictional Materials, *Int. J. Num. Anal. Meth. Geomech.*, in press.

[43] Nacar, A., Needleman, A. and Ortiz, M., A Finite Element Method for Analyzing Localization in Rate Dependent Solids at Finite Strains, submitted for publication.

[44] Tvergaard, V., *J. Mech. Phys. Solids* **30** (1982) 399.

[45] Zienkewicz, O.C., The Finite Element Method, 3rd ed., (McGraw-Hill, New York, 1977).

[46] Hughes, T.J.R. and Belytschko, T., *J. Appl. Mech.* **50** (1983) 1033.

[47] Asaro, R.J., *Acta Metall.* **27** (1979) 445.

[48] Piercy, G.R., Cahn, R.W. and Cottrell, A.H., *Acta Metall.* **3** (1955) 331.

[49] Harren, S., Elasto-Plastic Deformations in Rate-Dependent Metallic Polycrystals, Ph.D. Thesis, Brown University, 1988.

[50] Speich, G.R. and Spitzig, W.A., *Metall. Trans.* **13A** (1982) 2239.

[51] Becker, R. and Needleman, A., *J. Appl. Mech.* **53** (1986) 491.

[52] Needleman, A. and Becker, R., in Interdisciplinary Issues in Materials Processing and Manufacturing, Vol. 1, Samanta, S.K. et al., (eds.), (ASME, New York, 1987) p. 139.

[53] Becker, R., Needleman, A., Richmond, O. and Tvergaard, V., *J. Mech. Phys. Solids* **36** (1988) 317.

[54] Rice, J.R. and Johnson, M.A., in Inelastic Behavior of Solids, Kanninen, M.F., et al., (eds.), (McGraw-Hill, New York, 1970) p. 641.

[55] Rice, J.R., in The Mechanics of Fracture, Erdogan, F., (ed.), AMD Vol. 19 (ASME, New York, 1976) p. 23.

[56] Suresh, S., Vasudevan, A.K., Tosten, M. and Howell, P.R., *Acta Metall.* **35** (1987) 25.

[57] Becker, R., Needleman, A., Suresh, S., Tvergaard, V. and Vasudevan, A.K., An Analysis of Ductile Failure by Grain Boundary Growth, *Acta Metall.*, in press.

[58] Rice, J.R., *J. Appl. Mech.* **35** (1968) 379.

[59] Needleman, A., *J. Appl. Mech.* **54** (1987) 525.

[60] Nutt, S.R. and Needleman, A., *Scripta Metall.* **21** (1987) 705.

[61] Dubensky, E.M. and Koss, D.A., *Metall. Trans.* **18A** (1987) 1887.

[62] Becker, R., *J. Mech. Phys. Solids* **35** (1987) 577.

[63] Needleman, A. and Kushner, A.K., The Effect of Void Distribution on Plastic Flow in Porous Solids, in preparation.

[64] Li, F.Z., Needleman, A. and Shih, C.F., Creep Crack Growth by Grain Boundary Cavitation: Crack Tip Fields and Crack Growth Rates under Transient Conditions, *Int. J. Fract.*, in press.

[65] Chaboche, J.L., this volume.

[66] Needleman, A. and Tvergaard, V., *J. Mech. Phys. Solids* **35** (1987) 151.

[67] Paris, P., Tada, H., Zahoor, A. and Ernst, H., *ASTM STP* **668** (1979) 5.

[68] Tvergaard, V. and Needleman, A., *J. Mech. Phys. Solids* **34** (1986) 213.

Theoretical and Applied Mechanics
P. Germain, M. Piau and D. Caillerie (Editors)
Elsevier Science Publishers B.V. (North-Holland)
© IUTAM, 1989

FLOW OF GRANULAR MATERIALS

Stuart B. SAVAGE

Laboratory for Hydraulics,
Hydrology, and Glaciology
ETH, CH-8092 Zürich [a]

This paper deals with rapid deformations of bulk solids made up of discrete particles. The main focus is on flows at high concentrations and high deformation rates in which the interstitial fluid is relatively unimportant and the bulk behavior is governed largely by interactions between the solid constituents through interparticle collisions and Coulomb - frictional rubbing contacts of longer duration. Some distinctive characteristics of granular flows, and concepts that are key to their understanding are discussed. Three geophysical problems flow problems involving mass-movements are described briefly: (i) the motion of rockfalls, (ii) instabilities in mudflows, and (iii) particle size segregation.

1. INTRODUCTION

Rapid deformations of bulk solids such sand, ore, coal, crushed oil shale, grain, granular snow, pack ice, and metal and ceramic powders have been termed *granular flows*. During flow, the solids concentrations typically are quite high, and as a result the bulk behaviour is governed largely by interactions between the solid constituents arising from collisional impacts and longer term frictional rubbing contacts. In some instances, the interstitial fluid phase plays a minor role in determining the overall constitutive behavior. This might occur, for example, when the density and viscosity of the interstitial fluid and the mean relative velocity of the two phases are small. In other cases, the interactions between the fluid and solid phases are significant; the motion of the fluid can provide the driving force for the flow of the solid phase. The dynamical behavior of these materials can be very complex; its description involves aspects of traditional fluid mechanics, plasticity theory, soil mechanics and rheology.

An understanding of the mechanics of the flow of granular materials is of fundamental importance in connection with a very wide range of technological and scientific problems. Many of these have been the subject of study for many years. Examples of problems in the soil mechanics and rock mechanics area (that involve other than quasi-static deformations) include landslides, rockfalls, debris flows and stability of tailings dumps. Those in the hydraulics area include sediment transport, flows in slurry pipelines, hydraulic dredging, ice jams in rivers, mechanics of mush ice and sub-aqueous grain flows. Some related problems that are of concern to structural engineers are the stresses developed by granular materials on the walls of bins and hoppers, and the ice forces on bridge piers and artificial islands used for oil exploration. In the area of geo- and planetary physics, one can cite snow avalanches, drift of pack ice, turbidity currents, and the motion of planetary rings. The design of materials handling equipment would benefit greatly from a clearer insight

---

[a]on leave from Department of Civil Engineering & Applied Mechanics, McGill University, Montreal.

into bulk material behavior. There are numerous chemical engineering applications such as fluidised beds, spouted beds, manufacture of pharmaceuticals, etc. that involve granular flows. New methods of Xerography, powder metal forming processes and ultrastructural processing of ceramics are just a few high-tech examples which require knowledge about bulk solids flow. Although the examples mentioned above might appear to be of a very disparate nature they in fact have many similarities. For example, the flow pattern of pack ice driven by wind and ocean currents through a converging channel such as the Bering Straight on a scale of many kilometers is remarkably like that of the gravity flow of sand through a hopper on a laboratory scale of less than a meter [1].

During the past fifteen years, research into the fundamental mechanics of granular flows has increased both in pace and in scope. Many of these investigations, whether experimental, theoretical, or numerical simulations of granular flows, have been focused at the microstructural level. The reasoning behind this is that a proper understanding the gross flow behavior on the large scale will follow from detailed examinations of the interactions between individual particles. Some of this work has been made possible through the recent availability of supercomputers, and exploitation of theoretical techniques developed for the study of liquids at the molecular level. These studies have not only greatly improved our present qualitative understanding of granular flows in complex practical situations but should eventually lead to the capability of quantitative predictions of general flows.

It would be difficult to give even a brief review of the variety of current work on the mechanics of granular materials in the time available for this lecture. A presentation of this kind would perhaps also be repetitious since there have been several recent reviews of such work. Savage [2] has provided an overall review including both theoretical analyses and laboratory experiments. More recently, Jenkins [3] has presented a careful and detailed analytical review of kinetic theories of granular flows, while a more heuristic discussion concentrating on the physical aspects has been given by Haff [4]. Campbell [5] has described the various types of computer simulations of granular flows. A review of theoretical and experimental work with a particular emphasis on chemical engineering applications has been provided by Nedderman, Houlsby, Savage and Tuzun [6,7,8]. Somewhat earlier, Jenkins and Cowin [9] reviewed continuum theories of granular flows, and Wieghardt [10] described some of experimental work on hopper flows and plowing. I shall not attempt another review here, but instead, after some brief preliminary remarks, will describe some of my own current work that has applications to geophysical problems. These examples should illustrate a number of interesting features of granular flows, and as well bring out both the similarities and the differences that exist between such flows and those of more common kinds of fluids.

2. GENERAL DISCUSSION OF CONSTITUTIVE BEHAVIOUR

We shall begin with a brief description of some aspects of the constitutive behaviour of granular materials. To perform analyses of various flow situations, we eventually require continuum descriptions of the constitutive equations for the fluxes of mass, momentum and energy. However, to obtain a better physical understanding of the bulk behaviour, it is sometimes useful to focus at a finer scale and consider things at the level of individual particle interactions. This approach can in fact be pursued further to obtain analytical descriptions in terms of kinetic theories for this bulk behaviour.

2.1. Mechanisms for generation of stresses

For the sake of simplicity in these preliminary remarks, we shall consider the bulk to be made up of solid particles immersed in a gas of negligible density so that we can ignore any interstitial fluid effects. During deformation of the bulk material, individual particles typically acquire, in addition to their mean translational motion, more or less random translational fluctuation velocities and spins, as a result of collisions and particle overriding. For dry granular materials, there are three main mechanisms by which stresses can be generated during bulk deformations; these are (i) dry Coulomb-type rubbing friction, (ii) transport of momentum by particle translation, and (iii) momentum transport by collisional interactions. (Similar mechanisms are at work during the analogous transport of energy, etc.) Although there are instances where all three of these mechanisms are effective, there are limiting flow regimes in which a single one plays the dominant role. At high solids concentrations and low shear rates, the particles will be in close rubbing contact and the stresses are of the quasi-static, rate-independent Coulomb-type described in the soil mechanics literature. On the other hand at very low concentrations and high shear rates, the granular material will be in some ways analogous to a dilute gas. The mean free path is large compared to the particle diameter, and stresses result from the interchange of particles between adjacent layers moving at different mean transport velocities. If both the concentration and shear rate are large, then momentum transport occurs as a result of collisional interactions, since there rarely occur void spaces of sufficient size to permit particle transport betweeen adjacent shear layers.

2.2. Kinetic theories

The flow regime in which nearly instantaneous collisions are the dominant means of momentum and energy transport has been termed the *grain inertia* regime by Bagnold [11]. He argued that under these circumstances momentum transfer occurs by a succession of glancing collisions as the grains of one layer overtake those of the adjacent slower layer. Both the change in momentum during a single collision and the rate at which collsions occur are proportional to the relative velocity of the two layers. Hence, both the shear and normal stresses will vary as the square of the shear rate. This behaviour was verified at high shear rates in his own experiments (which involved neutrally buoyant wax spheres in Newtonian fluids) and by several subsequent workers [12,13,14,15,16].

Bagnold's basic ideas have been extended along the lines of the 'hard-sphere' kinetic theories used for the analysis of dense gases and liquids [17] - [22]. Some important results that emerge are the evolution equations for the particle velocity fluctuations and spins. The kinetic energy associated with the translational velocity fluctuations has been related to the 'granular temperature', which has an obvious analogy with the definition of the temperature in a fluid at the molecular level. This analytical work has been paralleled by computer simulations of granular flows that have been patterned after molecular dynamics calculations. Some of these [23,24,25] have involved the assumption of 'hard' particles such that the collisions can be treated as nearly instantaneous. Other work has considered 'softer' particles which result in more complex kinds of interparticle interactions of finite durations [26,27,28,34]. These simulations are deterministic in the sense that they keep track of the position and velocity of all the particles within the computation 'box' and use 'periodic' boundary conditions to effectively extend the region of computation. While they are the most direct and exact kinds of computations, they are also expensive in terms of computer time and memory. Computations can be carried out more rapidly by using

Monte-Carlo methods [29], but at the cost of loosing some details about the structure of the flow field, which must be supplied by the programmer. A combination of these deterministic and probabalistic approaches has been used in an adaptation of the 'direct Monte-Carlo simulation technique' for granular flows [30].

It has been recognized that one of the important aspects in the analysis of shear flows is the treatment of boundary conditions, both at a free surface and next to a solid wall. The determination of the velocity 'slip' and granular temperature 'jumps' at such 'surfaces' is the subject of current study [31,32,33,34,35].

Most engineering applications and other natural flow situations probably fit into a flow regime in which *both* nearly instantaneous collisional and longer time particle interactions occur and contribute to the generation of stresses. Although such cases can be treated by the 'soft' particle computer simulations, it is extremely difficult to extend the analytical kinetic theory models in a rigorous way to handle this kind of mixed flow regime. The only analytical attempts to do so have involved a rather crude patching together of results from analyses for the quasi-static and grain-inertia flow regimes mentioned above. Simple examples of this may be found in Savage [36], Norem, Irgens and Schieldrop [37] and in the more detailed analysis of Johnson and Jackson [38]. Savage [36], for example, suggested that, for the case of a shear flow in a gravitational field, one might represent the total stresses as the linear sum of a rate-independent, dry friction part plus a rate-dependent 'viscous' part obtained from the high shear rate granular flow kinetic theories. Because of its practical importance, this flow regime is obviously one that deserves increased attention.

Proper treatment of the boundary conditions and inclusion of the effects of long time frictional rubbing contacts are essential if one is to solve boundary value problems in a physically realistic way. Their importance is illustrated by comparing the success acheived by Johnson and Jackson in their shear cell simulations [38] and later chute flow simulations with the problems experienced by Szidarovszky, *et al.* [39] in analogous chute flow computations. The very limited range of conditions for which the latter workers were able to obtain solutions, and the extreme difficulties that were encountered in obtaining them, resulted from the failure to include the above effects.

2.3. Some distinctive characteristics of granular flows

2.3.1. Dilatancy

The concept of dilatancy was defined by Reynolds [40] as the property possessed by a mass of granular material to alter its volume in accordance with a change in the arrangement of its grains. If we start with a mass of grains in a fairly closed packed arrangement and subject it to a load to cause a shear deformation in at least portions of the bulk, then from geometrical considerations of the particle overriding it can be seen that an increase in volume of the bulk will occur. During these quasi-static deformations, dilatancy is a consequence of kinematic restrictions. However, under rapid shearing conditions analogous things occur. In Bagnold's simple grain inertia model it was found that the normal stress (which Bagnold termed the dispersive stress) increases with the square of the shear rate. Hence, during shearing motions, the development of these stresses will tend to expand the thickness of the shear layer. For this reason sheared granular materials are sometimes termed 'dilatant' materials [41].

2.3.2. Internal friction angle

Unlike a fluid, granular material can be piled in a heap. For a given material the free surface can be inclined at some maximum value corresponding to the material's *angle of repose.* This behaviour is similar to the sliding of a block on an inclined rough plane which will occur when the inclination angle is greater than the friction angle. Analogous things occur in the interior of a slowly deforming granular material. For a cohesionless material, the behaviour is well described by the Mohr-Coulomb yield criterion [42,43], which states that yielding will occur at a point on a plane element when

$$|S| = N \tan \phi, \tag{2.1}$$

where S and N are respectively the shear and normal stress acting on the element, and ϕ is the internal friction angle. For spherical particles, like glass beads, $\phi \simeq 24°$, and for more angular particles, like crushed sand, ϕ is around $37°$. What is perhaps surprising is that when (dry) granular materials are sheared quite rapidly, the ratio of shear to normal stress remains nearly constant, and close to the value observed during quasi-static deformations.

2.3.3. Effect of experimental test procedures on observed rate dependence

A number of experiments have been performed using annular shear cell devices to obtain information about the dependence of shear and normal stresses upon shear rate and particle concentration [12,13,14,15,16]. Typically, granular material was contained in an annular trough in the lower half of the shear cell and capped by an annular ring attached to the upper half of the shear cell. The horizontal surfaces which came in contact with the granular material were roughened, so that when the two halves of the shear cell were rotated with respect to one another about a vertical axis, the granular material was sheared. Quite different results are obtained, depending upon how the tests are performed. When tests are performed so as to maintain a constant volume (i.e., constant mean solids concentration), at high shear rates it is found that stresses vary with the square of the shear rate. Stresses also increase strongly with increase in solids concentration, but the *ratio* of shear to normal stress is *nearly* constant. These results are predicted by the kinetic theories. However, if shear rate is increased at a constant normal load, and the bed of dry particles is allowed to expand freely, it is found that the shear stress is nearly independent of shear rate. This is quite consistent with the nearly constant dynamic friction angle (ratio of shear to normal stress). But, is is quite different from what would occur with a Newtonian fluid, and it led some of the early investigators to infer a complete lack of rate dependence for sheared granular materials. Some typical results for tests at constant normal stress taken from Stadler [15] are shown in figure 1. At the higher velocities the flow is in the grain inertia regime. A detailed discussion of this behaviour and its relationship to flow down inclined chutes [44] is given in Savage and Hutter [45].

2.3.4. Large energy dissipation

During rapid flows the particles are in a highly agitated state, continually colliding with one another. Since the particles are inelastic and frictional, the kinetic energy associated with the velocity fluctuations is dissipated with each collision. Granular systems are extremely dissipative and in the absence of some driving mechanism, the fluctuations very quickly die away. This sizeable energy dissipation is one major difference between granular flows and gases at the molecular level. In the case of the granular material, energy can be degraded one level lower; the kinetic energy of the particle fluctuations is transformed, resulting in an increase in the temperature of the individual grains.

One means of maintaining the velocity fluctuations in a mass of granular material is to supply energy by vibrating a wall of the container. Solving the fluctuation kinetic energy

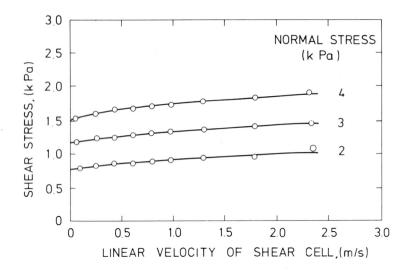

FIGURE 1. Shear stress versus linear velocity of shear cell at constant applied normal stress for 1.1 mm dry glass beads (after Stadler [15]).

equation obtained, for example, from the kinetic theory of Jenkins and Savage [18], one can determine the variation of the granular temperature, $T = \langle \mathbf{C}^2 \rangle / 3$, with distance y from the vibrating wall, where $\langle \mathbf{C}^2 \rangle$ is the mean square of the translational velocity fluctuation $\mathbf{C}$. It is found (see Savage [46]) that $T \propto \exp(-\alpha y/\sigma)$ where $\alpha = [6(1 - e)]^{1/2}$, σ is the particle diameter, and e is the coefficient of restitution of the particles. Thus, for typical materials the 'e-folding' distance for the granular temperature in this situation is of the order of a particle diameter.

For the case of a simple shear flow, the mean shear supplies the energy that maintains the fluctuations. Equilibrium (in terms of the velocity fluctuations) is reached when the shear work (the product of the shear stress and shear rate) is equal to the collisional rate of energy dissipation per unit volume. At higher concentrations, one finds that the shear layers are quite thin and typically of the order of 10 particle diameters thick; this is observed under both quasi-static and rapid flow conditions.

2.3.5. 'Granular Reynolds number'

Consider the case of a simple granular shear flow that might be generated by the relative velocity U between two rough parallel plates spaced a distance H_s apart. Let us express the shear stress as $\tau = \mu_{eff}(U/H_s)$ where μ_{eff} is the effective viscosity of the sheared granular material. Now define a *granular Reynolds number* for this flow as

$$Re_G = \frac{\rho U H_s}{\mu_{eff}}, \tag{2.2}$$

where $\rho = \rho_p \nu$ is the bulk density, ρ_p is the mass density of the individual particles, and ν is the solids volume fraction (volume of solids/total volume). From the kinetic theory of Lun, *et al.* [20] and the shear cell experimental data of Savage and Sayed [13] for $\nu = 0.5$, we find that $\tau \simeq 2\rho_p \sigma^2 (U/H_s)^2$. Hence, the granular Reynolds number is

$$Re_G \simeq \frac{1}{4}\left(\frac{H_s}{\sigma}\right)^2. \tag{2.3}$$

We have obtained the rather curious result that Re_G depends upon the ratio of shear layer thickness to particle diameter. For typical values of $H_s/\sigma \simeq 4$ to 10 we find that $Re_G \simeq 4$ to 25. Thus, even though on the microscale the mechanism of stress generation is dominated by grain inertia effects, the bulk behaviour is a very 'viscous' one in the normal fluid mechanics sense. It would be interesting to investigate whether for low particle concentrations and large layer thicknesses (say $\nu \simeq 0.1$ and $H_s/D = O(50)$) there will occur something akin to the laminar-turbulent flow transition in normal Newtonian fluids.

3. EXAMPLES INVOLVING GEOPHYSICAL MASS-FLOWS

We shall briefly describe three examples of granular flows that involve mass-movements in a geophysical context. They illustrate the application of some of the concepts described above. The choice of these particular examples has no special significance other than that they are subjects of some of my current research.

3.1. Rockfalls

In the Alpine regions, rockfalls pose a threat to life and property. When rock material is dislodged on steep slopes, it accelerates down the slope, reaches a maximum velocity, and then decelerates as the the angle of inclination of the bed decreases; eventually bed friction brings the material to rest. The moving streams of debris, in the form of numerous discrete blocks and fragments that are generated during the rockfall, have been termed "strutzstroms". Erisman [59] has made various comparisons to give one an appreciation for the enormous energies involved in these events. For example, the prehistoric Flims landslide in Switzerland had a volume of about $1.2 \times 10^{10} \, \mathrm{m}^3$ and probably took place over a period of a couple of minutes. Its motion resulted in the dissipation of an amount of energy equal to the *total present day energy consumption of the world for a period of 10 hours*.

Sturtzstroms often spread out into very thin layers and flow on surfaces that are much less inclined than the angle of repose of the debris material. For over 100 years, since Albert Heim [47,48] observed and described the Elm rockfall in Switzerland, attempts have been made to explain the apparent fluid-like behaviour and high mobility of these slides (see [49 - 63]). Various hypotheses have included fluidization by an upward current of air [49], a kind of hovercraft action [50,51,52], fluidization by high pressure steam [54], mechanical fluidization aided by the presence of interstitial dust [55], vibrational fluidization by large earthquakes [58], lubrication by a layer of molten rock [59,60], acoustic fluidization [63], and the presence of a thin layer of vigorously fluctuating particles beneath a densely packed overburden [64]. All of these hypotheses are at best controversial and none have been universally accepted. Few have been accompanied by detailed computations.

3.1.1. Size effects

It has also been noted by Heim [48], Scheidegger [53], Hsü [55], Lucchitta [57], Davies [61], and Li Tianchi [62], among others, that the ratio of the total vertical fall height, H, to the total horizontal travel, L, (both measured from top of scar to tip of rockslide) depends upon the total volume of the rock material. Figure 2a shows the ratio H/L, termed the "Fahrböschung" by Heim [48], versus total volume, V, for a number of field events, using data taken from Hsü [55], Lucchitta [57], and Li Tianchi [62]. It is seen that for rockfall volumes greater than about $0.5 \times 10^6 \, \mathrm{m}^3$, there is a decrease in H/L with increasing volume.

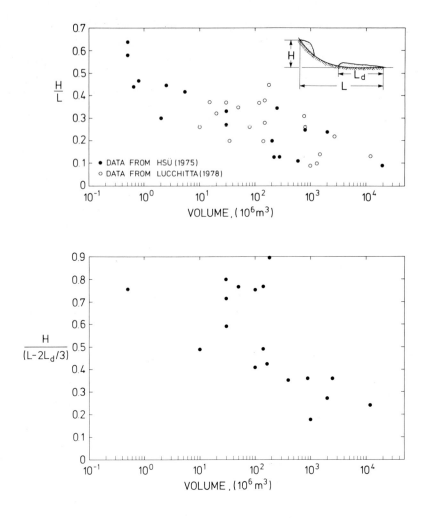

FIGURE 2. (a) Variation of H/L with rockfall volume; based on data from Hsü [55] and
Lucchitta [57], with additional information from Li Tianchi [62]. (b) Slope of line
connecting estimated inital and final centre of mass positions versus rockfall volume.

Such a plot appears to imply a reduction in the bed friction coefficient with volume
V. However, Davies [61] suggested that the apparently 'long' travel of the nose is largely
a geometrical consequence, and that a reduction of bed friction was not necessary to yield
the long runouts. If there is similarity in the final form of the deposit, then the end to
end length, L_d, should vary as the 1/3 power of the volume. By making use of published
landslide data for 24 events, Davies obtained the correlation that

$$L_d = 9.98V^{0.32}, \tag{3.4}$$

suggesting such a similarity relationship. In a similar analysis of 76 events Li Tianchi [62]
obtained somewhat later

$$L_d = 6.48V^{0.32}. \tag{3.5}$$

Davies [61] went on to propose that if the centre of mass was at the midpoint of the final deposit, then the total length of travel is

$$L = \frac{H}{\tan\phi} + \frac{L_d}{2}, \tag{3.6}$$

where ϕ is the friction angle of the granular material, which he assumed to be about $35°$. Although considerable scatter was present, Davies found reasonable agreement between (3.6) and Scheidegger's correlation of H/L versus V for fall heights between 400 m amd 2000 m.

I have reexamined data published in [55,57,62] in an attempt to determine the slope of the line from the initial to final centre of mass, $\tan\theta$, which, as Hsü [56] and Davies have suggested, is a more relevant friction slope than the ratio H/L. Examining results from laboratory model tests [65,66] and, for example, profiles for the Elm rockfall [55,56], it is seen that the shape of the pile is more wedge-like than symmetrical. Thus, the final centre of mass is roughly $2L_d/3$ from the tip rather than being at the centre of the final deposit. We then approximate the slope by

$$\tan\theta = \frac{H}{L - 2L_d/3}. \tag{3.7}$$

In the absence of volume effects, θ should be roughly constant (different bed curvatures will give rise to 'scatter' in θ). Figure 2b shows the results for $\tan\theta$ calculated from data given in [55,57,62]. Although the values of $\tan\theta$ are larger than H/L, and the scatter is increased from that in figure 2a, a decrease with volume is still evident in figure 2b, in contradiction with what would be inferred from Davies' [61] proposal. Thus, the volume dependence of the runout distance apparently remains an enigma. We restrict further discussion in what follows to volumes less than about $10^6\,\mathrm{m}^3$, such that the size effect is not of concern.

3.1.2. Granular flow analysis

It was of interest to see what would result if one analysed the motion of the rockfall as a granular flow. Savage and Hutter have made initial attempts to construct simple models of these kinds of geophysical flows by considering the two-dimensional motion of a finite mass of granular material down plane [45] and curved surfaces [67]. The granular material was treated as an incompressible, cohesionless continuum obeying a Mohr-Coulomb yield criterion. The pile of material was assumed to be long and thin, the streamwise velocity profile nearly uniform, and the curvature of the bed was taken to be moderate (cf. figure 3). By integrating over the thickness of the pile, the following (nondimensional) depth-averaged equations of motion were obtained

$$\frac{\partial h}{\partial t} + \frac{\partial(h\bar{u})}{\partial x} = 0, \tag{3.8}$$

and

$$\frac{\partial \bar{u}}{\partial t} + \bar{u}\frac{\partial \bar{u}}{\partial x} = \sin\zeta - \tan\delta\,\mathrm{sgn}(\bar{u})\left(\cos\zeta + \lambda\kappa\bar{u}^2\right) - \epsilon\,k_{actpass}\,\cos\zeta\,\frac{\partial h}{\partial x}, \tag{3.9}$$

where $\bar{u}$ is the depth average streamwise velocity defined by

$$\bar{u} = \int_0^h u\,dy, \tag{3.10}$$

h is the local depth, ζ is the local bed inclination angle, κ is the local curvature of the

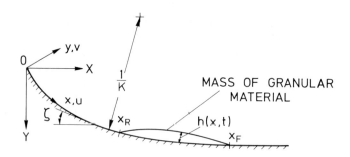

FIGURE 3. Definition sketch for the motion of a finite mass of granular material down a curved bed.

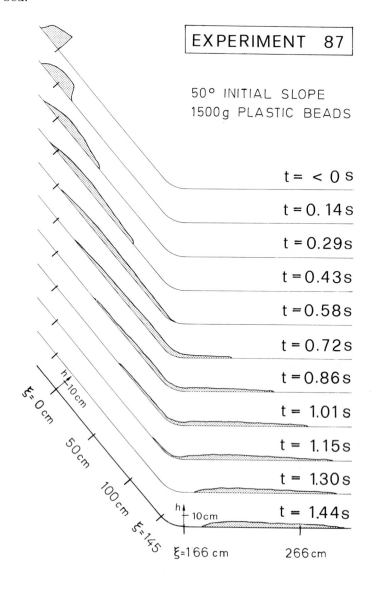

FIGURE 4. Evolution of a pile of plastic particles released from rest on a 50° incline. Drawings of depth profiles at various times taken from photographs (after Plüss [71]).

bed profile, t and x are the nondimensional time and streamwise distance, ϵ is the ratio of the characteristic depth to length of the pile, and λ is the ratio of the characteristic length to the bed radius of curvature. Following the assumption of Coulomb-like behaviour, the shear stress at the bed was related to the the normal stress there by means of a bed friction angle δ. In accordance with the behaviour discussed in §2.3.3 (cf. figure 1), the bed friction was approximated as a constant, independent of the velocity. Following a common soil mechanics practice, the normal stresses parallel and perpendicular to the bed were related to one another through an *earth pressure coefficient*, $k_{actpass}$, which may take on the active or passive values,

$$\left.\begin{array}{c} k_{act} \\ k_{pass} \end{array}\right\} = 2\left[1 \mp \sqrt{1 - (1 + \tan^2\delta)\cos^2\phi}\right]/\cos^2\phi - 1, \qquad \text{for } \partial\bar{u}/\partial x = \begin{array}{c} > 0 \\ < 0 \end{array}. \quad (3.11)$$

It was found quite difficult to obtain accurate and reliable numerical solutions to this set of equations; however, analytical similarity solutions [45,68] and perturbation solutions [69] have been obtained. The similarity solutions were found for cases of granular flow down inclined planes and gradually curved beds; in them the pile has the shape of a parabolic cap . The numerical approach that was found to be the most simple and accurate was a Lagrangian scheme in which the computational grid was advected with the material 'particles' [45,67]. Until recently, the only laboratory scale experiments that could be used to verify theoretical analyses were those of Huber [70]. To supplement them, more extensive measurements of the motion of finite masses of various granular materials released from rest on curved beds have been carried out by Hutter, and coworkers [71,65,66]. Some typical experimental results, in the form of snapshots at various times of the shape of a pile of plastic beads flowing down a rough inclined [71] bed, are shown in figure 4. Figures 5a and 5b compare the measured temporal evolution of the front, rear, and the total length of the moving pile with that predicted by the Lagrangian finite difference scheme.

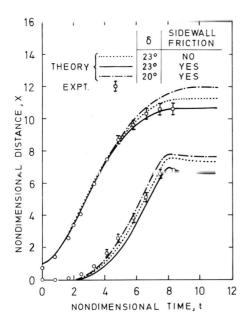
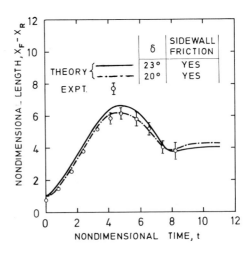

FIGURE 5. (a) Comparison of predicted and measured [71] positions of leading and trailing edge of granular avalanche shown in figure 4. (b) Similar comparison for total length of pile versus time.

Most notable is the rapid evolution into a very thin layer of material. The predicted positions of the front and rear of the pile, as well as its depth profile, are in good agreement with the measurements. The analysis has made use of no unusual fluidization mechanism. We have merely assumed a very simple and ordinary Coulomb-type behaviour; the spreading and fluid-like behaviour result directly from the dynamical equations of motion.

Work to consider physically more realistic cases of three-dimensional motions is presently underway [72,73].

3.2. Development of surge waves in mudflows

The sturtzstroms discussed in the previous section are rapid, gravity-induced, mass movements involving essentially dry materials. Mudflows [74 - 79] are another form of debris flow, differing from the sturtzstroms in that they have a significant water content in addition to granular solids of different sizes which range from fine clay particles, silt, to sand, pebbles and finally rocks and boulders having diameters of the order of meters. Mudflows commonly originate in mountainous regions and are responsible for considerable property damage and loss of life each year. Their great destructive power is due in large part to two characteristic features: (i) the ability of the flow to transport boulders, trees, and other large debris, and (ii) the tendency of the flow to develop into the form of surge waves, with steep fronts as much as several meters high. The largest boulders collect at the very front of the surge wave, rumbling and crashing into one another, scouring the channel bed, and destroying virtually anything in their path.

The mechanics of these flows is not well understood. The wide range of materials involved leads to a complex constitutive behaviour [75 - 86]. The interactions between the particles of different sizes is complicated by the presence of the interstitial 'fluid'. The viscosity and cohesion of this 'fluid' can be much larger than that of typical building mortar. One of the simpler assumptions is to approximate the behaviour by a Coulomb - viscous model [75,76]. The presence of the mud or clay leads to a finite cohesion or yield stress. The larger discrete granular solids give rise to a Coulomb-like behaviour in which shear stresses are related to the normal stresses in a nearly linear fashion (cf. §2.3). Finally, there are contributions to the stresses which depend upon bulk deformations in a viscous, rate-dependent manner. In addition to the problem of understanding and describing theoretically the details of the flow of this complex mixture under uniform, steady state conditions, two interesting questions arise:

1. What are the mechanisms which cause a more or less uniform, steady flow to develop into surge waves or pulse-like flow?

2. Why is there an accumulation of the large boulders at the leading edge of the surge wave?

We consider the first matter next, and discuss the second in §3.3.

3.2.1. Roll waves

Particularly striking photgraphs and descriptions of mudflow waves in Jiang-jia Ravine in China are given in the paper of Li Jian, *et al.* [87]. The continuous flow in this channel can develop into a train of surge waves, with intervals of very low, and even, in some cases, zero discharge between the individual surges. Curry [88], Broscoe and Thomson [89], and Li Jian and Luo Defu [90] have presented other discussions of the pulsations in mudflows. Davies [77,78] has recently reviewed these kinds of field observations of the mudflow type of debris flows and hypothesized that the pulsating flow in the Jiang-jia Gulley and other similar events in China were the result of *roll waves* [93,94,95,96,97] in laminar flows of the highly viscous mud.

There is evidence that roll waves develop in other kinds of granular flows. They have been observed during the flow of dry granular materials in the laboratory [44,78]. This suggests that roll waves *might* occur in nature during rockfalls [55,56,61]. Transverse ridges, that may have resulted from such instabilities, have been observed in sturtzstrom deposits [56,57,58]. Schaerer and Salway [91] observed the development of wave motions during dry snow avalanches analogous to roll waves observed in water flows. Hopfinger [92] has suggested that the transition from a dense flow snow avalanche to an airborne powder snow avalanche is the result of roll wave instabilities.

In typical turbulent open channel water flows, the uniform flow becomes unstable and roll waves begin to develop when the Froude number is greater than about 2. The Froude numbers for the flows in the field events examined by Davies were somewhat lower than this. It is interesting to consider how the material constitutive behaviour for mudflows, which is quite different from that used to model water flows, might affect the roll wave stability criteria and the quasi-steady solutions of periodic roll waves. In particular, how does material cohesion affect the critical Froude numbers? Savage [98] has used a generalization of the usual Chezy bed friction law to consider these questions and investigate the development of roll waves in mudflows, snow avalanches and sturtzstroms. Some results of this study are outlined below.

3.2.2. Stability analysis

Let us consider a uniform debris flow at the normal depth down a wide channel inclined at a constant angle ζ and examine its linear stability when subjected to small disturbances of wavelength λ (figure 6). In the hydraulics literature, gradually varied open channel flows

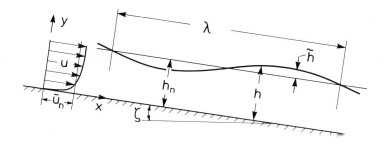

FIGURE 6. Definition sketch for roll waves down inclined channel.

are usually treated by using (empirical) bed friction laws that are based upon uniform flow at the normal depth. One such law is the Vedernikov formula, which is a generalization of the Chezy formula, i.e.

$$\overline{u} = (\beta \sin \zeta)^{1/a} h^{(1+b)/a}, \tag{3.12}$$

where a, b, and β are empirically determined coefficients. We can rearrange (3.12) to write it in terms of the bed shear stress, i.e.

$$\tau_b = \frac{\overline{\rho} g}{\beta} \frac{\overline{u}^a}{h^b}, \tag{3.13}$$

where $\bar{\rho}$ is the depth averaged mean density of the fluid-particle mixture, and g is the gravitational acceleration. Note that in the case of the Chezy formula (for a turbulent Newtonian flow in a fully rough channel) the constants in (3.12) are

$$\beta^{1/a} = C_f = \sqrt{8g/f}, \tag{3.14}$$
$$a - 2, \qquad b - 0, \tag{3.15}$$

where C_f is the Chezy flow resistance factor, and f is the usual Darcy friction factor.

In order to include the effects of the cohesive interstitial fluid, and the interactions between the large particulate materials, we use a generalization of (3.13) and express the shear stress in the following form

$$\tau_b = K + \frac{\bar{\rho}g}{\beta}\frac{\bar{u}^a}{h^b}, \tag{3.16}$$

where K is the shear strength, defined as the sum of the cohesion C and the granular interparticle friction, i.e.

$$K = C + \left(\frac{\bar{\rho} - \rho_f}{\bar{\rho}}\right)\bar{\rho}gh\cos\zeta\,\tan\delta_0, \tag{3.17}$$

ρ_f is the density of the interparticle fluid, and δ_0 is the quasi-static friction angle associated with the granular material. Equations (3.16) and (3.17) can be regarded essentially as a representation the Coulomb-viscous model [75,76]. The factor $(\bar{\rho}-\rho_f)/\bar{\rho}$ in the last term of (3.17) accounts for the presence of the pore fluid and its effect in determining the effective interparticle stress. In general, we anticipate that β in (3.16) may depend upon ζ.

Now carry out the depth-averaging in a manner similar to that used in §3.1. The depth averaged conservation of mass equation takes on the same form as (3.8). Because of the somewhat different nondimensionalization (which is based upon flow conditions at the normal depth), and the more general bed friction law, the depth averaged conservation of linear momentum equation has a form different from (3.9), i.e.

$$\epsilon F^2 \left(\frac{\partial \bar{u}}{\partial t} + \alpha\bar{u}\frac{\partial \bar{u}}{\partial x} \;-\; (\alpha - 1)\frac{\bar{u}}{h}\frac{\partial h}{\partial t}\right) = \tan\zeta - \frac{C^*}{h} - \left(\frac{\bar{\rho} - \rho_f}{\bar{\rho}}\right)\tan\delta_0$$
$$-\; \Gamma F^2 \frac{\bar{u}^a}{h^{b+1}} - \epsilon\,k_{actpass}\frac{\partial h}{\partial x} + \epsilon\frac{F^2}{hR}\frac{\partial}{\partial x}\left(h\frac{\partial \bar{u}}{\partial x}\right), \tag{3.18}$$

where now h and $\bar{u}$ represent the *nondimensional* depth and depth averaged velocity, ϵ is the ratio of normal depth h_n to wavelength λ, $F = u_n/\sqrt{gh_n\cos\zeta}$ is the Froude Number, and u_n is the (physical) depth-averaged normal velocity. The momentum coefficient α is defined by

$$\alpha = \frac{\overline{u^2}}{\bar{u}^2}, \qquad \text{where} \qquad \overline{u^2} = \frac{1}{h}\int_0^h u^2\,dy, \tag{3.19}$$

the nondimensional cohesion by

$$C^* = \frac{C}{\bar{\rho}gh_n\cos\zeta}, \tag{3.20}$$

and

$$\Gamma = \left(\frac{g}{\beta}\right)\frac{u_n^{a-2}}{h_n^b}. \tag{3.21}$$

In (3.18), R is a Reynolds number defined by

$$R = \frac{\bar{\rho}\,u_n\lambda}{\mu_L}. \tag{3.22}$$

where the effective longitudinal viscosity μ_L is similar to the term introduced by Needham and Merkin [96], and also used by Hwang and Chang [97]. This viscosity was included to permit the study of nonlinear, periodic waves; it provides a means of smoothly passing through the jump at the front of a well developed surge wave. An estimate of its magnitude could be determined by consideration of the depth profiles of fixed granular jumps such as those described in [44].

We can now examine the stability of uniform flow in the usual fashion [93,95,96]. Consider small disturbances to the normal velocity and depth such that

$$\bar{u} = 1 + \tilde{u}, \qquad \text{where} \qquad \tilde{u} \ll 1,$$
$$h = 1 + \tilde{h}, \qquad \text{where} \qquad \tilde{h} \ll 1. \tag{3.23}$$

Substituting (3.23) in (3.8) and (3.18) and retaining only linear terms, eliminating $\tilde{u}$, and considering solutions for $\tilde{h}$ in the form

$$\tilde{h} = H(k)e^{ikx-\omega t}, \tag{3.24}$$

where k and ω are the wavenumber and (complex) frequency of the depth perturbation, eventually leads to the following expression for the critical Froude number for instability of the flow at the normal depth

$$F^2 = \left\{ B\left[4\alpha C^{*2}(\alpha - 1) + 4C^* k_{actpass}(\alpha B - A) + B^2 k_{actpass}^2\right]^{1/2} - 2AC^* + 2\alpha C^* B \right.$$
$$\left. + B^2 k_{actpass} \right\} / \left\{ 2(A^2 - 2AB\alpha + \alpha B^2) \right\}, \tag{3.25}$$

where

$$A = \frac{\epsilon k^2}{R} + \Gamma(a + b + 1), \tag{3.26}$$

$$B = a\Gamma + \frac{\epsilon k^2}{R}. \tag{3.27}$$

It is found from (3.25) that the longest waves are the most unstable, and that they have the largest growth rates. Let us examine how the critical Froude number varies with the nondimensional cohesion C^*. Tabulated below are typical values of the momentum coefficient α, and the constants a and b, associated with certain kinds of flows and velocity distributions.

Flow type	velocity distribution	α	a	b
turbulent	power or log law	1.05	2	0
laminar	$1 - (1 - y)^2$	1.20	1	1
dilatant (Bagnold)	$1 - (1 - y)^{3/2}$	1.25	1	1/2

As viscosity and particle concentration increase, the profile of the sheared layer is likely to change from a blunt shape, typical of a pure Newtonian turbulent fluid, to one having a parabolic shape more characteristic of a laminar flow, and finally at high concentrations, to a linear shape or even one having an inflection point [80,84].

Li Jian, *et al.* [87] have presented details of mudflow measurements taken at the research station of Jiang-Jia Ravine, China where bursting and large surge waves frequently occur. The shear strength of the mudflow in this region is much higher than that in other districts. Using this information, Savage [98] estimated values of some of the parameters that appear in the stability analysis. The Froude number for the actual flow was 2.16. Assuming that $(\bar{\rho} - \rho_f)/\bar{\rho} \simeq 0.1$ and $\tan \delta_0 \simeq 0.3$, it was found that $\Gamma \simeq 0.0042$ and $C^* \simeq .01$.

Figure 7 shows the critical Froude number versus the dimensionless cohesion for various values of the parameters α, a, and b, and for $\Gamma = 0.00422$ and $k_{actpass} = 1$. It is seen that the critical Froude numbers are very sensitive to the values of the parameters α, a, and b. The flow corresponding to a typical turbulent flow (curve 1) is more stable than that corresponding to a laminar flow (curve 7). This is well known for fluids without cohesion, but the interesting feature is the strong reduction in stability with increasing cohesion. At the value of C^* ($\simeq .01$) estimated from the field data of [87], the predicted critical Froude numbers are seen to be well below the measured Froude number of 2.16 for this flow.

In summary, the analysis predicts the occurence of roll waves in the complex solid-fluid mixture making up the mudflow. Increased cohesion, increased viscosity, and increased effects of interactions between solid particles all reduce the critical Froude number and increase the likelihood of the initation of roll waves, which subsequently develop into the nonlinear surge waves associated with the pulsing mudflows.

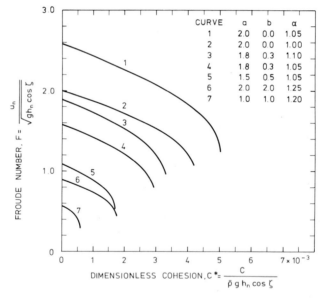

FIGURE 7. Effect of cohesion on critical Froude number. Values of coefficients $(a, b$ and $\alpha)$ for curves 1, 6 and 7 correspond respectively to turbulent, dilatant [11] and laminar (Newtonian) flow profiles. Calculations are for $k_{actpass} = 1, k = 0, \Gamma = 0.00422$.

3.3. Particle size segregation

As mentioned earlier, one of the interesting features of the surging mudflows was the collection of large stones and boulders at the front of the surge wave [77,90]. This is an example of particle size segregation that occurs in a free surface shear flow. In systems of granular materials, the individual grains usually differ from each other in terms of size, density, surface characteristics, shape, etc. When such a mass of material is agitated or deformed in the presence of a gravitational field, segregation or grading of the particles can occur and particles having the same or similar properties tend to collect together in one part of the system. Savage [99] has recently reviewed work on interparticle percolation and various segregation mechanisms in granular materials. Devices such as pinched sluices, Humphreys spirals and Reichert cones are widely used in the mineral processing industry for sorting materials of different sizes, materials of different densities, concentrates from tailings, etc. These devices take advantage of the mechanism of gravity separation when granular materials, in the form of slurries or in the dry state, are sheared as they flow down inclined surfaces. In the geological context there is a closely related phenomenon associated with the deposition of sediments that is termed 'reverse' or 'inverse' grading [100,101,102,103]. The result is the *opposite* of 'normal grading' in which the finer particles are found in the upper layers of a lake or river bed and the coarse ones are lower down. Normal grading is due to the differences in particle sedimentation velocities, i.e. when a mass of particles is discharged into the water, the faster falling (large) particles reach the bed first, and the slowest falling (fine) particles are deposited last and form the top of the bed.

3.3.1. Davies' surge wave experiments

Davies [78] has recently performed some very interesting laboratory experiments to investigate steady-state surge waves in both dry granular materials, and mixtures of solid particles and water or other more viscous liquids. He used a moving-bed channel, the inclined bed of which consisted of a corrugated nylon belt driven by a variable speed motor. For sufficiently large amounts of material, and for flow conditions less than some critical Froude number (cf. §3.2), steady uniform flows corresponding to the 'normal depth' could be developed. Above this, the flow was unstable and surface waves developed. At higher Froude numbers, a finite amount of the granular solid-fluid mixture, when placed on the upstream moving bed, would evolve into a steady, roughly 'tear-drop' shaped mass. As shown in figure 8, this has a sharply curved blunt 'nose', a uniformly sloping 'tail', and a constant depth 'body', the length of which depended upon the total amount of material.

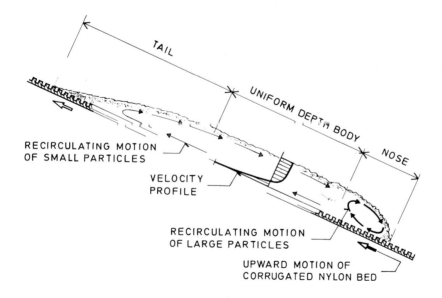

FIGURE 8. Schematic diagram of stationary granular surge wave in Davies' [78] moving-bed laboratory experiments. Circulatory paths of small and large particles and accumulation of large particles at nose of surge wave shown.

Since this surge wave was steady with respect to the laboratory reference frame, it could be studied at leisure to determine depth, velocity, and particle concentration profiles. The overall flow pattern was a recirculating motion, with velocities directed upstream for particles next to the bed, and downstream for particles near the free surface. Typically, a relatively thin shear layer next to the bed was observed along the 'nose' and 'body' region; above it was the remaining material, deforming at a relatively small shear rate. In the 'tail' region, the shear layer was thicker, and the shear rates smaller, than in the bed shear layer under the 'nose' and 'body'. The particle solids concentration in the 'tail' region was lower than that elsewhere. These facts are consistent with the shape of the pile. Examination of the depth averaged equilibrium equations for this steady flow shows that the decrease in the depth in the 'tail' region can only occur if the 'dynamic bed friction angle' (ratio of bed shear to normal stress) decreases in the upstream direction. Since the fluid-solid mixture will yield rate-dependent stress effects, lower particle concentrations and lower shear rates in the 'tail' will give rise to a lower ratio of bed shear to normal stress there. To examine the size segregation effects, Davies poured small amounts of larger particles onto the stationary (recirculating) surge wave. The larger coloured particles (about twice the size of those in the main body) soon made their way to the front of the surge wave and remained there. This phenomenon can be explained in terms of the mechanism of inverse grading and by consideration of the resulting individual particle kinematics.

3.3.2. Analysis of inverse grading in granular shear flows

Bagnold [11] has attempted to give a physical explanation for inverse grading based upon his 'grain-inertia' theory which predicted that stresses were proportional to $\sigma^2(du/dy)^2$. His concept was that the *larger* particles migrate to regions where the shear rate, du/dy, is small (i.e. to the free surface). Bagnold's suggestion has been treated in more detail by Takahashi [81]; but, as Iverson and Denlinger [79] have noted, the explanation is based upon circular arguments and is not entirely satisfactory.

A plausible, alternative mechanism of 'kinetic sieving' was proposed by Middleton [100]. This idea was developed in quantitative terms by Savage and Lun [104], who presented an analysis for flow of a binary mixture of small and large spherical particles of equal mass density down a inclined chute. Their treatment, which is an elaboration of the statistical mechanical interpretation of percolation proposed by Cooke and Bridgwater [105], is summarized below.

The model is restricted to relatively high particle concentrations, and slow shearing flows. By slow flows it is meant that the particles are in nearly continuous contact with one another, overriding each other during the shear process. Thus, we neglect diffusive processes which are of greater importance in the low density, rapid granular flow regime (cf. §2.2) in which particles interact vigorously and primarily through nearly instantaneous collisions. Assume that the flow takes place in layers which are in motion relative to one another as a result of the mean shear rate γ developed by the rough lower boundary (see figure 9). As a result of overriding of adjacent layers and the continual rearrangement of particles within a layer, the contact force network and the void spaces are undergoing continual random changes. At any instant, there will be a distribution of void spaces in a given layer (figure 10). If a void space is large enough, then a particle from the layer above can fall into it as the layers move relative to one another. For a given overall solids concentration, the probability of finding a hole that a small particle can fall into is larger than the probability of finding a hole that a large particle can fall into. Hence, there is a tendency for particles to segregate out, with fines at the bottom and coarse ones at the top. Savage and Lun [104] termed this the 'random fluctuating sieve' mechanism. Through the

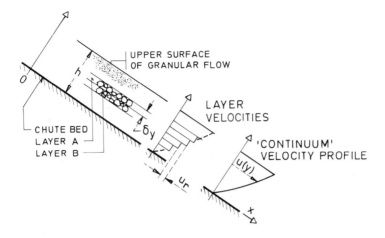

FIGURE 9. Flow of particles in layers down a rough inclined plane.

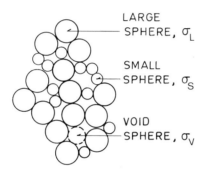

FIGURE 10. 'Random continuous network' for binary mixture of small and large spherical
particles in a layer (after Savage and Lun [104]).

use of information entropy concepts, they obtained the net percolation velocities, q_S and
q_L, of the small and large particles. Comparisons of their predicted percolation velocity
with the experiments of Bridgwater, *et al.* [106] showed reasonably good agreement for the
variation of the nondimensional percolation velocity, $q_S/(\gamma\sigma_L)$, with the diameter ratio,
σ_S/σ_L, of small to large particles. The model was applied to analyze the size segregation
process during the flow down a roughened inclined chute. Uniformly mixed coarse and
fine particles of equal mass density entered the upstream end of the chute. Besides the
percolation flux of particles, an opposing flux was imposed to satisfy the assumed condition
that there is no overall mass flux in the direction normal to the plane of the inclined
chute. The mass conservation equation for the small particles was solved by the method of
characteristics to determine concentration profiles and distances for complete separation
of the fine from the coarse particles. Savage and Lun [104] also performed experiments
with a binary sized mixture of spherical polystyrene beads, and concentration profiles were
determined at various downstream distances. Reasonable agreement was found between
the measured and predicted concentration profiles, and the fine from coarse separation
distances.

3.3.3. Particle segregation in surge waves

Given the above physical description of the 'inverse grading' process we can explain the accumulation of the larger size particulate material at the front of the surge wave as observed in Davies' laboratory experiments, and commonly seen in the field. Figure 8 shows a schematic diagram of both the large and small particle recirculating motions in the surge wave taken from the video recordings of the experiments. The recirculating loop traced out by the large particles is much smaller than that taken by the smaller particles. Consider the motion of a large particle starting at the nose near the bed. Because of the rough bed, the particle is moved rearward towards the tail. However, due to the kinetic sieving mechanism it quickly rises through the surrounding smaller particles to the surface layer, where the mean motion is towards the nose. Because of the upward rise, it never travels longitudinally very far, and so remains confined to the nose region of the surge wave. Davies found that when the maximum height of the wave was deep, the grains were dispersed more uniformly over the length of the pile. For these thicker surges, the upper layer of material is nearly a non-shearing plug region. Shear is essential for the operation of the kinetic sieve mechanism; thus, particles will not rise through such a plug region and are more likely to be carried further upstream from the nose region.

4. CONCLUSIONS

The subject of 'granular flows' encompasses a wide range of problems; some of these are of practical engineering importance in connection with industrial processes, and others relate to our understanding of natural geophysical phenomena. The behaviour of granular materials during rapid flows exhibits many features that are quite different from those typical of more normal fluids. While our understanding of these flows is slowly improving, the analyses of the material behaviour and the application of these results in the spirit and style of applied mechanics still present many challenges.

ACKNOWLEDGEMENTS

This paper was written while the author was on leave at the Laboratory of Hydraulics, Hydrology and Glaciology, ETH, Zürich. I am indebted to the Director, Professor D. Vischer for financial support and for the pleasant and stimulating surroundings which made this work possible. The research was also supported by the Natural Sciences and Engineering Research Council of Canada (NSERC) through an operating grant for which I am extremely grateful. The work on granular avalanches described here is part of a continuing project in collaboration with Prof. K. Hutter. I am grateful to R. Svendsen for his careful reading of the first draft of this paper.

REFERENCES

[1] Sodhi, D.S., Ice Arching and the Drift of Pack Ice Through Restricted Channels, CRREL Rept. 77-18, Cold Region Research and Engr. Lab., U.S. Army Corps of Eng. Hanover New Hampshire, (1977) 11 pp.

[2] Savage, S.B., The Mechanics of Rapid Granular Flows, in: Wu, T.Y. and Hutchinson, J., (eds.), Advances in Applied Mechanics **24** (Academic, 1984) 289-366.

[3] Jenkins, J.T., Balance Laws and Constitutive Relations for Rapid Flows of Granular Materials, in: Chandra, J. and Srivastav, R., (eds.), Proc. Army Research Office Workshop on Constitutive Relations, Philadelphia (1987).

[4] Haff, P.K., A Physical Picture of Kinetic Granular Flows, J. Rheology **30** (1986).

[5] Campbell, C.S., Computer Simulation of Rapid Granular Flows, Proc. 10th National Congress on Appl. Mech., Austin, Texas (1986).

[6] Nedderman, R.M., Tuzun, U., Savage, S.B., and Houlsby, G.T., The Flow of Granular Materials, I. Discharge Rates From Hoppers, Chem. Engnrg. Sci. **37** (1982) 1597-1609.

[7] Tuzun, U., Houlsby, G.T., Nedderman, R.M., and Savage, S.B., The Flow of Granular Materials, II. Velocity Distributions in Slow Flows, Chem. Engnrg. Sci. **37** (1982) 1691-1709.

[8] Savage, S.B., Nedderman, R.M., Tuzun, U. and Houlsby, G.T., The Flow of Granular Materials, III. Rapid Shear Flows, Chem. Engnrg. Sci. **38** (1983) 189-195.

[9] Jenkins, J.T. and Cowin, S.C., Theories for Flowing Granular Materials, in: Cowin, S.C., (ed.), Mechanics Applied to the Transport of Bulk Materials, ASME AMD-31 (1979) 79-89.

[10] Wieghardt, K., Experiments in Granular Flow, Ann. Rev. Fluid Mech. **7** (1975) 89-114.

[11] Bagnold, R.A., Experiments on a Gravity Free Dispersion of Large Solid Spheres in a Newtonian Fluid Under Shear, Proc. R. Soc. London, Ser. A **225** (1954) 49-63.

[12] Savage, S.B. and McKeown, S., Shear Stresses Developed During Rapid Shear of Dense Concentrations of Large Spherical Particles Between Concentric Rotating Cylinders, J. Fluid Mech. **127** (1983) 453-472.

[13] Savage, S.B. and Sayed, M., Stresses Developed by Dry Cohesionless Granular Materials Sheared in an Annular Shear Cell, J. Fluid Mech. **142** (1984) 391-430.

[14] Hanes, D.M. and Inman, D.L., Observations of Rapidly Flowing Granular-Fluid Mixtures, J. Fluid Mech. **150** (1985) 357-380.

[15] Stadler, R., Stationäres, schnelles Fliessen von dicht gepackten trockenen und feuchten Schüttgütern. Dr.-Ing. Dissertation, Univ. Karlsruhe, Karlsruhe, West Germany (1986).

[16] Stadler, R. and Buggisch, H., Influence of the Deformation Rate on Shear Stress in Bulk Solids, Theoretical Aspects and Experimental Results, in: Reliable Flow of Particulate Solids, (EFCE Publication Series No. 49, Bergen, Norway, 1985) pp. 15.

[17] Savage, S.B. and Jeffrey, D.J., The Stress Tensor in a Granular Flow at High Shear Rates, J. Fluid Mech. **110** (1981) 255-272.

[18] Jenkins, J.T. and Savage, S.B., A Theory for the Rapid Flow of Identical, Smooth Nearly Elastic Particles, J. Fluid Mech. **130** (1983) 187- 202.

[19] Haff, P.K., Grain Flow as a Fluid-Mechanical Phenomenon, J. Fluid Mech. **134** (1983) 401-430.

[20] Lun, C.K.K., Savage, S.B., Jeffrey, D.J. and Chepurniy, N., Kinetic Theories for Granular Flow: Inelastic Particles in Couette Flow and Slightly Inelastic Particles in a General Flowfield, J. Fluid Mech. **140** (1984) 223-256.

[21] Jenkins, J.T. and Richman, M.W., Grad's 13-Moment System for a Dense Gas of Inelastic Spheres, Arch. Rat. Mech. Anal. **87** (1985) 355-377.

[22] Jenkins, J.T. and Richman, M.W., Kinetic Theory for Plane Flows of a Dense Gas of Identical, Rough,Inelastic, Circular Disks, Phys. Fluids **28** (1985) 3485-3494.

[23] Campbell, C.S. and Brennen, C.E., Computer Simulation of Granular Shear Flows, J. Fluid Mech. **151** (1985) 167-188.

[24] Campbell, C.S. and Brennen, C.E., Chute Flows of Granular Material: Some Computer Simulations, J. Appl. Mech. **52** (1985) 172-178.

[25] Campbell, C.S. and Gong, A., The Stress Tensor in a Two-Dimensional Granular Shear Flow, J. Fluid Mech. **164** (1986) 107-125.

[26] Walton, O.R. and Braun, R.L., Stress Calculations for Assemblies of Inelastic Spheres in Uniform Shear, Acta Mechanica **63** (1986) 73-86.

[27] Walton, O.R. and Braun, R.L., Viscosity and Temperature Calculations for Assemblies of Inelastic Frictional Disks, J. Rheology **30** (1986) 949-980.

[28] Walton, O.R., Braun, R.L. and Cervelli, D.M., Flow of Granular Solids: 3-Dimensional Discrete-Particle Computer Modelling of Uniform Shear, Preprint UCRL-95232, Lawrence Livermore Nat'l. Lab. (1986).

[29] Hopkins, M. and Shen, H., A Monte-Carlo Simulation of a Rapid Simple Shear Flow of Granular Materials, preprint for U.S.-Japan Seminar on Micromechanics of Granular Materials, Sendai, Japan, October 26-30 (1987) 313-331.

[30] Basik, B. and Savage, S.B., Direct Monte-Carlo Simulation of Granular Flows, ASCE Eng. Mech. Div. Specialty Conf., Buffalo, New York, May 20-22 (1987) pp. 212.

[31] Hui, K., Haff, P.K., Ungar, J.E., and Jackson, R., Boundary Conditions for High Shear Grain Flows, J. Fluid Mech. **145** (1984) 223-233.

[32] Jenkins, J.T. and Richman, M.W., Boundary Conditions for Plane Flows of Smooth, Nearly Elastic, Circular Disks, J. Fluid Mech. **171** (1986) 53-69.

[33] Campbell, C.S. and Gong, A., Boundary Conditions for Two-Dimensional Granular Flows, preprint for U.S.-Japan Seminar on Micromechanics of Granular Materials, Sendai, Japan, October 26-30 (1987) 157-162.

[34] Walton, O.R., Braun, R.L., Mallon, R.G. and Cervelli, D.M., Particle-Dynamics Calculations of Gravity Flow of Inelastic, Frictional Spheres, preprint for U.S.-Japan Seminar on Micromechanics of Granular Materials, Sendai, Japan, October 26-30 (1987) 156.1-156.9.

[35] Gutt, G.M. and Haff, P.K., Boundary Conditions on Continuum Theories of Granular Flow, Brown Bag Preprint Series BB-70, Division of Physics, Mathematics, and Astronomy, Caltech (1988) 24 pp.

[36] Savage, S. B., Granular Flows Down Rough Inclines - Review and Extension, in: Jenkins, J.T. and Satake, M., (eds.), Mechanics of Granular Materials: New Models and Constitutive Relations (Elsevier, 1983) pp. 261-82.

[37] Norem, H., Irgens, F. and Schielhop, B., A Continuum Model for Calculating Snow Avalanches, in: Salm, B. and Gubler, H., (eds.), Avalanche Formation, Movement and Effects (IAHS Publ. No. 126, 1987) pp. 363-379.

[38] Johnson, P.C. and Jackson, R., Frictional-Collisional Constitutive Relations for Granular Materials, with Application to Plane Shearing, J. Fluid Mech. **176** (1987) 67-93.

[39] Szidarovszky, F., Hutter, K. and Yakowitz, S., A Numerical Study of Steady Plane Granular Chute Flows Using the Jenkins-Savage Model and Its Extension, Int. J. Num. Meth. Engrg. **24** (1987) 1993-2015.

[40] Reynolds, 0., On the Dilatancy of Media Composed of Rigid Particles in Contact, Phil. Mag. Ser. 5 **20** (1885) 469-481.

[41] Takahasi, T., 1981 Debris flow, Ann.Rev.Fluid Mech. **13** (1981) 57-77.

[42] Scott, R.F., Principles of Soil Mechanics (Addison-Wesley, Reading, 1963).

[43] Schofield, A.N. and Wroth, C.P., Critical State Soil Mechanics (McGraw-Hill, New York, 1968).

[44] Savage, S. B., Gravity Flow of Cohesionless Granular Materials in Chutes and Channels, J. Fluid Mech. **92** (1979) 53-96.

[45] Savage, S.B. and Hutter, K., The Motion of a Finite Mass of Granular Material Down a Rough Incline, J. Fluid Mech., (1988) in press.

[46] Savage, S.B., Streaming Motions in a Bed of Vibrationally Fluidized Dry Granular Material, J. Fluid Mech. **194** (1988) 457-478.

[47] Heim, A., Der Bergsturz von Elm, Deutsch Geol. Gesell. Zeitschrift **34** (1882) 74-115.

[48] Heim, A., Bergsturz und Menscheleben, Beiblatt zur Vierteljahresschrift der Natf. Ges. Zurich **20** (1932) 1-218.

[49] Kent, P.E., The Transport Mechanism in Catastrophic Rockfalls, J. Geol. **74** (1965) 79-83.

[50] Shreve, R.L., Sherman Landslide, Alaska, Science **154** (1966) 1639-1643.

[51] Shreve, R.L., The Blackhawk Landslide, Geol. Soc. Am., Spec. Paper **108** (1968) 47 pp.

[52] Shreve, R.L., Leakage and Fluidization in Air-Lubricated Avalanches, Geol. Soc. Am. Bull. **79** (1968) 653-658.

[53] Scheidegger, A.E., Physical Aspects of Natural Catastrophics (Elsevier, 1975).

[54] Goguel, J., Scale Dependent Rockslide Mechanisms, in: Voight, B., (ed.), Rockslides and Avalanches, *Vol. 1* (Elsevier, 1978) 167-180.

[55] Hsü, K., On Sturzstorms - Catastrophic Debris Streams Generated by Rockfalls, Geol. Soc. Am. Bull. **86** (1975) 129-140.

[56] Hsü, K., Albert Heim: Observations on Landslides and Relevance to Modern Interpretations, in: Voight, B., (ed.), Rockslides and Avalanches *Vol. 1* (Elsevier, 1978) 69-93.

[57] Lucchitta, B.K., A Large Landslide on Mars, Geol. Soc. Amer. Bull. **80** (1978) 1601-1600.

[58] McSaveney, M.J., Sherman Glacier Rock Avalanche, Alaska, U.S.A., in: Voight, B., (ed.), Rockslides and Avalanches, *Vol. 1* (Elsevier, 1978) 197-258.

[59] Erismann, T.H., Mechanisms of Large Landslides, Rock Mechanics **12** (1979) 15-46.

[60] Erismann, T.H., Flowing, Rolling, Bouncing, Sliding: Synopsis of Basic Mechanisms, Acta Mechanica **64** (1986) 101-110.

[61] Davies, T.R.H., Spreading of Rock Avalanche Debris by Mechanical Fluidization, Rock Mech. **15** (1982) 9-29.

[62] Li Tianchi, A Mathematical Model for Predicting the Extent of a Major Rockfall, Z. Geomorph. **27** (1983) 473-482.

[63] Melosh, J., The Physics of Very Large Landslides, Acta Mechanica **64** (1986) 89-99.

[64] Dent, J.D., Flow Properties of Granular Materials Large Overburden Loads, Acta Mechanica **64** (1986) 111-122.

[65] Hutter, K., Plüss, Ch. and Maeno, N., Some Implications Deduced from Laboratory Experiments on Granular Avalanches, Mitteilung No. 94 der Versuchsanstalt für Wasserbau, Hydrologie und Glaziologie an der ETH (1988) 323-344.

[66] Hutter, K., Plüss, Ch. and Savage, S.B., Dynamics of Avalanches of Granular Materials from Initiation to Runout, Part II: Laboratory experiments. (1988) in preparation.

[67] Savage, S.B. and Hutter, K., Dynamics of Avalanches of Granular Materials from Initiation to Runout, Part I: Analysis, (1988) in preparation.

[68] Savage, S.B. and Nohguchi, Y., Similarity Solutions for Avalanches of Granular Materials Down Curved Beds, Acta Mechanica (1988) 20 pp., in press.

[69] Savage, S.B., Symbolic Computation of the Flow of Granular Avalanches, preprint for Symposium on Symbolic Computation, ASME Winter Annual Meeting, Chicago, Nov. 27 - Dec. 2 (1988) 7 pp.

[70] Huber, A., Schwallwellen in Seen als Folge von Felssturzen, Mitteilung No. 47 der Versuchsanstalt für Wasserbau, Hydrologie und Glaziologie an der ETH (1980) 122 pp.

[71] Plüss, Ch., Experiments on Granular Avalanches, Diplomarbeit, Abt X, Eidg. Techn. Hochschule, Zürich (1987) 113 pp.

[72] Hutter, K., Siegel, M., Savage, S.B., and Nohguchi, Y., Two-dimensional Spreading of a Granular Avalanche Down an Inclined Plane, Part I: Theory, (1988) in preparation.

[73] Lang, R., and Hutter, K., The Three-Dimensional Simulation of Laminar Flow Region Granular Avalanches, Symp. on Snow and Glacier Research Relating to Human Living Conditions, Lom, Norway, September 4-9, (1986).

[74] Innes, J.L., Debris Flows, Progr. Phys. Geogr. **7** (1983) 469-501.

[75] Costa, J.E., Physical Geomorphology of Debris Flows, in: Costa, J.E. and Fleisher, P.J., (eds.), Developments and Applications of Geomorphology (Springer, 1984) 268-317.

[76] Johnson, A.M., Debris Flow (with contributions from J.R. Rodine), in: D. Brunsden, D. and Prior, D.B., (eds.), Slope Instability (Wiley, 1984) 257-361.

[77] Davies, T.R.H., Large Debris Flows: A Macro-Viscous Phenomenon, Acta Mechanica **63** (1986) 161-178.

[78] Davies, T.R.H., Debris Flow Surges - A Laboratory Investigation, Mitteilung No. 96 der Versuchsanstalt für Wasserbau, Hydrologie und Glaziologie an der ETH (1988) 122 pp.

[79] Iverson, R.M. and Denlinger, R.P., The Physics of Debris Flows - A Conceptual Assessment. in: Erosion and Sedimentation in the Pacific Rim (Proceedings of the Corvallis Symposium), IAHS Publ. No. 165. (1987) 155-165.

[80] Hashimoto, H. and Tsubaki, T., Characteristics of Debris Flow With a Plug, Memoirs Faculty Engrg. 44 (1984) 273-289.

[81] Takahashi, T., Debris Flow on Prismatic Channel, Journ. Hydr. Div., ASCE **106** (1980) 381-395.

[82] Takahashi, T., Debris flow, Ann. Rev. Fluid Mech. **13** (1981) 57-77.

[83] Takahashi, T., Debris Flow and Debris Flow Deposition, in: Shahinpoor, M., (ed.), Advances in the Mechanics and Flow of Granular Materials *Vol. II.* (1983) 57-77.

[84] Tsubaki, T. and Hashimoto, H., Interparticle Stresses and Characteristics of Debris Flow, J. Hydrosci. & Hydraulic Engrg. **1** (1983) 67-82.

[85] Chen, C.-L., Generalized Viscoplastic Modeling of Debris Flow, Journ. Hydr. Div., ASCE **114** (1988) 237-258.

[86] Chen, C.-L., General Solutions for Viscoelastic Debris Flow, Journ. Hydr. Div., ASCE **114** (1988) 259-282.

[87] Li Jian, Yuan Jianmo, Bi Cheng & Luo Defu, The Main Features of the Mudflow in Jiang-Jia River, Zeit. Geomorph. **27** (1983) 325-341.

[88] Curry, R.R., Observation of Alpine Mudflows in the Ten Mile Range, Central Colorado, Geol. Soc. Amer. Bull. **77** (1966) 771-776.

[89] Broscoe, A.J. and Thomson, S., Observations on an Alpine Mudflow, Steele Creek, Yukon, Canad. Jour. Earth Sci. **6** (1969) 219-229.

[90] Li Jian and Luo Defu, The Formation and Characteristics of Mudflow and Flood in the Mountain Area of the Dachao River and its Prevention, Zeit. Geomorph. **25** (1981) 470-484.

[91] Schaerer, P.A. and Salway, A.A., Seismic and Impact-Pressure Monitoring of Flowing Avalanches, Journ. Glaciology **26** (1980) 179-187.

[92] Hopfinger, E.J., Snow Avalanche Motion and Related Phenomena, Ann. Rev. Fluid Mech. **15** (1983) 47-76.

[93] Dressler, R.F., 1949 Mathematical Solution of the Problem of Roll Waves in Open Inclined Channels, Comm. Pure Applied Math. **2** (1949) 149-194.

[94] Ishihara, T., Iwagaki, Y. and Iwasa, Y., Discussion of "Roll Waves and Slug Flows in Inclined Open Channels", Journ. Hydr. Div., ASCE **86** (1960) 45-60.

[95] Brock, R.R., Periodic Permanent Roll Waves, Journ. Hydr. Div., ASCE **96** (1970) 2565-2580.

[96] Needham, D.J. and Merkin, J.H., On Roll Waves Down an Open Inclined Channel, Proc. R. Soc. Lond. A**394** (1984) 259-278.

[97] Hwang, S.-H. and Chang, H.-C., Turbulent and Inertial Roll Waves in Inclined Film Flow, Phys. Fluids **30** (1987) 1259-1268.

[98] Savage, S. B., Roll Waves in Coulomb-Viscous Materials: Applications to Mud Flows, Snow Avalanches, and Sturtzstroms, in preparation.

[99] Savage, S.B., Interparticle Percolation and Segregation in Granular Materials: A Review, in: Selvadurai, A.P.S., (ed.), Developments in Engineering Mechanics (Elsevier, Amsterdam, 1987) 347-363.

[100] Middleton, G.V., Experimental Studies Related to Problems of Flysch Sedimentation, in: Lajoie, J., (ed.), Flysch Sedimentology in North America (Business and Economics Science Ltd., Toronto, 1970) 253-272.

[101] Middleton, G.V. and Hampton, M.A., Subaqueous Sediment Transport and Deposition by Sediment Gravity Flows, in: Stanley, D.J. and Swift, D.J.P., (eds.), Marine Sediment Transport and Environmental Management (Wiley, New York, 1976) 197-218.

[102] Sallenger, A.H., Inverse Grading and Hydraulic Equivalence in Grain-Flow Deposits, J. Sedimentary Petrology **49** (1979) 553-562.

[103] Naylor, M.A., The Origin of Inverse Grading in Muddy Debris Flow Deposits - A Review, J. Sedimentary Petrology **50** (1980) 1111-1116.

[104] Savage, S.B. and Lun, C.K.K., Particle Size Segregation in Inclined Chute Flow of Dry Cohesionless Granular Solids, J. Fluid Mech. **189** (1988) 311-335.

[105] Cooke, M.H., and Bridgwater, J., Interparticle Percolation: A Statistical Mechanical Interpretation, Ind. Engng. Chem. Fundam. **18** (1979) 25-27.

[106] Bridgwater, J., Cooke, M.H., and Scott, A.M., Inter-particle Percolation: Equipment Development and Mean Percolation Velocities, Trans. Instn Chem. Engrs. **56** (1978) 157-167.

Theoretical and Applied Mechanics
P. Germain, M. Piau and D. Caillerie (Editors)
Elsevier Science Publishers B.V. (North-Holland)
© IUTAM, 1989

INTERACTIONS IN BOUNDARY-LAYER TRANSITION

Frank T. SMITH

Department of Mathematics
University College London
Gower Street
London WC1E 6BT, U.K.

Certain theoretical studies of boundary-layer transition are described, based on high Reynolds numbers and with attention drawn to the various nonlinear interactions and scales present. The article concentrates in particular on theories for which the mean-flow profile is completely altered from its original state. Two- and three-dimensional flow theory and conjectures on turbulent-boundary-layer structures are included. Specific recent findings noted, and in qualitative agreement with experiments, are: nonlinear finite-time break-ups in unsteady interactive boundary layers; strong vortex/wave interactions; and prediction of turbulent boundary-layer displacement- and stress sublayer-thicknesses.

1. INTRODUCTION

Our concern in this article is with transition to turbulence in boundary layers, incompressible or compressible, with stress laid in particular on the importance of certain types of interactions, e.g. viscous/inviscid, two-/three-dimensional, small-/large-scale, wave/vortex. There appear to be many types of boundary-layer transition observable in practice, depending on the disturbance environment, e.g. on how rough the solid surface supporting the boundary layer is or on how unsteady the oncoming free stream is.

Two possible extremes of transition are the following. First there is the "faster", savage, transition that is epitomised by the wake-passing effect, from an upstream row of rotor blades, on the boundary layers on stator blades in turbines. Broadly, a characteristic measure of the stator boundary-layer properties, such as the heat transfer at a fixed station on the surface, shows a very rapid change taking place there, apparently from a fully laminar to a fully turbulent state almost immediately as the highly turbulent wake of a typical rotor hits the stator blade. When the wake moves on there is a slightly less rapid adjustment, back to laminar conditions, until the violent wake of the next rotor blade hits the stator and the rapidly changing temporal cycle repeats itself. The other possible extreme of boundary-layer transition is "slower" (civilised) transition. This is exemplified by extensions of the classical experiments by Schubuaer & Skramstad in the 1940's, on tiny unsteady controlled disturbances introduced upstream into a basic planar laminar boundary layer, the extensions being carried out in the recent experiments of Saric, Gaster, Kachanov and their co-workers, and others, who examine the effects of gradually increasing the size of the small input disturbance upstream. Typically, say for an input sizeable disturbance due to a ribbon, vibrating at a fixed frequency (approximately) and situated ahead of the lower-branch linear neutral position corresponding to that frequency, the Tollmien-Schlichting travelling disturbance observed in the boundary-layer flow decays spatially in x up to the neutral position, then it exhibits relatively slow growth in amplitude just beyond the neutral position, and (unless the disturbance is infinitesimal) that is followed by the gradual development of a nonlinear travelling-wave state, still predominantly two-dimensional in nature. For sufficiently large, although still small, amplitudes of disturbance the occurrence of the nonlinear planar state just mentioned is itself then followed often by very significant three-dimensional action appearing in the unsteady motion downstream with the three-dimensionality in some cases developing a preferred cross-stream wavenumber dependence, which corresponds to oblique and quite fast-growing waves at angles of about 55°-65° from the free-stream direction, and in other cases producing a sustained vortex-like pattern of quite long streamwise length scale. Associated with this, the amplitude spectrum sometimes exhibits a strong growth in its sub-harmonic components, particularly the half harmonic, then multiples thereof, and thereafter the spectrum starts to fill up relatively rapidly. In addition, intermittency may occur in any of the above stages experimentally, again depending on the disturbance environment. Turbulent flow ensues some way downstream and, although it may be difficult to determine exactly how far downstream, the nominally full turbulent state eventually reached may be regarded as virtually the same as that reached via the faster type of by-pass transition described earlier.

The aim in the present article is to report on some theoretical research in progress which is directed towards theoretically understanding the transition process somewhat more, be it faster or slower transition, and ultimately of course towards a rational account of turbulent flow and turbulence-modelling if possible, granted that the latter step is without doubt a formidable one. We concentrate here mostly on the slower types of transitions, although the faster by-pass types are referred to later on, and attention is drawn to non-linear interactions which completely alter the mean-flow profile.

As a guide for the theory we should remark here that the Schubauer-Skramstad experimental results on infinitesimal disturbances agree well on the whole with classical linear-Orr-Sommerfeld theory; that linear triple-deck theory based on assuming the Reynolds number to be large also agrees fairly well with the above experiments; that the typical spatial regions of interest in the earlier noted experiments on slow transition straddle the linear neutral curve but they are often especially concerned with the nonlinear features emerging as the unsteady flow passes between the linear branches of the neutral curve; and that the characteristic Reynolds number (local R_δ or global Re) is relatively large in those transition experiments. The above aspects of the experiments motivate the present rational theory, which addresses nonlinear effects arising as the unsteady boundary-layer flow progresses downstream. The only major assumption in the theory is that the typical Reynolds number is large.

Section 2 below considers the theoretical <u>two</u>-dimensional nonlinear flow properties, as a starting point and to gain some overall ideas on the downstream progress of non-linear travelling disturbances. Emphasis is given to two main aspects, namely the high-frequency theory [1] and the recent findings of nonlinear finite-time break-ups (Brotherton-Ratcliffe & Smith [2], Smith [3]). Then section 3 moves on to the more realistic case of <u>three</u>-dimensional flow theory. Certain features of the planar case cross over directly to the three-dimensional case, but the latter also brings in some very significant responses of its own. In particular, nonlinear interactions can occur sooner, i.e. at lower amplitude, in the three- compared with the two-dimensional version and special mention is made of the recent findings on vortex/Tollmien-Schlichting nonlinear interactions (Hall & Smith [4,5]). It is found that these theoretical nonlinear three-dimensional interactions agree well qualitatively and sometimes quantitatively with the experimental findings mentioned previously. Section 4 gives a tentative view on turbulent flow [1,6]. Further the theory seems to be quite promising so far, overall, although there is still a very long way to go in the required three-dimensional context.

2. TWO-DIMENSIONAL NONLINEAR INTERACTION PROPERTIES

To start the nonlinear unsteady flow theory we consider two-dimensional features first, as a guideline. Given that the Reynolds number can be taken to be a large parameter we address the Navier-Stokes equations but scale the nondimensional velocities $\bar{u}$, $\bar{v}$ in the x, y directions, and the nondimensional pressure $\bar{p}$, in the Tollmien-Schlichting (TS) fashion, which is equivalent to the triple deck scalings

$$(\bar{u}, \bar{v}) = (\text{Re}^{-1/8}\lambda^{1/4}u, \text{Re}^{-3/8}\lambda^{3/4}v) + \ldots, \tag{2.1a}$$

$$\bar{p} = \text{Re}^{-1/4}\lambda^{1/2}p(X,T) + \ldots, \tag{2.1b}$$

$$(x, y, t) = (x_0 + \text{Re}^{-3/8}\lambda^{-5/4}X, \text{Re}^{-5/8}\lambda^{-3/4}Y, \text{Re}^{-1/4}\lambda^{-3/2}T). \tag{2.1c}$$

Here the scalings (2.1a–c) apply in the so-called lower deck, a viscous sublayer close to the surface, while the factor $\lambda = \lambda(x_0)$ is the reduced skin friction of the oncoming undisturbed boundary layer, e.g. in Blasius flow $\lambda(x_0) \propto x_0^{-\frac{1}{2}}$, at the typical $O(1)$ station $x = x_0$. With (2.1) acting, the flow problem comes down to the unsteady viscous-inviscid interactive one of solving the nonlinear boundary-layer equations

$$\frac{\partial u}{\partial X} + \frac{\partial v}{\partial Y} = 0 \ , \tag{2.2a}$$

$$\frac{\partial u}{\partial T} + u \frac{\partial u}{\partial X} + v \frac{\partial u}{\partial Y} = - \frac{\partial p}{\partial X} + \frac{\partial^2 u}{\partial Y^2} \ , \tag{2.2b}$$

subject to the conditions

$$u = v = 0 \quad \text{at} \quad Y = 0 \ , \tag{2.2c}$$

$$u \sim Y + A(X,T) \quad \text{as} \quad Y \to \infty \ , \tag{2.2d}$$

$$p(X,T) = \frac{1}{\pi} \int_{-\infty}^{\infty} \frac{\partial A}{\partial \xi} (\xi, T) \frac{d\xi}{(X - \xi)} \ , \tag{2.2e}$$

representing in turn the no-slip constraint at the surface, the displacement effect $(\propto - A)$ on the motion, and the pressure-displacement interaction between the unknown pressure p and the unknown displacement $(-A)$ caused by the outer quasi-steady potential flow. Other rather obvious initial or spatial boundary conditions are assumed depending on the flow configuration. Further, (2.2e) applies for subsonic two-dimensional flow whereas in the supersonic counterpart (2.2e) is replaced by Ackeret's law

$$p(X,T) = - \frac{\partial A}{\partial X} (X,T) \tag{2.2e'}$$

[see also a note in section 3 on compressibility], and other replacements for the pressure-displacement interaction yield applications to transonic flows, hypersonic flows, internal motions, water layers and so on.

The two main alternatives to the nonlinear triple-deck version (2.2) of the unsteady flow at high Reynolds numbers are the Navier-Stokes equations and interactive boundary-layer versions, both requiring numerical treatments. The former tends to be hindered more in general by grid-resolution difficulties, among others, whereas interactive boundary-layer methods, which are preferable and involve sensible interpretation of asymptotic properties such as

(2.1),(2.2) at finite Reynolds numbers, have still to be developed seriously for unsteady flows (a start is made in [7]). With the present rational approach beginning with the nonlinear TS problem (2.2) and its extensions later on the hope is that the small- and large-scale structures of transitional boundary layers can be understood in a more definite way, including the interaction between those scales and the possibility of coherent structures emerging in line with experimental findings in transitional and turbulent boundary layers.

Some properties of the viscous-inviscid unsteady nonlinear system (2.2) are quite well known (see, e.g., in Smith & Burggraf [1]). The linearized version, linearized about the original steady flow $u = Y$, $v = p = A = 0$, compares well with Orr-Sommerfeld results as noted before. It gives the neutral value $\Omega = \Omega_N \approx 2.30$, with Ω denoting the disturbance frequency i.e. $\partial/\partial T \to -i\Omega$ or, as an approximate generalization for subsequent working,

$$\Omega = |\partial/\partial T| \quad , \qquad\qquad (2.3)$$

schematically. At $\Omega = \Omega_N$ there is a supercritical bifurcation, which yields stable nonlinear travelling states for Ω just exceeding Ω_N (equivalent to $x_0 > x_{0N}$ or $R_\delta > R_{\delta N}$, in view of the λ factor, "N" standing for neutral), as opposed to the basic steady flow which is unstable there. Computations of the nonlinear travelling states for a range of values of Ω greater than Ω_N are being finalized [8] by O.R. Burggraf, A.T. Conlisk and the author, while Duck [9] and Smith [10] have recently presented numerical time-marching solutions of (2.2) starting from initial conditions at a scaled time T of zero: see also near the end of this section.

Of probably most concern next is what happens downstream either in an initial-value problem (see below) or as the station x_0 increases. This latter corresponds to increasing Ω since the skin-friction factor $\lambda(x_0)$ in (2.1) accentuates the time-dependence; the same effect of increasing our parameter Ω results, at a fixed station x_0 , from an increase in the input frequency upstream, again from the scalings in (2.1). In such cases, our interest is steered first towards investigating the properties of the nonlinear TS system (2.2) when the reduced frequency parameter defined in (2.3) is large,

$$\Omega \gg 1 \; . \qquad\qquad (2.4)$$

In that regime it is found that an initially small disturbance passes through at least two stages I, II as it progresses downstream. These successive stages are conveniently defined by their representative pressure levels,

$$p = O(\Omega^{1/2})(\text{stage I}) \quad \text{and} \quad p = O(\Omega)(\text{stage II}). \qquad (2.5a,b)$$

During stage I, the scaled form of the unsteady pressure response turns out to be governed by a generalized cubic Schrödinger equation,

$$\frac{\partial P}{\partial \hat{T}} - i \frac{\partial^2 P}{\partial \hat{X}^2} = \frac{P}{\sqrt{2}} - iP|P|^2 \qquad (2.6)$$

from analysis of (2.2) (Smith [10]), in slow variables $\hat{X}$, $\hat{T}$ and in a suitable frame moving fast downstream with the group velocity. The principal features of the solution to (2.6) are that the disturbance broadens spatially in an exponentially fast form, as $\hat{T}$ increases, and the maximum amplitude also grows exponentially rapidly, with the amplitude solution taking on an elliptical shape.

In stage II, further downstream say, where the pressure p rises to $0(\Omega)$, the system (2.2) becomes fully nonlinear but viscous forces appear at first sight to become negligible. As a result the unsteady motion there is governed by the Benjamin-Ono equation

$$\frac{\partial A}{\partial T} + A \frac{\partial A}{\partial X} = -\frac{1}{\pi} \int_{-\infty}^{\infty} \frac{\partial^2 A}{\partial \xi^2} \frac{(\xi, T)d\xi}{(X - \xi)} \quad , \qquad (2.7)$$

for $A(X,T)$. This matches with stage I earlier on and can admit nonlinear travelling waves and solitary waves as solutions. The flow structure here is subject to the existence of a viscous unsteady sublayer, however of thickness $\sim \Omega^{-\frac{1}{2}}$ close to the surface, and that sublayer can itself break down in a singular form to provide eruptions of vorticity into the outer inviscid zone where (2.7) applies. Again, the recent study of reversed-flow singularities by Smith (11) is relevant here, along with the work in (2.9)ff. below.

The other major features so far in the development of the sizeable high-frequency disturbance above are: (a) the amplitude of the nonlinear disturbance is much greater than that of the basic steady flow (we recall that this is for the flow nearest to the surface); (b) the mean shear stress at the surface is increased by an order of magnitude above its original steady laminar value; (c) there are large oscillations of the unsteady shear stress about that mean; (d) secondary instability of the inflexion-point kind may well also be present. These features (a) – (d) continue to hold in the subsequent development of the disturbance even further downstream.

The next stage is the Euler stage, which finds the disturbance stronger still, further downstream, and the entire mean-flow profile is altered [1,6]. Formally the larger disturbance amplitude and increased frequency parameter

Ω ($\rightarrow Re^{1/4}$) make the dominant spatial scales of the triple deck structure compress, so that now typically x, y, t are all fast, of order $Re^{-1/2}$, while the velocities $\bar{u}$, $\bar{v}$ and the pressure $\bar{p}$ rise to $0(1)$. Hence the current Euler structure is controlled predominantly by the inviscid nonlinear system

$$\frac{\partial \bar{u}}{\partial x} + \frac{\partial \bar{v}}{\partial y} = 0 \ , \tag{2.8a}$$

$$\frac{\partial \bar{u}}{\partial t} + \bar{u} \frac{\partial \bar{u}}{\partial x} + \bar{v} \frac{\partial \bar{u}}{\partial y} = - \frac{\partial \bar{p}}{\partial x} \ , \tag{2.8b}$$

$$\frac{\partial \bar{v}}{\partial t} + \bar{u} \frac{\partial \bar{v}}{\partial x} + \bar{v} \frac{\partial \bar{v}}{\partial y} = - \frac{\partial \bar{p}}{\partial y} \tag{2.8c}$$

and appropriate initial and boundary conditions. The unsteady Euler equations (2.8) allow the necessary merging with the earlier stage II, since the equation (2.7) for instance is a reduced-amplitude long-wave case of a solution to (2.8). Like the earlier stage, though, the validity of (2.8) is subject to the behaviour of the nonlinear viscous unsteady sublayer, now of $0(Re^{-3/4})$ thickness, adjoining the surface and that sublayer can act as a source of vorticity bursts, occurring possibly in a random fashion, which can rejuvenate the nonlinear disturbance as it continues to move downstream. This combined nonlinear structure, featuring both larger-scale properties governed mainly by the Euler equations and smaller-scale properties due to viscous action and bursting, has some promising connections with the full computations and/or modelling and/or experiments on transitional or turbulent boundary layers by Walker & Abbott [12], Fasel [13], Orszag & Patera [14] and Blackwelder and co-workers, among others, especially if the extension to three-spatial dimensions is allowed for (see sections 3,4 below). Again, we may also expect to be able to describe a "savage" transition (see introduction) as one containing sufficiently high amplitudes [0(1)] and frequency ranges that it by-passes the earlier stages I, II and enters the fully nonlinear Euler stage (2.8) directly.

Much more needs to be known about this stage, possibly from further computational studies, to increase our structural understanding, and the same goes for related interactive boundary-layer and composite methods aimed at incorporating the important viscous effects present also. Computational studies by the author with Dr. D.J. Doorly and Prof. S.C.R. Dennis are in progress: see also work by H. Fasel and M.E. Ralph and co-workers. Some encouragement can be drawn nevertheless from the two-dimensional theory so far, because of the connections with other approaches as noted previously and because the various stages yield both spectral focussing, from (2.6), and spectral broadening, from inflexional instabilities and/or sublayer bursting, aspects that are observed in practice. The major extra element to be acknowledged in all the high-frequency nonlinear theory is three-dimensionality, which is considered in the next section.

Most recently interest has returned to the general range $\Omega = 0(1)$ where, for finite amplitudes, the unsteady interactive-boundary-layer system (2.2) applies in full. The principal finding there is that localized <u>finite-time break-ups</u> can occur (Brotherton-Ratcliffe & Smith [2], Smith [3]). These break-ups take the form

$$X - X_s = c(T - T_s) + (T_s - T)^N \xi \;,\; \frac{\partial p}{\partial X} \sim (T_s - T)^{-1} p_1{'}(\xi) \qquad (2.9)$$

near the break-up position X_s and time T_s- , with the local coordinate ξ of $0(1)$, the phase speed c of $0(1)$ and $N = 3/2, 5/4, 7/6, 9/8, \ldots, 1$. For <u>moderate</u> break-ups $N > 1$ and then the inviscid Burgers equation for the pressure function $p_1(\xi)$,

$$-\left[p_1 - \left(\frac{N}{N-1} \right) \xi p_1{'} \right] + p_1 p_1{'} = 0 \;, \qquad (2.10a)$$

describes the local terminal behaviour, in scaled terms, yielding the appropriate smooth solution

$$\xi = (1 - N)p_1 - e_1 p_1{}^{N/(N-1)} \qquad (2.10b)$$

implicitly, where e_1 is an arbitrary positive constant. For <u>severe</u> break-ups where $N = 1$, on the other hand, the local governing equations become

$$u_1 = \frac{\partial \psi_1}{\partial Y} \;,\; (\xi + u_1) \frac{\partial u_1}{\partial \xi} - \frac{\partial \psi_1}{\partial \xi} \frac{\partial u_1}{\partial Y} = - p_1{'}(\xi) \qquad (2.11)$$

for $(u_1, \psi_1)(\xi, Y), p_1(\xi)$, in normalized form, with c set to zero without loss of generality, and solutions are given in [3]. Both sorts of break-up provoke increasingly large reversed wall-shear responses in the local motion.

Various unsteady interactive-boundary-layer computations appear to support (Smith [3]) the singular description in (2.9) – (2.11), and there is also fair agreement with the Navier-Stokes computations of Fasel [13]. Moreover, the break-ups (2.9) – (2.11) apply to all the unsteady interactive flows known to date.

Following the break-up above, new physical effects come into play locally associated mainly with the unsteady Euler equations (an appropriate computational approach in principle then is in [7]). Intuition would suggest that this break-up process, when repeated again and again, may be connected with the occurrence of <u>intermittency</u> in practice.

3. THREE-DIMENSIONAL NONLINEAR INTERACTIVE FEATURES

The aim now is to extend the above Re-large theory to allow for three-dimensional effects, as experimental findings tell us we should. Some of the above theory and conclusions can go straight through, in the various stages, but much more volatile events can take place even at relatively low amplitudes, in three dimensions.

Once again we start with the triple-deck version since it allows for linear and nonlinear TS disturbances, but now in the three-dimensional form

$$\frac{\partial u}{\partial X} + \frac{\partial v}{\partial Y} + \frac{\partial w}{\partial Z} = 0 \quad , \tag{3.1}$$

$$\frac{\partial u}{\partial T} + u\,\frac{\partial u}{\partial X} + v\,\frac{\partial u}{\partial Y} + w\,\frac{\partial u}{\partial Z} = -\,\frac{\partial p}{\partial X} + \frac{\partial^2 u}{\partial Y^2} \quad , \tag{3.1b}$$

$$\frac{\partial w}{\partial T} + u\,\frac{\partial w}{\partial X} + v\,\frac{\partial w}{\partial Y} + w\,\frac{\partial w}{\partial Z} = -\,\frac{\partial p}{\partial Z} + \frac{\partial^2 v}{\partial Y^2} \quad , \tag{3.1c}$$

with

$$p(X,Z,T) = -\,\frac{1}{2\pi} \int_{-\infty}^{\infty}\int_{-\infty}^{\infty} \frac{\partial^2 A}{\partial \xi^2}\,(\xi,\eta,T)\,\frac{d\xi\,.\,d\eta}{[(X-\xi)^2 + (Z-\eta)^2]^{\frac{1}{2}}} \quad , \tag{3.1d}$$

$$u \sim Y + A(X,Z,T), \quad w \to 0 \quad \text{as} \quad Y \to \infty \quad , \tag{3.1e}$$

$$\text{no-slip at} \quad Y = 0 \quad , \tag{3.1f}$$

as generalized from (2.1), (2.2) , in the incompressible regime. Computations of (3.1) are given in [20], while the compressible regime is studied in [21].

One of the lowest-amplitude three-dimensional responses occurs at high frequencies (Ω as defined in (2.3)) through the "resonant triad" mechanism (Craik [15]), again as the disturbance moves downstream, in line with experiments (see introduction). When (2.4) holds the triad can come into play in the form of the pressure expansion

$$p = P_1 \exp{[i\alpha X - i\Omega T]}$$

$$+ P_2 \exp \left[\frac{i\alpha X}{2} + i\beta Z - \frac{i\Omega T}{2} \right]$$

$$+ P_3 \exp \left[\frac{i\alpha X}{2} - i\beta Z - \frac{i\Omega T}{2} \right] + 0(\Omega^{-\frac{1}{2}}) \ . \tag{3.2}$$

and similarly for the other flow variables, all expanded in inverse powers of Ω . Here the P_1-wave is the incoming two-dimensional disturbance, whereas the P_2- and P_3-waves are three-dimensional parasitic ones which can start up provided that $\beta = \sqrt{3} \ \alpha$, i.e. the three-dimensional waves are inclined at angles

$$\theta_{res} = 60° \tag{3.3}$$

to the free-stream direction. This angle compares favourably with some of the observed values mentioned in the introduction. At second order the nonlinear interaction between the three waves takes effect and yields from (3.1) the governing equations (Smith & Stewart [16])

$$\frac{\partial P_1}{\partial T} + 2 \frac{\partial P_1}{\partial X} = (1 - i) \frac{P_1}{\sqrt{2}} - 2iP_2P_3 \quad , \tag{3.4a}$$

$$\frac{\partial P_2}{\partial T} + \frac{5}{4} \frac{\partial P_2}{\partial X} + \frac{\sqrt{3}}{4} \frac{\partial P_2}{\partial Z} = (1 - i) \frac{P_2}{2} - \frac{i}{4} P_1P_3{}^* \ , \tag{3.4b}$$

$$\frac{\partial P_3}{\partial T} + \frac{5}{4} \frac{\partial P_3}{\partial X} - \frac{\sqrt{3}}{4} \frac{\partial P_3}{\partial Z} = (1 - i) \frac{P_3}{2} - \frac{i}{4} P_1P_2{}^* \ , \tag{3.4c}$$

the nonlinearity coming from the coupling of P_2, P_3 which reproduces P_1, for example. In (3.4) the asterisk denotes the complex conjugate and X, Z, T are to be regarded as relatively slow coordinates, compared with the faster dependence in the waves of (3.2). The terms involving (1 − i) in (3.4) are effects due to viscosity and are responsible for all the significant growth in the system. Some further details including certain computational and analytical nonlinear properties are given by Smith & Stewart [16] and extended by Stewart [17]. Comparisons given in [16] between the above three-dimensional theory and the experiments of Kachanov & Levchenko [18] are rather encouraging, yielding both small- and large-scale agreement in quantitative terms.

It is worth observing here that the three-dimensional nonlinear interaction in (3.2) occurs at scaled pressure levels p of 0(1), much less than in the corresponding two-dimensional version summarized in (2.5). Moreover, the interaction is occurring theoretically in the same range, viz. Ω increasing, station x_0 increasing, as is observed experimentally. To maintain the above relevance to experimental findings and go further into transition requires theoretical investigations of pressure levels slightly higher than in (3.2). Some such investigations for three-dimensional effects have been started by the author and co-workers and the current state of the investigations is summarized briefly in [19] and elsewhere.

Next, at $0(1)$ typical frequencies (Ω) finite-amplitude disturbances governed by (3.1) can lead to a three-dimensional extension of the finite-time break-ups described in section 2 (equations (2.9) – (2.11)), which is of much interest.

Also at $0(1)$ typical frequencies, however, vortex/TS interactions can arise at comparatively small amplitudes. There are several such nonlinear interactions possible, depending again on the disturbance environment, and much research is currently in progress on these. The research stems from the Hall & Smith [4] nonlinear interaction for channel flow, governed in normalized form by

$$\left[\frac{\partial^2}{\partial y^2} - \frac{\partial}{\partial t}\right] U_v = V_v \bar{u}' \quad , \tag{3.5a}$$

$$\left[\frac{\partial^2}{\partial y^2} - \frac{\partial}{\partial t}\right] \frac{\partial^2}{\partial y^2} V_v = G\bar{u}U_v - 2\rho\bar{u}\bar{u}', \tag{3.5b}$$

$$U_v = V_v = \frac{\partial}{\partial y} V_v = 0 \quad \text{at} \quad y = 0, 1 \tag{3.5c}$$

for the vortex flow (subscript v), with $\bar{u} = y(1 - y)$ being the original Poiseuille-flow profile, G the scaled Görtler number for curved channels, while the TS amplitude (squared) $\rho(t)$ is controlled by

$$\frac{d\rho}{dt} = \rho \left\{\alpha_1 + \alpha_2 \int_0^1 \bar{u}U_v dy + \alpha_3 \left[\frac{\partial U_v}{\partial y}(0,t) - \frac{\partial U_v}{\partial y}(1,t)\right]\right\} \quad , \tag{3.5d}$$

thus yielding the nonlinear interaction. Here the α's are constants. The principal features of the solutions possible are described in the last-named reference, both for curved channels $(G \neq 0)$ and for the straight-walled case $(G = 0)$. A sample application made subsequently to boundary layers [5], for instance, leads to the following vortex/TS-interaction system, in suitably scaled terms:

$$\frac{\partial U_v}{\partial X} + \frac{\partial V_v}{\partial Y} + i\beta W_v = 0 \quad , \tag{3.6a}$$

$$Y \frac{\partial U_v}{\partial X} + V_v = \frac{\partial^2 U_v}{\partial Y^2} \quad , \tag{3.6b}$$

$$Y \frac{\partial W_v}{\partial X} + iK\rho Y^{-2} = \frac{\partial^2 W_v}{\partial Y^2} \quad , \tag{3.6c}$$

$$U_v \to 0, \quad V_v \to 0, \quad W_v \sim \rho\{-iK\ell nY + \phi\} \text{ as } Y \to 0 \quad , \tag{3.6d}$$

$$U_v \to A, \quad V_v \sim -AY, \quad W_v \propto Y^{-3} \text{ as } Y \to \infty \tag{3.6e}$$

for the induced vortex motion, with

$$\frac{\partial P}{\partial X} + \left[c_1\lambda_b X + c_2 \frac{\partial U_v}{\partial Y}(X,0)\right]P = 0 \quad \& \quad \rho = |P|^2 \tag{3.6f}$$

for the TS pressure amplitude. Here $K = \beta(1 - \beta^2/4\alpha^2)$ where α, $\pm \beta/2$ are the TS streamwise and spanwise wavenumbers respectively, while ϕ, c_1, c_2 are constants for given α, β and the term $\propto \lambda_b$ gives the nonparallel-flow effect. Solution properties are described in [5]. In both of the interactions above there is some qualitative agreement with experiment and, in both, the ultimate behaviour of the nonlinear solutions, with increasing time or distance, can lead on either to the full system (3.1) above, and thence possibly to break-up, or to a subsequent interactive stage of much increased vortex streamwise-length scale, but reduced TS amplitude, in which the whole mean-flow profile is completely altered from its original steady state. This latter interaction is a short-scale/long-scale one and it appears to be the strongest type of nonlinear vortex/TS interaction found to date.

Related research on compressibility effects (see also [21]) and on pipe flows is under way by Mr. N.D. Blackaby and Mr.A.G. Walton.

4.   TOWARDS THE TURBULENT BOUNDARY LAYER

It may be speculated that certain of the above features whether in two or three spatial dimensions provide the building blocks for the structure of turbulent boundary layers, although we should stress the tentative nature of the theory here.   The theory, again, is aimed at understanding scales, structures and the dominant dynamics. This is especially for nonlinear interactions in which the mean-flow profile is affected substantially.

First, the combined highly nonlinear properties of sections 2,3 are probably connected with K-type [22] breakdown, we feel. This is especially for the finite-frequency theory mentioned above and its associated cascade of length and times scales locally.

Next, the Euler stage yields two major properties that quite resemble two major properties of turbulent boundary layers in practice, namely (a) the large thickening compared with the steady laminar version and (b) the emergence of a very thin wall layer, in the mean flow.

The property (a) is derived from an examination [6] addressing the two-dimensional nonlinear neutral curve for typical boundary layers which suggests that the highest "reach" y of the nonlinearly disturbed boundary-layer solution is

$$O(Re^{-\frac{1}{2}} \ell n \, Re). \tag{4.1}$$

At such distances from the wall nonlinear solutions exist which travel along almost at the freestream speed, yet with only a minor perturbation of the freestream velocity $u = 1$ (again this is as in experiments [23,24] or turbulence-modelling). The distance (4.1) is dependent on the vorticity distribution, we note, with exponential decay assumed here.

The property (b) concerning the wall-layer extent comes from the following argument [6] on the dynamics. The Euler-stage flow described in section 2 creates a new inviscid slip velocity of order unity and thence a viscous $O(R^{-3/4})$ sublayer, as observed before. In effect a new, extra, boundary layer of thickness $O(Re^{-3/4})$ is therefore set up increasing the typical wall-shear factor $\partial u/\partial y$ from the original laminar value $O(Re^{\frac{1}{2}})$ to $O(Re^{3/4})$. This new thinner layer is then susceptible to new TS waves of shortened time and length scales, in view of the λ factors given in (2.1), and these new TS waves in turn can either break up or form a new Euler stage. The latter then induces a new inviscid slip velocity of order unity and hence a new sublayer, this time of y-extent $O(Re^{-7/8})$, and hence yet another new, thinner, boundary-layer profile. The dynamic process continues indefinitely in principle, generating ever shorter length and time scales, until all the localized structure becomes affected by viscosity, that is, when the length scale reduces to $O(Re^{-1})$. The "reach" of this last localized structure is multiplied by a logarithmic factor just as in (4.1), however, since each new boundary-layer profile created exhibits exponential decay in the y direction.

So the predicted final wall-layer thickness is

$$O(Re^{-1} \, \ell n \, Re) \tag{4.2}$$

for the thinnest layer in which unsteady inertial and viscous forces matter equally.   The two features (4.1),(4.2) should be compared with the corresponding features of empirical-turbulence modelling, e.g. Cebeci-Smith [25], the latter yielding the thicknesses

$$O(u_\tau), \; O(Re^{-1}u_\tau^{-1}) \; [\text{where } u_\tau \sim (\ell n \, Re)^{-1}] \tag{4.3a,b}$$

in turn.   The first thicknesses (4.1),(4.3a) disagree but that may well be due to the two-dimensionality of the theory so far.   The second thicknesses (4.2),(4.3b), in contrast, are in exact agreement, and this is encouraging.

Again, the dynamic wall-shear-stress values predicted from the above agree well with the experimental/empirical sizes, as do the spectral broadening and the cascade of scales implied.   Clearly, the extension to three spatial dimensions should also be made in this Euler-stage work and that may well yield many extra features of relevance including an account of the thickness (4.3a).   There are additional connections with "slugs" and "puffs" [26,27] in channel and pipe flows, as described in [6].

ACKNOWLEDGEMENTS

Thanks are due to the Science & Engineering Research Council, U.K., to the United Technologies Independent Research Program, E. Hartford, U.S.A., and to ICASE, NASA Langley, U.S.A., for their support of various parts of this research, and to many colleagues.

REFERENCES

[1]   Smith, F.T. and Burggraf, O.R.,   Proc. Roy. Soc. A399 (1985) 25-55.

[2]   Brotherton-Ratcliffe, R.V. and Smith, F.T., Mathematika 34 (1987) 86-100.

[3]   Smith, F.T., Mathematika (1988), to appear.

[4]   Hall, P. and Smith, F.T., Proc. Roy. Soc. A417 (1988) 255-282.   Also, ICASE Rept. (1987).

[5]   Hall, P. and Smith, F.T., ICASE Rept. (1988).

[6]   Smith, F.T. and Doorly, D.J., J. Fluid Mech. (1988), to appear.   Also, Rept. (1987) UTRC 87-43.

[7]   Smith, F.T., Papageorgiou, D.T. and Elliott, J.W., J. Fluid Mech. 146 (1984) 313-330.

[8] Conlisk, A.T., Burggraf, O.R. and Smith, F.T., Trans. A.S.M.E. (1987), Forum on Unsteady Separation, Cincinnati, Ohio, U.S.A., June 1987.

[9] Duck, P.W., J. Fluid Mech. 160 (1985) 465.

[10] Smith, F.T., United Tech. Res. Cent., E. Hartford, Ct., U.S.A., Rept. UTRC85-36 (1985) (see also A.I.A.A. paper no. 84-1582 (1984) and J. Fluid Mech. 169 (1986) 353-377).

[11] Smith, F.T., Proc. Roy. Soc. A (1988), in press. Also, Rept. (1988) UTRC 88-12.

[12] Walker, J.D.A. and Abbott, D.E., in Turb. in Internal Flows (ed. S.N.B. Murthy), Hemisph. Pub. Corp., Wash. (1977).

[13] Fasel, H., Proc. Symp. on Turb. & Chaotic Phen. in Fluids (ed. T. Tatsumi), Elsevier (1984).

[14] Orszag, S.A. and Patera, A.T., J. Fluid Mech. 128 (1983) 347.

[15] Craik, A.D.D., Proc. IUTAM Symp. on Lam.-Turb. transition (ed. V.V. Kozlov), Springer-Verlag (1985).

[16] Smith, F.T. and Stewart, P.A., United Tech. Res. Cent. E. Hartford, Ct., U.S.A., Rept. UTRC 86-26 (1986) (also to appear J. Fluid Mech. 1987)).

[17] Stewart, P.A., Ph.D. thesis, Univ. of London (1988), in preparation.

[18] Kachanov, Y.S. and Levchenko, V.Y., J. Fluid Mech. 138 (1984) 209.

[19] Smith, F.T., On Transition to Turbulence in Boundary Layers, in: European Conference on Turbulence (Springer-Verlag 1987). Also, Utd. Tech. Cent., E. Hartford, Ct., U.S.A., Rept. (1986) UTRC 86-10.

[20] Smith, F.T., Computers & Fluids, Volume dedicated to R.T. Davis (1988), to appear; and presentation, R.T. Davis Memorial Symposium, Cincinnati (1987).

[21] Smith, F.T., J. Fluid Mech. (1988), in press. Also, Rept. UTRC 87-52.

[22] Klebanoff, P.S., Tidstrom, K.D. and Sargent, L.M., J. Fluid Mech. 12 (1962) 1.

[23[Gad-el-Hak, M., Blackwelder, R.E. and Riley, J.J., J. Fluid Mech. 110 (1981) 73.

[24] Riley, J.J. and Gad-el-Hak, M., in Frontiers in Fluid Mechanics (eds. S.H. Davis and J.L. Lumley, 1985).

[25] Cebeci, T. and Smith, A.M.O., Turbulent Shear Flows (Academic Press, New York, 1974).

[26] Wygnanski, I.J. and Champagne, F.H., J. Fluid Mech. 59 (1973) 281.

[27] Wygnanski, I.J., Sokolov, M. and Friedman, D., J. Fluid Mech. 69 (1975) 283.

Theoretical and Applied Mechanics
P. Germain, M. Piau and D. Caillerie (Editors)
Elsevier Science Publishers B.V. (North-Holland)
© IUTAM, 1989

STOCHASTIC MODELLING OF FATIGUE ACCUMULATION

Kazimierz SOBCZYK

Institute of Fundamental Technological Research

Polish Academy of Sciences

Warsaw, Poland

1. INTRODUCTION

An important deterioration process which takes places in structures operating under time-varying actions is fatigue. It has been recognized as a frequent cause for failure of engineering structures. On the other hand, fatigue phenomenon itself is still far from being completely understood and it remains a challenging task today.

The fatigue phenomenon is now deemed to originate in local yield in the material or, in other words, in the sliding of atomic layers. This sliding is caused by combination of dislocations and local stress concentration. Under time-varying stress conditions there is migration of dislocations and localized plastic deformation. Microscopic cracks are created which can grow and join together to produce major cracks. However, during the course of propagation the crack encounters various types of metallurgical structures and imperfections, so that the rate of growth is variable in time. Today, it is widely accepted that fatigue phenomenon takes place via the formation and growth of cracks in material.

Fatigue process is a complex phenomenon and can be (and usually is) considered from different points of view. It is of interest for *physics*, since it is intriguing to recognize the true physical (e.g. atomic and molecular) mechanisms of fatigue deterioration process on microlevel which might be responsible for generating macroscopic fatigue. While the existing efforts in this direction are of great importance (cf. the kinetic theory of fracture by S.N.Zhurkov; cf. also paper [1] in which the dislocation theory is used to determine a type of fatigue crack growth equation, or paper [2] in which the connection between submicrocracks and the structure of solid is studied experimentally) they can not yet give a basis for consistent and quantitative treating of macroscopic fatigue fracture.

Fatigue is obiously a subject of *mechanics*; for mechanical engineer there are of interest why, when and where fatigue might occur and consequences it could have on overall stress distribution in structure.

Fatigue is an important issue in *metallurgy* since the manufacturer always wants to produce structural elements that will be free from fatigue as long as possible; one of the basic questions for a metallurgist is: what should be the basic metallurgical factors (such as: alloy composition, heat treatment, mechanical working) to produce elements of high fatigue resistance.

From very engineering point of view fatigue is a problem in the design of many structures and this leads to the questions commonly considered within *reliability theory*. For a structure required to perform safely during its entire service life, it is necessary to have an appropriate probability distribution of ultimate fatigue failure.

In spite of the above problems there exists an important issue which seriously determine the direction of recent research on fatigue. This is the fact that our basic knowledge on fatigue comes from experiments. The experimental data constitute basic source of information on fatigue of various materials subjected to various loading conditions. In such situation it is important to look at fatigue phenomenon from the point of view of the *mathematical modelling*. However, experimental test results as well as field data indicate that fatigue process in real materials involves considerable random variability. This variability varies with respect to many parameters such as material properties, type of load, environmental conditions etc. The random nature of the fatigue process is most obvious if a structure is subjected to randomly varying loading. But even in very well controlled laboratory conditions (under deterministic cyclic loading) results obtained show considerable statistical dispersion. Because of this inherent randomness in fatigue data a *stochastic modelling* seems to be most appropriate one.

In the last two decades an increasing amount of attention has been devoted to various statistical and probabilistic aspects of fatigue damage. A significant number of studies are concerned with question of statistics of dispersed fatigue data and applications of the results to estimation of fatigue reliability of parts in service. As for as stochastic models for fatigue are concerned they are primarily intended to describe the probabilistic mechanism of macroscopic fatigue accumulation in order to draw conclusions concerning fatigue life prediction. Due to inherent complexity of fatigue phenomenon the existing results are, however, far from satisfactory. The question of how to perform a stochastic modelling and analysis of random fatigue process which would lead to a consistent theory of random fatigue still constitutes an important and challenging problem in recent research.

The objective of this lecture is twofold. First, we wish to review shortly the basic existing approaches to stochastic modelling of fatigue process. Next, we describe in more details a cumulative random compound model for fatigue crack growth in which crack length is represented as a random sum of random components. Parameters of the model can be identified from the appropriate fatigue experiments. They can also be related to the quantities occurring in the existing empirical fatigue crack growth equations.

2. MAIN EXISTING APPROACHES TO STOCHASTIC MODELLING OF FATIGUE

2.1 Fatigue life models (distributions)

From engineering point of view, for a structure required to perform safely during its service life, it is necessary to have an appropriate probability distribution of ultimate fatigue failure, or life-time distribution. This need leads to problems analogous to those in general reliability theory, where estimation of the life-time distribution of the system considered is a major concern.

Let us denote a random life-time of a structural element by T. A number of papers have appeared in which various probability laws for random variable T have been proposed (cf.[3,4,5]). Let denote the (cumulative) probability distribution of T, that is P{T<t} by F(t). The function 1-F(t) is called the survival function. Weibull (cf.[3]) introduced a distribution of the extreme value type (cf.[6]) of the form

$$F(t) = 1 - \exp\left[-\left(\frac{t-\varepsilon}{\nu}\right)^k\right], \qquad t \geq \varepsilon \qquad (2.1)$$

with three parameters (ε, ν, k), where $\varepsilon > 0$ is called the threshold. If the lower bound of random variable T is assumed to be equal to zero $(\varepsilon = 0)$, then one obtains two parameter Weibull distribution. The Weibull distribution function often provides a reasonable fit to empirical distribution functions of life-time that have been obtained from fatigue tests.

In practice, the scattered data of fatigue life time is most often described by exponential distribution, i.e.

$$F(t) = 1 - \exp(-at), \qquad t \geq 0, \qquad a > 0 \qquad\qquad (2.2)$$

and log-normal distribution:

$$f(t) = \frac{dF}{dt} = \frac{1}{t\sigma\sqrt{2\Pi}} \exp\left[-\frac{1}{2\sigma^2}(\log t - \mu)^2\right], \qquad (2.3)$$

where μ, σ are constants $\sigma > 0$, $-\infty < \mu < \infty$.

It is worthy noting that (cf.[4]) the usefulness of the exponential distribution is essentially limited due to the fact that the previous use of structural element does not effect its future life-time. This means that if an element has not failed up to a time t_1, then the probability distribution of its future life-time $t-t_1$ is the same as if the element was quite new.

Failure of structural elements may be related to causes that depend on the smallest or the largest value in the sample of data. For example, the failure may depend on the strength of the weakest of elementary elements, or it may depend on the size of the largest crack-like defect. In such situations the distribution of extreme value in a sample taken from some initial distribution may constitute a good model for a life-time. The distribution of extreme values in a statistical sample depend, in general, on the sample size n and on the type of initial distribution. If, however, n becomes large enough and if the initial distribution belongs to so called "exponential type" (cf.[6]) then it has been shown that the smallest or largest values converge (as $n \Rightarrow \infty$) to so called type I asymptotic distribution (of extreme values) known also as Gumbel distribution. The Gumbel distribution of the smallest extreme has the form

$$F(t) = 1 - \exp\left[-\exp\left[\frac{t-\nu}{\beta}\right]\right] \qquad\qquad (2.4)$$

or

$$f(t) = \frac{1}{\beta}\exp\left[\frac{t-\nu}{\beta} - \exp\left[\frac{t-\nu}{\beta}\right]\right] \qquad\qquad (2.4')$$

where $\beta > 0$, $-\infty < \nu < +\infty$.

Birnbaum and Saunders (cf.[4]) have proposed a cumulative fatigue model based on renewal theory and derived a two-parameter family of distribution functions for a life-time which provides good fit to empirical distributions obtained from extensive fatigue test data. Other recently proposed stochastic models for fatigue process lead to other distributions for a life-time. For example,

Ditlevsen [7] and Sobczyk [8] derived independently and by different reasonings the inverse Gaussian distribution for the life-time which has the form:

$$f(t) = f(t;\mu,\lambda) = \left[\frac{\lambda}{2\Pi\ t^3}\right]^{1/2} \exp\left[-\frac{\lambda(t-\mu)^2}{2\mu^2 t}\right] \qquad (2.5)$$

where t>0, μ>0, λ>0. This distribution has been used earlier in reliability theory as a life-time model.

It is clear that a general and mathematically natural approach to the formulation of fatigue life distribution is as follows. The accumulation of fatigue in real materials subjected to realistic actions as a function of time t is a certain stochastic process $X(t,\gamma)$ defined for $t \in [t_0, \infty)$, $\gamma \in \Gamma$ where $(\Gamma, \mathcal{F}, P)$ is the basic probability space that is, the triple consisting of a space of elementary events Γ, σ - field $\mathcal{F}$ of subsets of Γ and probability P defined on $\mathcal{F}$.

Fatigue failure occurs at such time t=T, being a random variable, for which $X(t,\gamma)$ crosses, for the first time, a fixed critical level x^*. Therefore, the fatigue life-time distribution is the first-passage time distribution for the process $X(t, \gamma)$, i.e. the distribution of random variable T defined as

$$T = \sup\left\{t: X(\tau,\gamma) < x^*\right\}, \qquad t_0 \le \tau < t. \qquad (2.6)$$

Now, we shall describe briefly how process $X(t,\gamma)$ can be constructed using the existing information (mostly, in the form of experimental data) about fatigue process.

2.2. Evolutionary probabilistic models

A general probabilistic approach to the modelling of fatigue is based on the following reasoning. A sample of material together with a fatigue process taking place in it is regarded as a certain dynamical system whose evolution in time is described by a stochastic vector process $\underline{X}(t)=[X_i(t),\ldots,X_n(t)]$. The component processes $X_i(t)$, i=1,2...,n characterize specific features of the fatigue accumulation, and $0 \le X_i(t) \le x_i^*$ denotes a critical or limiting value of $X_i(t)$. In order to obtain a model for random fatigue one may try to describe the evolution in time of the probabilistic structure of the process $\underline{X}(t)$. The infinitesimal characteristics of such an evolution (depending on all basic factors provoking fatigue) should be deduced from the knowledge of fatigue process in real materials.

Denoting the joint probability density of the transition of $\underline{X}(t)$ from the state $\underline{x}_0$ at time t by $p(\underline{x},t) = p(\underline{x},t;\underline{x}_0,t_0)$ we can postulate the evolution equation in the form

$$\frac{\partial p(\underline{x},t)}{\partial t} = \mathcal{L}\,[p(x,t);\ S(t),\ Z(t)] \qquad (2.7)$$

where $\mathcal{L}$ can be termed the fatigue operator which must be constructed; it can be a differential, integral or differential-integral operator. $S(t)$ denotes an appropriate parameter (or parameters) characterizing external loading and $Z(t)$ is used to denote all other possible quantities important in fatigue problem. In such a formulation, which was first indicated by Bolotin (cf.[9]) the modelling problem lies in proper construction of the fatigue operator $\mathcal{L}$ on the basis of data.

In order to make the idea described above efficient, one has to introduce some assumptions and hypotheses. A hypothesis which is often introduced reguires the fatigue process to be Markovian. This makes it possible to use a large variety of mathematical schemes elaborated in the theory of Markov stochastic process theory, and leads to interesting results. This approach is represented in a series of recent works (cf.[10,11,12,13]). There are two classes of Markovian fatigue accumulation models: discrete - based on Markov chain theory (cf.[10,12,13]) and continuous ones formulated in framework of diffusion Markov processes (cf.[11]). In the Markov chain scheme process $\underline{X}(t)$ characterizing a random evolution of fatigue is a sequence of fatigue states $E_0,\ E_1,\ldots,\ E_n=E^*$ where E_0 denotes an "ideal" state and $E_n=E^*$ the state of ultimate damage. Randomness in transition from one state to another is accounted for by the probabilistic (Markovian) mechanism of the transition (by infinitesimal transition intensities). The corresponding Markov chain can be one-dimensional (if a measure of fatigue is characterized by a scalar quantity) - cf.[12], or multi-dimensional - as in the model elaborated by Bogdanoff and his collaborators (cf.[10]); cf. also paper of Cox and Morris [13] in which multidimensional Markov chain model is presented with a special emphasis on the role of the randomness of microstructure in the growth of short fatigue cracks.

In the Markovian models just mentioned the fatigue equation (2.7) takes the specified form which is the Kolmogorov equation for transition probability.

2.3. Randomized differential equation models

As we have already said, the fatigue failures in components and structures results from the nucleation and propagation of cracks. So, the fatigue accumulation is most often characterized by the length of a dominant crack. To estimate total fatigue life-time of an element one should, therefore, consider a nucleation phase and growth phase. The nucleation process of fatigue crack is very complex (and very difficult for experimental observations) and it is not yet fully recognized. Though exist now some efforts to model probabilisticly this fatigue stage, they are not mature enough to provide a reliable information for fatigue analysis. In modelling, most often the initial (macroscopic) crack length is hypothesized (and experimentally verified) and the efforts are focused on modelling of fatigue crack growth.

In modelling of fatigue crack growth it is primarily important to relate random factors provoking fatigue to the mechanisms of crack growth. A possible way to do that is to make use of the fracture mechanics laws and introduce random quantities to them (randomization of empirically motivated equations). It seems that such a methodology dominates the recent efforts in stochastic modelling of fatigue.

Studies of fatigue crack growth in elastic materials have provided a number of empirical equations which relate the crack length with material and loading parameters (cf.[14]). in general, they can be represented by the following nonlinear equation

$$\frac{dL(t)}{dt} - F\ [L,\ \Delta K,\ K_{max},\ R,\ S] \qquad (2.8)$$

where $L(t)$ is the crack length at time t, F is non-negative function of its arguments, ΔK - stress intensity factor range, K_{max} - maximum stress intensity factor, R - stress ratio, and S - maximum stress level in the loading spectrum. The most common example of eg.(2.8) is the Paris-Erdogan equation

$$\frac{dL}{dt} = C\ (\Delta\ K)^n \qquad (2.9)$$

where C and n are empirical constants.

In the papers of Yang and his collaborators (cf.[20] the so called hyperbolic sine model is employed, i.e.

$$\frac{dL}{dt} = C_1 \sin h\ [C_2(\log \Delta K + C_3] + C_4 \qquad (2.10)$$

where C_1, C_2, C_3, C_4, are constants to be determined from crack propagation data. For crack propagation in fastener holes under variable loading the following empirical equation has recently been proposed

$$\frac{dL}{dt} = Q\ L^b(t), \qquad Q = CS^\mu \qquad (2.11)$$

where C, b and μ are constants depending on the characteristics of loading and material.

In order to take into account the statistical dispersion of the crack growth rate, eg.(2.8) is randomized as follows

$$\frac{dL(t)}{dt} = \xi(t,\gamma)\ F\ [L,\ \Delta K,\ K_{max},\ K,\ R,\ S] \qquad (2.12)$$

where $\xi(t,\gamma)$ is, in general, non-negative stochastic process. Of course, eg.(2.12) is supplemented by the initial condition (deterministic or random)

$$L(t_0) = L_0 \qquad (2.13)$$

characterizing the "initial" size of the crack in question.

Stochastic differential equation (2.12) together with condition (2.13) constitutes a general framework for various specific randomized crack growth models; cf. Kozin, Bogdanoff [15], Sobczyk [8,12], Madsen [16], Lin, Yang [17], Ditlevsen [7], Doliński [18], Spencer [19] and references therein. In the papers cited above the stochastic model have been formulated by randomization of equation (2.9). As it has already been mentioned, in papers [7] and [8] an inverse Gaussian distribution has been obtained for the life-time of a specimen.

Since process $\xi(t,\gamma)$ should be non-negative, it was proposed to be a stationary lognormal random process (cf.[20]). Process $\xi(t,\gamma)$ is defined by the fact that its logarithm i.e., $Z(t,\gamma)=\log\xi(t,\gamma)$ is a normal process. Making use of equation (2.11) we have the following randomized equation

$$\frac{dL}{dt} = \xi\ (t,\gamma)\ Q\ L^b\ (t) \qquad (2.14)$$

or

$$Y = b\ U + q + Z(t,\gamma) \qquad (2.15)$$

where

$$Y = \log \frac{dL}{dt}, \qquad U = \log L\ (t), \qquad q = \log Q. \qquad (2.16)$$

Since $Z(t,\gamma)$ at time t is a normal random variable, it follows from equation (2.15) that the log crack growth rate $Y=\log\left[\dfrac{dL}{dt}\right]$ is also a normal random variable, conditional on a given crack size $L(t)$. Integration of equation (2.14) from zero to t gives

$$L(t) = \frac{L_0}{\left[1 - c\ Q\ L_0^c\ \Lambda\ (t)\right]^{1/c}} \qquad (2.17)$$

in which $L_0=L(0)$ and $c=b-1$ and $\Lambda(t,\gamma)=\displaystyle\int_0^t \xi(\tau,\gamma)d\tau$. For the lognormal random variable model, i.e. when $\xi(t,\gamma)=\xi(\gamma)$ the crack size distribution can be derived analytically. Comparison of the results of the above model with data provided in [20] show very satisfactory agreement.

A stochastic differential equation model has been studied for a wide class of stationary noises $\xi(t,\gamma)$ by use of the central limit theorem for dependent random variables (cf.Bolotin [21,22].

2.4 Stochastic cumulative jump models

According to many experimental observations fatigue crack grows intermittently and goes through active and dormant periods. On the other hand, it is reasonable to assume that the growth of the crack is due mainly to a sequence of shocks occurring randomly in time. This leads to modelling by random jump processes (cf. the models considered by Sobczyk [23], Kogaiev and Lebiedinsky [24] and Ditlevsen and Sobczyk [25]). If $L(t)$ denotes the length of a dominant crack at an arbitrary moment $t \geq t_0$ then it is characterized by the following random sum of random components

$$L(t) = L_0 + \sum_{i=1}^{N(t)} \Delta L_i, \qquad N(t)>0 \qquad (2.18)$$

where $N(t)$ is a random number of jumps in crack length in the interval $(t_0,t]$. Quantities ΔL_i characterize random elementary increments in fatigue crack growth and are assumed to be identically distributed independent non-negative random variables. In what follows we shall show how such a model can be made useful (cf. Sobczyk, Trębicki [26]).

3.   CUMULATIVE COMPOUND POISSON MODEL

3.1 Basic probability distributions

In general, the counting stochastic process $N(t)$ can have quite different properties, but here it is assumed that $N(t)$ is a Poisson homogeneous process, that is probability $P_k(t)$ that within time interval $[0,t]$ the number of events (crack's jumps) is equal to k is

$$P_k(t) = e^{-\lambda_0 t} \frac{(\lambda_0 t)^k}{k!} \tag{3.1}$$

Let T be a positive random variable characterizing random time of reaching by process $L(t,\gamma)$ of a fixed, critical level ξ. Probability distribution of T is $\left[Y_i(\gamma) = \Delta L_i(\gamma)\right]$:

$$P(T>t) = P\{L(t,\gamma) < \xi\}$$

$$= P\left\{L_0 + \sum_{i=1}^{N(t)} Y_i(\gamma) < \xi\right\} = \sum_{k=0}^{\infty} P\left\{L_0 + \sum_{i=1}^{k} Y_i(\gamma) < \xi\right\} P_k(t)$$

$$= \sum_{k=0}^{\infty} P_k(\xi) P_k(t).$$

where $P_k(\xi)$ is the probability that after k increments the crack size is less than critical value ξ.

The probability density of T is

$$f_T(t) = -\frac{d}{dt} P(T>t) = -\sum_{k=0}^{\infty} \frac{P_k(\xi)}{k!} \frac{d}{dt}\left[(\lambda_0 t)^k e^{-\lambda_0 t}\right]$$

After transformations

$$f_T(t) = \lambda_0 e^{-\lambda_0 t} \sum_{k=0}^{\infty} \frac{(\lambda_0 t)^k}{k!}\left[P_k(\xi) - P_{k+1}(\xi)\right]. \tag{3.2}$$

Of course, $P_0(\xi)=1$ and $P_k(\xi) \geq P_{k+1}(\xi)$.

If random variables $Y_i(\gamma)$ are exponentially distributed (with parameter a) then (cf.[26])

$$P_k(\xi) - P_{k+1}(\xi) = \frac{a^k(\xi-L_0)^k}{k!} e^{-a(\xi-L_0)}$$

and we finally obtain

$$f_T(t) = \lambda_0 e^{-\lambda_0 t - a(\xi-L_0)} I_0 \left[2\sqrt{\lambda_0 ta(\xi-L_0)} \right] \qquad (3.3)$$

where $I_0( )$ is a modified Bessel function of order zero.

Another important characteristics of process $L(t)$ is the probability distribution of a crack size at arbitrary time t. Using the apparatus of moment generating functions one obtains that $F_{L(t)}(1)=F_{L_1(t)}(1-L_0)$, where

$$F_{L_1(t)}(1) = \sqrt{\frac{\lambda_0 at}{1}} e^{-\lambda_0 t - al} I_1 \left[2\sqrt{\lambda_0 alt} \right]. \qquad (3.4)$$

If the intensity of the Poisson process $N(t)$ is a time-varying function, i.e. $\lambda_0=\lambda(t)>0$, then by transformation of variable

$$\lambda(t) = \frac{d\eta(t)}{dt}, \qquad \eta(t) = \int_0^t \lambda(u) du \qquad (3.5)$$

we obtain that

$$P_k(t) = \frac{[\eta(t)]^k}{k!} e^{-\eta(t)} \qquad (3.6)$$

and finally, after application of formula (3.2), the distribution of T is expressed as $(t \geq 0)$:

$$f_T(t) = \frac{d\eta}{dt} e^{-\eta(t)-a(\xi-L_0)} \sum_{k=0}^{\infty} \frac{a^k(\xi-L_0)^k}{k!} \frac{[\eta(t)]^k}{k!}. \qquad (3.7)$$

3.2 Relation to experimental data

The model just presented and the above formulae form the framework in which a real fatigue process can be embedded. To do this the parameters occurring in the model (λ_0, a, ξ) should be obtained from the appropriate fatigue tests. Since such experiments are of future concern, it is reasonable to relate the model parameters to those occurring in empirical fatigue crack growth equations and to characteristics of random loading process.

Let us assume that structural element is subjected to random loading characterized by stationary stochastic process $S(t,\gamma)$. The modified Paris-Erdogan equation (adapted to random loading) can be represented as (cf.[8])

$$\frac{dL}{dt} = \mu_0 C \ (K_{rms})^n, \qquad L(t_0)=L_0 \qquad (3.8)$$

where μ_0 is a (constant) average number of maxima of process $S(t,\gamma)$, and K_{rms} its root mean square. Different criteria can be adopted to relate the model parameters to the parameters of loading and material. One of the possibilities is as follows:

$$\lambda_0 = \mu_0(s_0) \qquad (3.9)$$

where $\mu_0(s_0)$ is the average number of maxima of the process $S(t,\gamma)$ above level s_0 (where s_0 is appropriately selected; for example s_0 might be the fatigue limit of the material considered).

In order to "approximate" parameter a it is reasonable to use the following condition

$$F(a) = \int_0^{\bar{t}} \left\langle \ [L_{P-E} \ (t) - L(t,\gamma)]^2 \ \right\rangle \ dt = \min_a \qquad (3.10)$$

where $L_{P-E}(t)$ is a solution of empirical equation (3.8) and $L(t,\gamma)$ - is the process represented by our model (2.18); $\bar{t}$ - is the time in which the curve $L_{P-E}(t)$ crosses the critical level ξ; $\left\langle \cdot \right\rangle$ denotes the mean value.

To approximate the parameter ξ, a limiting critical condition (from fracture mechanics) has to be used. We use the fact that in the analysis of fatigue under periodic loading it is commonly accepted that unstable crack growth takes place when K, the stress intensity factor, assumes its critical value K_{cr}. In the case of infinite sheet it means that

$$K_{cr} = S_{cr} \sqrt{\pi \xi} \, .$$

Let, for a given load, $S_{cr} = S_{max}$. Then

$$\xi = \frac{K_{cr}^2}{\pi S_{max}^2}$$

where K_{cr} is a constant determined experimentally. The above suggests that in the case of random loading $S(t,\gamma)$ the following value of ξ can be assumed.

$$\xi = \frac{K_{cr}^2}{\pi \left\langle S_{max} \right\rangle^2} \qquad (3.11)$$

where for stationary and Gaussian process

$$\left\langle S_{max} \right\rangle = \langle S(t,\gamma) \rangle + S_{rms} \sqrt{\frac{\pi(1-\eta^2)}{2}} \qquad (3.12)$$

$$\eta^2 = 1 - \frac{\omega_2^2}{\omega_0 \omega_4} \, , \qquad \omega_k = \int_{-\infty}^{+\infty} \omega^k \, g_S(\omega) \, d\omega \qquad (3.13)$$

and $g_S(\omega)$ is the spectral density of $S(t,\gamma)$.

Numerical calculations for real data allows us to visualize a dependence of the probability distributions of the life-time and that of crack size on various parameters involved in the problem (e.g. on the level s_0, on the correlation radius of process $S(t,\gamma)$ etc.) - cf. [26].

4. FATIGUE CRACK GROWTH WITH RETARDATION; CUMULATIVE BIRTH MODEL

Of particular interest in the study of crack growth under variable - amplitude loading is the decrease in growth rate called the crack *growth retardation*, which normally follows a high overload (cf. [27]).

Recently, Ditlevsen and Sobczyk [25] presented an attempt to construct a probabilistic model for fatigue crack growth with retardation of the form (2.18) where N(t) is the birth process whose infinitesimal growth intensity is appropriately constructed to account for the retardation.

The idea steems from the conviction that the crack growth intensity should be treated as state -dependent (also in the case of homogeneous cyclic loading). This suggest the usage of the pure birth process as $N(t)$ whose intensity is

$$\lambda_k = \lambda_0 k, \qquad k = 1, 2 \ldots \qquad \lambda_0 > 0 \qquad\qquad (4.1)$$

Since a peak overload causes a decrease in crack growth rate, it is postulated that:

$$\lambda = \lambda_k(t) = \lambda_0 k - \lambda_{OL}(t, k; t_1) \qquad\qquad (4.2)$$

where $\lambda_{OL}(t, k; t_1)$ describes a retarding effect of an overload which has occurred at time t_1. It depends on several variables such as: the overloading ratio ρ defined as

$$\rho = \frac{K_{max,o}}{K_{max}} ,$$

the crack size, metallurgical properties, etc.

In order to obtain practicable results, λ_{OL} is specified to have the form

$$\lambda_{OL}(t, k; t_1) = k \, \lambda_{OL}(t; t_1)$$

$$\lambda_{OL}(t, k; t_1) = \mu(t, \zeta) \, e^{-\alpha(t-t_1)} H(t-t_1) \qquad\qquad (4.3)$$

where

$$H(t-t_1) = \begin{cases} 1, & t > t_1 + \theta, \\ 0, & \text{otherwise.} \end{cases}$$

In (4.3) $\mu(t, \zeta)$ is a suitably bounded retardation magnitude function depending on time t and on the collection of relevant variables denoted symbolically by ζ (e.g. the overloading ratio); $\frac{1}{\alpha}$ may be interpreted as the characteristic retardation time, whereas the parameter θ reflects a delay of the start of the retardation.

In the case of periodic loading with multiple (separated) overloads occurring at $t_1, t_2, \ldots, t_n$ the infinitesimal "overloading intensity" λ_{OL} is postulated to have the form:

$$\lambda_{OL}(t;t_1,t_2\ldots,t_n) = \sum_{i=1}^{n} \mu_i(t,\zeta)\, e^{-\alpha_i(t-t_i)}\, H(t-t_i) \qquad (4.4)$$

where the quantities $\mu_i(t,\zeta)$, α_i and θ_i are associated with overload which has occurrent at t_i.

In paper [25] the probability distribution of crack size [represented by (2.13) with N(t) being a pure birth process with intensity (4.2), (4.3), (4.4) has been derived and then the basic parameters of the model have been related to measurable quantities; in particular the retardation time α^{-1} has been related to the Wheeler empirical model.

5. CONCLUDING REMARKS

1. It would be extremely desirable to formulate stochastic theory of fatigue that deals with all physical and chemical process on microscopic level that might be responsible for generating macroscopic fatigue process we observe. While the existing physical theories are helpful in explaining qualitatively some aspects of fatigue, they can not give yet a basis for a quantitative treatment of fatigue problems and obtaining results for practical needs. In view of these difficulties it is important to build macroscopic (phenomenological) models by introducing intuitively plausible assumptions justified by experiments.

2. Main objective in building stochastic (phenomenological) models for fatigue is to provide a consistent base for accurate prediction of the fatigue deterioration process for purposes of reliability estimation.

   Since modelling is a transformation of data into a model this will be acceptable if it is consistent with data. Additionally. it would be very desirable if a model could provide greater insight and understanding of the phenomenon.

   An important question is: how well a model describes all features of the phenomenon revealed by the analysis of existing data ?. It is very difficult (if possible at all) to build a stochastic model for fatigue process which would be consistent with all informations contained in data. Up to now, no existing model is completely satisfactory. But, the modelling which has already been performed shows the possible methodologies along with their advantages and deficiencies.

3. In order to gain better insight into stochastic nature of fatigue there is a strong need for new type of fatigue experiments.

On the one hand the experiments are required which would be designed and performed to verify the existing stochastic models. On the other hand, the experiments are needed in which fatigue process is viewed as multidimensional one (characterized by several parameters; e.g. by the crack length and the characteristics of the plastic zone in the crack's tip). The results of such experiments would allow us to construct more adequate stochastic models.

4. Stochastic modelling of fatigue stimulates new statistical (unconventional) problems. They are associated with statistical inference and estimation of parameters of stochastic fatigue models. New developments in statistical analysis of point stochastic processes (e.g. estimation of the intensity of Poissson and birth processes) as well as in estimation theory associated with stochastic differential equations can be very helpful in modelling of fatigue (cf.[28]).

REFERENCES

[1]   Yokobori T., Yokobori A.T., Kamei A., Intern. J. Fracture, vol.11, 5 (1975), 781.

[2]   Kuksenko V.S. et al.,Intern.J.Fracture vol.11, 5 (1975), 829.

[3]   Weibull W., ASME J. Appl. Mech., vol.18, (1951), 293.

[4]   Birnbaum Z.W., Sunders S.C., J. Amer. Stat. Assoc., vol.53 (1958), 151.

[5]   Birnbaum Z.W., Sunders S.C., J. Appl. Probability, vol.6, (1969), 319.

[6]   Gumbel E., Statistics of Extremes (Columbia Univ. Press , N.York, 1958).

[7]   Ditlevsen O., Eng. Fracture Mech., vol.23 (1986), 467.

[8]   Sobczyk K., Eng. Fracture Mech., vol.24 (1986), 609.

[9]   Bolotin V.V., Problems of Strength (in Russian), vol.2 (1971).

[10] Bogdanoff J.L., Kozin F., Probabilistic Models of Cumulative Damage (John Wiley and Sons, 1985).

[11] Oh K.P., Proc. Roy. Soc. London, A367 (1979), 47.

[12] Sobczyk K., J. Mec. Theor. Appl. (Num. Spec.), 1982, 147.

[13] Cox B.N., Morris W.L., Fatigue and Fract. of Eng. Materials vol.10, 5 (1987), 419.

[14] Hoeppner D.W., Krupp W.E., Eng.Fract.Mech., vol.6 (1974), 47.

[15] Kozin F., Bogdanoff J.L., Eng.Fract.Mech., vol.14 (1981), 1.

[16] Madsen H.O., DIALOG 6-1982 (Danish Eng. Academy, Lyngby, 311.

[17] Lin Y.K., Yang J.N., Eng. Fract. Mech., vol. 18 (1983), 243.

[18] Doliński K., Eng. Fract. Mech., vol. 25 (1986) 809.

[19] Spencer B.F., Tang J., J. Eng. Mech. (to appear 1988).

[20] Yang J.N., Salivar G.C., Annis C.G., Eng.Fract.Mech., vol.18,(1983),257.

[21] Bolotin V.V., Zurn. Prikl. Mekh. Tekhn. Fiz. (in Russian), 5, 1980.

[22] Bolotin V.V., Prediction of Safety of Machines and Structures (in Russian), Moskow, Mashinostrojenije, 1984.

[23] Sobczyk K., On the reliability models for random fatigue damage, in: Brulin O. Hsieh R., New Problems in Continuum Mechanics (Waterloo Univ. Press, 1983).

[24] Kogajev V.H., Lebiedinsky S.G., Mashinoviedienije, (in Russian), 4, 1983.

[25] Ditlevsen O., Sobczyk K., Eng.Fract.Mech.,vol.24 (1986), 861.

[26] Sobczyk K., Trębicki J., (in preparation)

[27] Jones R.E., Eng. Fract. Mech., vol.5 (1973), 585.

[28] Sobczyk K., Adv. Appl. Probability, vol.19 (1987), 652.

Theoretical and Applied Mechanics
P. Germain, M. Piau and D. Caillerie (Editors)
Elsevier Science Publishers B.V. (North-Holland)
© IUTAM, 1989

FLUID-DRIVEN FRACTURES

David A. SPENCE

Department of Mathematics
Imperial College of Science & Technology
London SW7 2BZ, UK.

Fluid-driven fractures, both natural and man-made, occur in a
variety of contexts in the earth's crust: (i) in connection with the
transport of magma and the formation of intrusive dykes and sills,
(ii) in geothermal energy extraction (iii) in explosive volcanism,
(iv) in hydraulically induced fracture as a mechanism for enhanced
recovery of hydrocarbons;   (v) hydrofracture is also a feasible
mechanism in the deposition of economic ore deposits.   In the
lecture theoretical attempts to model the propagation of a crack
filled with viscous fluid and surrounded by deformable host rock
will be described, the fluid being driven either by buoyancy forces
as in (i), (ii) and (iii) above, or by an externally supplied
pressure as in (iv) and (v).   These begin with the work of
Barenblatt [20] and Weertman [5].   They give rise to complex
fluid-elastic problems, which must be treated simultaneously.
Similarity solutions exist for certain hydrofracture problems, and
reference will be made to recent work on time-dependent solutions
for unsteady magma transport. Theoretical treatments are aimed at
gaining fundamental understanding, and make use of simplifying
assumptions, usually treating the host rock as elastic and the fluid
viscosity as constant.   These, and phenomena such as "leak-off" and
solidification, will be discussed.

1. INTRODUCTION

Fluid-driven fracture is an important mechanism for the transport of magmas
in the Earth's crust and upper mantle.   By propagating upwards through host
rock as a result of buoyancy forces magmas produce dykes and sills that can
be observed in outcrops and inferred from geodetic data (Shaw [1], Pollard
[2], Delaney & Pollard [3]).   In Iceland and Hawaii dykes a few kilometers
high and tens of kilometers long are intruded laterally from a central magma
reservoir with adjacent rift zones, and give rise episodically to local
eruptions.

In recent years there have been a number of attempts at mathematical
modelling.   Pollard [4] has recently reviewed applications of fracture
mechanics to the structural interpretation of dykes. Earlier, Weertman [5,6]
postulated a steady-state shape for a dyke rising as a result of magma
buoyancy, and attempted to calculate a steady-state velocity; solutions were
only found however for cases where this velocity was either zero, or close to
the speed of sound in the solid.   Stevenson [7] pointed out that dykes
containing a finite volume of magma cannot propagate upward with a steady
shape: thus a dyke will possess an "umbilical cord" extending to the magma
source, rather than a "tadpole tail" that pinches shut.   In each case such
thin tails are likely to solidify and isolate the fluid magma from its
source.

Spence, Sharp and Turcotte [8] addressed the vertical propagation of dykes in the Earth's crust in which crack growth is driven not by a specified fluid flux but by the differential pressure gradient between solid and fluid. They find a steady-state solution for an infinite dyke whose top migrates upwards at a constant velocity and whose tail asymptotically approaches a constant thickness. For specified values of the rock fracture toughness, elastic moduli, and differential pressure gradient between host rock and magma, a steady-state solution is found for only a single value of the asymptotic dyke thickness (the maximum is somewhat less than twice this value). For a fracture toughness typical of basalt in the upper mantle a solution is found for dyke thickness of order 2m, but the length over which this is approached is somewhat deeper than most magmas are presumed to have originated.

Mathematically, a closely-related problem to that of magma transport is presented by the technique of hydrofracture, used for petroleum recovery, whereby fluid (possibly containing proppants) is forced down a bore hole under pressure to create a reservoir into which oil can drain from the surrounding shale. A state-of-the-art survey of this field up to 1982 is contained in the volume of conference proceedings edited by Nemat-Nasser, Abé and Hirakawa [11]. Semi-empirical theories for predicting width and extent of hydraulically induced fractures are reviewed by Geertsma and Haafkens [12]. Advani, Lee and Khattab [1,3] (who give many other references) have used Lagrangian methods to treat an ellipsoidal cavity, and Spence and Sharp [14] have obtained similarity solutions for two-dimensional and axisymmetric cavities with power-law filling rates. The relationship of this work to magma-driven crack propagation and dyke injection has been explored in references [15], [16], [17]. Under typical conditions the propagation of the dyke is limited by the viscosity of the magma; the fracture resistance of the elastic medium can be neglected. As an example, it is found that a magma with a viscosity of 10^2Pa.s can be injected into a crack with a length of 2km and a maximum width of 0.5m in a period of 15 minutes.

In the present paper, we treat first the problem of the axisymmetric lens-shaped cavity in a homogeneous elastic medium, being filled by a line source on the axis r = 0 so that its perimeter is a crack of radius R(t) and the volume of fluid contained in the cavity is $Q(t) = At^\alpha$. In this case a self-similar solution exists. This was found in [14] for the corresponding 2-dimensional problem. With axial symmetry the eigenfunctions (Popov [18], [19]) are Legendre polynomials. The stress singularity at the tip of the crack could be identified in the 2-dimensional case from an expression due to Barenblatt [20]. In the axi-symmetric case this has an analogue, and is again a parameter characterising the solution.

In section 3 a similar mathematical approach is outlined for the case of a cavity containing a constant volume of buoyant fluid, and some computations indicating the change in shape with time as the fluid moves to the top of the cavity are presented.

One of the most spectacular examples of buoyancy-driven magma fracture is the Kimberlite eruption responsible for the emplacement of diamonds at the earth's surface (Anderson [21]). These eruptions require high velocities in order to quench diamonds; Pasteris [22] estimates the velocities to lie in the range 0.5-5m/s. The mechanism is discussed by McGetchin and Ulrich [23].

If the cavity shape is given by h(y,t), then its evolution is described by the equation

$$\frac{\partial h}{\partial t} + \frac{1}{3\eta_m} \frac{\partial}{\partial y} h^3 \left[-\frac{\partial p}{\partial y} + g\Delta\rho \right] = 0$$

where η_m is the viscosity of the fluid contained, (magma in most contexts), $\Delta\rho = \rho_s - \rho_m$ is the density difference between the magma and surrounding host rock and p is the deviatoric stress. An approximation that may be relevant in the case of relatively rapid buoyancy-induced flow is to neglect the pressure term in the equation, in which case it becomes a form of the kinematic wave equation for which explicit solutions can be found. These show a shock front propagating upward into an undisturbed medium. This approach is explored in section 4.

2.   FILLING OF A LENS-SHAPED CAVITY FROM A CENTRAL SOURCE

We consider a cavity of instantaneous radius R(t) lying in a horizontal plane of cleavage between two isotropic rock layers, and filling with fluid of (large) viscosity η from a line source on the axis r = 0 as indicated in Figure 1.   The surrounding rock is treated as elastic, with modulus

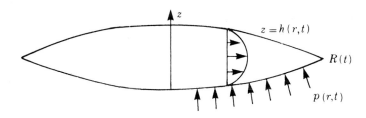

FIGURE 1.   Schematic diagram of cavity flow.

$\varepsilon = (1-\nu)^{-1}$ (shear modulus), where ν is Poisson's ratio   The shear stress on the internal walls of the cavity will be neglected in comparison with the normal stress.   The pressure $p(r,t)(=-\sigma_{zz})$ and the cavity height $2h(r,t)$ are connected by an integral relation derived by solving the equations of linear elasticity in a half space, with zero shear stress on the circle $|r| < R$:

$$p(r,t) \;=\; -\frac{\varepsilon}{r}\frac{\partial}{\partial r} \int_0^{R(t)} r\; \tilde{K}\!\left(\frac{r}{s}\right) \frac{\partial}{\partial s} h(s,t)\; ds \qquad\qquad (1)$$

The kernel for shear-free deformation is

$$\tilde{K}\!\left(\frac{r}{s}\right) \;=\; s \int_0^{\infty} J_1(\xi r)\; J_1(\xi s)\; d\xi \;\equiv\; s\, K_{22}(r,s)$$

They are also connected by Reynolds' equation of hydrodynamic lubrication theory, in which the flow is assumed to be so slow that inertia terms are negligible. For the axisymmetric geometry it takes the form

$$\frac{\partial h}{\partial t} \;=\; \frac{1}{3\eta r}\frac{\partial}{\partial r}\!\left[r\, h^3\, \frac{\partial p}{\partial r}\right] \qquad\qquad (2)$$

The total volume of fluid in the cavity is

$$Q(t) = 4\pi \int_0^{R(t)} r\, h\,(r,t)\, dr \tag{3}$$

The ratio

$$T = 3\eta/\varepsilon$$

represents a viscoelastic scale time for the balance between elastic deformation and viscous flow.

2.1 Self-similar solution

In the case

$$Q = A\left(\frac{t}{T}\right)^{\alpha} \tag{4}$$

where α, A are constant, the equations possess a self-similar solution given by

$$\left. \begin{array}{c} p = \varepsilon \left(\dfrac{t}{T}\right)^{-1/3} P(\rho) \qquad h = k\left(\dfrac{t}{T}\right)^{\beta} H(\rho) \\[2em] \text{in terms of a similarity variable} \\[1.5em] \rho = \dfrac{r}{R(t)}, \end{array} \right\} \tag{5}$$

$R(t)$ being given by $R = k(t/T)^{\lambda}$, with $\beta = (1/3)\alpha - 2/9$, $\lambda = (1/3)\alpha + 1/9$. Equations (1) and (2) become

$$P(\rho) = -\frac{1}{\rho}\frac{d}{d\rho} \int_0^1 \rho\, \tilde{K}\left(\frac{\rho}{\sigma}\right) H'(\sigma)\, d\sigma \tag{6}$$

and

$$\beta H(\rho) - \lambda\rho\, H'(\rho) = \frac{1}{\rho}\frac{d}{d\rho}\left[\rho\, H^3\frac{dP}{d\rho}\right] \tag{7}$$

When these have been solved, the normalising constant k is found by quadrature as

$$k = A^{1/3}\left[4\pi \int_0^1 \rho\, H(\rho)\, d\rho\right]^{-1/3} \tag{8}$$

The solution must be such that the cavity height remains positive, so that

$$H(\rho) > 0, \qquad 0 \le \rho < 1$$

and the tip is defined by

$$H(1) = 0$$

Integration of (7) using this condition gives a form of the equation that is useful in numerical work:

$$\alpha \int_{\rho}^{1} \sigma \, H(\sigma) \, d\sigma + \lambda \, \rho^2 \, H(\rho) = -\rho \, H^3 \, \frac{dP}{d\rho} \qquad (9)$$

2.2 Eigenfunction expansion

In terms of the kernel K_{22} in equation (1), the relation between P and H′ is

$$P(\rho) = -\frac{1}{\rho} \frac{d}{d\rho} \int_{0}^{1} \rho\sigma \, K_{22}(\rho, \sigma) \, H'(\sigma) \, d\sigma \qquad (10)$$

The eigenfunctions of this kernel (Popov [18], [19]) are

$$U_m(\rho) = \rho \, C_{2m}^{3/2} \, (\sqrt{1-\rho^2}) \qquad (11)$$

where $C_{2m}^{3/2}$ are Gegenbauer polynomials (Erdelyi, Magnus, Oberhettinger and Tricomi [24] p176: $C_r^{\nu}(z)$ is the coefficient of h^r in the expansion of $(1 - 2hz + h^2)^{-\nu}$). Using standard results it can be shown that

$$\int_{0}^{1} K_{22}(\rho, \sigma) \, U_m(\sigma) \, \frac{\sigma d\sigma}{\sqrt{1-\sigma^2}} = k(m) \, U_m(\rho) \qquad (12)$$

where

$$k(m) = \left[\frac{\pi}{2}\right] (2m + 1)!! (2m)! / (2m+2)!! 4^m (m!)^2$$

The formula

$$\frac{d}{dz} C_n^{\nu}(z) = 2\nu \, C_{n-1}^{\nu+1}(z) \qquad (13)$$

can be used to express $C_{2m}^{3/2}$ as the derivative of $C_{2m+1}^{\frac{1}{2}}$ which is the Legendre polynomial P_{2m+1}. This forms the basis of a numerical treatment of the problem. If $H(\rho)$ is expanded in a series of Legendre polynomials of argument $(1-\rho^2)^{\frac{1}{2}}$, namely

$$H(\rho) = \sum_m A_m \, P_{2m+1} \, (\sqrt{1-\rho^2}) \qquad (14)$$

the pressure gradient $\frac{dP}{d\rho}$ can be calculated explicitly in terms of Gegenbauer polynomials. Differentiation of (14) gives

$$H'(\rho) = -(1-\rho^2)^{-\frac{1}{2}} \sum_m A_m \, U_m(\rho)$$

and use of (12) then gives

$$P(\rho) = \frac{1}{\rho} \frac{d}{d\rho} \sum_m A_m \, k(m) \, \rho \, U_m(\rho) \qquad (15)$$

Differentiating we find

$$\frac{dP}{d\rho} = -\sum_m A_m \, k(m) \, \Phi_m(\rho) \qquad (16)$$

$$\Phi_m(\rho) = \frac{\rho}{(1-\rho^2)^{\frac{1}{2}}} \left[\left(12 + \frac{3\rho^2}{1-\rho^2} \right) C_{2m-1}^{5/2}(\sqrt{1-\rho^2}) - \frac{15\rho^2}{\sqrt{1-\rho^2}} C_{2m-2}^{7/2} \right]$$

As $\rho \to 1$ $\Phi_m(\rho)$ is singular like $(1-\rho^2)^{-1}$, but the singularity does not enter equation (9), being suppressed by the zero in H^3. However as $\rho \to 0$ $\Phi_m(\rho) \to 0$, and all the terms of the expansion (16) are zero. But (9) shows that

$$\lim_{\rho \to 0} \rho \frac{dP}{d\rho} = -\frac{\alpha}{H^3(0)} \int_0^1 \rho \, H(\rho) \, d\rho \tag{17}$$

and the right hand side is non-zero. Therefore the expansion (14) must be augmented by a term which will produce a contribution to $\rho \frac{dP}{d\rho}$ that does not vanish as $\rho \to 0$.

Accordingly we seek a further function $H(\rho) = \psi(\rho)$ such that

$$\int_0^1 \rho \, \sigma \, K_{22}(\rho, \sigma) \, \psi'(\sigma) \, d\sigma = \rho^2 \ln \rho, \tag{18}$$

for then $P(\rho) = -\frac{1}{\rho} \frac{\rho}{d\rho} (\rho^2 \ln \rho) \sim -2\ln \rho$, and $\rho \frac{dP}{d\rho} \to -2$ as $\rho \to 0$. (18) is a first kind integral equation for $\psi'(\sigma)$ which can be solved by use of the Wiener–Hopf technique (or with hindsight by direct substitution), giving

$$\psi'(\rho) = -\frac{4}{\pi} \left[\cos^{-1}\rho - (\ln 2 - \frac{1}{2}) \frac{\rho}{\sqrt{1-\rho^2}} \right]$$

whence on integration with $\psi(1) = 0$

$$\psi(\rho) = -\frac{4}{\pi} \left[\rho \cos^{-1}\rho - .80685 \sqrt{1-\rho^2} \right] . \tag{19}$$

2.3 Numerical procedure

We now write

$$H^N(\rho) = \sum_{m=0}^{N-2} A_m P_{2m+1}(\sqrt{1-\rho^2}) + B\psi(\rho) \tag{20}$$

so that the solution is characterised by an N-vector $\underset{\sim}{A}^{(N)} = (A_0 .. A_{N-2}, B)^T$ of coefficients, and choose $\underset{\sim}{A}^{(N)}$ so as to minimise the squared error sum

$$\sum_{i=1}^M \left[\text{LHS} - \text{RHS} \right]^2_{\rho_i} \tag{21}$$

of the left and right hand sides of (9) evaluated at M points ρ_i, $i = 1...M$, subject to the positivity constraint

$$H^N(\rho_i) > 0.$$

The computational details are similar to those used by Spence and Sharp [14] for the two-dimensional version of this problem, and will not be repeated here.

2.4. Stress viscosity near crack-tip

In general the cavity shape $H(\rho)$ behaves like $A\sqrt{(1-\rho^2)}$ near the tip $\rho = 1$, where from (20)

$$A = \sum_m A_m C_m - \left[\frac{4}{\pi} \right] (\ln 2 - \tfrac{1}{2})B, \qquad (22)$$

C_m being the coefficient of x in $P_{2m-1}(x)$, namely $C_m = (-1)^m (2m+1)!!/2^m m!$. This behaviour results in a square-root singularity in tensile stress in the plane $z = 0$ just outside the crack. The kernel $K_{22}(\rho,\sigma)$ can be written $\frac{\pi}{2\rho} \left\{ K\left[\frac{\sigma}{\rho}\right] - E\left[\frac{\sigma}{\rho}\right] \right\}$ and using standard results on the differentiation of elliptic integrals we find near $\rho = 1$

$$P(\rho) \sim - (A\pi) \int_0^1 \frac{\sigma}{\sqrt{(1-\sigma^2)}} \frac{d\sigma}{\rho^2 - \sigma^2} = - \frac{A\pi^2}{4\sqrt{(\rho^2-1)}} \qquad (\rho > 1) \qquad (23)$$

the negative sign indicating that the stress is tensile.

The singularity represents a disposable parameter in the solution, which must be chosen to match the known fracture strength of the matrix within which the crack is propagate. As discussed by Spence and Turcotte (1985), we have in most cases suppressed the singularity by ensuing that

$$A = 0 \qquad (24)$$

3. CAVITY CONTAINING BUOYANT FLUID

In this section the mathematical treatment of an idealised model of a magma plume will be given. Spence, Sharp and Turcotte [8] treated the case of a crack propagating vertically upwards and fed at a constant rate from a reservoir; but interest also attaches to the case of a cavity of width $2h(y,t)$ and constant volume extending from $y = 0$ to $y = 2\ell(t)$. As in the previous section, the slow viscous flow in the interior satisfies the equation of mass conservation

$$2 \frac{\partial h}{\partial t} + \frac{\partial Q}{\partial y} = 0. \qquad (25)$$

Q is the flux of fluid across a surface $y = $ constant and is given by Reynolds' equation

$$Q = - \frac{2}{3\eta} h^3 \left[\frac{\partial p_0}{\partial y} - g \, \Delta\rho \right] \qquad (26)$$

where $\eta = $ viscosity, $\Delta\rho = $ density difference between fluid and surrounding rock, $p_0 = $ deviatoric pressure given by

$$p_0(y, t) = - \frac{\varepsilon}{\pi} \int_0^{2\ell(t)} \frac{\partial h}{\partial z} \frac{dz}{z-y} . \qquad (27)$$

The integral is a Cauchy principal value, and ε as before is the shear modulus μ of the rock divided by $(1-\nu)$. The cavity volume per unit width is

$$V = \int_0^{2\ell} 2h(z, t)dz \tag{28}$$

and will be treated as constant in this section, although this restriction is not essential. If time is scaled with $T = \eta/\varepsilon$ and y with $L = \varepsilon/g\Delta\rho$ the constants disappear from the equations leaving

$$2\frac{\partial h}{\partial t} + \frac{\partial Q}{\partial y} = 0, \quad Q = -\frac{2}{3}h^3\left[\frac{\partial p}{\partial y} - 1\right], \quad p = -(1/\pi)\int_0^{2\ell}\frac{\partial h}{\partial z}\frac{\partial z}{z-y} \tag{29}$$

To treat these numerically, we introduce a trigonometrical variable by means of $y = \ell(1 + \cos\theta)$ and seek a solution in which

$$h = (1/\ell)\sum_1^N A_n \sin n\theta \tag{30}$$

Then when the Cauchy integral for p is evaluated using a standard result we obtain

$$p = \left(\frac{1}{\ell^2}\right)\sum_1^N nA_n\frac{\sin n\theta}{n \sin\theta} \tag{31}$$

and

$$\frac{\partial p}{\partial y} = -(1/\ell \sin\theta)^3 \sum nA_n (n \cos n\theta \sin\theta - \sin n\theta \cos\theta) \equiv P(\theta) \text{ say}$$

Therefore

$$Q = -(2/3)h^3(P(\theta) - 1) \tag{32}$$

is known in terms of the $\{A_n\}$ as a function of θ. By use of the Fast Fourier Transform we are able to express Q in the form

$$Q = \sum E_n \sin n\theta, \tag{33}$$

where E_n depends on $\underset{\sim}{A}$.

The mass conservation equation then leads to a set of linear equations for the time derivatives $\{\dot{A}_n\}$ of the form

$$L \underset{\sim}{\dot{A}} - \frac{1}{2}\left[\frac{\dot{\ell}}{\ell}\right] M \underset{\sim}{A} + \underset{\sim}{E} = 0 \tag{34}$$

where L and M are simple $(N-1)\times(N-1)$ banded matrices. The volume of the cavity is $V = \pi A_1$ and may be assumed known, in which case the remaining unknowns are $\dot{A}_2...\dot{A}_N$ and $\dot{\ell}/\ell$, and to close the problem it is necessary to prescribe one further linear relationship among them. This is provided by the stress intensity factor at $y = 2\ell$, which as in the last section depends on the curvature at this point, namely

$$K = A_1 + 2A_2 + ...NA_N \tag{35}$$

If K is determined by the fracture mechanics of the host rock, then $\dot{K}$ is known, and we have N-1 equations for N-1 unknowns from which the evolution of

the cavity can be studied by time-stepping.   An example of the evolution of
an elliptic cavity of constant vertical extent is given in Figure 2.

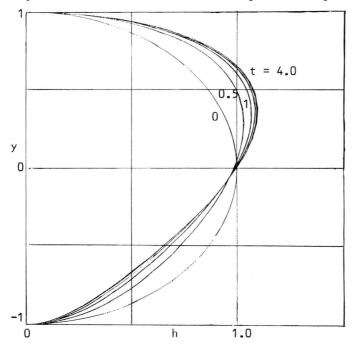

FIGURE 2:   Development of h(y,t) for elliptic cavity of fixed height.

4.   BUOYANCY-DOMINATED FLOWS

As remarked in the introduction, a possible approximation for the upward
spread of a cavity dominated by buoyancy is to disregard the
deviatoric-pressure term $\dfrac{\partial p_0}{\partial y}$ in comparison with the buoyancy term $g\Delta p$ in
(26). If time is then rescaled with $g\Delta p/\eta$, the mass conservation equation
(29) reduces to

$$\frac{\partial h}{\partial t} + \frac{\partial}{\partial y}\left[\frac{1}{3}h^3\right] = 0 \tag{36}$$

or

$$\frac{\partial h}{\partial t} + h^2\frac{\partial h}{\partial y} = 0 \tag{37}$$

This is a form of the kinematic wave equation.   The solution h(y,t) can be
found in terms of an initial profile

$$h(y,0) = f(y) \tag{38}$$

by using characteristics as

$$h = f(y-th^2) \tag{39}$$

For example, for a cavity that is initially elliptical

$$f(y) = (1-y^2)^{\frac{1}{2}}$$

and we find

$$h = \left\{(1+4ty+4t^2)^{\frac{1}{2}} - (1-2ty)\right\}^{\frac{1}{2}} 2^{-\frac{1}{2}}t^{-1}$$

As time increases from 0 to ½ the cavity lies entirely on the interval
-1 < y < 1, the fluid migrating towards the upper end so that it
progressively becomes pear-shaped. At t = ½ a point of inflexion develops at
y = 1 and a shock forms. This then begins to migrate, at a rate determined
by the condition that the volume within the cavity remains constant. This
condition can be written

$$0 = \frac{d}{dt} \int_0^{y_s(t)} h \, dy = h_s \frac{dy_s}{dt} + \int_0^{y_s} \frac{\partial h}{\partial t} \, dy$$

and the last term, by (36), equals $-\frac{1}{3} h_s{}^3(t)$. Here the suffix s denotes
quantities at the shock front. Therefore the migration of the cavity is
given by

$$\frac{dy_s}{dt} = \frac{1}{3} h_s{}^2 \tag{40}$$

This is precisely the velocity at which a discontinuity in a weak solution of
(37) can propagate when the diffusion term represented by the pressure
gradient is of the form that occurs in (26). The solution for y_s as a
function of t is most easily found by writing h = sinθ. Then y = cosθ +
tsin²θ and the volume is

$$2 \int_{-1}^{y_s} h \, dy = \pi - (\theta_s - \tfrac{1}{2}\sin 2\theta_s) + \frac{4}{3}t \sin^3\theta_s = \pi,$$

π being the constant volume. From this equation θ_s is found as a function of

t: for t ≫ 1, $\theta_s \sim \pi - \left(\frac{3\pi}{4t}\right)^{1/3}$ and $y_s \sim \left(\frac{3\pi}{4}\right)^{2/3} t^{1/3}$. Curves illustrating

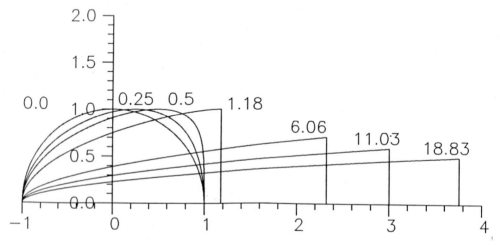

FIGURE 3:   Development of elliptic profile from kinematic wave equation.
            Shock first forms at t = 0.5.

the development of such a cavity are shown in Figure 3. Figure 4 illustrates
the case when the initial profile is h(y,0) = 1-y² so that the ends are
pointed. In this case a shock first forms at t = .65 at an internal point,
and reaches y = 1 when t = 1, subsequently advancing in a similar manner to
that in the first case.

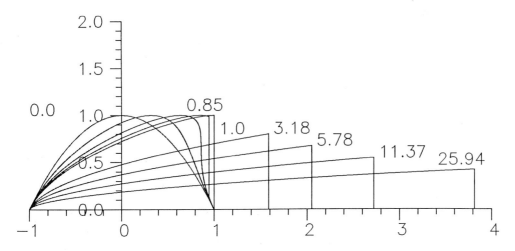

FIGURE 4:   Development of biconvex cavity.   In this case a shock first forms
at $y = .866$, $t = .6495$ but the upper limit of cavity remains
at $y = 1$ up to $t = 1$.

Asymptotically the rate of propagation and the size of the shock jump depend
only on the initial volume and the solution (39) shows that $y = th^2 + g(h)$,
$(g(h) = f^{-1}(h))$ so the volume is

$$V = \frac{1}{3} th_s^3 + \int_0^{h_s} hg'(h)dh = \frac{1}{3} th_s^3 + O(h_s^2)$$

and for $t \gg 1$, reverting to dimensional quantities, the asymptotic
expressions are

$$h_s \sim \left[\frac{\eta}{g\Delta\rho} \frac{3V}{4t} \right]^{1/3} , \qquad y_s \sim \left[\frac{3V}{4} \right]^{2/3} \left[\frac{g\Delta\rho t}{\eta} \right]^{1/3} \qquad (41)$$

giving rise to shapes such as those of Figure 5.   These solutions exhibit
unsteady time-dependent behaviour at constant volume. Without the pressure
term in (26) it seems impossible to reproduce a steady rising plume of magma.

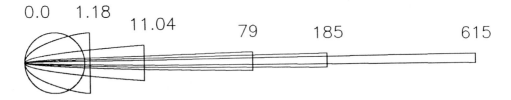

FIGURE 5:   Asymptotic development of constant-volume cavity (equations (41))
Schematic.

However at long times the pressure term is of order $t^{-1/2}$ in comparison with
the buoyancy term, so the buoyancy is the dominant term, and the
approximation of neglecting the deviatoric pressure is justified except at
the shock front.   At the shock front the gradient $\partial h/\partial y$ is infinite and the
pressure term cannot be neglected.   But if we introduce a new variable

$$x = \frac{y_s - y}{h_s^{\frac{1}{2}}} \tag{42}$$

to measure distance from the shock front scaled with the (jump in h)$^{\frac{1}{2}}$, and write

$$h = h_s(t)H(x) \tag{43}$$

in the full equation (26), including the deviatoric pressure term, then

$$p(y,t) = h_s^{\frac{1}{2}}(t)P(x), \qquad P(x) = -\frac{1}{\pi}\int_0^\infty \frac{H'(s)ds}{s-x} \tag{44}$$

and H satisfies the integro-differential equation

$$\frac{1}{\pi}\frac{d}{dx}\int_0^\infty \frac{H'(s)ds}{s-x} = 1 - 1/H^2(x) \ . \tag{45}$$

This is exactly the equation that was solved numerically in [8]. It shows that the shock layer has a universal structure, in which pressure plays a significant role, and the ratio

$$\frac{\text{Shock layer thickness}}{\text{Distance travelled}} = O(t^{-\frac{1}{2}})$$

The solution in [8] showed that for points just outside the layer

$$P(x) \sim -\frac{\lambda}{(-x)^{\frac{1}{2}}} \ , \qquad \lambda = 1.308$$

From this we deduce that the physical pressure has a singularity of the form

$$p \sim -K/[2(y-y_s(t))]^{\frac{1}{2}}$$

where

$$K = (\lambda \sqrt{2}) \left[\frac{\mu}{1-\nu}\right]^{3/4} \left[\frac{3V\eta}{4t}\right]^{1/4} \tag{46}$$

so the stress-intensity factor decreases like $t^{-1/4}$, and might ultimately drop to a level at which the crack could no longer propagate in the surrounding medium. We now have a picture of an unsteady solution with a narrow front in which the thickness rapidly increases, so that $H(x) = 2\lambda \sqrt{x}$ at the tip, forming a bulge followed by a long tail in which the volume is conserved, the rate of propagation decreasing like $t^{-2/3}$.

As an example to illustrate the crack growth and stress intensity from the formulae (41), (46) we take $V = 100m^2$, $\eta = 3$ Pa s, $\mu = 2 \times 10^{10}$Pa, $\nu = .25$, $\Delta\rho = 300$kg m^{-3} and obtain the following values

t (secs)	h_s (m)	y_s (m)	$\frac{dy_s}{dt}$ (ms^{-1})	K (MNm$^{-3/2}$)
1	.422	178	59.3	472
10	.196	383	12.8	266
10^2	.091	825	2.75	149
10^3	.042	1778	.59	84
10^4	.020	3832	.13	47
10^5	.009	8255	.03	27
10^6	.004	17784	.006	15

The flow rate per unit width at the crack tip is V/2t.

Laboratory values of K for granites and basalts are of order 1-3MNm$^{-3/2}$, but the values at depth are likely to be higher: however the solution suggests that the strength of surrounding rock would not limit the rise of such an inclusion at the times considered.

The use of lubrication theory assumes that inertia times are small compared with viscous terms: the ratio is of order $y_s/4t^2$ and is certainly small for t > 100s in the example quoted, although the rounded shape of the wave head raises other difficulties of approximation and these may affect the validity of the calculated stress singularity.

ACKNOWLEDGEMENT

The author is grateful for support under grant 0524/87 of the NATO Collaborative Research Grants Programme. Astrid Holstad and Panaghis Vergottis did the calculations for Figures 2-5.

REFERENCES

[1] Shaw, H.R. The fracture mechanisms of magma transport from the mantle to the surface. In Physics of Magmatic Processes, R.B. Hargraves (Ed.), (Princeton University Press, 1980) pp.201-264.
[2] Pollard, D.D., Geophys. Res. Lett., 3 (1973) 513-516.
[3] Delaney, P.T. and Pollard, D.D., American Journal of Science, V282 (1982) 856-885.
[4] Pollard, D.D., Elementary fracture mechanics applied to the structural interpretation of dykes: In Mafic dyke swarms, Halls, H.C. & Fahrig W.F. (Eds.) Geol. Soc. of Canada, Special paper 34 (1987) pp.5-24.
[5] Weertman, J., Journal of Geophysical Research, V.76 (1971a) 1171-1183.
[6] Weertman, J., Journal of Geophysical Research, V.76 (1971b) 8544-8553.
[7] Stevenson, D.J., Lunar and Planetary Sciences 13th (1982) 768-769.
[8] Spence, D.A., Sharp, P.W. and Turcotte, D.L. J. Fluid Mech., v174 (1987) 135-153.
[9] Rubin, A.M. and Pollard, D.D., Origins of blade-like dykes in volcanic rift zones: In Decker, R.W. and Wright, T. (Eds.) Volcanism in Hawaii, U.S. Geological Survey Professional Paper 1350 Ch. 53 (1987) pp.1449-1470.
[10] Sobolev, S.V. Cyclic model of deep fissure magmatism. Intl. Union of Geodesy and Geophysics, XIX General Assembly, Vancouver, (August 1987) Abstracts V2.

[11] Nemat-Nasser, S., Abé, H. and Hirakawa, S. (Eds.) Hydraulic fracturing and geothermal energy. Proceedings of first Japan-US Joint Seminar. November 1982. (Martinus Nijhoff 1983, The Hague).

[12] Geertsma, J. and Haafkens, R. J. Energy Res. Technol. 101 (1979) 8-19.

[13] Advani, S.H., Lee, J.K. and Khattab, H., Fluid flow and associated response modelling of hydraulic stimulation processes, In Finite Elements in Fluids, Vol. 5 (Gallagher, R.H., Oden, J.T., Zienhiewicz, O.C., Kawai, T. and Kawahara, M., (Eds.) (John Wiley, 1984).

[14] Spence, D.A., and Sharp, P. Proc. R. Soc. Lond. A400 (1985) 289-313.

[15] Spence, D.A. and Turcotte, D.L., Journal of Geophysical Research, v90 (1985) 575-580.

[16] Emerman, S.H., Turcotte, D.L. and Spence, D.A., Transport of Magma and Hydrothermal Solutions by Laminar and Turbulent Fluid Fracture: Physics of the Earth and Planetary Interiors, v41 (1986) 249-259.

[17] Turcotte, D.L. Emerman, S.H. and Spence, D.A., Mechanics of dyke injection, In Mafic dyke swarms, Halls, H.C. and Fahrig, W.F. (Eds.) Geological Association of Canada Special Paper 34 (1987) pp.25-29.

[18] Popov, G. In Prikl. Math. Mekh., 27(5) (1963).

[19] Popov, G. In Prikl. Math. Mekh., 28(3) (1964).

[20] Barenblatt, G.I., Advances in Applied Mechanics, v7 (1962) 55-129.

[21] Anderson, O.L., The role of fracture dynamics in Kimberlite pipe formation, In Kimberlites, Diatremes and Diamonds: Their Geology, Petrology and Geochemistry Boyd, F.R. and Meyer, H.O.A. (Eds.) Vol. 1, (American Geophysical Union, 1979) pp.344-353.

[22] Pasteris, J.D., Ann. Rev. Earth Planet. Sci., 12 (1984) 133-153.

[23] McGetchin, T.R. and Ulrich, G.W., J. Geophys. Res., 78(11) (1973) 1833-1853.

[24] Erdelyi, A. (Ed), Magnus, W., Oberhettinger, F., Tricomi, F.G., Higher Transcendental Function, v.II, New York, McGraw-Hill (1953).

Theoretical and Applied Mechanics
P. Germain, M. Piau and D. Caillerie (Editors)
Elsevier Science Publishers B.V. (North-Holland)
© IUTAM, 1989

NONLINEAR ASPECTS OF DELAMINATION IN STRUCTURAL MEMBERS

Bertil Storåkers

Department of Strength of Materials and Solid Mechanics
The Royal Institute of Technology
S-100 44 Stockholm, Sweden

Interlayer crack propagation or delamination may cause considerable degradation of stiffness and strength of composite structures. The phenomenon has already received appreciable attention within the realms of linear fracture mechanics, in particular regarding free-edge effects, but in nonlinear situations, be they of kinematic or constitutive origin, progress is more recent. Some methods and results are reviewed as regards nonlinear delamination events induced mainly by buckling or peeling. Structural members such as beams, plates and shells are given a unified treatment with illustrations aimed at providing insight into stable and unstable crack growth. Special attention is given to energy theorems in various formulations and their power as a basis for computational determination of energy release rates is discussed. Experimental techniques and observations, as regards crack growth, are discussed especially with respect to self-similarity. Some particular issues of constitutive and environmental origin are briefly commented upon and certain matters warranting further investigation are identified.

1.   INTRODUCTION

The introduction of layered composites or laminates in design has markedly increased the potentiality to tailor structural elements for special purposes and more efficient use. To serve as intended, in particular with regard to safety, it is essential that laminates are free from defects or at least that the presence of defects, occasionally invisible, may be detected and the consequences analysed. The manufacture of composites requires involved procedures and there are many potential sources for defects to occur in the finished product, [1]. Interlayer cracks or delaminations may also result from events during service life such as low velocity impacts with particular implications for aerospace applications, [10]. Related aspects have received considerable attention and delamination problems are prominent in the many and frequent symposia arranged by the American Society for Testing and Materials, [2] to [7], of which one, [6], was entirely devoted to the discipline. Likewise numerous contributions to major conferences, [8], deal with related matters and their significance has recently been emphasized in a comprehensive survey by Garg [9].

In linear situations delamination crack behaviour has in the past mostly been analysed with three-dimensional fracture mechanics as a basis, [11]. A situation, frequently met in practice and of fundamental interest in the current context, is when the presence of delaminations in compressed structural members might promote buckling combined with crack growth. For prediction and analysis of such events, methods based upon stability theory and, if proper, fracture mechanics must be combined. Such nonlinear problems have been addressed mainly during the latest decade. It is the present intention to summarize some of the progress made mainly based on beam, plate and shell theory together with associated experimental techniques and attempt to provide a common ground for some salient features. The approach is mechanistic and consciously biased towards analytical modelling and methods together with computational and experimental techniques. No historical account will be given and a complete survey, soon to be out-dated, is not attempted; reference will be given only to some earlier contributions representative of the context. The topics to be dealt with have earlier been illuminated in a general fashion by Kachanov [12], Bolotin et al. [13] and more recently by Garg [9] and Yin [14].

2.   BUCKLING AND GROWTH OF A SURFACE LAYER DELAMINATION

Impact on laminates often gives rise to delamination. In Fig. 1 is shown, schematically after Dorey [15], cross–sections of damaged panels the delaminations resulting from drop–weight tests on carbon fibre plastics.

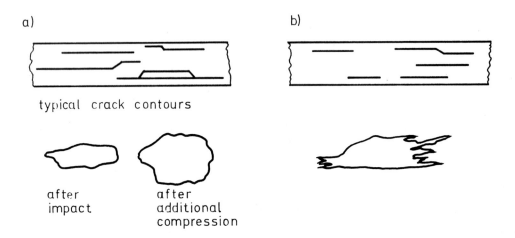

FIGURE 1
Delamination damage after impact in a) carbon fibre/epoxy, b) carbon fibre/PEEK.

It is obvious by a mere glance at the crack patterns illustrated in Fig. 1 that determination of local stress fields even under simple loading and linear behaviour poses a formidable computational problem. When nonlinear kinematics becomes significant it seems natural to initially adopt a two–dimensional formulation and model the mechanical behaviour by aid of plate theory, or when relevant, shell theory, and as a consequence confining the analysis to flaws with characteristic lengths of greater order than a typical thickness.

Before discussing deformation and eventual crack growth in a general situation, it seems advisable to first consider a simple illustration that contains some but by far not all features to be discussed. To this end the combined buckling and delamination growth of a thin surface layer under compression is singled out for attention. This case in various formulations has attracted the interest of many writers, Chai et al. [16] being pioneering ones for a rectilinear case. Apart from a methodological point of view the problem is of interest per se as related to the adherence of preventive surface coatings, [17], to promote e.g. corrosion and wear resistance. The matter is also actual in electronic devices.

A delaminated strip, of assumed infinite width, is depicted in Fig. 2 and is part of a layer attached to a half–space subjected remotely to uniaxial compression resulting in a longitudinal strain ϵ_0. If for simplicity and following Chai et al. [16] homogeneity and isotropy is assumed such that the layer and the half–space are Hookean, (E, ν), the strip will buckle at a strain

$$\epsilon_b = \frac{\pi^2}{3(1-\nu^2)} \left(\frac{t}{\ell}\right)^2 \tag{2.1}$$

by ordinary plate theory.

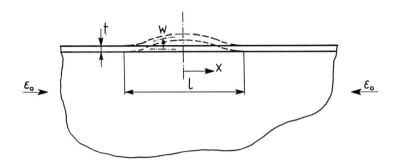

FIGURE 2
Delaminated strip under compression.

In the postbuckling range under increasing ϵ_0, to first order the strip will undergo a deflection

$$w = \frac{w_0}{2}\left(1 + \cos\frac{2\pi x}{\ell}\right) \qquad (2.2)$$

and membrane strain

$$\epsilon_m = (1-\nu^2)\epsilon_b + \nu^2\epsilon_0 \qquad (2.3)$$

under constant longitudinal membrane force.

The strain energy per unit length in the strip may then be expressed as

$$W = E(1-\nu^2)\,t\ell\left(2\epsilon_0\epsilon_b - \epsilon_b^2 + \frac{\nu^2}{1-\nu^2}\epsilon_0^2\right). \qquad (2.4)$$

Should delamination progress, the resulting energy release rate in the system is then

$$G = -\frac{\partial W}{\partial \ell} + E\epsilon_0^2 t/2 \qquad (2.5)$$

where the last term, perhaps introduced somewhat ad hoc, originates from the strip surroundings. If a simple crack growth criterion $G = G_c$ is adopted then by (2.1), (2.4) and (2.5) the continued relation between remote strain, now denoted ϵ_c, and crack length, ℓ, is given by

$$(\epsilon_c - \epsilon_b)(\epsilon_c + 3\epsilon_b) = \frac{2G_c}{Et(1-\nu^2)}. \qquad (2.6)$$

The sequence of events conceivable is illustrated in Fig. 3, which is based on parameter values $G_c/(Et) = 10^{-5}$, $\nu = 0.3$, thought to be representative of situations met in practice. Thus for a slender strip with $\ell > \ell_s$, the path commences at 1 and buckling occurs at 2. Under increasing external load further delamination occurs at 3 progressing in a (weakly) stable manner. On the contrary when $\ell < \ell_s$ progressing delamination will be unstable until possible arrest. At very stubby strips events are catastrophic since no arrest is predicted.

By (2.6) it is readily found that the transition strain is $\epsilon_s = 3\epsilon_b$ and further by (2.1) that

$$\frac{\ell_s}{t} = \pi\left[\frac{2}{3}\frac{Et}{G_c(1-\nu^2)}\right]^{\frac{1}{4}}. \qquad (2.7)$$

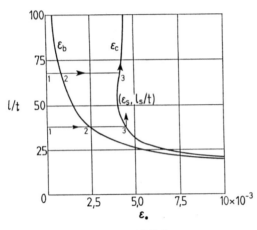

FIGURE 3

Buckling (ϵ_b) and crack growth (ϵ_c)
characteristics for a surface strip.

It is evident then that whenever $\ell > \ell_s$ initially, the remote compressive strain required for crack growth to occur will always exceed the buckling value by a factor of three and delamination will progress in a stable manner at least within the model assumptions. For shorter delaminations the strain difference between buckling and growth will diminish as has been found also for the case of a circular disbond, [18].

As regards method of approach, in a pioneering investigation Kachanov [12] set out to determine final delaminated states for some particular systems by in principle seeking the minimum of the potential energy complemented with a term accounting for the fracture energy. In this term, $-G_cS_c$, the total cracked area, S_c, constitutes an additional variable to be determined. Though somewhat questionable when irreversibility aspects are taken into account, for the case just discussed certainly Kachanov's procedure may be utilized to generate identical governing equations. From a fundamental point of view it must be concluded though that not even at simple crack geometries is a unique solution to be expected and in particular at irregular contours an incremental procedure seems preferable. Kachanov's philosophy has, however been pursued by Bottega [19] for truly two–dimensional situations though augmented by heuristic arguing to cover also local crack growth.

The dominant features of the strip problem, as illustrated in Fig. 3, will prevail also in the corresponding axisymmetric case, with a circular crack contour and balanced biaxial compression, as may be concluded from results given by Yin [18]. This is not self–evident though as the postbuckling behaviour of a circular delamination is qualitatively different from that of a rectilinear strip. For one thing, as emphasized by Yin, full nonlinearity must be retained in the postbuckling equations in order to arrive at reliable results.

Analytical simplicity is lost once a finite thickness of the substrate is introduced in the so–called thin film problem as merely to determine the buckling strain requires solution of an eigenvalue problem of several degrees of freedom. Chai et al. [16] did analyse several particular cases and tentatively it may be concluded that the relative thickness of the film should be less than 5% or so for the model to apply well.

3.   DELAMINATION ANALYSES FOR VARIOUS COMPOSITE BEAMS, PLATES AND SHELLS

Having achieved some insight from the prototype thin film problem, it is to be anticipated that analyses of delaminated structural members having several characteristic finite dimensions might be technically quite involved. However, it is clearly of technical importance though to determine stiffness and strength degradation. To this end several contributions have been made with the aim of determining buckling characteristics of various delaminated elements and also in some cases the interaction with crack propagation. In the majority of cases, however, attention has been confined to crack contours allowing a one–parameter description such as rectilinear and circular ones.

Already Chai et.al. [16] dealt with some cases of plates of finite thickness and length. More detailed analyses of delaminated plates with rectilinear contours, as in Fig. 4a, have been carried out by Simitses, Yin and Sallam, [20] to [24]. In spite of the simple geometry studied, the first step in the analysis of combined postbuckling and delamination growth, that is determination of the load to initiate buckling, requires solution of an eigenvalue problem involving 24 degrees of freedom if no symmetry is present.

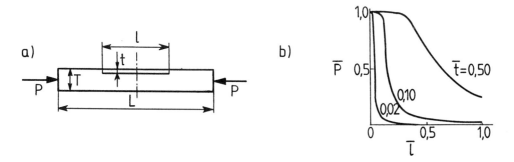

FIGURE 4

a) Plate section with a rectilinear delamination b) reduction of buckling load as function of delamination length $\bar{l} = \ell/L$ for various thicknesses $\bar{t} = t/T$.

The reduction of buckling load due to a symmetrically (lengthwise) situated delamination as found by Simitses et al. [20] for clamped conditions is reproduced in Fig. 4b. For relatively short and thin delaminations, $\bar{l} < \bar{t} < 0.2$ (say), the critical load is not very much affected and buckling is essentially global. The same circumstances prevail for simple support though rather when $\bar{l}/2 < \bar{t} < 0.2$. Thin film analysis was found to be fairly well applicable at least when $\bar{t} < 0.2$ and $\bar{t} < \bar{l}$ for clamped supports and $\bar{t} < \bar{l}/2$ for simple ones. Among further findings Simitses et al. concluded that a lengthwise symmetrically situated delamination causes the highest reduction of buckling load when $\bar{t} < \bar{l}$ but not necessarily so otherwise.

However, the initiation of buckling in a delaminated member by no means necessarily implies that the overall load carrying capacity is exhausted. Instead to predict further events a postbuckling analysis with due regard to the possibility of progressing delamination is required. In this spirit further studies of transversely delaminated plates under uniaxial compression have been carried out by Yin et al. [22].

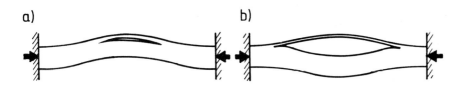

FIGURE 5

Crack openings modes for short and long delaminations.

Fig. 5 illustrates the qualitative postbuckling behaviour of a wide column containing a stubby and a slender delamination respectively. In the first case the overall behaviour is not very much affected. The delamination will merely trigger buckling at a load slightly below that of a perfect structure. Thus from a design point of view this case is perhaps of less concern although Yin et al. [22] point out some interesting features in the postbuckling range where a peak load may be attained and the two detached segments might resume contact.

For a slender delamination, as in Fig. 5b, postbuckling is triggered essentially by localized buckling of the thinner segment at a load substantially reduced with respect to that of a perfect column. Initially in the postbuckling range the configuration is qualitatively similar to that of a short crack, Fig. 5a, but a transition state is arrived at implying a reversal of sign of curvature of the thicker segment Fig. 5b. Again in this case a load maximum may occur and the crack start to close partially. An additional unknown boundary is then to be determined as part of an already technically complex analysis.

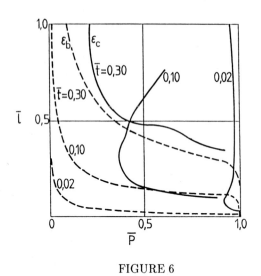

FIGURE 6

Buckling and crack growth characteristics
for a delaminated plate.

As was clear from the introductory thin film problem, delamination may progress in different ways. For the rectilinear plates studied by Yin et al. [22] it was predicted that at relatively small values of the critical energy release rate, $\bar{G} = G_c \ell^4 / Et^5 < 10^{-3}$ say, crack propagation will always be catastrophic and essentially so also for large values, $\bar{G} > 10^{-1}$ say. In Fig. 6, constructed from results given in [20], [22] and valid for an intermediate value, $\bar{G} = 10^{-2}$, delamination length is shown as a function of external loading for three values of the delamination thickness. The buckling and crack growth curves should be interpreted in the same spirit as those of Fig. 3 associated with the thin film case although in Fig. 6 the external loading is normalized with respect to the buckling load of an undamaged member. It is evident that conceivable events depend in a quite complex manner on the characteristic parameters involved and as a consequence results should be used with due consideration.

All matters discussed so far have been based upon assumed homogeneity and at least in—plane isotropy. The effect of unsymmetric cross–ply has, however, been studied by Sallam and Simitses [21] and when analysing cylindrical buckling of laminates Yin [23], [24] discusses how isotropy solutions for delaminations may be drawn upon to generate solutions for bending–stretching coupling. Qualitatively the results are similar for the two cases though quantitatively the coupling effect might drastically reduce critical loads.

The case of a circular crack contour has been dealt with for axisymmetric situations by among others Kachanov [12], Bottega and Maewal [25] and Evans and Hutchinson [26] based on first order postbuckling theory. The load required for initiation of buckling of a circular plate with a concentric delamination has been determined by Yin and Fei [27] for a range of geometric parameters. The results proved to be very similar to those for the rectilinear case, as in Fig. 4b, and Yin and Fei contemplate whether buckling essentially may be governed by a particular slenderness parameter also for contours of irregular shapes.

For a delaminated segment of relative thickness 0.1 Yin and Fei [28] have also analysed postbuckling behaviour and crack growth. Although the phenomena involved resemble those for a rectilinear delamination, there are qualitative and quantitative differences due to dissimilar postbuckling characteristics.

Some major contributions to the understanding of the behaviour of compressed delaminated panels have been given by Whitcomb together with Shivakumar [29] to [35]. For one thing Whitcomb [32] has analysed the individual contributions to the energy release rate from different modes of crack propagation and also investigated the effects of geometric imperfections and thermal strains.

Within the fracture mechanics philosophy adopted so far and originating from Griffith, it has been tacitly assumed that the entire energy release rate governs progressing delamination. When partitioning into individual contributions from crack growth modes of opening, G_I, shearing, G_{II}, and tearing, G_{III}, then within an energy balance concept necessarily

$$G_I + G_{II} + G_{III} = G_c \tag{3.1}$$

would govern mixed–mode crack growth.

There is ample evidence, [36], however, that critical energy release rates in pure Mode II are higher than those of Mode I and on occasion substantially so. The effect has been found even more pronounced, [37], when it comes to growth in pure Mode III. Such features may not be accommodated by (3.1) and for mixed–mode crack growth alternative criteria of the general form

$$f(G_I, G_{II}, G_{III}) = 0 \qquad\qquad (3.2)$$

have been proposed.

For compressed panels of essentially the same geometry as depicted in Fig. 4, though including four different laminate types, Whitcomb [32] determined, by aid of a finite element technique, the separate contributions of G_I and G_{II} to the energy release rate in the post-buckling range. As a notable feature Whitcomb found that the ratio G_{II}/G_I varied considerably, G_I dominating in the initial stages but then diminishing while G_{II} was monotonically increasing. Three different growth criteria of the form (3.2) were adopted and leaving out details the resulting predictions turned out to be quite disparate. The influence of different mixed–mode criteria has further been elaborated upon by Donaldson [38] for Whitcomb's model. Thus when adopting a modified fracture surface morphology based criterion, [39], and some phenomenological ones, Donaldson found that for cases when predictions based on different criteria vary, delamination growth may be self–arresting.

Whitcomb [32] further brought out some relevant features as regards the role of geometric imperfections in the present context. Thus an initial deflection of a delaminated segment may grow monotonically under compression and progressing delamination may very well occur before the theoretical buckling load is attained. It is to be expected that imperfection effects will be of particular importance for shorter delaminations corresponding to the lower parts of the delamination curves in Figs 3 and 6 above. This is confirmed by Whitcomb's results, which further show that imperfection effects are almost exclusively confined to opening crack modes. Also thermal mismatch is included in Whitcomb's analysis by way of a superposed thermal load. When this effect is strong enough to cause thermal buckling the consequences might be particularly severe.

Appealing as it may be to the analyst, a treatment of a through–width delamination does not generate many results applicable to more general situations. Whitcomb and Shivakumar [33] have, however, dealt with postbuckling of square and rectangular delaminations. It was found then at uniaxial compression that the strain energy release rate exhibited a substantial variation along the delamination contour and as a natural consequence that self–similar growth was not to be expected. In case of isotropy and a square contour the predicted direction for initiation of growth was transverse to the loading direction while for rectangular contours the matter becomes ambigous.

For embedded circular and elliptic delaminations Whitcomb [34], [35] found large gradients of the energy release rate along the contours under nominal uniaxial compression. The implications are that transverse growth is to be expected at least for smaller aspect ratios in case of elliptic delaminations. The results were generated by a 3D finite element technique allowing for mode partitioning and as emphasized by Whitcomb [35], the location of initiation of growth will in general depend on the particular growth criterion adopted. Whitcomb also did encounter crack closure in his analysis but the implications were left for future study.

Although only flat panels have been considered so far difficulties have arisen both of a fundamental and a technical nature. They will still increase when attention will be directed to curved members with more complex geometry. Remembering the pains involved already in buckling analysis of undamaged composite shells, as recently discussed by Zhou [40], it is perhaps no surprise that related aspects for delaminated shells have been treated very sparsely. Some preliminary insight has been gained, however, by linear buckling analyses of cylindrical shells, [41], [42], and panels, [43], [44], containing longitudinal delaminations.

As was mentioned above delaminations, especially due to impact, seldom come singly. Still though buckling of members containing multiple delaminations has received little attention due to its complexity. Bolotin et al. [13] have analysed the case of a column with n equidistant delaminations of equal length as illustrated in Fig. 7a. Determination of the buckling load in general requires the solution of an eigenvalue problem of 6(n+2) degrees of freedom and as concluded by Bolotin et al. the eigenmodes to be expected are fundamentally different from the case of a single delamination as illustrated in Fig. 5. In particular this causes difficulties in efforts to construct some equivalent stiffness for the delaminated part of the column.

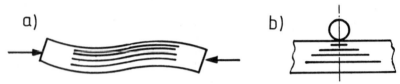

FIGURE 7
Rectilinear and circular multiple delaminations.

Other analyses of multiply delaminated columns, [16], [45], have dealt with cases when one delaminated ply is thick enough to resist significant bending. A conceivable delamination pattern due to impact, as sketched in Fig. 7b, is still open for exploration.

In all cases discussed this far it has been assumed that delamination is present already when external loading is applied to a structural member. In practice the sequence of events might be reversed, e.g. at impact during service, and either the structure may sustain its capacity to carry load or catastrophic failure may occur as recorded by Williams and Rhodes [46]. Some implications have been discussed qualitatively by Chai and Babcock [47]. Likewise the role of residual stresses, due to fabrication or other reasons, has not been much elucidated although some analyses exist for idealized situations [17], [26], [48], [49].

4. THE ENERGY RELEASE RATE AT GENERAL DELAMINATION GROWTH

Crack contours discussed so far have mostly been of simple shape and the standard methods used to determine energy release rates are not applicable in general situations. Instead an approach to energy balance corresponding to well established ones in 3D fracture mechanics, [50], but cast in a 2D nonlinear formulation is warranted. This matter has recently been dealt with by Storåkers and Andersson [51]. Following these writers and to gain in simplicity without fundamental loss of generality, a flat composite plate containing a single interlayer crack of arbitrary contour, Fig. 8, will be considered first.

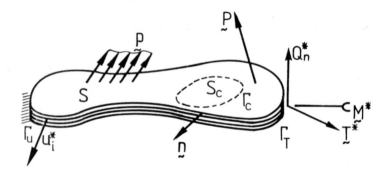

FIGURE 8
Composite plate with one delamination.

With the crack present the system may be decomposed into three plates each one characterized by individually homogeneous, though otherwise arbitrary, strain energy functions

$$W = W(e_{\alpha\beta}, \kappa_{\alpha\beta}) \tag{4.1}$$

where $e_{\alpha\beta}$, $\kappa_{\alpha\beta}$ are middle surface strains and curvatures respectively in obvious index notation, $\alpha = 1,2$.

For the present purpose it suffices to adopt the von Karman approximation

$$e_{\alpha\beta} = \tfrac{1}{2}(u_{\alpha,\beta} + u_{\beta,\alpha} + u_{3,\alpha} u_{3,\beta}) \tag{4.2}$$

and

$$\kappa_{\alpha\beta} = -u_{3,\alpha\beta} \tag{4.3}$$

referred to in-plane coordinates x_α.

If displacement gradients are adopted as primary stretching variables, then by (4.1), (4.2) the conjugate, nominal, membrane forces are

$$S_{\alpha\beta} = \frac{\partial W}{\partial e_{\alpha\beta}} \quad, \tag{4.4}$$

$$S_{\alpha 3} = \frac{\partial W}{\partial e_{\alpha\beta}} u_{3,\beta} \tag{4.5}$$

together with moments

$$M_{\alpha\beta} = \frac{\partial W}{\partial \kappa_{\alpha\beta}} \; . \tag{4.6}$$

Whenever (4.4) and (4.6) are invertible (4.5) is too, save for pathological situations, and as a consequence by a Legendre transformation a truly complementary strain energy function

$$W^c = W^c (S_{\alpha i}, M_{\alpha\beta}) \tag{4.7}$$

based on nominal variables may be introduced, $i = 1,2,3$.

The advantage of this somewhat unorthodox formulation, with the view of determining energy release rates, is that variation principles and also bounds for the potential energy may be constructed. This was evidently first observed by Stumpf [52], [53] for Hookean plates, the generalization to arbitrary strain energy functions, however, being immediate.

Thus variation of the potential

$$U = \int (W - p_i u_i) dS - \int (T_\alpha^* u_\alpha + Q_n^* u_3 + \epsilon_{\alpha\beta} M_\alpha^* u_{3,\beta}) d\Gamma_T , \tag{4.8}$$

$\epsilon_{\alpha\beta}$ being the permutation tensor, generates in the notation of Fig. 8 the equilibrium equations

$$S_{\alpha\beta,\alpha} + p_\beta = 0 \;\bigg]$$
$$M_{\alpha\beta,\alpha\beta} + S_{\alpha3,\alpha} + p_3 = 0 \;\bigg]$$

(4.9)

and the boundary conditions

$$n_\alpha S_{\alpha\beta} = \overset{*}{T}_\beta \;\Bigg]$$

$$n_\alpha Q_\alpha + \frac{\partial M_{nt}}{\partial x_t} = \overset{*}{Q}_n + \frac{\partial \overset{*}{M}_{nt}}{\partial x_t} \;\Bigg]$$

$$M_{nn} = \overset{*}{M}_{nn} \;\Bigg]$$

(4.10)

where

$$Q_\alpha = M_{\beta\alpha,\beta} + S_{\alpha3}$$

(4.11)

is an effective shear force, n and t denoting the normal and tangential directions at the boundary contour respectively.

Likewise the dual potential, $U^c = -U$, reads

$$U^c = \int W^c dS - \int [T_\alpha u^*_\alpha + (Q_n + \frac{\partial M_{nt}}{\partial x_t}) u^*_3 - M_{nn} \frac{\partial u^*_3}{\partial x_n}] d\Gamma_u$$

(4.12)

generating compatibility equations and kinematic boundary conditions.

Presently, however, the main interest is focused on the potential energy change δU when local crack growth occurs extending the cracked area by δS_c. Thus by appeal to (4.8)

$$\delta U = \int (\delta W - p_i \delta u_i) dS - \int \| W \| d\delta S_c - \int (\overset{*}{T}_\alpha \delta u_\alpha + \overset{*}{Q}_n \delta u_3 + \epsilon_{\alpha\beta} \overset{*}{M}_\alpha \delta u_{3,\beta}) d\Gamma_T$$

(4.13)

where $\| \; \|$ denotes the jump over the crack front. Then by straight–forward use of the principle of virtual work, (4.13) may be shown, [51], to be equivalent to

$$-\delta U = \int \| P_{\alpha\beta} \| n_\alpha n_\beta \, \delta a \, d\Gamma_c$$

(4.14)

where δa is the local crack advance and

$$P_{\alpha\beta} = W \delta_{\alpha\beta} - S_{\alpha\gamma} u_{\gamma,\beta} + M_{\alpha\gamma} u_{3,\gamma\beta} - Q_\alpha u_{3,\beta}$$

(4.15)

is a plate analogue of Eshelby's energy momentum tensor satisfying the balance equations

$$P_{\alpha\beta,\alpha} - p_i u_{i,\beta} = 0 \quad .$$

(4.16)

The complementary analogues to (4.15) and (4.16) are

$$P^c_{\alpha\beta} = -W^c \delta_{\alpha\beta} + S_{\alpha\gamma,\beta} u_\gamma - M_{\alpha\gamma,\beta} u_{3,\gamma} + Q_{\alpha,\beta} u_3 \qquad (4.17)$$

and

$$P^c_{\alpha\beta,\alpha} - p_{i,\beta} u_i = 0 \qquad (4.18)$$

respectively.

In the light of (4.14) it seems appropriate to interpret some earlier proposals to determine energy release rates locally for a simple situation. To this end the split beam illustrated in Fig. 9 is singled out for attention.

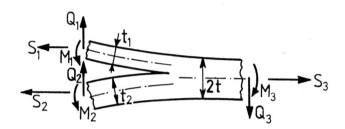

FIGURE 9
Split beam with load resultants at crack front.

Then by (4.14), (4.15) and considerations of continuity for the three beam elements, the energy release rate becomes

$$G = -\frac{b}{2} \left\| \frac{S^2}{EA} + \frac{M^2}{EI} \right\| \qquad (4.19)$$

where the beam elements have been assumed to be Hookean and customary engineering notation has been used. Further b is the beam width and thus total forces and bending moments are Sb and Mb respectively. The energy release rate according to (4.19) is then to be determined with due attention to the further dynamic continuity conditions

$$\left. \begin{aligned} S_3 &= S_1 + S_2 \\ M_3 &= M_1 + M_2 + (S_1 t_2 - S_2 t_1)/2 \end{aligned} \right\} \qquad (4.20)$$

in the notation of Fig. 9.

At a beam buckling problem Yin and Wang [54] heuristically combined nonlinear kinematics with the small–strain version of the J–integral and arrived at an expression for the energy release rate, which may readily be shown to be equivalent to (4.19). In a direct approach based on first principles Williams [55] derived for a split beam an expression in conformity with (4.19) and also demonstrated its ease of application for a variety of flexural tests involving also large rotations, [56], in contrast to the moderate ones considered presently. Based on crack closure arguments, Withcomb and Shivakumar [33] have also derived a strain energy release rate expression for linear elastic plates in conformity with (4.14), (4.15) .

Furthermore Yin [23] has discussed also path–independence of the J–integral when applied to nonlinear plate problems. The approach is not without ambiguity though when using non–conjugate variables but requires heuristic arguing and physical insight. In the present setting based on primary plate variables, path–independence is inherent in the balance equations (4.16) or alternatively (4.18), which may be applied in a straight—forward manner.

It seems desirable to extend the general theory outlined above for plates to apply also to nonlinear shell problems. To this end it may first be established that it has been shown by Stumpf [57] that the principles laid down by him for Hookean plates apply also to shells provided displacement gradients and curvature changes as in (4.2), (4.3) are replaced by

$$u_{\alpha,\beta} \rightarrow u_{\alpha,\beta} - b_{\alpha\beta}u_3 \tag{4.21}$$

$$u_{3,\alpha} \rightarrow u_{3,\alpha} + b_{\alpha\beta}u_\beta \tag{4.22}$$

$$u_{3,\alpha\beta} \rightarrow u_{3,\alpha\beta} + b_{\gamma\alpha,\beta}u_\gamma + b_{\gamma\alpha}u_{\gamma,\beta} + b_{\gamma\beta}u_{\gamma,\alpha} - b_{\gamma\alpha}b_{\gamma\beta}u_3 \tag{4.23}$$

In (4.21) to (4.23) $b_{\alpha\beta}$ is the shell middle surface curvature tensor and on the right–hand sides a comma denotes differentiation with respect to curvilinear Gaussian surface coordinates, no distinction having been made between contravariance and covariance for brevity. Frequently analyses of shallow shells rest on Donnell–Mushtari–Vlasov theory and when this approximation is relied upon, only (4.21) should be introduced as a replacement. Any details as regards application to the present delamination theory are, however, open for exploration. The same goes for the question of stability of crack growth which has only been briefly discussed above in particular circumstances.

So far only methods to determine the entire energy release rate have been outlined and there remains the question of mode partitioning. The matter has recently been dealt with by Williams [55] for the split beam just discussed. If for no other reason than to avoid algebraic complexity, generality will not be sought for presently but the approach by Williams followed as an illustration.

Referring to Fig. 9 the task is to partition the energy release rate with respect to variables contributing to crack propagation in Modes I and II respectively in ordinary fracture mechanics sense. Based on the kinematics of beam theory, for assumed homogenous material properties Williams proposes partitioning of membrane forces and bending moments according to

$$\left. \begin{array}{ll} S_1 = S_I + S_{II}, & S_2 = \left[\dfrac{1-\alpha}{\alpha}\right] S_I + S_{II} \\[2ex] M_1 = M_{II} - M_I, & M_2 = \left[\dfrac{1-\alpha}{\alpha}\right]^3 M_{II} + M_I \end{array} \right\} \tag{4.24}$$

where $\alpha = t_1/2t$.

Remembering the continuity conditions (4.20), the energy release rate according to (4.19) may be evaluated to yield

$$G = \frac{3}{4Ebt^3} \left[\frac{1+\beta}{(1-\alpha)^3} M_I^2 + \frac{(1-3\alpha+3\alpha^2)(1-2\alpha)^2}{3\alpha(1-\alpha)}(S_{II}t)^2 + \frac{3(1-\alpha)(1+\beta)}{\alpha^2} M_{II}^2 \right.$$

$$\left. -2(1-2\alpha)(1+\beta) S_{II}t M_{II} \right] \tag{4.25}$$

where $\beta = (1-\alpha)^3 / \alpha^3$.

Evidently there is no coupling in (4.25) between the conjectured mode–specified variables according to (4.24) and G_I and G_{II} may be identified separately. Incidentally at non–vanishing membrane forces, (4.25) is at variance with the corresponding result in

[55] as apparently moment continuity, eq. $(4.20)_2$ presently, was not enforced. From the viewpoint of economy at computations of delamination growth, it seems of great interest to investigate the accuracy of a partitioning technique based on 2D variables relative to a conventional 3D approach.

5. COMPUTATIONAL TECHNIQUES AND PARTICULAR PROBLEMS

In practice it is rare that analytical solutions may be found when the particular nonlinearities under discussion predominate. Definitely there is a need for reliable computational methods to predict the growth rate of arbitrarily shaped defects and the resulting residual strength of damaged structural composite members.

From a computational point of view the problem may be considered as a fully three–dimensional one. In general also internal boundaries need to be determined as part of the overall solution due to possible contact between delaminated parts. This will further add to existing nonlinearities of primarily kinematic origin. In case of anisotropic materials all three modes of deformation are in general present at a delamination front and techniques are needed for partitioning of energy release rates. While problems of such generality still await their solutions, in cases dealt with different simplifying assumptions have been introduced most commonly as regards dimensionality.

Geometries of the simple kind illustrated in Fig. 4a above are not typical of cases met in practice. They may, however, characterize carefully designed experimental situations aimed at an increased understanding of fundamental questions related to the problem at hand. Access to associated computational solutions are accordingly of importance.

In an early contribution Whitcomb [29] analysed through–width delaminations in compressed coupons of graphite/epoxy–aluminium, the system being essentially as illustrated in Fig. 4a. Postbuckling solutions were derived using a finite element technique based on 2D continuum modelling. The separate contributions of G_I and G_{II} to the energy release rate were determined by aid of a virtual crack closure technique. A notable feature is that the ratio G_I/G_{II} was found to vary considerably in the postbuckling range, G_I dominating in the initial stages but then diminishing while G_{II} increased monotonically with applied compressive strain. In subsequent work, [30], for essentially the same system, a superposition technique was introduced in order to simplify computations. Such a procedure, used also to advantage by others, [18], [54], is perhaps somewhat ad hoc when nonlinear kinematics is at hand but may be justified within the plate theory outlined above.

For delamination contours of more complex shape a few attempts have been made to apply plate theory. Elliptic crack contours in particular have been given some attention. For a graphite/epoxy–aluminium laminate Shivakumar and Whitcomb, [31], have carried out linear buckling analyses for different delamination shapes, orientations, material anisotropy and lay–ups. Save for highly anisotropic situations, satisfactory agreement was found between buckling loads computed by aid of the finite element code STAGS and those obtained using a Rayleigh–Ritz procedure although only three degrees of freedom were utilized in the later approach.

Postbuckling and growth of an elliptic delamination has been considered by Chai and Babcock [47] for a film of orthotropic material bonded to an isotropic substrate. These writers rely on a Rayleigh–Ritz procedure involving five degrees of freedom and a continued two–parameter description of the crack contour at growth. For different anisotropy properties Chai and Babcock were then able to make predictions for a variety of crack characteristics regarding growth directions, instability and arrest.

A similar problem, though for a panel of finite thickness and allowing also for bending–stretching coupling, has been recently dealt with by Kassapoglou [58]. Truncated power series, having seven degrees of freedom, were adopted as displacement trial functions in the postbuckling range. The governing partial differential equations were solved directly by identification of equal powers. At comparison with results by Chai and Babcock [47] substantial disagreement was found in some situations already for the load required to initiate buckling and as one conclusion Kassapoglou [58] questions the procedure to analyse a geometrically non–trivial problem relying on just a few degrees of freedom. The doubts may be well–founded as even in a fully linear problem, a square plate subjected to a central transverse force, a four degrees of freedom Rayleigh–Ritz procedure applied by Whitcomb and Shivakumar [33] resulted in surprisingly poor accuracy. In case of postbuckling the situation might be even worse as has also recently been indicated by Yin [14].

It goes without saying that reliability of computed data is a question of primary importance in any numerical approach. Accuracy and computational efficiency when determining energy release rates at combined stretching and bending action has recently been dealt with by Storåkers and Andersson [51]. The system analysed by these writers was an isotropic elastic layer bonded to a rigid foundation save for a circular delamination subjected to transverse pressure. When using the extremum principles for von Karman plates as above, it was found that to determine energy release rates within 10 % accuracy by aid of a Rayleigh–Ritz procedure based on eqs. (4.14), (4.15) and their complementary analogues above, at least ten degrees of freedom were needed in both kinematic and dynamic trial functions in this essentially one–dimensional problem. It was possible though to improve the situation a little by utilizing the balance equations, (4.16), (4.18), for the energy momentum tensor. It is unavoidable to conclude then that for a truly two–dimensional situation, it must be expected that application of a simple Rayleigh–Ritz procedure will in general result in poor accuracy. On the other hand it was found in [51] that by using more recent developments of the finite element method such as the p–version, [59], there was no difficulty in determining energy release rates within 1 % say by utilizing only 100 degrees of freedom or so for very nonlinear situations exhibiting strong boundary layers.

In two–dimensional situations the potential of the hp–version of the finite element method still remains to be explored when it comes to solving general delamination problems within plate approximations. It must be remembered though that for non–smooth crack contours and for certain boundary conditions different plate models may generate results that deviate considerably from those of the corresponding three–dimensional solution also in cases where characteristic length to thickness ratios are large, [60]. For such cases fully three–dimensional solutions are of necessity.

When it comes to partitioning of the energy release rate, there is a need to investigate the accuracy when using plate variables, as sketched above, related to three–dimensional formulations. At nonlinear kinematics the matter seems to have been dealt with only for beam models as in the finite element analysis of a cracked lap shear specimen by Dattaguru et al. [61].

In very recent work Whitcomb [35] gave postbuckling solutions for relatively thick embedded delaminations of elliptical shape in panels under uniaxial compression. A three–dimensional finite element method was employed and strain energy release rates for different modes were obtained by use of a virtual crack closure technique. Perhaps it is fair to say that this work constitutes the only computationally reliable one, which is available for a more general practical situation.

Some main findings by Whitcomb [35] showed that energy release rates G_I and G_{II} were large as compared to G_{III} and varied considerably along the crack contour. Peak values of G_I and G_{II} might occur at different locations and depend on the imposed strain and the specific delamination shape. As crack faces were occasionally found to overlap there is a further need for analysis of this phenomenon.

6.   EXPERIMENTAL ASPECTS

In order to determine interlayer fracture characteristics for laminates it is a common proce- dure to perform tests involving flexure. Some standard arrangements are illustrated in Fig. 10. There exists a volumnious literature on the way of interpreting test results and also on fracture data determined for specific materials. The latter aspect will not be dwelt upon further but details, also including fatigue properties, may be found in summaries such as that by Jones et al. [62].

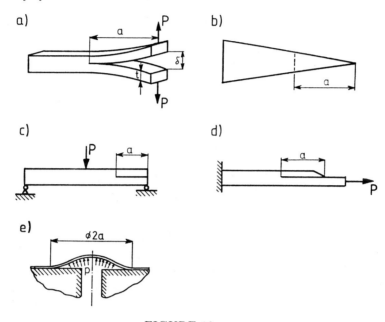

FIGURE 10

Test arrangements and specimens to determine energy release rates,
a) DCB, b) WTDCB, c) ENF, d) CLS, e) blister.

Test setups shown in Fig. 10a–d constitute statically determinate systems when beam mo- dels are adopted. With reference to Fig. 9 above the evaluation of energy release rates is then immediate by eq. (4.19) which furthermore, as should be emphasized, is general in the sense that the beam elements involved must not necessarily have equal material properties. It is required though that individual stiffnesses are known and one alternative way to de- termine critical energy release rates in single load tests is by the standard formula

$$G = \frac{P^2}{2} \frac{dC}{dS}_c \qquad (6.1)$$

where C is the compliance of the specimen. The variation of compliance with cracked area, S_c, is then to be determined experimentally. When the compliance variation with crack size is known theoretically, application of (6.1) leads to expressions equivalent to (4.19). If (6.1) is used to analyse the cracked lap shear test, Fig. 10d, the common procedure of "ignoring bending effects" will result in an error of significance. In the present setting this amounts to using (4.19) while ignoring (4.20).

As regards alternative ways to determine critical energy release rates by aid of double can- tilever beams, Fig. 10a, a variety of methods exists as recently summarized by Gue'dra et al. [63]. The end notched flexure specimen, Fig. 10c, has been given special attention in se- veral contributions by Carlsson, Gillespie, Pipes and others [64], [65], [66]. The stability of flexural delamination tests will in general depend on the way of load introduction and the matter has been elaborated upon by Williams [55] and Carlsson et al. [65]. The present concern is, however, with nonlinearities and in particular kinematical ones may become an important issue when slender specimens are used. It proves useful also as regards testing to draw upon the energy theorems laid down above.

It has been shown earlier, [51], in connection with an axisymmetric blister test as proposed by Williams [67] and illustrated in Fig. 10e, that deformation is governed by a single load parameter

$$\bar{p} = \frac{pa^4}{Et^4} \tag{6.2}$$

where E is a representative modulus limiting then the discussion to the case of a quadratic strain energy function. As a consequence the potential energy of the system may be expressed as

$$U = \frac{Et^5}{a^2} f\left(\frac{pa^4}{Et^4}\right) \tag{6.3}$$

and, by some energy considerations, at crack propagation the energy release rate becomes simply

$$G = \frac{1}{S_c}\left(pV + \int_0^{} pdV\right) \tag{6.4}$$

without any further assumptions. In (6.4), which contains only variables directly measurable, S_c is the delaminated area and V the volume under the deformed blister.

The analysis sketched for the blister test rests on self–similarity of crack growth. These circumstances prevail also when using a width–tapered beam, illustrated in Fig. 10b and perhaps better classified as a plate, the crack length being the only characteristic in–plane dimension.

In analogy with the reasoning above, for the WTDCB–test the characteristic load parameter, is

$$\bar{P} = \frac{Pa^2}{Et^4} \tag{6.5}$$

and accordingly

$$U = \frac{Et^5}{a^2} f\left(\frac{Pa^2}{Et^4}\right) \ . \tag{6.6}$$

Now as the sum of the dual potentials introduced above, (4.8), (4.12), vanishes identically,

$$\delta\left(\int WdS - P\delta + U^c\right) = 0 \ , \tag{6.7}$$

then by the virtual work theorem

$$\frac{\partial U^c}{\partial P} = \delta \ , \tag{6.8}$$

which in passing may be interpreted as a nonlinear version of Castigliano's second theorem.

Again as $U^C = - U$, by (6.6)

$$\left(\frac{\partial U^C}{\partial a}\right) = \frac{2}{a}\left(- U^C + \frac{\partial U^C}{\partial P} P\right) \tag{6.9}$$

and further by (6.8)

$$\frac{\partial U^C}{\partial a} = \frac{2}{a}\left(- \int_0 \delta dP + P\delta\right) . \tag{6.10}$$

For the WTDCB–specimen the cracked area may be expressed as

$$S_c = \beta a^2, \tag{6.11}$$

the parameter β depending only on the tapering of the specimen.

Then by partial integration of (6.10) and introduction of (6.11), the energy release rate reduces to the simple expression

$$G = \frac{\partial U^C}{\partial S_c} = \frac{1}{S_c} \int_0 P d\delta . \tag{6.12}$$

This result applies as well to a blister test, when the pressure load has been replaced by a central point load as proposed by Malyshev and Salganik [68].

In uniform beams, such as the double cantilever one, crack propagation will not be self–similar but there is no difficulty in showing that, similarly as above, the complementary energy potential may be written

$$U^C = \frac{Et^5 b}{a^3} f\left(\frac{Pa^3}{Et^4 b}\right) \tag{6.13}$$

where b is the beam section width.

In analogy with (6.12) the energy release rate at crack growth reduces to a simple expression

$$G = \frac{3}{S_c} \int_0 P d\delta . \tag{6.14}$$

The results dicussed so far are valid in general when rotations are moderate, that is when displacements are within one tenth say of a characteristic in–plane dimension. Occasionally when testing very slender beams large rotations might evolve though strains may still be small. Such situations have been analysed by Devitt et al. [69], Wang et al. [70] and Williams [56], [71]. While Devitt et al. and Williams utilize solutions involving elliptic integrals deriving from the elastica, Wang et al. prefer a Rayleigh–Ritz procedure involving two degrees of freedom. Williams, [71], takes the matter a little further and arrives at an energy release rate expression containing only variables directly measurable. As an alternative the approach outlined above may be modified to account also for large rotations but the details will be dealt with elsewhere.

Leaving the mechanics of testing there is a large variety of experimental findings warranting at least brief comments in the present context. Numerous test results clearly indicate that the energy release rate at initation of crack growth in general varies with the mode of propagation and critical values of increasing magnitude have been found for opening (I), shearing (II) and tearing (III) respectively in pure modes. By experimental observations and correlation with FEM–analysis of local stress fields, e.g. Corleto et al. [72] have attributed this difference to disparate sizes and shapes of the damage zone observed ahead of interlaminar crack tips in graphite/epoxy. This being so at initiation, progressing delamination might require higher release rates, [73], [74], the occurrence of this so–called R–curve phenomenon being well known for crack growth in metals. For composites the effect might be due to fiber bridging in Mode I while it might be absent in Mode III as reported by Donaldson [37] for unidirectional graphite/epoxy. Further it goes without saying that the state of anisotropy will affect crack growth properties. As one observation Marom et al. [75] found, for three kinds of woven reinforced composites, that the strain energy release rate for propagation in Mode I was less sensitive to the angle between delamination directions and fabric axes than that of Mode II. Discussion of physical and micromechanical aspects of these important matters is the main theme of several contributions to [7].

The presence of rate effects has been reported by several investigators. Thus Devitt et al. [69] found, when carrying out DCB–tests a weak power law dependence of the energy release rate on crack speed. In similar experiments Dharan [76] found a common increase of 15 per cent or so for a tenfold increase of the loading rate. Effects on the contrary in Mode II tests of graphite reinforced composites have, however, been reported by Chapman et al. [77].

Some other effects that influence the mechanical behaviour of laminates are well known to the materials science and engineering communities but deserve to be at least mentioned in the present context if only as a reminder to analysts. The significance of environmental effects, such as arising from moisture and temperature, has been surveyed by Springer [78]. Again mode sensitivity might be predominant as found at flexural testing e.g. by Davies and de Charantenay [79]. Thus these investigators found that propagation in Mode II was far less sensitive to temperature, in the interval -30^0 C to 120^0 C, than in Mode I, the energy release rate associated with the latter showing a substantial increase at higher temperatures.

Though delamination growth has been investigated in multitude by standard flexural tests, for natural reasons tests involving combined buckling and growth are much less frequent. A summary of experimental studies on the influence of delamination damage on compression strength has recently been given by Baker et al. [80] and only some additional particular investigations will be commented upon briefly for reasons of flavour.

As regards the issue of initiation of buckling Wang et al. [45] investigated local and global symmetric buckling of through–width delaminated short–fiber composites under uniaxial compression. The study was conducted both by analytical and experimental means and when varying a variety of geometrical parameters, being the main object, predictions were found to be in very good agreement with test results. Gillespie and Pipes [81] investigated the effect of through–width delaminations by way of four–point static loading of sandwich beams with graphite/epoxy face sheets. Critical loading, leading to fast interlaminar fracture, was substantially reduced by disbonds and varied in a linear way with delamination length. Whitcomb [29] tested delamination growth in five graphite/epoxy aluminium specimens by compression fatigue and, guided by an accompanying FEM–analysis, concluded that growth was dominated by the Mode I component. In similar testing of six specimens of graphite/epoxy, Whitcomb [30] again found high growth rates only at large G_I –values though rapidly decreasing with delamination size.

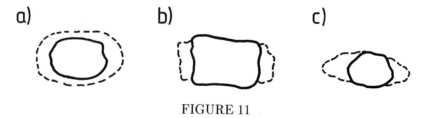

FIGURE 11

Delamination contours after combined buckling and growth under uniaxial compression, — initial, – – – current, after a) Chai and Babcock [47], b) Mousley [82], c) Konishi and Johnston [83].

Experiments regarding the simultaneous effects of low–velocity impact and compression loading have been carried out by Chai [84] and further discussed by Chai and Babcock [47]. Thus high–speed photography and moiré technique was used to record the combined events of buckling and delamination growth in graphite/epoxy laminates. A typical growth mode, essentially orthogonal to the direction of compression and unstable as found by Chai [84], is sketched in Fig. 11a. Predominantly lateral growth was also found by Mousley [82] when investigating delamination growth in 24 ply carbon fibre/epoxy coupons under static compression as illustrated in Fig. 11b. Growth of elliptic delaminations in sandwich panels, [58], was found to be of a catastrophic kind, which, within the test arrangements, inhibited determination of the location of growth initiation. Likewise unstable growth of embedded circular delaminations has been recorded by Ramkumar [85]. Konishi and Johnston [83] studied static and fatigue growth (R=–1) of built–in circular delaminations developing essentially into elliptical shapes. A growth pattern recorded under static load is illustrated in Fig. 11c and also in fatigue major growth occurred laterally relative to the loading direction.

It must be concluded though that these numerous and highly interesting experimental findings as regards combined buckling and delamination growth, are still not matched by analyses of reliable accuracy.

7. CONCLUDING REMARKS

An effort has been made to elucidate some of the challenges involved when investigating nonlinear delamination phenomena. The main features of this branch of mechanics are understood at least in principle but to achieve a deeper insight several issues, some of them emphasized above, need to be further addressed. These include theoretical foundations and modelling as well as analytical and computational methods preferably in close interaction with experimental investigations of both basic and applied kinds. The serious investigator will find it rewarding that additional insight of any generality will most probably be of direct practical value.

ACKNOWLEDGEMENT

The writer is indebted to his collegue Professor Börje Andersson for helpful advice and valuable discussions in particular regarding computing techniques.

REFERENCES

[1] Johnson, W. and Ghosh, S.K., *Some physical defects arising in composite material fabrication,* J. Mater. Sci. **16**, (1981), pp. 285–301.
[2] Tsai, S.W., (ed.), *Composite materials: Testing and design,* ASTM STP 674, American Society for Testing and Materials, Philadelphia, (1979).
[3] Reifsnider, K.L., (ed.), *Damage in composite materials,* ASTM STP 775, (1982).
[4] Daniel, I.M., (ed.), *Composite materials: Testing and design,* ASTM STP 787, (1982).
[5] Wilkins, D.J., (ed.), *Effects of defects in composite materials,* ASTM STP 836, (1984).
[6] Johnson, W.S., (ed.), *Delamination and debonding of materials,* ASTM STP 876, (1985).
[7] Johnston, N.J., (ed.), *Toughened composites,* ASTM STP 937, (1987).
[8] Matthews, F.L., Buskell, N.C.R., Hodgkinson, J.M. and Morton, J., (eds.), Proc. ICCM–VI, ECCM–2, **1–5**, (Elsevier, 1987).
[9] Garg, A.C., *Delamination – a damage mode in composite structures,* Engng Fracture Mech. **29**, (1988), pp. 557–584.
[10] Dorey, G., *Impact damage in composites–development, consequences and prevention,* in: [8] **3**, pp. 1–26.
[11] Wang, S.S., *Fracture mechanics for delamination problems in composite materials,* J. Compos. Mater. **17**, (1983), pp. 210–223.

[12] Kachanov, L.M., *Separation failure of composite materials,* Polymer Mech. **12,** (1976), pp. 812–815.

[13] Bolotin, V.V., Zebel'yan, Z.Kh. and Kurzin, A.A., *Stability of compressed components with delamination–type flaws,* Probl. Proch. **7,** (1980), pp. 813–819.

[14] Yin, W.–L., *Recent analytical results on delamination buckling and growth,* in: Armanios, E.A., (ed.), *Interlaminar fracture in composites,* (Trans. Tech. Publ., to appear).

[15] Dorey, G., *Impact and crashworthiness of composite structures,* in: Davies, G.A. O., (ed.), *Structural impact and crashworthiness,* **1,** (Elsevier, 1984), pp. 155–192.

[16] Chai, H., Babcock, C.D. and Knauss, W.G., *One dimensional modelling of failure in laminated plates by delamination buckling,* Int. J. Solids Structures **17,** (1981), pp. 1069–1083.

[17] Argon, A.S., Gupta, V., Landis, H.S. and Cornie, J.A., *Intrinsic toughness of interface between SiC coatings and substrates of Si or C fiber,* J. Mater. Sci., (to appear).

[18] Yin, W.–L., *Axisymmetric buckling and growth of a circular delamination in a compressed laminate,* Int. J. Solids Structures **21,** (1985), pp. 503–514.

[19] Bottega, W.J., *A growth law for propagation of arbitrary shaped delaminations in layered plates,* Int. J. Solids Structures **19,** (1983), pp. 1009–1017.

[20] Simitses, G.J., Sallam, S. and Yin, W.–L., *Effect of delamination of axially loaded homogeneous laminated plates,* AIAA J. **23,** (1985), pp. 1437–1444.

[21] Sallam, S. and Simitses, G.J., *Delamination buckling and growth of flat cross–ply laminates,* Composite Structures **4,** (1985), pp. 361–381.

[22] Yin, W.–L., Sallam, S. and Simitses, G.J., *Ultimate axial load capacity of a delaminated plate,* AIAA J. **24,** (1986), pp. 123–128.

[23] Yin, W.–L., *Cylindrical buckling of laminated and delaminated plates,* Proc. AIAA/ASME/ASCE/AHS 27th SDM Conf **1,** San Antonio, (1986), pp. 165–179.

[24] Yin, W.–L., *The effects of laminated structure on delamination buckling and growth,* J. Compos. Mater., (to appear).

[25] Bottega, W.J. and Maewal, A., *Delamination buckling and growth in laminates,* J. Appl. Mech. **50,** (1983), pp. 184–189.

[26] Evans, A.G. and Hutchinson, J.W., *On the mechanics of delamination and spalling in compressed films,* Int. J. Solids Structures **20,** (1984), pp. 455–466.

[27] Yin, W.–L. and Fei, Z., *Buckling load of a circular plate with a concentric delamination,* Mech. Res. Comm. **11,** (1984), pp. 337–344.

[28] Yin, W.–L. and Fei, Z., *Delamination buckling and growth in a clamped circular plate,* AIAA J., (to appear).

[29] Whitcomb, J. D., *Finite element analysis of instability related delamination growth,* J. Compos. Mater. **15,** (1981), pp. 403–426.

[30] Whitcomb, J.D., *Strain–energy release rate analysis of cyclic delamination growth in compressively loaded laminates,* in: [5], (1984), pp. 175–193.

[31] Shivakumar, K.N. and Whitcomb, J.D., *Buckling of a sublaminate in a quasi–isotropic composite laminate,* J. Compos. Mater. **19,** (1985), pp. 2–18.

[32] Whitcomb, J.D., *Parametric analytical study of instability–related delamination growth,* Compos. Sci. Technol. **25,** (1986), pp. 19–48.

[33] Whitcomb, J.D. and Shivakumar, K.N., *Strain–energy release rate analysis of a laminate with a postbuckled delamination,* in: Luxmoore, A.R., Owen, D.R.J., Rajapakse, P.S. and Kanninen, M.F., (eds.), *Numerical methods in fracture mechanics,* (Pineridge, 1987). pp. 581–605.

[34] Whitcomb, J.D., *Mechanics of instability–related delamination growth,* NASA Tech. Memo., No. 100662, (1988).

[35] Whitcomb, J.D., *Three–dimensional analysis of a postbuckled embedded delamination,* NASA Technical Paper 2823, (1988).

[36] Tay, T.E., Williams, J.F. and Jones, R., *Characterization of pure and mixed mode fracture in composite laminates,* Theor. Appl. Fracture Mech. **7,** (1987), pp. 115–123.

[37] Donaldson, S.L., *Interlaminar fracture due to tearing (Mode III),* in: [8] **3,** (1987), pp. 274–283.

[38] Donaldson, S.L., *The effect of interlaminar fracture properties on the delamination buckling of composite laminates,* Comp. Sci. Technol. **28,** (1987), pp. 33–44.

[39] Hahn, H.T. and Johannesson, T., *A correlation between fracture energy and fracture morphology in mixed–mode fracture of composites,* in: Carlsson, J. and Ohlson, N.G., (eds.), Proc. *Mechanical behaviour of materials – IV* **1,** (Pergamon, 1984), pp. 431–438.

[40] Zhou, C.-T., *Some basic problems in the analysis of instability for composite cylindrical shells,* in: [8] **5**, (1987), pp. 101–112.

[41] Troshin, V.P., *Effect of longitudinal delamination in a laminar cylindrical shell on the critical external pressure,* Mech. Comp. Mat. **17**, (1983), pp. 563–567.

[42] Sallam, S. and Simitses, G.J., *Delamination buckling of cylindrical shells under axial compression,* Composite Structures. **6**, (1987), pp. 83–101.

[43] Simitses, G.J. and Chen, Z.Q., *Delamination buckling of pressure–loaded thin cylinders and panels,* in: Marshall, I.H., (ed.), *Composite structures* 1, (Elsevier, 1987), pp. 294–308.

[44] Simitses, G.J. and Chen, Z., *Buckling of delaminated, long, cylindrical panels under pressure,* Computers & Structures **28**, (1988), pp. 173–184.

[45] Wang, S.S., Zahlan, N.M. and Suemasu, H., *Compressive stability of delaminated random short–fiber composites, Part I– Modeling and methods of analysis, II– Experimental and analytical results,* J. Compos. Mater. **19**, (1985), pp. 296–316, 317–333.

[46] Williams, J.G. and Rhodes, M.D., *Effect of resin on impact damage tolerance of graphite/epoxy laminates,* in: [4], (1982), pp. 450–480.

[47] Chai, H. and Babcock, C.D., *Two–dimensional modelling of compressive failure in delaminated laminates,* J. Compos. Mater. **19**, (1985), pp. 67–98.

[48] Marshall, D.B. and Evans, A.G., *Measurement of residually stressed thin films by indentation, I. Mechanics of interface delamination,* J. Appl. Phys. **56**, (1984), pp. 2632–2638.

[49] Rossington, C., Evans, A.G., Marshall, D.B. and Khuri-Yakub, B.T., *Measurements of adherence of residually stressed thin films by indentation, II. Experiments with ZnO/Si,* J. Appl. Phys. **56**, (1984), pp. 2639–2644.

[50] Shih, C.F., Moran, B. and Nakamura, T., *Energy release rate along a three–dimensional crack front in a thermally stressed body,* Int. J. Fracture **30**, (1986), pp. 79–102.

[51] Storåkers, B. and Andersson, B., *Nonlinear plate theory applied to delamination in composites,* J. Mech. Phys. Solids, (to appear).

[52] Stumpf, H., *Die Extremalprinzipe der nichtlinearen Plattentheorie,* ZAMM **55**, (1975), pp. T110–T112.

[53] Stumpf, H., *Dual extremum principles and error bounds in the theory of plates with large deflections,* Arch. Mech. **27**, (1975), pp. 485–496.

[54] Yin, W.-L. and Wang, J.T.S., *The energy–release rate in the growth of a one–dimensional delamination,* J. Appl. Mech. **51**, (1984), pp. 939–941.

[55] Williams, J.G., *On the calculation of energy release rates for cracked laminates,* Int. J. Fracture **36**, (1988), pp. 101–119.

[56] Williams, J.G., *Large displacement effects in the DCB test for interlaminar fracture in modes I and II,* in: [8] **3**, (1987), pp. 233–241.

[57] Stumpf, H., *The derivation of dual extremum and complementary stationary principles in geometrical nonlinear shell theory,* Ing.-Arch. **48**, (1979), pp. 221–237.

[58] Kassapoglou, C., *Buckling, post–buckling and failure of elliptical delaminations in laminates under compression,* Composite Structures **9**, (1988), pp. 139–159.

[59] Guo, B.Q. and Babuska, I., *The hp–version of the finite element method. Part 2: General results and applications,* Comp. Mech. **1**, (1986), pp. 203–220.

[60] Babuska, I. and Scapolla, T., *Benchmark computation and performance evaluation for a rhombic plate bending problem,* Int. J. Num. Meth. Engng, (to appear).

[61] Dattaguru, B., Everett, Jr., R.A., Whitcomb, J.D. and Johnson, W.S., *Geometrically nonlinear analysis of adhesively bonded joints,* J. Engng. Matrls Techn. **12** (1984), pp. 59–65.

[62] Jones, R., Paul, J., Tay, T.E. and Williams, J.G., *Assessment of the effect of impact damage in composites: Some problems and answers,* Theor. Appl. Fracture Mech. **9**, (1988), pp. 83–95.

[63] Gue'dra, D., Lang, D., Rouchon, J., Marais, C. and Sigety, P., *Fracture toughness in Mode I: A comparison exercise of various test methods,* in: [8] **3**, (1987), pp. 346–357.

[64] Carlsson, L.A., Gillespie, J.W. and Trethewey, B.R., *Mode II interlaminar fracture of graphite/epoxy and graphite/PEEK,* J. Reinforced Plastics Compos. **5**, (1986), pp. 170–187.

[65] Carlsson, L.A., Gillespie, Jr., J.W. and Pipes, R.B., *On the analysis and design of the end notched flexure (ENF) specimen for Mode II testing,* J. Compos. Mater. **20**, (1986), pp. 594–603.

[66] Smiley, A.J. and Pipes, R.B., *Rate sensitivity of Mode II interlaminar fracture toughness in graphite/epoxy and graphite/PEEK composite materials,* Compos. Sci. Technol. **29**, (1987), pp. 1–15.

[67] Williams, M.L., *The continuum interpretation for fracture and adhesion,* J. Appl. Pol. Sci. **13**, (1969), pp. 29–40.

[68] Malyshev, B.M. and Salganik, R.L., *The strength of adhesive joints using the theory of cracks,* Int. J. Fracture **1**, (1965), pp. 114–128.

[69] Devitt, D.F., Schapery, R.A. and Bradley, W.L., *A method for determining the Mode I delamination fracture toughness of elastic and viscoelastic composite materials,* J. Compos. Mater. **14**, (1980), pp. 270–285.

[70] Wang, S.S., Suemasu, H. and Zahlan, N.M., *Interlaminar fracture of short–fiber SMC composite,* J. Compos. Mater. **18**, (1984), pp. 574–594.

[71] Williams, J.G., *Large displacement and end block effects in the 'DCB' interlaminar test in Modes I and II,* J. Compos. Mater. **21**, (1987), pp. 330–347.

[72] Corleto, C., Bradley, W. and Henriksen, M., *Correspondence between stress fields and damage zones ahead of crack tip of composites under Mode I and Mode II delamination,* in: [8] **3**, pp. 378–387.

[73] Hashemi, S., Kinloch, A.J. and Williams, J.G., *Interlaminar fracture of composite materials,* in: [8] **3** (1987), pp. 254–264.

[74] O'Brien, T.K., *Characterization of delamination onset and growth in a composite laminate,* in: [3], (1982), pp. 140–167.

[75] Marom, G., Roman, I., Harel, M., Rosensaft, M., Kenig, S. and Moshonov, M., *The characterization of Mode I and Mode II delamination failures in fabric–reinforced laminates,* in: [8] **3**, (1987), pp. 265–273.

[76] Dharan, C.K.H., *Delamination fracture and acoustic emission in carbon, aramid and glass–epoxy composites,* in: [8] **1**, (1987), pp. 405–414.

[77] Chapman, T.J., Smiley, A.J. and Pipes, R.B., *Rate and temperature effects on Mode II interlaminar fracture toughness in composite materials,* in: [8] **3**, (1987), pp. 295–304.

[78] Springer, G.S., *Environmental effects on epoxy matrix composites,* in: [2], (1979), pp. 291–312.

[79] Davies, P. and de Charantenay, F.X., *The effect of temperature on the interlaminar fracture of tough composites,* in: [8] **3**, (1987), pp. 284–294.

[80] Baker, A.A., Jones, R. and Callinan, R.J., *Damage tolerance of graphite/epoxy composites,* Composite Structures **4**, (1985), pp. 15–44.

[81] Gillespie, Jr., J.W. and Pipes, R.B., *Compressive strength of composite laminates with interlaminar defects,* Composite Structures **2**, (1984), pp. 49–69.

[82] Mousley, R.F., *In–plane compression of damaged laminates,* in: Morton. J., (ed.), *Structural impact and crashworthiness,* **2**, (Elsevier, 1984), pp. 494–503.

[83] Konishi, D.Y. and Johnston, W.R., *Fatigue effects on delaminations and strength degradation in graphite/epoxy laminates,* in: [2], (1979), pp. 597–619.

[84] Chai, H., *The growth of impact damage in compressively loaded laminates,* Ph.D. Thesis, California Institute of Technology, (1982).

[85] Rakumar, R.L., *Fatigue degradation in compressively loaded composite laminates,* NASA Contractor Report 165681, (1981).

Theoretical and Applied Mechanics
P. Germain, M. Piau and D. Caillerie (Editors)
Elsevier Science Publishers B.V. (North-Holland)
© IUTAM, 1989

FRACTAL APPLICATIONS TO COMPLEX CRUSTAL PROBLEMS

Donald L. TURCOTTE

Department of Geological Sciences
Cornell University
Ithaca, New York 14853 U.S.A.

Complex scale-invariant problems obey fractal statistics. The basic definition of a fractal distribution is that the number of objects with a characteristic linear dimension greater than r satisfies the relation $N \sim r^{-D}$ where D is the fractal dimension. Fragmentation often satisfies this relation. The distribution of earthquakes satisfies this relation. The classic relationship between the length of a rocky coast line and the step length can be derived from this relation. Power law relations for spectra can also be related to fractal dimensions. Topography and gravity are examples. Spectral techniques can be used to obtain maps of fractal dimension and roughness amplitude. These provide a quantitative measure of texture analysis. It is argued that the distribution of stress and strength in a complex crustal region, such as the Alps, is fractal. Based on this assumption, the observed frequency-magnitude relation for the seismicity in the region can be derived.

1. INTRODUCTION

The scale invarience of geological phenomena is one of the first concepts taught to a student of geology. It is pointed out that an object with a scale, i.e. a coin, a rock hammer, a person, must be included whenever a photograph of a geological feature is taken. Without the scale it is often impossible to determine whether the photograph covers 10 cm or 10 km. For example, it is impossible to determine the altitude from which a photograph of a rocky coast line is taken.

The concept of fractals was introduced in this geological context by Mandelbrot [1]. Noting that the length of a rocky coast line increased as the length of the measuring rod decreased according to a power law, he associated the power with a fractal (fractional) dimension. The basis of this observation is that the rocky coast line is scale invariant.

The basic definition of fractals is a statistical distribution in which the number of objects is proportional to the inverse of their size raised to a power, the power is the fractal dimension. The power law distribution is the only distribution that is scale invarient. It was recognized empirically in a wide variety of fields long before the concept of fractals was defined.

Fractal distributions are found in geology under a variety of circumstances. Examples include the distribution of islands, of rock fragments, of earthquakes, and of faults. In each case, the distribution can be associated with scale invarience.

The concept of fractals can also be applied to continuous variables in one or more dimensions. Any time series would be an example, specifically the earth's magnetic field at a point on the surface. If the measurement is made over a period T, a spectral decomposition can be carried out. The result is often given in terms of the spectral energy density as a function of frequence. If the spectral energy density has a power law dependence β on the frequency the distribution is a fractal for a specified range of β. The analysis can be extended to two dimensions and it is found that the earth's topography is closely approximated as a fractal under a wide variety of circumstances. Mandelbrot [2] has used this approach to generate synthetic topography landscapes that look remarkably similar to actual landscapes. Details have been given by Voss [3,4].

Although landscapes in general exhibit fractal statistics to a very good approximation, regularities associated with natural scales are also apparent. An important scale is the thickness of the elastic lithosphere, the plates of plate tectonics. This scale determines the characteristic dimensions of flexures of the lithosphere under loads. Examples that exhibit this characteristic length scale are sedimentary basins and mountain ranges. However, on smaller scales there is little evidence of characteristic lengths and the general topography in a wide variety of terranes, both young and old, appears to obey fractal statistics.

Topography is often the result of displacements on faults. Since the distribution of earthquakes and faults is fractal, it should not be suprising that the resulting topography is also fractal. Several authors have given quantitative treatments of displacements on fractal distributions of faults [5,6]. Topography is also generated by the folding of rocks. Folds occur on all scales and also are found to be generally scale invarient.

However, much topography is generated by erosional processes. Erosional processes are poorly understood but river patterns, centers of deposition of sediments, and the spectral distribution of floods must play important roles. Culling [7-9] has suggested that erosion and deposition satisfy the heat equation. Some alluvial fans and river deltas are consistent with this model. But solutions of the heat equation will not produce a fractal topography; in fact, no linear theory will produce a fractal topography. There is considerable evidence that the largest storms produce the greatest erosion. Great storms create new gullies and river valleys and thereby renew the large scale erosional features that have been removed by smaller storms. Since floods appear to obey fractal statistics it is not unreasonable to relate this distribution to fractal landscapes.

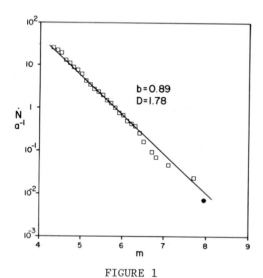

FIGURE 1

Number of earthquakes occurring per year N with a surface wave magnitude greater than m. The open squares are data for southern California from 1932-1972 [16]; the solid circle is the rate of occurrence of great earthquakes in southern California [13].

2. DEFINITION OF A FRACTAL DISTRIBUTION

Fractals and fractal distributions occur in a wide variety of physical problems. The most general and useful definition of a fractal distribution is

$$N = \frac{C}{r^D} \qquad (2-1)$$

where N is the number of "objects" with a characteristic linear

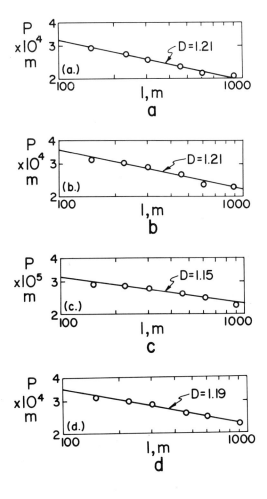

a

b

c

d

FIGURE 2

Fractal dimensions for specified topographic contours in several mountain belts: (a) 3,000 ft. contour of the Cobblestone Mountain Quadrangle, Transverse Ranges, California; (b) 5,400 ft. contour of the Tatooh Buttes quadrangle Cascade Mountains, Washington; (c) 10,000 ft. contour of the Byers Peak quadrangle, Rocky Mountains, Colorado; (d) 1,000 ft. contour of the Silver Bay Quadrangle, Adirondack Mountains, New York.

dimension greater than r, D is the fractal dimension, and C is a constant of proportionality. For any physical application there will be upper and lower limits on the applicability of the fractal distribution. The essential feature of the fractal distribution is its scale invariance. No characteristic length scale enters into the definition (2-1). If scale invariance extends over a sufficient range of length scales then the fractal dimension provides a useful measure of the relative importance of large versus small objects.

Under some circumstances different fractal dimensions may be applicable at different scales. Also, other definitions of fractal distributions can be derived from (2-1). However, in a number of applications the basic definition can be applied directly. For example, the Korcak empirical relation for the number of islands with area greater than a specified value fits (2-1) with D = 1.30 [10]. Another example of direct applicability is fragmentation [11]. Under a wide variety of circumstances sieve analyses fit (2-1) which is known empirically in fragmentation as Rosin's law [12].

3. APPLICATION TO SEISMICITY

The location, timing, and magnitudes of earthquakes can be accurately determined from seismic arrays. The completeness of coverage decreases with decreasing magnitude but a relative large range of coverage is available globally.

Under many circumstances the number of earthquakes N with a magnitude greater than m satisfies the empirical relation [13]

$$\log N = -bm + a \qquad (3\text{-}1)$$

where a and b are constants. This relationship has been found to be applicable both regionally and on a world-wide basis. The b-value is widely used as measure of regional seismicity. Aki [14] showed that (3-1) is equivalent to the definition of a fractal distribution.

The moment of an earthquake is defined by:

$$M = \mu \, \delta \, A \qquad (3\text{-}2)$$

where μ is the shear modulus, A the area of the fault break and δ is the mean displacement on the fault break. The moment of the earthquake can be related to its magnitude by

$$\log M = cm + d \qquad (3\text{-}3)$$

were c and d are constants. Kanamori and Anderson [15] have established a theoretical basis for taking c = 1.5. These authors have also shown that it is a good approximation to take

$$M = \alpha r^3 \qquad (3\text{-}4)$$

where $r = A^{1/2}$ is the linear dimension of the fault break. Combining (3-1), (3-3), and (3-4) gives

$$\log N = -2b \log r + \beta \qquad (3\text{-}5)$$

with

$$\beta = \frac{bd}{1.5} + a - \frac{b}{1.5} \log \alpha \qquad (3\text{-}6)$$

and (3-5) can be rewritten as

$$N = \beta' \, r^{-2b} \qquad (3\text{-}7)$$

A comparison with the definition of a fractal given in (2-1) shows that

$$D = 2b \qquad (18)$$

Thus the fractal dimension of regional or world-wide seismic activity is simply twice the b value.

As an example of a regional frequency-magnitude distribution we consider seismicity in southern California as summarized by Main and Burton [16]. Based on data from 1932 to 1972 the number of earthquakes per year N with magnitude greater than m is given as a function of m in Figure 1. In the magnitude range $4.25 < m < 6.5$ the data are in excellent agreement with (3-1) taking b = 0.89 or D = 1.78. With c = 1.5, d = 16, and $\alpha = 3.27 \times 10^7$ dyne/cm^2 the corresponding range for r from (3-4) and (3-3) is 0.9 km $< r < 12$ km. Data for smaller amplitude earthquakes; m < 4.25, would be expected to fit this relationship but in broad regions instrumented coverage is inadequate.

Also included in Figure 1 is the N associated with great earthquakes on the southern section of the San Andreas fault as given by Sieh [17]; m = 8.05 and $N = 0.006 a^{-1}$ (a repeat time of 163 years). An extrapolation of the fractal relation for regional seismicity appears to make a reasonable prediction of great earthquakes on this section of the San Andreas fault.

It should be emphasized that considerable caution must be exercised in using the fractal relation as a predictor for large earthquakes. Although the slope (b-value) in a particular region is likely to remain nearly constant, the level of seismicity may vary with time making any extrapolation subject to considerable error. For example, if a great earthquake reduces the stress level over a large region the level of seismicity following the earthquake would be expected to decrease. Also, a largest characteristic earthquake would be expected in each region. The size of this earthquake would be related to the regional tectonics. The seismic hazard would be strongly related to the magnitude of this largest earthquake, however, there is no reason to believe that the size would be related to other characteristics of the fractal distribution. Nevertheless, without the concept of fractals, a number of authors have used the fractal relation (2-1) to determine seismic hazards in a region [18-25].

4. PERIMETER RELATIONS

Although we suggest that the basic definition of a fractal is the number-size relation given in (2-1), the original definition of a fractal given by Mandelbrot [1] was the length of a trail or perimeter as a function of the step length. If the length of the step is r_i and if N_i is the number of steps required to obtain the length of the trail or perimeter P_i, we have

$$P_i = N_i \; r_i \tag{4-1}$$

Substitution of (2-1) gives

$$P_i = Cr_i^{1-D} \tag{4-2}$$

Mandelbrot [1] showed that the west coast of Britain satisfied (4-2) with $D \approx 1.25$.

Fractal dimensions for topography can be easily obtained from topographic maps. Several examples are given in Figure 2. The length along specified contours P is obtained using dividers of different lengths r. The results generally satisfy (4-2) and a fractal dimension can be obtained. An important question is whether there are systematic variations in the fractal dimension, for example, does it depend systematically on the age of the geological province. It is seen in Figure 2 that quite diverse geological provinces have very similar fractal dimensions. Since most fractal dimensions are near $D = 1.2$ it does not appear possible to use fractal dimensions obtained by this method to characterize geological terrains.

5. GENERATION OF SYNTHETIC FRACTAL TOPOGRAPHY

5.1   Fractal Probabilities

In order to relate the definition of a fractal distribution given in (2-1) to probabilities it is appropriate to consider the Sierpinski carpet illustrated in Figure 3. The solid square of unit dimension has a square with dimensions $r = 1/3$ removed from its center. Eight solid squares with dimensions $r = 1/3$ remain so that $N = 8$. Thus from (2-1) $D = \ln 8/\ln 3 = 1.8928$. The process is repeated. The fractal dimension lies between 1 (the Euclidian dimension of a line) and 2 (the Euclidian dimension of an area). The probability that a square with dimensions r includes solid is $Pr = 1$ when $r = 1$, $Pr = 8/9$ when $r = 1/3$, and $Pr = 68/81$ when $r = 1/9$. This is generalized to

$$Pr_i = N_i \; r_i^2 \tag{5-1}$$

and substitution of (2-1) gives

$$Pr_i = r_i^{2-D} \tag{5-2}$$

Although   derived   for   a
deterministic   fractal,   Figure   3,
this   relation   can   be   applied   to
two-dimensional random processes.

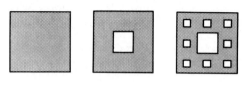

5.2   Application   to   Self-Affine
           Fractals

FIGURE 3

Applications   of   fractal   Illustration   of   the   Sierpinski
distributions   in   the   previous   carpet.
sections   were   to   self-similar
fractals.   In   this   section   we
extend the   analyses   to self-affine
fractals.   Topography is an example of a self-affine fractal, the vertical
coordinate of the earth surface is not self-similar to the horizontal
coordinates.

Consider a single valued function of a variable, an example would be the
topography h(x) along a track.   We assume that the variable is random but
has a specified power law dependence of the spectral energy density on
wave number.   We consider the specific case in which the increments h(x +
L) - h(x) satisfy the condition

$$\Pr\left[\frac{h(x + L) - h(x)}{L^H} < h'\right] = F(h') \qquad (5\text{-}3)$$

where H is a constant.   In many examples F(h') is a Gaussian.   If this is
the case and if $0 < H < 1$ then the random signal is known as fractional
Brownian motion.   For $H = 1/2$ we have Brownian motion.   Comparing (5-2)
and (5-3) we define

$$H = 2\text{-}D \qquad (5\text{-}4)$$

This is the basic definition of the fractal dimension for fractional
Brownian motion.

The variance of the increment is defined by

$$V(L) = <[h(x + L) - h(x)]^2> \qquad (5\text{-}5)$$

where the brackets < and > denote averages over many samples of x(t).
From (5-3), (5-4), and (5-5) we have

$$V(L) \sim L^{2H} \sim L^{(4-2D)} \qquad (5\text{-}6)$$

Random functions in time are often characterized by their spectral energy
densities S(k).   If one defines h(k,x) as the Fourier transform of h(x)
for

$0 < x < L,$

$$\bar{h}\,(k,t) = \frac{1}{L} \int_o^L h(x)e^{2\pi ikx}dx \qquad (5\text{-}7)$$

then

$$S(k) \sim L\,\bar{h}^{\,2}(k,x) \qquad (5\text{-}8)$$

as $L \to \infty$. In many cases of interest the spectral energy has a power law dependence on frequency

$$S(k) \sim k^{-\beta} \tag{5-9}$$

The relationship between β and the fractal dimension is obtained from (5-5), (5-6), (5-8) and (5-9) with the result

$$S \sim k^{-\beta} \sim L^{\beta} \sim LV \sim L^{2H+1} \sim L^{5-2D} \tag{5-10}$$

or

$$\beta = 5 - 2D \tag{5-11}$$

For a true fractal $1 < D < 2$ and $1 < \beta < 3$. For Brownian motion $\beta = 2$ and $D = 1.5$.

An alternative derivation of this result uses the basic definition of fractals (2-1) to relate the number of boxes of size L required to cover the profile. The standard deviation is related to the length of the profile L by

$$\sigma(L) = [V(L)]^{1/2} \sim L^{H} \tag{5-12}$$

For fractional Brownian motion, H is in the range $0<H<1$. In order to define the relevant fractal dimension a reference "box" is introduced with a width L and a height σ. If the fractal was self-similar the box would be square; however, for self-affine fractals arbitrary rectangular boxes must be used. Consider a set of n-th order smaller boxes with width $L_n=L/n$ and height $h_n=\sigma/n$, n is an integer. The number of n-th order boxes N_n required to cover a width L and a height $\sigma_n=\sigma(L/n)$ is

$$N_n = \frac{L\sigma_n}{L_n h_n} = n^2 \frac{\sigma_n}{\sigma} \tag{5-13}$$

and using (5-12) we find

$$N_n = n^2 \frac{(L/n)^H}{L^H} = n^{2-H} = ((L)/L_n)^{2-H} \tag{5-14}$$

Comparing (5-14) with the definition of the fractal dimension in (2-1) we obtain (5-4).

5.3   Synthetic Profiles

Examples of synthetically generated fractional Brownian motion are given in Figure 4. The method used to generate these random time series was as follows:

1.   The time interval was divided into a large number of increments. Each increment was given a random number based on a Gaussian probability distribution. This is then a Gaussian white noise sequence.
    2.   The Fourier transform of the sequence was taken.
    3.   The resulting Fourier coefficients were filtered by multiplying by $f^{-\beta/2}$.
    4.   An inverse Fourier transform was taken using the filtered coefficients.
    5.   In order to remove edge effects (periodicities) only the central portion of the time series was retained.

Brownian motion is illustrated in Figure 4d.

5.4   Synthetic Topography

The method used above to generate a synthetic fractal profile has been
extended to generate synthetic topography. Two-dimensional white noise
was generated on an NxN grid using a Gaussian distribution of values.
Then, a fractal filtering techniques was applied to the synthetic white
noise using the two-dimensional Fourier spectra approach:

    1)   By carrying out a 2-D discrete Fourier transform, an NxN array of
complex Fourier transform coefficients H_{st} was obtained using the relation

$$H_{st} = \sum_{n=0}^{N-1} \sum_{m=0}^{N-1} h_{nm} \exp \left[- \frac{2\pi i}{N} (sn + tm) \right] \qquad (5\text{-}15)$$

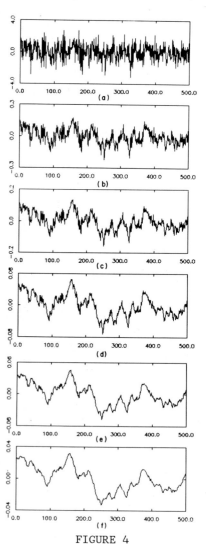

    2)   Each pair of indices (s,t)
in the tranform domain was assigned
an equivalent radial index k_{st}
according to the relation

$$k_{st} = (s^2 + t^2)^{1/2} \qquad (5\text{-}16)$$

    3)   A fractal dimension D was
specified and the corresponding
value for β was obtained from [4]

$$D = \frac{7 - \beta}{2} \qquad (5\text{-}17)$$

This is the two-dimensional
equivalent of (5-11). A new set of
complex coefficients H'_{st} was
obtained by filtering the original
H_{st} according to the relation

$$H'_{st} = H_{st}/k_{st}^{\beta/2} \qquad (5\text{-}18)$$

    4)   An inverse 2-D Fourier
transform was carried out to
generate the synthetic topography.

Examples of synthetic topography
are given in Figure 5. Black areas
are regions of higher then average
topography and white areas are
regions of lower than average
topography. The initial white
noise base is given in Figure 5a,
Figure 5b gives results for $\beta=2$ and
D=2.5, and Figure 5c gives results
for $\beta=2.6$ and D=2.2. The latter
corresponds quite well to real
topography.

FIGURE 4

Synthetically generated fractional
Brownian motion. (a) Gaussian white
noise. (b) $\beta = 1.2$, D = 1.9. (c) $\beta=$
1.6, D = 1.7. (d) $\beta = 2.0$, D = 1.5
(Brown noise). (e) $\beta = 2.4$, D = 1.3.
(f) $\beta = 2.8$, D = 1.1.

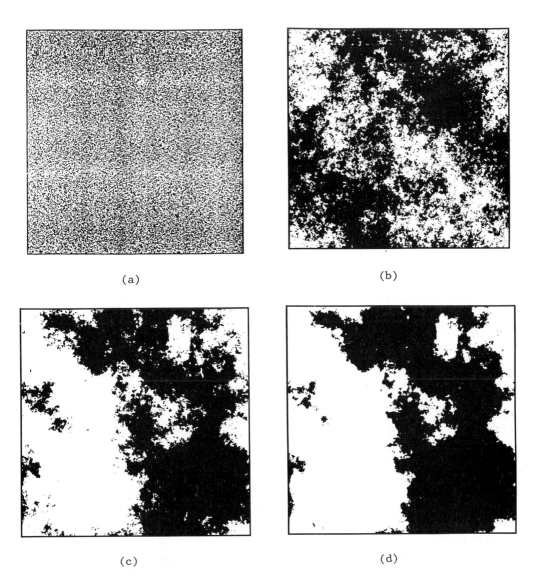

FIGURE 5
Synthetically generated images.   (a) Gaussian white noise.   (b) $\beta = 2$, $D = 2.5$.   (c) $\beta = 2.6$, $D = 2.2$.   (d) $\beta = 3$, $D = 2.0$.

5.5   Analysis of Actual Topography

Using digitized topography for Arizona, fractal dimensions and roughness
amplitudes were obtained for each 4.5 x 4.5 km subregion in the state;
each subregion contained 32 x 32 data points.   Examples for four randomly
selected subregions in Arizona are given in Figure 6.   It is seen that the
linear trend of the data is quite good.   The mean 2-D dimension for all of
Arizona is $D = 2.09$.   It is also of interest to carry out a study of the
1-D spectral behavior of the Arizona topography.   Average 1-D value for a
large number of tracks of Arizona data is $D = 1.52$.

6. Discussion

In previous sections it has been shown that the earth's topography obeys fractal statistics under a wide variety of circumstances. The implication is that the governing physics is scale invariant and nonlinear. these same statements can be made about turbulence, but at least for turbulence the governing equations are known. Two processes dominate the creation of topography. The first is mountain building; the earth's crust is deformed by discontinuous displacements on faults and continuum plastic deformations. The second is erosion. In neither case can the basic equations be written. Displacements on pre-existing faults plays an important role in mountain building. Friction must play an important role but the basic friction laws are not known. Also, deformation occurs on a fractal distribution of faults that occur on a wide range of scales. Faults appear to terminate in complex zones of deformation. Also, there is strong observational evidence that the stress field in deformation zones is strongly heterogeneous.

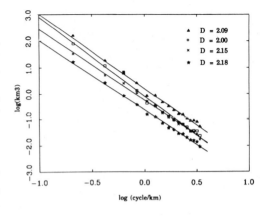

FIGURE 6
Dependence of the spectral energy on the wave number for four 32 x 32 point subregions in Arizona.

The understanding of erosional processes is even worse than the understanding of mountain building processes. Erosion during great floods clearly dominates. But models for scouring out river channels and the creation of gulleys are simply not available.

Given the difficulties discussed above, is the association with fractal statistics of any utility? The answer is yes in two ways. The first is that it may lead the way to the development of applicable, scale-invariant physical laws. The second is that it allows the use of fractal interpolation of data. Take bathymetry in the oceans as an example. Bathymetry is known along ship tracks and must be interpolated to obtain bathymetric charts. Since the bathymetry obeys fractal statistics a two-dimensional fractal Fourier expansion can be used. The relative amplitudes are specified by the fractal dimension and the ship track data are used to determine the phases. Hewett [26] has used this technique to obtain the three-dimensional porosity distribution in an oil field from a limited number of well logs.

ACKNOWLEDGEMENTS

This research was supported by grant NGR-33-010-108 from the National Aeronautics and Space Administration. This is contribution 841 of the Department of Geological Sciences, Cornell University.

REFERENCES

[1]   Mandelbrot, B. B., Science 156, (1967), 636-638.
[2]   Mandelbrot, B.B., The Fractal Geometry of Nature. (Freeman, San Francisco, 1982).
[3]   Voss, R.F., Pynn, R. and Skjeltorp, A., (eds.), (Plenum Press, New York, 1985), pp. 1-11.
[4]   Voss, R.F., Earnshaw, R.A., (ed.), (NATO ASI Series, Vol. F17, Springer-Verlag, Berlin, 1985), pp. 805-835.
[5]   King, G., Pure Ap. Geophys. 121, (1983), 761-815.
[6]   Turcotte, D.L., Tectonophys., 132, (1986), 261-269.
[7]   Culling, W.E.H., J. Geol., 71, (1963), 336-344.
[8]   Culling, W.E.H., J. Geol., 71, (1963), 127-161.
[9]   Culling, W.E.H., J. Geol., 73, (1965), 230-254.
[10] Mandelbrot, B.B., Proc. Natl. Acad. Sci. U.S.A., 72, (1975), 3825-3828.
[11] Turcotte, D.L., J. Geophys. Res. 91, (1986), 1921-1926.
[12] Rosin, P. and Rammler, E., J. Insti. Fuel, 7, (1933), 29-36.
[13] Gutenberg, B. and Richter, C.F., (Princeton University Press, Princeton, 1954).
[14] Aki, K., Simpson, D.W. and Richards, P.G, (eds.), (American Geophysical Union, Washington, D.C., 1981), pp. 566-574.
[15] Kanamori, H. and Anderson, D.L., Bull. Seis. Soc. Am., 65, (1975), 1073-1096.
[16] Main, I.G., and Burton, P.W., Seis. Soc. Am. Bull. 76, (1983), 297-304.
[17] Sieh, K.E., J. Geophys. Res. 83, (1978), 3907-3939.
[18] Smith, S.W., Geophys. Res. Let. 3, (1976), 351-354.
[19] Molnar, P., Seis. Soc. Am. Bull. 69, (1979), 115-133.
[20] Anderson, J.G., Seis. Soc. Amer. Bull. 69, (1979), 135-158.
[21] Anderson, J.G., Seis. Soc. Am. Bull. 76, (1986), 273-290.
[22] Anderson, J.G., and Luco, J.E., Seis. Soc. Am. Bull., 73, (1983), 471-496.
[23] Youngs, R.R., and Coppersmith, K.J., Seis. Soc. Am. Bull., 75, (1985), 939-964.
[24] Hyndman, R.D. and Weichert, D.H., Geophys. J. Roy. Astron. Soc., 72, (1983), 59-82.
[25] Singh, S.K., Rodriquez, M., and Esteva L., Seis. Soc. Am. Bull., 73, (1983), 1779-1796.
[26] Hewett, T.A., Fractal distributions of reservoir heterogeneity and their influence on fluid transport, Soc. Petrol. Eng. Paper 15386, 1986.

Theoretical and Applied Mechanics
P. Germain, M. Piau and D. Caillerie (Editors)
Elsevier Science Publishers B.V. (North-Holland)
© IUTAM, 1989

PLASTICITY AND CREEP AT FINITE STRAINS

Viggo Tvergaard

Department of Solid Mechanics
The Technical University of Denmark
Lyngby, Denmark

The equations governing large deformations of a solid are specified
for elastic-plastic as well as elastic-viscoplastic material behaviour. Conditions for uniqueness and bifurcation of the incremental
solution are discussed, emphasizing the strong dependence on the constitutive model. Some interest is devoted to tensile instabilities
that usually occur at large strains, including both necking instabilities and localization of plastic flow in a shear band. As an example
of a material model based on detailed micromechanical damage studies a
set of constitutive relations for high temperature creep are presented. This material model is applied to analyse creep failure in a tensile test, including the effect of necking. In this paper simple model
analyses are used to illustrate basic finite strain behaviour without
going into elaborate numerical modelling.

1. INTRODUCTION

Much interest in finite strain formulations has been related to the elastic behaviour of rubber, where the existence of a potential energy function is helpful
in the solution of boundary value problems, in spite of large geometry changes
(e.g. see Ogden [1]). In the presence of plastic or viscoplastic material behaviour the path dependence of the solutions is a significant further complication,
but numerical analysis has had a strong influence. Thus, since the early
1970'ies there has been an increasing amount of research in the area of finite
strain inelasticity.

The purpose of the present paper is to give an introduction to a number of basic
formulations, relating to inelastic material behaviour in a finite strain context. First, the field equations are outlined, including a description of geometric relations, a formulation of the equilibrium conditions, and an indication
of the structure of the incremental stress-strain relationships. In addition to
time-independent plasticity the discussion will cover viscous material behaviour, which plays an important role in descriptions of material strain-rate sensitivity at room temperature as well as creep at elevated temperatures. For the
presentation of the governing equations a convected coordinate formulation is
chosen here, as in Green and Zerna [2] and Budiansky [3].

Tensile instabilities in ductile materials are an important failure mode that
usually involves large straining. Alternatively, failure occurs by the development of damage on the micro-level, and in some cases final failure involves the
interaction of instabilities and damage. Here, conditions for uniqueness and
bifurcation in inelastic solids are given. Central in this context is Hill's
[4,5] theory of uniqueness and bifurcation in time-independent elastic-plastic
solids, but also extensions to elastic-plastic material models not covered by
Hill's theory are discussed. Furthermore, the conditions for uniqueness and bifurcation in elastic-viscoplastic materials are given.

Localization of plastic flow is a special type of instability, which is associated with loss of ellipticity of the governing differential equations (Rice [6]). In the three dimensional solid the corresponding mode of instability is a shear band, while under plane stress conditions, as those prevailing in a biaxially stretched thin sheet, a localized necking mode results. The strong dependence of localization predictions on the constitutive description is an important issue in finite strain studies, which will be illustrated here by a few examples.

In the development of constitutive relations that account for the micro-mechanics of damage basic model studies are usually needed, regarding the behaviour of voids, cracks, inclusions, etc. This is illustrated here by a constitutive model for creep of polycrystalline materials at elevated temperatures, incorporating the effect of grain boundary cavitation by the combined influence of diffusion and dislocation creep (Tvergaard [7]). Some of the approximations made in the model are emphasized and predictions of material failure are discussed.

2.   FORMULATION OF BASIC EQUATIONS

Several different formulations of finite strain theory have been used by different authors. In the present paper a Lagrangian, convected coordinate formulation of the field equations is chosen (e.g. see Green and Zerna [2] and Budiansky [3]), which has been found convenient in several analyses of inelastic behaviour at finite strain [8,9,10].

Relative to a fixed Cartesian frame, the position of a material point in the reference configuration is denoted by the vector $\underset{\sim}{r}$, and the position of the same point in the current configuration is $\underset{\sim}{\bar{r}}$. The displacement vector $\underset{\sim}{u}$ and the deformation gradient $\underset{\sim}{F}$ are given by

$$\underset{\sim}{u} = \underset{\sim}{\bar{r}} - \underset{\sim}{r} \quad , \quad \underset{\sim}{F} = \frac{\partial \underset{\sim}{\bar{r}}}{\partial \underset{\sim}{r}} \tag{2.1}$$

In many cases the reference configuration is identified with the initial undeformed configuration, but there are also cases where another choice of reference is more convenient.

Convected coordinates ξ^i are introduced, which serve as particle labels. The convected coordinate net can be visualized as being inscribed on the body in the reference state and deforming with the material. The displacement vector $\underset{\sim}{u}$ is considered as a function of the coordinates ξ^i and a monotonically increasing time-like parameter t . Covariant base vectors $\underset{\sim}{e}_i$ and $\underset{\sim}{\bar{e}}_i$ of the material net in the reference configuration and the current configuration, respectively, are given by

$$\underset{\sim}{e}_i = \frac{\partial \underset{\sim}{r}}{\partial \xi^i} \quad , \quad \underset{\sim}{\bar{e}}_i = \frac{\partial \underset{\sim}{\bar{r}}}{\partial \xi^i} \tag{2.2}$$

The metric tensors in the reference and current configurations are given by the dot products of the base vectors

$$g_{ij} = \underset{\sim}{e}_i \cdot \underset{\sim}{e}_j \quad , \quad G_{ij} = \underset{\sim}{\bar{e}}_i \cdot \underset{\sim}{\bar{e}}_j \tag{2.3}$$

and their determinants are denoted g and G , respectively, while the inverse of the two metric tensors are denoted by g^{ij} and G^{ij}. Latin indices range from 1 to 3, and the summation convention is adopted for repeated indices.

Components of vectors and tensors on the embedded coordinates are obtained by dot products with the appropriate base vectors. Thus, the displacement components on the reference base vectors satisfy

$$u_i = \underset{\sim}{e}_i \cdot \underset{\sim}{u} \quad , \quad u^i = \underset{\sim}{e}^i \cdot \underset{\sim}{u} \quad , \quad \underset{\sim}{u} = u^i \underset{\sim}{e}_i \tag{2.4}$$

Substituting (2.1a) in (2.2b) and using (2.4c) gives

$$\overline{\underset{\sim}{e}}_i = \underset{\sim}{e}_i + u^k_{,i} \underset{\sim}{e}_k \tag{2.5}$$

where $(\ )_{,i}$ denotes the covariant derivative in the reference frame. The Lagrangian strain tensor $\eta_{ij} = \frac{1}{2}(G_{ij} - g_{ij})$, expressed in terms of displacement components, is then found using (2.3) and (2.5)

$$\eta_{ij} = \frac{1}{2}\left[u_{i,j} + u_{j,i} + u^k_{,i} u_{k,j}\right] \tag{2.6}$$

Large geometry changes are conveniently visualized in terms of *principal fibres*. According to (2.1) the material line segment (fibre) with initial components dr_j on a fixed Cartesian frame has the current components $\overline{dr}_i = F_{ij} dr_j$, and the squared final length of the fibre is

$$\overline{dr}_i \overline{dr}_i = F_{ij} F_{ik} dr_j dr_k \tag{2.7}$$

The stretch λ of a fibre is the ratio of the current and initial lengths, $\overline{dr}_i \overline{dr}_i = \lambda^2 dr_i dr_i$, and the principal fibre directions are those at which the squared stretch is stationary with respect to varying fibre orientation

$$F_{ij} F_{ik} dr_j = \lambda^2 dr_k \tag{2.8}$$

Thus, the squared principal stretches are the eigenvalues of a symmetric, positive definite matrix $F_{ij} F_{ik}$ (see (2.7)) and the associated eigenvectors dr_j specify the orientations of the principal fibres in the reference frame (see Hill [11]). If (2.8) is multiplied by F_{nk} the eigenvalue problem can be rewritten in the form

$$F_{nk} F_{ik} \overline{dr}_i = \lambda^2 \overline{dr}_n \tag{2.9}$$

where the eigenvectors $\overline{dr}_i$ give the current orientations of the principal fibres. With the principal stretches λ_r, the principal logarithmic strains are $\epsilon_r = \ell n(\lambda_r)$.

The true stress tensor $\underset{\sim}{\sigma}$ in the current configuration (the Cauchy stress tensor) has the contravariant components σ^{ij} on the current base vectors, where

$$\underset{\sim}{\sigma} = \sigma^{ij}\bar{\underset{\sim}{e}}_i\bar{\underset{\sim}{e}}_j \quad , \quad \sigma^{ij} = \bar{\underset{\sim}{e}}^i \cdot \underset{\sim}{\sigma} \cdot \bar{\underset{\sim}{e}}^j \tag{2.10}$$

The contravariant components τ^{ij} of the Kirchhoff stress tensor on the current base vectors are defined by

$$\tau^{ij} = \sqrt{G/g}\,\,\sigma^{ij} \tag{2.11}$$

where $\sqrt{G/g} = d\bar{V}/dV = \rho/\bar{\rho}$, expressed in terms of the volume element dV and the density ρ . From the Cartesian components $\underset{\sim}{\sigma}$ of the true stress tensor (2.10a), the values of the principal true stresses and the corresponding orientations in the current configuration are found by standard techniques.

The requirement of equilibrium can be specified in terms of the principle of virtual work

$$\int_V \tau^{ij}\delta\eta_{ij}\,\,dV = \int_S T^i\delta u_i\,\,dS + \int_V \rho f^i\delta u_i\,\,dV \tag{2.12}$$

where V and S are the volume and surface, respectively, of the body in the reference configuration, and $\underset{\sim}{f} = f^i\underset{\sim}{e}_i$ and $\underset{\sim}{T} = T^i\underset{\sim}{e}_i$ are the specified body force per unit mass and surface tractions per unit area in the reference frame. The Euler equations of (2.12) express the same requirement directly in terms of the equilibrium equations and the corresponding boundary conditions

$$- \left[\tau^{ij} + \tau^{kj}u^i_{,k}\right]_{,j} = \rho f^i \tag{2.13}$$

$$u_i = 0 \quad \text{on} \quad S_U \quad , \quad \left[\tau^{ij} + \tau^{kj}u^i_{,k}\right]n_j = T^i \quad \text{on} \quad S_T \tag{2.14}$$

where displacements and tractions are specified on the surface parts S_U and S_T , respectively, and $\underset{\sim}{n} = n_j\underset{\sim}{e}^j$ is the surface normal in the reference state.

For the solution of boundary value problems incremental equilibrium equations are needed, due to the material path dependence. When the current values of all field quantities, e.g. stresses τ^{ij} , displacements u_i and tractions T^i , are assumed known, an expansion of (2.12) about the known state gives to lowest order

$$\Delta t \int_V \left\{\dot{\tau}^{ij}\delta\eta_{ij} + \tau^{ij}\dot{u}^k_{,i}\delta u_{k,j}\right\}dV = \Delta t \int_S \dot{T}^i\delta u_i\,\,dS + \Delta t \int_V \rho\dot{f}^i\delta u_i\,\,dV$$

$$- \left[\int_V \tau^{ij}\delta\eta_{ij}\,\,dV - \int_S T^i\delta u_i\,\,dS - \int_V \rho f^i\delta u_i\,\,dV\right] \tag{2.15}$$

Here, $(\dot{\,}) \equiv \partial(\,\,)/\partial t$ at fixed ξ^i , and Δt is the prescribed increment of the "time" t , so that $\Delta T^i = \Delta t\,\dot{T}^i$ are the components of the prescribed traction increments. The terms bracketed in (2.15) vanish according to (2.12), if the current state satisfies equilibrium. However, in linear incremental analyses the solution tends to drift away from the true equilibrium path, due to incrementation errors, and the bracketed terms in (2.15) can be included to avoid such drifting.

The *constitutive relations* to be considered here are based on the assumption that the total strain-rate is the sum of the elastic and plastic parts, $\dot{\eta}_{ij} = \dot{\eta}^E_{ij} + \dot{\eta}^P_{ij}$. Thus, with an elastic relationship of the form $\overset{\triangledown}{\sigma}^{ij} = R^{ijkl}\dot{\eta}^E_{kl}$, as is often assumed (Hutchinson [12]), the constitutive relations can be written as

$$\overset{\triangledown}{\sigma}^{ij} = R^{ijkl}(\dot{\eta}_{kl} - \dot{\eta}^P_{kl}) \tag{2.16}$$

Here, the Jaumann (co-rotational) rate of the Cauchy stress tensor $\overset{\triangledown}{\sigma}^{ij}$ is related to the convected rate by

$$\overset{\triangledown}{\sigma}^{ij} = \dot{\sigma}^{ij} + \left[G^{ik}\sigma^{jl} + G^{jk}\sigma^{il}\right]\dot{\eta}_{kl} \tag{2.17}$$

An incremental stress-strain relationship in terms of $\dot{\tau}^{ij}$, needed in (2.15), is obtained from (2.16) by substituting (2.17) and the incremental form of (2.11)

$$\dot{\tau}^{ij} = \sqrt{G/g}\,\dot{\sigma}^{ij} + \tau^{ij}G^{kl}\dot{\eta}_{kl} \tag{2.18}$$

The type of elastic relationship considered here is actually hypo-elastic, since it cannot be derived from a work potential [12,13]. However, in the limit of small stresses relative to Young's modulus it reduces to a standard small-strain elastic stress-strain relation. In elasticplastic analyses the elastic contribution to the total straining is usually very small so that the use of this hypo-elastic relationship rather than a truly elastic one is a reasonable approximation.

For *time-independent plasticity* the plastic part of the strain increment $\dot{\eta}^P_{ij}$ is homogeneous of degree one in $\overset{\triangledown}{v}^{kl}$. Using this in (2.17) and solving with respect to the stress increments leads to constitutive relations of the form

$$\dot{\tau}^{ij} = L^{ijkl}\dot{\eta}_{kl} \tag{2.19}$$

where L^{ijkl} is the tensor of instantaneous moduli.

Viscoplastic material models are time-dependent, so that here the parameter t denotes time. In this type of material model the plastic part of the strain rate is a function of the current stresses and strains, but not of the stress-rate. The relationship may be of the form

$$\dot{\eta}^P_{ij} = F(\sigma_e , \epsilon_e)\frac{\partial\Phi}{\partial\sigma^{ij}} \tag{2.20}$$

where Φ is a plastic potential function, while σ_e and ϵ_e are the effective stress and strain. Using this in (2.16) leads to constitutive relations of the form

$$\dot{\tau}^{ij} = L_*^{ijkl}\dot{\eta}_{kl} + \dot{\tau}_*^{ij} \tag{2.21}$$

where $\dot{\tau}_*^{ij}$ acts as an initial stress increment that represents the viscous terms.

3.   UNIQUENESS AND BIFURCATION

For time-independent plasticity, with a constitutive law of the form (2.19), boundary value problems are solved incrementally, using the incremental form of the equilibrium equations (2.12) or (2.13) and (2.14). A theory for the uniqueness and bifurcation in such solids has been developed by Hill [4,5,11]. At the current point of the loading history it is assumed that there are at least two distinct solution increments $\overset{\cdot a}{u_i}$ and $\overset{\cdot b}{u_i}$ corresponding to a given increment of the prescribed load (or displacement). The difference between these two solutions is denoted by $(\tilde{\ }) = (\overset{\cdot}{\ })^a - (\overset{\cdot}{\ })^b$, so that subtraction of the incremental principle of virtual work for the two solutions gives

$$\int_V \left\{ \tilde{\tau}^{ij} \delta\eta_{ij} + \tau^{ij} \tilde{u}^k_{,i} \delta u_{k,j} \right\} dV - \int_S \tilde{T}^i \delta u_i \ dS - \int_V \rho \tilde{f}^i \delta u_i \ dV = 0 \qquad (3.1)$$

where τ^{ij} are the current stresses. Thus, if the solution of the incremental boundary value problem is non-unique, (3.1) has a non-zero solution $(\tilde{\ })$.

For elastic-plastic solids obeying normality Hill [4,5] has made use of the expression

$$I = \int_V \left\{ \tilde{\tau}^{ij} \tilde{\eta}_{ij} + \tau^{ij} \tilde{u}^k_{,i} \tilde{u}_{k,j} \right\} dV - \int_S \tilde{T}^i \tilde{u}_i \ dS - \int_V \rho \tilde{f}^i \tilde{u}_i \ dV \qquad (3.2)$$

to prove uniqueness. A comparison solid is defined by choosing fixed instantaneous moduli $L_c^{ijk\ell}$, which are equal to the current plastic moduli for every material point currently on the yield surface, and the elastic moduli elsewhere. For this comparison solid with fixed moduli (3.2) reduces to the quadratic functional

$$F = \int_V \left\{ L_c^{ijk\ell} \tilde{\eta}_{k\ell} \tilde{\eta}_{ij} + \tau^{ij} \tilde{u}^k_{,i} \tilde{u}_{k,j} \right\} dV - \int_S \tilde{T}^i \tilde{u}_i \ dS - \int_V \rho \tilde{f}^i \tilde{u}_i \ dV \qquad (3.3)$$

A smooth yield surface and normality of the plastic flow rule are often used to model metal plasticity. Then the instantaneous moduli in (2.19) are of the form

$$L^{ijk\ell} = \mathcal{L}^{ijk\ell} - \mu M^{ij} M^{k\ell} \qquad (3.4)$$

where μ is zero or positive for elastic unloading or plastic loading, respectively, while M^{ij} is normal to the yield surface. For this type of material model it can be proved that the relation

$$\tilde{\tau}^{ij} \tilde{\eta}_{ij} \geq L_c^{ijk\ell} \tilde{\eta}_{k\ell} \tilde{\eta}_{ij} \qquad (3.5)$$

is satisfied at every material point, and thus $F \leq I$. Therefore, since any non-trivial solution of (3.1) gives $I = 0$, the requirement $F > 0$ is a sufficient condition for uniqueness.

Equality in (3.5) and thus $I = 0$ for $F = 0$ requires that both solution increments, $\overset{\bullet a}{u}_i$ and $\overset{\bullet b}{u}_i$, give plastic loading at all material points currently on the yield surface. Then, if $\overset{\bullet a}{u}_i$ is identified with the prebifurcation solution and $\tilde{u}_i$ with the bifurcation mode, the variation of the prescribed load (or deformation) parameter λ with the bifurcation mode amplitude ξ initially after bifurcation can be written on the form

$$\lambda = \lambda_c + \lambda_1 \xi + \ldots , \qquad \xi \geq 0 \qquad (3.6)$$

with λ_1 chosen sufficiently large. Generally the minimum value of λ_1 is positive. The relation (3.6) has been extended into an actual post-bifurcation expansion by Hutchinson [14], accounting for elastic unloading zones that spread in the material.

Necking in a uniaxial tensile test specimen is the most well known example of a tensile instability, shown by Considere [15] to occur at the maximum load point, where the uniaxial true tensile stress equals the current tangent modulus, $\sigma = E_t$. Hutchinson and Miles [16] have used the general theory based on eqs. (3.1)-(3.4) to show that 3D-effects delay bifurcation beyond the maximum load point, so that Considere's one dimensional result corresponds to the limit of a long thin bar.

Configuration dependent loading plays a role in some bifurcation analyses, e.g. the rotating turbine disk analysed by Tvergaard [17]. In a co-rotating cylindrical coordinate system with ξ^1 , ξ^2 and ξ^3 denoting the initial radial, angular and axial coordinates, respectively, the external loading is the centrifugal force ρf^i . Here, the body force term in (3.3) takes the form

$$\tilde{f}^i = \omega^2 \tilde{\phi}^i + 2\omega\tilde{\omega}\phi^i \qquad (3.7)$$

$$\phi^1 = \xi^1 + u^1 , \quad \phi^2 = u^2 , \quad \phi^3 = 0 , \quad \tilde{\phi}^\alpha = \tilde{u}^\alpha , \quad \tilde{\phi}^3 = 0 \qquad (3.8)$$

where ω is the angular velocity. If ω is the prescribed quantity, the constraint $\tilde{\omega} = 0$ has to be substituted into (3.7); but this still leaves the first term non-zero. In the more realistic case where the angular momentum is taken to be prescribed, both terms in (3.7) are non-zero, and here the axisymmetric solution can remain unique beyond the maximum angular velocity. Bifurcation into sinusoidal modes with wave number m around the circumference has been investigated [17] for a disk with a central bore, having the initial inner and outer radii R_i and R_0 and the thickness h . Bifurcation predictions in Fig. 1, for different cases and material models, show that the $m = 2$ mode is first critical in all cases, which agrees very nicely with experimental observations by Percy *et al.* [18] who found necking at two opposite sides of the bore before bursting. This necking mode is illustrated by results of a plane stress post-bifurcation analysis shown in Fig. 2.

Non-normality of the plastic flow rule is an important feature of elastic-plastic models describing the frictional dilatant behaviour of rocks or soils [17] and is also important during void nucleation in ductile metals [20]. Here, the instantaneous moduli are of the form [21]

$$L^{ijk\ell} = \mathcal{L}^{ijk\ell} - \mu M_G^{ij} M_F^{k\ell} \qquad (3.9)$$

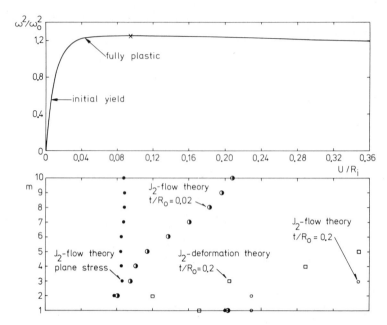

Fig. 1. Angular velocity ω vs. radial bore expansion U for a ductile turbine disk with $R_i/R_0 = 0.1$. Lower diagram shows bifurcation mode number m vs. critical bore expansion (from [17]).

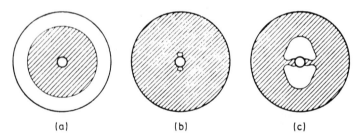

(a)   (b)   (c)

Fig. 2. Current plastic zone (hatched area) for a ductile turbine disk with $R_i/R_0 = 0.1$ (plane stress result). (a) $U/R_i = 0.024$, (b) $U/R_i = 0.078$, (c) $U/R_i = 0.110$.

where M_F^{ij} represents the yield surface normal, and $M_G^{ij} \neq M_F^{ij}$. In this case the relation (3.5) cannot be proved for the usual comparison solid. Raniecki and Bruhns [22] have proposed an alternative comparison solid, in which both M_G^{ij} and M_F^{ij} are replaced by $\frac{1}{2}(M_G^{ij} + rM_F^{ij})/\sqrt{r}$, for $r > 0$. These alternative comparison moduli satisfy (3.5), so that a lower bound to the first critical bifurcation point is obtained, while the usual comparison solid gives an upper bound. Uniqueness is only guaranteed up to the highest lower bound, but various solutions indicate that the actual bifurcation occurs much closer to the upper bound, which tends to be well above the lower bound [22,21].

The formation of a *vertex on the yield surface* is implied by physical models of polycrystalline metal plasticity, based on the concept of single crystal slip [23,24]. At a vertex the instantaneous moduli are functions of the stress-rate direction, e.g. of the form

$$L^{ijk\ell} = L^{ijk\ell}(\overset{\triangledown}{s}{}^{mn}) \qquad\qquad (3.10)$$

where s^{ij} is the stress deviator. For a pyramidal vertex Sewell [25] has shown that the total loading moduli (all slip systems active) satisfy the inequality (3.5). For a phenomenological corner theory of plasticity, J_2 corner theory [26], similar conditions apply [27]. If the prebifurcation solution satisfies total loading everywhere, so that this condition can also apply to the bifurcated solution, the initial post-bifurcation behaviour is of the form (3.6), with the minimum value of λ_1 determined by the requirement of total loading. If total loading is not satisfied everywhere, application of the total loading moduli in (3.3) will only give a lower bound, while application of the moduli corresponding to the prebifurcation solution gives an upper bound [28].

A smooth bifurcation, i.e. $\lambda_1 \rightarrow \infty$ in (3.6), occurs in the case where the total loading region at the vertex shrinks to a single stress rate direction, as has been shown by Needleman and Tvergaard [27]. Such smooth bifurcations can also occur for the classical elastic-plastic solid (3.4). The comparison solid does show a normal bifurcation, but due to the extra requirement of plastic loading (or total loading) at all material points currently at the yield surface, the elastic-plastic solid may not allow for a non-zero $(\tilde{\ })$ solution. At smooth bifurcations the linear lowest order ξ contribution in (3.6) will be replaced by a term with exponent smaller than unity [27]. A treatment of smooth bifurcation based on a special continuum formulation for small strains has been given recently by Nguyen and Triantafyllidis [29].

For *elastic-viscoplastic material behaviour* constitutive relations of the form (2.21) are considered, with the inelastic part of the strain-rate given by a purely viscous expression such as (2.20). It is again assumed that there are two distinct solutions $\overset{\bullet}{u}{}^a_i$ and $\overset{\bullet}{u}{}^b_i$ (displacement rates) corresponding to a given rate of change of the prescribed load (or displacement), and the difference between the two velocity fields is denoted $(\tilde{\ }) = (\overset{\bullet}{\ })^a - (\overset{\bullet}{\ })^b$. Now, since the current stress and strain fields are known, the inelastic part of the current strain rate is uniquely determined by (2.20) so that the only possible difference between the two total strain rates $\overset{\bullet}{\eta}{}^a_{ij}$ and $\overset{\bullet}{\eta}{}^b_{ij}$ is an elastic contribution. Thus, $\tilde{\eta}_{ij}$ is an elastic strain rate field, $\tilde{u}_i$ is compatible with this field, and $\tilde{\tau}^{ij}$ is the corresponding elastic stress rate field. Consequently, (3.1) shows directly that uniqueness and bifurcation is entirely governed by the elastic part of the material response.

Often the function F in (2.20) is taken to be proportional with $[\sigma_e/g(\epsilon_e)]^{1/m}$, where the rate-hardening exponent m is a measure of the strain-rate sensitivity, while the function $g(\epsilon_e)$ incorporates the strain hardening. Thus, for a rate-sensitive version of J_2 flow theory (2.20) takes the form

$$\overset{\bullet}{\eta}{}^P_{ij} = \overset{\bullet}{\epsilon}{}^P_e \frac{3}{2} \frac{s_{ij}}{\sigma_e} \quad , \quad \overset{\bullet}{\epsilon}{}^P_e = \overset{\bullet}{\epsilon}_0 \left[\frac{\sigma_e}{g(\epsilon_e)} \right]^{1/m} \tag{3.11}$$

where s_{ij} is the stress deviator and $\overset{\bullet}{\epsilon}_0$ is a reference strain-rate. In the limit $m = 0$ (3.11) reduces to the corresponding expression for time independent plasticity, and bifurcation is then governed by the instantaneous elastic-plastic moduli as discussed in connection with (3.4); but for any positive value of m (even very small) bifurcation is governed by the elastic moduli. Usually, the elastic bifurcation occurs much later than the plastic bifurcation, so that only imperfections can explain the instabilities observed in the case of elastic-viscoplastic solids. However, for small rate-sensitivities, such as $m = 0.001$, it is found that even extremely small imperfections give essentially identical response of the rate-sensitive and the time independent solids in the vicinity of the elastic-plastic bifurcation point (e.g. see [30]).

4.   LOCALIZATION OF PLASTIC FLOW

The equations given in the previous section are concerned with bifurcation into a diffuse mode, which will often occur while the governing differential equations are elliptic. Localization of plastic flow in a narrow shear band is a different type of instability, which is observed in a wide variety of materials as a rather sudden change from a smooth deformation pattern.

The basic phenomenon of localization can be studied by a relatively simple model problem for solids subject to uniform straining, as illustrated in Fig. 3. The localized shearing is assumed to occur in a thin slice of material with reference normal n_j , while the strain fields outside this band are assumed to remain uniform throughout the deformation history. The quantities inside and outside the band are denoted by $(\ )^b$ and $(\ )^o$, respectively, and a Cartesian reference coordinate system is used. Since uniform deformation fields are assumed both inside and outside the band, equilibrium and compatibility inside the solid are automatically satisfied, apart from the necessary conditions at the band interface. These conditions are

$$u^b_{i,j} = u^o_{i,j} + c_i n_j \tag{4.1}$$

$$(T^i)^b = (T^i)^o \tag{4.2}$$

where c_i are parameters to be determined, while T^i are the nominal tractions on the interface.

The incremental form of (4.2) can be written as (see (2.14))

$$\left[(\dot{\tau}^{ij} + \dot{\tau}^{kj}u^i_{,k} + \tau^{kj}\dot{u}^i_{,k})n_j\right]^b = \left[(\dot{\tau}^{ij} + \dot{\tau}^{kj}u^i_{,k} + \tau^{kj}\dot{u}^i_{,k})n_j\right]^o \tag{4.3}$$

When the constitutive relations (2.19) are substituted herein, using the incremental form of (4.1), the result is the following set of incremental algebraic equations for the unknown parameters c_ℓ

$$(A^{iq\ell})_b\, n_q\dot{c}_\ell = \left[(A^{iq\ell})_o - (A^{iq\ell})_b\right]\dot{u}^o_{\ell,q} \tag{4.4}$$

where $A^{iq\ell}$ denotes the expression

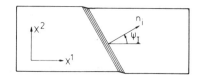

Fig. 3. Plastic flow localization in a uniformly strained solid.

$$A^{iq\ell} = \left[L^{kjpq}(g^i_k + u^i_{,k})(g^\ell_p + u^\ell_{,p}) + \tau^{qj}g^{\ell i} \right] n_j \tag{4.5}$$

If there is an initial inhomogeneity inside the band, such as a lower yield strength or a higher degree of damage, the incremental equations (4.4) are inhomogeneous, since $(A^{iq\ell})_b$ differs from $(A^{iq\ell})_o$, and thus the equations describe the gradual evolution of localization. In such cases localization is said to occur when straining stops outside the band (elastic unloading), and the interest is usually focussed on determining the initial angle of inclination of the band, ψ_I, that gives first localization.

If there is no initial inhomogeneity, equations (4.4) are homogeneous, so that a non-trivial solution for c_ℓ can only occur at a bifurcation point. The first such bifurcation into a shear band mode coincides with the loss of ellipticity of the equations governing incremental equilibrium, and the corresponding critical band inclination is that of the characteristics (Hill [31], Rice [6]). Thus, the analysis of this material instability follows the theoretical framework due to Hadamard [32].

The classical elastic-plastic solid with a smooth yield surface and normality of the plastic flow rule is very resistant to localization, unless the strain hardening level is very low. However, several investigations of shear localization have shown that localization at more realistic strain levels can be found, when deviations from the classical material model are accounted for, such as plastic dilatation, non-normality of the plastic flow rule, or the formation of a vertex on subsequent yield surfaces. Fig. 4 shows an example of localization predictions for a material that forms a vertex on the yield surface, with an inhomogeneity in the form of a lower initial yield stress inside the band [33]. The initial post-bifurcation behaviour is stable, analogous to $\lambda_1 > 0$ in (3.6). For small imperfections it is found that the localized flow saturates in this case, while larger imperfections give failure by the development of large strains inside the band.

The model problem described here is an important tool for the understanding of finite strain behaviour. The incremental solution of the three simultaneous algebraic equations (4.4) for c_ℓ is so simple that parameter studies are feasi-

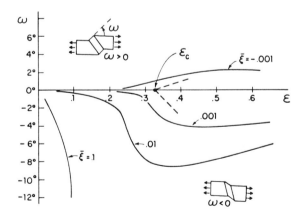

Fig. 4. Growth of a shear band in a solid that develops a vertex on the yield surface. Dashed lines show initial post-bifurcation slopes for a homogeneous solid, and the imperfection amplitude $\bar{\xi}$ represents a material inhomogeneity (from [33]).

ble, e.g. regarding the effect of stress state on loss of ellipticity, the degree of imperfection sensitivity, and the effect of the choice of constitutive
equations. In fact, such model studies should be recommended prior to any elaborate numerical study for non-uniformly strained solids, in which plastic flow
localization could become an issue.

In metal forming problems involving deep drawing the onset of *localized necking
in thin sheets* is an important failure mode that limits the sheet metal formability. In terms of three dimensional theory bifurcation occurs into a diffuse
mode that should be analysed based on eqs. (3.1)-(3.6); but often a plane stress
formulation is used as a simplified model. When a plane stress state is assumed
both inside and outside the band, sheet necking is the plane stress analog of
the shear localization problem described above. Here, equations (4.1) and (4.2)
are replaced by

$$u^b_{\alpha,\beta} = u^o_{\alpha,\beta} + c_\alpha n_\beta \qquad (4.6)$$

$$h^b(T^\alpha)^b = h^o(T^\alpha)^o \qquad (4.7)$$

where Greek indices range from 1 to 2 , and the initial sheet thicknesses
inside and outside the band are denoted by h^b and h^o , respectively. The incremental algebraic equations for the two unknown parameters c_α resulting from
(4.6) and (4.7) are analogous to (4.4).

The simple plane stress (M-K) analysis for the effect of a thickness inhomogeneity on the onset of localized necking in biaxially stretched sheets was first
introduced by Marciniak and Kuczynski [34], while the strong effect of the constitutive model was shown by Stören and Rice [35]. In the sheet metal forming
literature results are presented in terms of a forming limit diagram, which
shows the principal logarithmic strains ϵ^o_1 and ϵ^o_2 outside the band at localization failure. Fig. 5 shows such a forming limit diagram, which illustrates
the effect of the yield surface curvature (b = 0 and b = 1 denote kinematic
and isotropic hardening, respectively) for a power hardening material with N =
0.1 [36]. A porous ductile material model has been used in Fig. 5, with no initial inhomogeneity, but more void nucleation inside the band. The increased
yield surface curvature (b = 0) has the effect of a rounded vertex, which
tends to reduce the strains at failure. As in most forming limit diagrams the
curves in Fig. 5 correspond to proportional straining; but it should be empha-

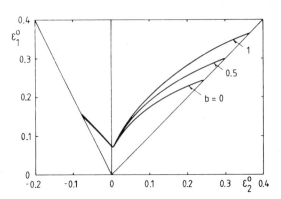

Fig. 5. Forming limit curves for thin sheet under proportional biaxial straining, with a material inhomogeneity represented by more void nucleation inside
the band. Isotropic and kinematic hardening denoted by b = 1 and b = 0 (from
[36]).

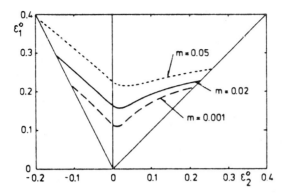

Fig. 6. Forming limit curves for thin sheet under proportional biaxial strain-ing, for various rate hardening exponents. The same material inhomogeneity is used in all cases, represented by more void nucleation inside the band than outside (from [37]).

sized that deviations from this assumption have a strong effect on localization.

In the case of an *elastic-viscoplastic material* the equations (4.1) and (4.2) for localization in a shear band and the equations (4.6) and (4.7) for sheet necking are still valid. Also the incremental algebraic equation (4.4) for c_ℓ retains its form, apart from two additional terms on the right-hand side that result from the viscous terms $\overset{\bullet}{\tau}{}^{ij}_{*}$ in (2.21). Since the elastic moduli $L^{ijk\ell}_{*}$ replace the instantaneous moduli in (4.5), it is clear that bifurcation into a localized mode is entirely governed by elasticity. Thus, bifurcation is not pre-dicted at realistic stress levels, and therefore localization in visco plastic solids relies on the gradual amplification of initial inhomogeneities, analogous to the behaviour discussed in section 3 for diffuse bifurcation modes.

When the stress dependence of the viscous part of the strain rate follows a power law analogous to (3.11), with a rate hardening exponent m, the predic-tions for the corresponding time-independent plastic solid appear in the limit $m \rightarrow 0$. Relative to these predictions, viscosity tends to delay the onset of localization, both in the case of shear bands and sheet necking. This delay is illustrated by the forming limit diagram in Fig. 6, for a power hardening mate-rial with $N = 0.2$ [37]. As in Fig. 5, a porous ductile material model has been used, with more void nucleation inside the band than outside. It is interesting that the delay found for a given value of the rate hardening exponent m is essentially independent of the strain-rate applied, as was also found by Hutchinson and Neale [38].

5. A MATERIAL MODEL FOR CREEP RUPTURE

In the previous sections the constitutive relations have only been mentioned briefly, and results have been presented without specifying the details of the material models employed. Here, a set of constitutive relations for creep in a polycrystalline metal at elevated temperatures will be presented, as an example of a material model that relies on detailed micromechanical studies relating to the development of damage. Subsequently, an analysis of creep failure in a ten-sile test specimen is used to illustrate the material description, thus relating to the discussion of tensile instabilities in the previous sections.

5.1 Material Model

Experimental observations show that microscopic cavities nucleate and grow at the grain boundaries, mainly on grain boundary facets normal to the maximum principal tensile stress direction [39,40,41]. Coalescence of these cavities leads to micro-cracks, and the final intergranular creep fracture occurs as the micro-cracks link up. In addition to dislocation creep of the grains, grain boundary diffusion plays a significant role, and in some cases the diffusive cavity growth is so rapid that the growth rate is constrained by the rate of dislocation creep of the surrounding material [42].

The set of constitutive relations for creep with grain boundary cavitation to be discussed here has been proposed by Tvergaard [7], as an extension of work by Rice [43] and Hutchinson [44]. It is assumed that the grains deform by dislocation creep, represented as power law creep, and that the cavities nucleate and grow on a certain number of grain boundary facets. On each facet the cavities are assumed to be uniformly distributed, with the current average spacing $2b$ (see Fig. 7). Then, the average separation between the two grains adjacent to a cavitating facet is $\delta = V/\pi b^2$, and the rate of growth of this average separation is

$$\dot{\delta} = \frac{\dot{V}}{\pi b^2} - \frac{2V}{\pi b^2}\frac{\dot{b}}{b} \qquad (5.1)$$

where V, $\dot{V}$ and $\dot{b}$ are the volume of one cavity, the rate of growth of this volume, and the rate of change of the cavity half-spacing, respectively.

If a cavitating facet is modelled as a penny-shaped crack, as suggested by Rice [43], the average rate of separation $\dot{\delta}$ given in (5.1) can also be specified as the average rate of opening of the crack. For a single crack in an infinite solid an expression found by He and Hutchinson [45] for this average rate of opening gives

$$\dot{\delta} = \beta \frac{S-\sigma_n}{\sigma_e} \dot{\epsilon}_e^C \, 2R \qquad (5.2)$$

when modified to account for a non-zero normal tensile stress σ_n on the crack surfaces [46]. Here, R is the current radius of the crack, and $S = \sigma^{ij} \bar{n}_i \bar{n}_j$ represents the value of the normal stress on the facet in the absence of cavitation, where σ^{ij} is the macroscopic Cauchy stress tensor and $\bar{n}_i$ is the facet

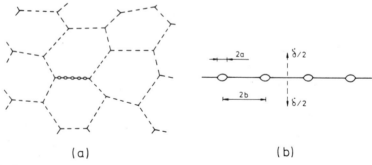

(a)   (b)

Fig. 7.   (a) An isolated, cavitated grain boundary facet in a polycrystalline material.   (b) Equally spaced cavities on a grain boundary.

normal in the current configuration. The macroscopic effective Mises stress is σ_e, and the effective creep strain rate for an uncracked material is taken to be $\dot{\epsilon}_e^C = \dot{\epsilon}_T (\sigma_e/\sigma_0)^n$, where σ_0 is a reference stress quantity, n is the creep exponent, and $\dot{\epsilon}_T$ is a temperature dependent reference strain rate. For the power law creeping material He and Hutchinson [45] have given an asymptotic expression for the value of the parameter β in (5.2)

$$\beta \simeq \frac{4}{\pi}\left[1 + \frac{3}{n}\right]^{-\frac{1}{2}} \tag{5.3}$$

The requirement that the value of $\dot{\delta}$ must be the same in (5.1) and (5.2) determines the value of the normal stress σ_n on the facet and the cavity growth rate $\dot{V}$, which is a function of σ_n.

The macroscopic creep strain rates in a material containing a certain density of penny-shaped cracks have been determined by Hutchinson [44] for a dilute concentration of cracks. For the material subject to grain boundary cavitation the same expression is used, modified to account for a non-zero traction σ_n on the crack surfaces [7],

$$\dot{\eta}_{ij}^C = \dot{\epsilon}_T \left[\frac{\sigma_e}{\sigma_0}\right]^n \left[\frac{3}{2}\frac{s_{ij}}{\sigma_e} + \rho\left\{\frac{3}{2}\frac{n-1}{n+1}\frac{s_{ij}}{\sigma_e}\left(\frac{S-\sigma_n}{\sigma_e}\right)^2 + \frac{2}{n+1}\frac{S-\sigma_n}{\sigma_e}m_{ij}\right\}\right] \tag{5.4}$$

Here, s_{ij} is the macroscopic stress deviator, while $m_{ij} = \bar{n}_i\bar{n}_j$ so that $S = \sigma^{ij}m_{ij}$. The parameter ρ reflects the density of cavitating facets, which may be expressed by [44]

$$\rho \simeq 4R^3 \Lambda (n+1)\left[1 + \frac{3}{n}\right]^{-\frac{1}{2}} \tag{5.5}$$

where R is the crack radius, and Λ is the number of cracks per unit volume.

The total strain-rate is taken to be the sum of the elastic part, the creep part and the thermal expansion part, $\dot{\eta}_{ij} = \dot{\eta}_{ij}^E + \dot{\eta}_{ij}^C + \dot{\eta}_{ij}^T$. Then, using the elastic relationship $\overset{\triangledown}{\sigma}{}^{ij} = \mathcal{R}^{ijk\ell}\dot{\eta}_{k\ell}^E$, the macroscopic stress strain relationship can be written as

$$\overset{\triangledown}{\sigma}{}^{ij} = \mathcal{R}^{ijk\ell}(\dot{\eta}_{k\ell} - \dot{\eta}_{k\ell}^C - \dot{\eta}_{k\ell}^T) \tag{5.6}$$

This relation is transformed into the standard form (2.21) by application of (2.17) and (2.18).

So far, equations (5.1) to (5.6) are independent of the particular mechanisms for the nucleation and growth of cavities to be considered. When surface diffusion is sufficiently rapid relative to grain boundary diffusion, the cavities tend to grow in a quasi-equilibrium spherical-caps shape, while slower surface diffusion favours crack-like cavity shapes. Furthermore, observations indicate continuous nucleation of cavities, with the number of cavities growing as a function of the accumulated inelastic strain. Whatever the mechanisms are, they will result in expressions for $\dot{V}$ and $\dot{b}$ to be substituted in (5.1).

Cavity growth in the spherical-caps shape was considered in [7], making use of detailed micro-mechanical studies that account for the combined influence of diffusion and dislocation creep. Without repeating the detailed expressions here, it is noted that

$$\dot{V} = \dot{V}\left[\sigma_n \ , \ \sigma_e \ , \ \sigma_m \ , \ \mathcal{D} \ , \ n \ , \ \dot{\epsilon}_0 \ , \ \sigma_0 \ , \ a \ , \ b\right] \tag{5.7}$$

where σ_n , σ_e and σ_m are the average normal stress, Mises stress and mean stress, respectively, in the vicinity of the void, $\mathcal{D}$ is the grain boundary diffusion parameter, n , $\dot{\epsilon}_0$ and σ_0 are power law creep parameters, a is the cavity radius and 2b is the average cavity spacing. Cavity coalescence on a grain boundary facet occurs at $a/b \simeq 1$ in cases where the normal stress σ_n is very small, while for larger values of σ_n failure may occur at somewhat smaller values of a/b , by ductile tearing or cleavage of the remaining ligaments [40]. In applications of these constitutive relations cavity coalescence on grain boundary facets has been used as failure criterion, thus neglecting the last part of tertiary creep where the micro-cracks link up.

The complete material model, with the cavity growth expression (5.7) substituted in (5.1), results in creep constrained cavitation ($\sigma_n/S \ll 1$) when the applied stresses are sufficiently low, while unconstrained cavitation ($\sigma_n/S \simeq 1$) is predicted for higher applied stress levels. The behaviour is also strongly affected by the temperature, which enters through the temperature dependence of the material parameters.

5.2 Failure in Tensile Test

Uniaxial tensile tests at constant load are often used to determine the rate of creep and the creep rupture time for metals at high temperatures. The constant load $P = \sigma_N A_0$ is applied at time t = 0 , and due to creep straining the cross-sectional area is gradually reduced below the initial value A_0 so that the true tensile stress grows above the nominal stress σ_N , resulting in an increase of the creep strain rate. Experiments show that localized necking as well as intergranular fracture are important failure mechanisms.

In the absence of cavitation failure occurs when the cross-sectional area is reduced to zero. The corresponding life time expression

$$t_c^0 = \left[n \ \dot{\epsilon}_0 \left(\frac{\sigma_N}{\sigma_0}\right)^n\right]^{-1} \tag{5.8}$$

is readily determined, if the thickness of the bar is assumed to remain uniform, with incompressible material behaviour and no elasticity (Hoff [47]). Generally, grain boundary cavitation will take place during the creep process, and thus the test will be interrupted by intergranular fracture prior to the time specified by (5.8).

Necking can be studied by a simple model, in which the initial cross-sectional area is taken to be A_0 , apart from a slightly thinner region with the initial cross-sectional area $(1 - \bar{\xi})A_0$. A uniaxial stress state is assumed both inside and outside the thinner region. This model, illustrated in Fig. 8, is essentially a one-dimensional version of the M-K type analysis for sheet necking discussed in section 4. For this model the failure time (5.8) in the absence of cavitation is reduced to the value

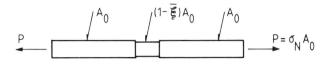

Fig. 8. One-dimensional model of tensile test specimen.

$$t_c = (1 - \bar{\xi})^n \, t_c^0 \qquad (5.9)$$

which corresponds to vanishing cross-sectional area in the neck region, as has been shown by Hutchinson and Obrecht [48].

It is clear from the analysis leading to (5.9) that necking occurs only as an amplification of an initial inhomogeneity. If there is no initial imperfection the thickness remains uniform throughout the life time, and the failure time reduces to that given by (5.8). It is noted that this lack of a bifurcation into a localized mode agrees completely with the results obtained for elastic-visco-plastic materials in sections 3 and 4. For these materials the possibility of elastic bifurcation at unrealistically high stress levels was mentioned; but such bifurcations are excluded in the present model analysis, since elastic deformations are neglected.

The interaction between necking and grain boundary cavitation can be studied by applying the material model of section 5.1 to the model problem in Fig. 8 (see Tvergaard [49]). Results of such studies are shown in Fig. 9 for a material with $n = 5$ and a density of cavitating facets corresponding to $\rho = 0.2$. The cavities are assumed to be present from the beginning, with the initial radius and spacing specified by $(a/b)_I = 0.1$ and $(b/R)_I = 0.1$. The value of the grain boundary diffusion parameter $\mathcal{D}$ is taken to be relatively small, so that dislocation creep gives a noticeable contribution to cavity growth and cavitation is not creep constrained.

Fig. 9 shows the time development of the difference ΔA between the cross-sectional areas inside and outside the neck, the logarithmic strain ϵ_1 in the tensile direction, and the damage parameter a/b, for three different levels of the initial imperfection $\bar{\xi}$. In all three cases failure is predicted to occur by intergranular fracture $(a/b = 1)$ inside the neck, at a stage where significant thinning has taken place corresponding to rather large strains in the neck, $\epsilon_1 \simeq 0.9$. For the larger imperfection a significant amount of necking has oc-

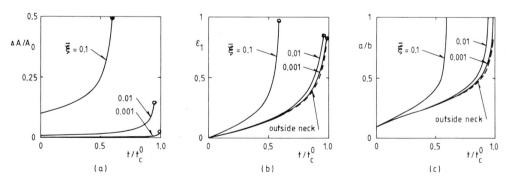

Fig. 9. Behaviour of uniaxial tensile creep test specimens under constant load for $n = 5$, $(a/b)_I = 0.1$ and $(b/R)_I = 0.1$. (a) Difference between cross-sectional areas. (b) Strains. (c) Grain boundary damage (from [49]).

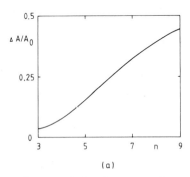
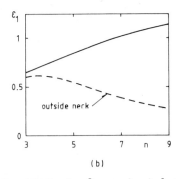

(a)   (b)

Fig. 10. Dependence of failure on the creep exponent, for uniaxial tensile creep tests with $\bar{\xi} = 0.01$, $(a/b)_I = 0.1$ and $(b/R)_I = 0.1$. (a) Difference between cross-sectional areas at failure. (b) Strains at failure (from [49]).

curred $(\Delta A/A_0 \simeq 0.5)$, and the critical time given by (5.9) is essentially reached when failure occurs by coalescence of grain boundary cavities. On the other hand, for the smallest imperfection considered nearly no necking occurs prior to grain boundary cracking; but still the failure time (5.9) is nearly reached, since failure occurs at relatively large strains. It is noted that a smaller value of the nominal stress σ_N , or a larger value of $\mathcal{D}$, increases the relative influence of diffusion on cavity growth, and when this effect is strong enough to give creep constrained cavitation, failure by grain boundary cracking occurs at small strains long before the critical time (5.9) is reached.

As has been mentioned previously, an elastic–viscoplastic material model with a power law expression of the form (3.11) for the stress dependence of the viscous terms tends towards time independent plasticity in the limit $m \to 0$. Since power law creep is of the form (3.11), with $n = 1/m$ and a constant $g = \sigma_0$ (no strain hardening), the value $m = 0.2$ used in Fig. 9 is clearly far from the time independent limit, where bifurcation into a necking mode occurs. Therefore, a stronger effect of necking would be expected for a larger value of the creep exponent, although the largest realistic value of the creep exponent, of the order of $n \simeq 9$, is still far from the values relevant to rate dependent plasticity at room temperature. The effect is illustrated in Fig. 10 by model calculations identical to that in Fig. 9 for $\bar{\xi} = 0.01$, with the values of the grain boundary diffusion parameter $\mathcal{D}$ chosen such that the relative influence of diffusion on cavity growth is kept fixed [49]. In all cases final failure occurs by intergranular cracking in the neck, but clearly much more necking takes place prior to failure for $n = 9$ than for $n = 3$.

6. CONCLUDING REMARKS

An introduction has been given here to the equations governing large deformations of solids. Conditions for uniqueness of the incremental solution have been specified, both for elastic–plastic materials and elastic–viscoplastic materials, and bifurcation as well as plastic flow localization has been discussed in some detail. Furthermore, basic micro-mechanical considerations needed to develop material models that account for failure have been illustrated for the case of creep damage.

Material modelling based on experimental or micro-mechanical studies is an important part of research on damage at finite strains, and due to the strong non-linearities involved analyses have to rely heavily on numerical methods. These aspects will be treated in detail in the following two lectures by Chaboche [50] and Needleman [51].

REFERENCES

[1] Ogden, R.W., Non-Linear Elastic Deformations (John Wiley & Sons, 1984).
[2] Green, A.E. and Zerna, W., Theoretical Elasticity (Oxford University Press, Oxford, 1968).
[3] Budiansky, B., Remarks on Theories of Solid and Structural Mechanics, in: Lavrent'ev, M.A. et al., (eds.), Problems of Hydrodynamics and Continuum Mechanics (SIAM, Philadelphia, 1969) pp. 77-83.
[4] Hill, R., J. Mech. Phys. Solids 6 (1958) 236.
[5] Hill, R., Bifurcation and Uniqueness in Nonlinear Mechanics of Continua, in: Problems of Continuum Mechanics (SIAM, Philadelphia, 1962) pp. 155-164.
[6] Rice, J.R., The Localization of Plastic Deformation, in: Koiter, W.T., (ed.), Theoretical and Appl. Mech. (North-Holland, 1977) pp. 207-220.
[7] Tvergaard, V., Acta Metall. 32 (1984) 1977.
[8] Needleman, A., J. Mech. Phys. Solids 20 (1972) 111.
[9] Tvergaard, V., J. Mech. Phys. Solids 24 (1976) 291.
[10] Needleman, A. and Tvergaard, V., Finite Elements, Special Problems in Solid Mechanics, in: Oden, J.T. and Carey, G.F., (eds.), Finite Elements, Special Problems in Solid Mechanics, Vol. V (Prentice-Hall, Inc., 1984) pp. 94-157.
[11] Hill, R., Aspects of Invariance in Solid Mechanics, in: Adv. Appl. Mech. 18 (1978) 1.
[12] Hutchinson, J.W., Finite Strain Analysis of Elastic-Plastic Solids and Structures, in: Hartung, R.F., (ed.), Numerical Solution of Nonlinear Structural Problems (ASME, New York, 1973) 17.
[13] McMeeking, R.M. and Rice, J.R., Int. J. Solids Structures 11 (1975) 601.
[14] Hutchinson, J.W., J. Mech. Phys. Solids 21 (1973) 163.
[15] Considere, M., Annales des Ponts et Chaussees 9 (1885) 574.
[16] Hutchinson, J.W. and Miles, J.P., J. Mech. Phys. Solids 22 (1974) 61.
[17] Tvergaard, V., Int. J. Mech. Sci. 20 (1978) 109.
[18] Percy, M.J., Ball, K. and Mellor, P.B., Int. J. Mech. Sci. 16 (1974) 809.
[19] Rudnicki, J.W. and Rice, J.R., J. Mech. Phys. Solids 23 (1975) 371.
[20] Gurson, A.L., Porous Rigid-Plastic Materials Containing Rigid Inclusions - Yield Function, Plastic Potential, and Void Nucleation, in: Taplin, D.M.R. (ed.), Proc. Int. Conf. Fracture 2A (1977) pp. 357-364.
[21] Tvergaard, V., J. Mech. Phys. Solids 30 (1982) 399.
[22] Raniecki, B. and Bruhns, O.T., J. Mech. Phys. Solids 29 (1981) 153.
[23] Hill, R., J. Mech. Phys. Solids 14 (1966) 95.
[24] Hutchinson, J.W., Proc. R. Soc. Lond. A318 (1970) 247.
[25] Sewell, M.J., A Survey of Plastic Buckling, in: Leipholz, H., (ed.), Stability (University of Waterloo Press, 1972) 85.
[26] Christoffersen, J. and Hutchinson, J.W., J. Mech. Phys. Solids 27 (1979) 465.
[27] Needleman, A. and Tvergaard, V., Aspects of Plastic Post-Buckling Behaviour, in: Hopkins, H.G. and Sewell, M.J., (eds.), Mechanics of Solids, The Rodney Hill 60th Anniversary Volume (Pergamon Press, Oxford, 1982) 453.
[28] Tvergaard, V., Int. J. Thin-Walled Struct. 1 (1983) 139.
[29] Nguyen, S.Q. and Triantafyllidis, N., Plastic Bifurcation and Postbifurcation Analysis for Generalized Standard Continua. Ecole Polytechnique, Palaiseau, France (1988).
[30] Tvergaard, V., Rate-Sensitivity in Elastic-Plastic Panel Buckling, in: Dawe, D.J. et al., (eds.), Aspects of the Analysis of Plate Structures, A volume in honour of W.H. Wittrick (Clarendon Press, Oxford, 1985) 293.
[31] Hill, R., J. Mech. Phys. Solids 10 (1962) 1.
[32] Hadamard, J., Lecons sur la propagation des ondes et les équations de l'hydrodynamique. Libraire Scientifique A. (Hermann, Paris, 1903).
[33] Hutchinson, J.W. and Tvergaard, V., Int. J. Solids Structures 17 (1981) 451.
[34] Marciniak, K. and Kuczynski, K., Int. J. Mech. Sci. 9 (1967) 609.
[35] Stören, S. and Rice, J.R., J. Mech. Phys. Solids 23 (1975) 239.

[36] Tvergaard, V., J. Mech. Phys. Solids 35 (1987) 43.

[37] Needleman, A. and Tvergaard, V., Limits to Formability in Rate-Sensitive Metal Sheets, in: Carlsson, J. and Ohlson, N.G., (eds.), Mechanical Behavior of Materials – IV (Pergamon Press, Oxford, 1984) pp. 51-65.

[38] Hutchinson, J.W. and Neale, K.W., Sheet Necking – III. Strain Rate Effects, in: Koistinen, D.P. and Wang, N.-M., (eds.), Mechanics of Sheet Metal Forming (Plenum Publ. Corp., New York, 1978) pp. 111-126.

[39] Hull, D. and Rimmer, D.E., Phil. Mag. 4 (1959) 673.

[40] Cocks, A.C.F. and Ashby, M.F., Progress in Materials Science 27 (1982) 189.

[41] Argon, A.S., Mechanisms and Mechanics of Fracture in Creeping Alloys, in: Wilshire, B. and Owen, D.R.J., (eds.), Recent Advances in Creep and Fracture of Engineering Materials and Structures (Pinerage Press, Swansea, U.K., 1982) 1.

[42] Dyson, B.F., Metal Science 10 (1976) 349.

[43] Rice, J.R., Acta Metall. 29 (1981) 675.

[44] Hutchinson, J.W., Acta Metall. 31 (1983) 1079.

[45] He, M.Y. and Hutchinson, J.W., J. Appl. Mech. 48 (1981) 830.

[46] Tvergaard, V., J. Mech. Phys. Solids 32 (1984) 373.

[47] Hoff, N.J., J. Appl. Mech. 20 (1953) 105.

[48] Hutchinson, J.W. and Obrecht, H., Tensile Instabilities in Strain-Rate Dependent Materials, in: Proc. 4th Int. Conf. on Fracture, Vol. 1 (Waterloo, Canada, 1977) 101.

[49] Tvergaard, V., Acta Metallurgica 35 (1987) 923.

[50] Chaboche, J.L., Phenomenological Aspects of Continuum Damage Mechanics, this volume.

[51] Needleman, A., Computational Micromechanics, this volume.

Theoretical and Applied Mechanics
P. Germain, M. Piau and D. Caillerie (Editors)
Elsevier Science Publishers B.V. (North-Holland)
© IUTAM, 1989

SOME PHYSICAL PHENOMENA ASSOCIATED WITH CAVITATION

J.H.J. VAN DER MEULEN

Maritime Research Institute Netherlands
Wageningen, The Netherlands

The collapse of cavitation bubbles near a boundary may be
accompanied by jet or counterjet formation, the generation of a
vortex cavity of toroidal shape, rebound shock waves and other
physical phenomena. The important parameter appears to be the
proximity of the bubble to the boundary. Physical effects associated
with bubble collapse in acoustic or hydrodynamic cavitation are
luminescence, noise and erosion. It is shown that all of these
effects can be used as measures of cavitation intensity for
hydrodynamic cavitation.

1. INTRODUCTION

Acoustic and hydrodynamic cavitation are related to sound fields or low
pressure regions in a flow field. In general, cavitation can be regarded as a
process in which different phases can be discerned. These phases are:
inception, growth, motion and collapse. In recent years a considerable amount
of both theoretical and experimental work has been done to predict and
describe the implosion behaviour of single cavitation bubbles, or clusters of
cavities, near a boundary. In section 2 of the present paper new developments
in this field are reviewed.

The collapse of bubbles may produce physical effects, such as noise,
erosion and luminescence. In the field of acoustic cavitation, luminescence
has been studied in great detail. Some work on the origin of this phenomenon
is discussed in section 3 of this paper. In section 4, it is shown that
hydrodynamic cavitation may also produce luminescence. The assessment of the
intensity of hydrodynamic cavitation can thus be based on measuring the
intensity of noise, erosion or luminescence. Some new developments in this
field are presented.

2. CAVITATION BUBBLE COLLAPSE

2.1. Review

Most studies on the collapse of cavitation bubbles are aimed at
explaining the damage of components subjected to cavitation. Therefore, these
studies are usually related to bubbles collapsing near a boundary. In 1917,
Lord Rayleigh (1) presented an analysis on the collapse of a spherical cavity
in an infinite mass of fluid and derived a formula for the pressure at
collapse. The simple formula derived for the time of complete collapse
appeared to be surprisingly accurate, when compared to more recent
computations. Initially, the damage was merely attributed to shock waves
generated by the collapse and transmitted through the liquid upon an adjacent
surface. In that case the site of the collapse should be very close to the
damage surface, but then the collapse would no longer be spherical. For a
cavity in contact with a solid boundary, Naudé and Ellis (2) derived a theory
which predicted the occurrence of a jet. They also produced experimental
evidence for the damage capability of the jet by using high-speed

cinematography and by applying a soft surface technique. In an experimental study Shutler and Mesler (3) confirmed the occurrence of the jet, but they also noticed that the cavity attained a toroidal shape at reaching its minimum volume. The observed damage (on a soft surface) was attributed to pressure pulses caused by rebound bubbles originating from this torus. A most interesting analysis on the physics of imploding bubbles near a boundary was presented by Benjamin and Ellis (4) in 1966. Their reasoning was based on the concept of the Kelvin impulse. Thus, they explained the appearance of circulation in the liquid and the formation of a vortex system during the final stage of collapse, resulting in a toroidal shape of the cavity. This process is preceded by the formation of a jet. They also mentioned the migration of a collapsing bubble towards the wall and the fact that an initially spherical cavity collapsing near a wall at first becomes elongated in the normal direction.

Kling and Hammitt (5) studied the collapse of bubbles near a wall in a flowing system. When the initial distance between the bubble and the wall was sufficiently small, a jet was observed and a damage pit was created in the soft aluminium surface. The primary effect appeared to be the substantial movement of the centroid of the bubble towards the wall during the final stage of collapse and the initial stage of rebound.

In most of the experimental studies mentioned so far the bubbles were generated by electric-spark discharge. Gibson (6) has shown that the heat produced by the spark is not affecting the bubble motion during the subsequent collapse, unless the ambient pressure is lower than one tenth of a bar. Another technique of producing bubbles was introduced by Lauterborn (7, 8). The technique is based upon the focussing of a ruby laser pulse in water with the formation of a plasma and the subsequent formation of a bubble. This optical technique of producing cavitation has been called optic cavitation. Bubble dynamics was studied by high-speed cinematography with up to 900,000 frames per second. The studies incorporated bubble collapse and jet formation near a wall and interaction of bubbles. Besides the observation of jets, Lauterborn and Bolle (9) also observed the occurrence of counterjets on the first and second implosion of a bubble near a wall. Since these phenomena can also been observed in the photographs in Ref. (5), the authors concluded that counterjet formation is a distinct feature of the dynamics of a bubble near a wall.

Another mechanism for generating high pressures is the interaction between bubbles in a cloud during a collective collapse. In a theoretical study, Van Wijngaarden (10) showed that this mechanism is capable for causing the trailing edges of ship propellers to bend towards the pressure side, a phenomenon which had been observed frequently. Hansson and Mørch (11) introduced the concept of energy transfer in analyzing the collapse of a hemispherical cluster of cavities. The most important result was that a cavity cluster collapses from its outer boundary towards the centre and that the pressure at the centre increases considerably, causing the central cavities to be much more erosive than the outer cavities. Similar results are found in a theoretical study on cloud cavitation by Chahine (12).

2.2. Recent Developments

During the past two years several publications appeared which have contributed largely to our understanding of the physics of cavitation bubble collapse. Tomita and Shima (13) reported on a detailed experimental study on the mechanism of impulsive pressure generation from a single bubble collapsing near a boundary. For bubbles situated very close to the boundary they concluded that the liquid jet acts like a tiny bubble generator, while the high pressure developing in the final stage of collapse of the original bubble creates the driving pressure to collapse individual bubbles. Progress on theoretical calculations of the growth and collapse of bubbles near a rigid boundary or a free surface has been reported by Blake, Taib, and Doherty (14, 15) and Blake and Gibson (16). These authors were able to obtain in greater detail numerical results on (a) successive bubble shapes during collapse, (b) particle pathlines through growth and collapse and (c) pressure contours late in the collapse phase. In general, it was found that collapsing bubbles migrate towards a rigid boundary and away from a free surface, unless buoyancy forces become dominant (for extremely large bubbles only) in which case an opposite behaviour was found. Vogel (17) performed a comprehensive experimental study on the implosion behaviour of laser-induced bubbles near a (rigid) boundary in which the parameter γ, representing the ratio of the distance to the wall and the bubble radius, was systematically varied. Different optical and acoustical techniques were used to obtain an understanding of the process of jet, counterjet and vortex formation during collapse. The work included a study on shock waves transmitted by the imploding bubbles. The time duration of the shock wave passage appeared to be in the order of 100 ns. A new approach in studying the collapse of arrays of cavities is due to Dear and Field (18). The method consists of producing a thin layer of a water/gelatine mixture in which a number of required cavities are cut out. The layer is placed between glass blocks and is hit by a shock wave with a strength of 2.6 kbar. The implosion behaviour of the (two-dimensional) cavities is visualized using schlieren photography. Based on the results of their study, the authors concluded that jet production by shock interaction is a key damage mechanism in cavitation erosion.

Van der Meulen and Van Renesse (19, 20) studied the implosion behaviour of bubbles near a body in a static fluid and in real flow conditions. Some of the results, i.e. those that pertain to the static fluid, will be presented here.

The experiments were made in the MARIN high speed water tunnel. The test section had a 40 mm x 80 mm rectangular cross section. The body consisted of a Brüel en Kjaer Type 8100 hydrophone with a hemispherical nose and a 21.0 mm dia. cylindrical shaft. This hydrophone acted as a boundary for nearby imploding bubbles, whereas the hydrophone signal was used to assess the instants of bubble formation, first implosion and second implosion. A single stage Q-switched ruby laser induced cavitation bubbles above the cylindrical part of the hydrophone at an axial distance of 16.1 mm from the nose. The maximum bubble diameter was usually 2.5-3 mm. The optical system is schematically represented in Figure 1. The ruby laser flash had a time duration of 25 ns, while the energy was adjusted between about 0.15 and 1 J/pulse, depending on the tunnel pressure. This resulted in 6-40 MW pulses which were sufficiently powerful to cause a breakdown in the focal spot in water. The optical recording set-up consisted of a 35 mm camera and a Xenon flash tube which generated 1 μs single light flashes. The trigger signal that fired the ruby laser flash tube was fed into a variable time delay which, in turn, triggered the driver circuit of the Xenon flash tube. Single exposures of the cavitation bubbles could thus be made at any arbitrary moment after their initiation. Series of such exposures with increasing time delay rendered an overall picture of the process of bubble growth and collapse in time.

The water used in the experiments had a temperature of 20 C and an air content of 12 cm^3/l. Without the addition of any energy absorbing particles in the water, the repeatability of the bubble generation by the ruby laser pulse appeared to be rather poor. To achieve adequate information on the implosion behaviour of the bubbles it was necessary to improve on the repeatability. An acceptable level was obtained by dispersing a small amount of cream in the tunnel water, just prior to performing a test series, so that a concentration of about 50 ppm was reached. The experiments in still water were made at a static pressure of 82 kPa.

The implosion behaviour of bubbles near a boundary is mainly determined by the parameter $γ=2 h/d_{max}$, where h is the initial distance of the bubble centroid to the boundary and d_{max} the maximum bubble diameter. The total test program, discussed in (20), comprised 18 test series: 6 values of the parameter γ and 3 values of the velocity V_o ahead of the body (0, 10 and 20 m/s). Each test series consisted of about 40-60 runs; at each run a photograph was made at a set time delay t (t=0 corresponds to the beginning of the bubble

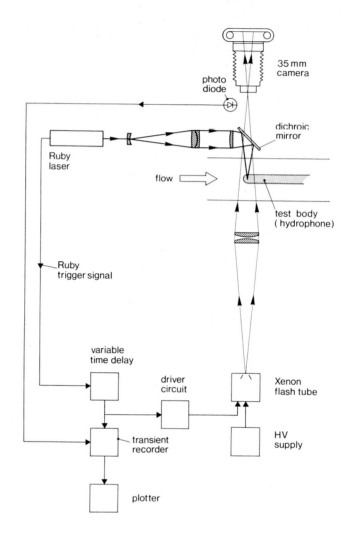

Fig. 1.   Schematic diagram of optical system.

growth at the instant of the laser pulse generation). From the recording of
the hydrophone signal, the instant of the first implosion t_1 and of the
second implosion t_2 were derived. Since t_1, and thus the maximum bubble
size, still showed considerable scatter, a selection of about 25-30 runs was
made for which the variation in t_1 was within $\pm$ 6 per cent. From this
selection the maximum bubble diameter d_{max} and average values of t_1 and
t_2 were obtained. A survey of the test conditions and results for V_0=0 m/s
is presented in Table 1. In this table $\bar{t}_1$ and $\bar{t}_2$ refer to averaged values.
t_R refers to the time of collapse according to Lord Rayleigh's (1) formula:

$$t_R = 0.457 \, d_{max} \, \sqrt{\rho/P},$$

where ρ is the liquid density and P the pressure.

For γ=3.09 the implosion behaviour of the bubble is rather weakly
influenced by the presence of the wall. The sequence observed is: (1) growth
to a spherical bubble of maximum size at t/t_1=0.5, (b) spherical collapse,
(c) elongation of the bubble in the normal direction during the final stage of
collapse (β=1.05 for t/t_1=0.95, where β is the ratio between the bubble
diameter in the normal direction and the bubble diameter in a direction

Table 1. Survey of test conditions and experimental
results for V_o=0 m/s

h	d_{max}	γ	$\bar{t}_1$	$\bar{t}_2$	$\bar{t}_2/\bar{t}_1$	$\bar{t}_1/2t_R$
mm	mm		μs	μs		
4.46	2.89	3.09	286	437	1.53	0.98
2.93	2.81	2.09	285	471	1.65	1.01
2.58	3.02	1.71	313	550	1.75	1.03
1.79	2.89	1.24	314	536	1.71	1.08
1.36	2.72	1.00	305	474	1.56	1.11
0.71	2.60	0.55	303	–	–	1.15

parallel to the wall), (d) shortly after the first implosion occurrence of
weak jet and weak counterjet and (e) after the second implosion occurrence of
small bubbles, most propably corresponding to the first jet and counterjet.
When γ becomes smaller, the movement of the bubble towards the wall becomes
progressively larger. This is shown in Figure 2, where the distance of the
bubble centroid to the wall y relative to the initial distance h is plotted
against $t/\bar{t}_1$ for all six cases considered.

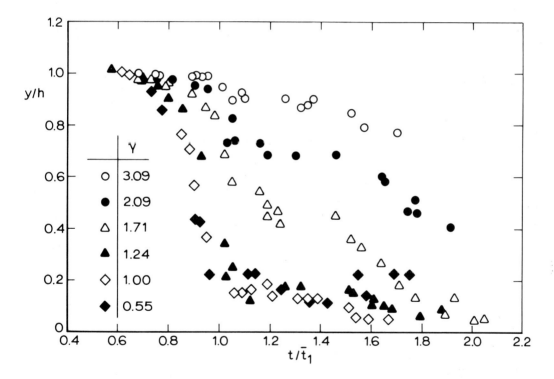

Fig. 2.   Relative distance of bubble centroid to wall versus relative time
for different values of γ.

For γ=2.09 the influence of the proximity of the wall on the implosion
behaviour is much stronger. The bubble is more elongated during the final
stage of collapse (β=1.08 for t/t_1=0.95). After the first implosion a
counterjet is observed, followed by a jet. At the second implosion small
bubbles are observed, most probably corresponding to a second counterjet and
jet. For γ=1.71 the bubble attaches to the wall between the first and second

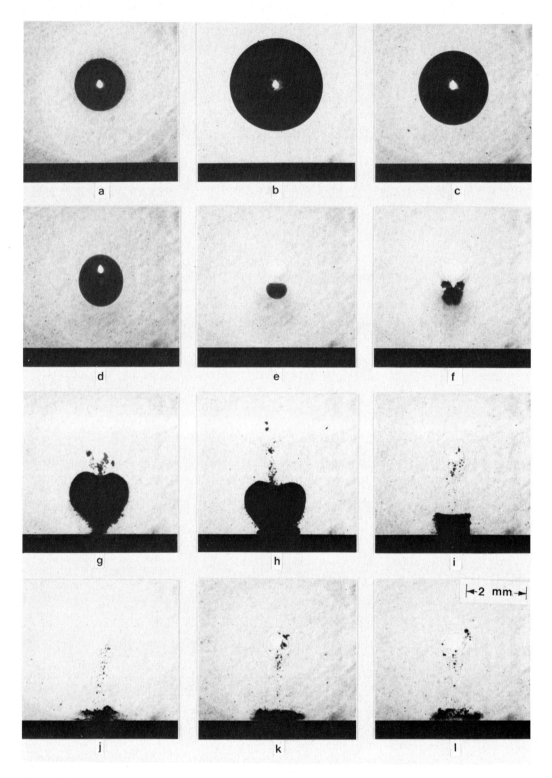

Fig. 3.　Stages of bubble growth, implosion, rebound and second implosion for
γ=1.71.
(a) t/t$_1$=0.06, (b) t/t$_1$=0.49, (c) t/t$_1$=0.80,
(d) t/t$_1$=0.95, (e) t/t$_1$=1.00, (f) t/t$_1$=1.02,
(g) t/t$_1$=1.19, (h) t/t$_1$=1.52, (i) t/t$_2$=0.99,
(j) t/t$_2$=1.01, (k) t/t$_2$=1.06, (l) t/t$_2$=1.19.

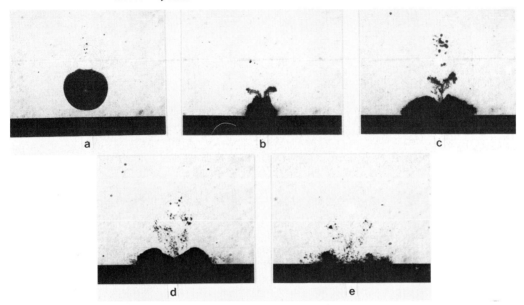

Fig. 4.   Stages of bubble implosion, formation of vortex cavity and second
           implosion for $\gamma = 1.24$.
           (a) $t/t_1 = 0.95$, (b) $t/t_1 = 1.03$, (c) $t/t_1 = 1.12$,
           (d) $t/t_1 = 1.60$, (e) $t/t_2 = 1.11$.

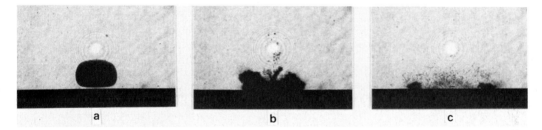

Fig. 5.   Stages of bubble implosion, formation of vortex cavity and second
           implosion for $\gamma = 1.00$.
           (a) $t/t_1 = 0.95$, (b) $t/t_1 = 1.09$, (c) $t/t_2 = 1.00$.

implosion. The series of photographs shown in Figure 3 gives a good
representation of the sequence of events. The bubble grows (Figure 3a,
$t/t_1 = 0.06$) and reaches a maximum size (Figure 3b, $t/t_1 = 0.49$). Next, the
bubble starts to implode and obtains an elongated shape (Figure 3c,
$t/t_1 = 0.80$, $\beta = 1.04$ and Figure 3d, $t/t_1 = 0.95$, $\beta = 1.14$). Figure 3e refers to
the instant of the first implosion ($t/t_1 = 1.00$), whereas Figure 3f
($t/t_1 = 1.02$) shows the beginning of a counterjet. The jet appears somewhat
later. In Figure 3g ($t/t_1 = 1.19$) the jet has struck the wall and in Figure 3h
($t/t_1 = 1.52$) the bubble has moved closer to the wall. A number of small
bubbles can still be seen, corresponding to the first counterjet. Figure 3i
($t/t_2 = 0.99$) was taken just prior to the second implosion, and Figure 3j
($t/t_2 = 1.01$) just after the second implosion. Small bubbles appear
corresponding to a second counterjet. These bubbles are also found in Figure
3k ($t/t_2 = 1.06$). Finally, in Figure 3l ($t/t_2 = 1.19$), the bubble starts to
disintegrate.
        For $\gamma = 1.24$ the bubble attaches to the wall shortly after the first

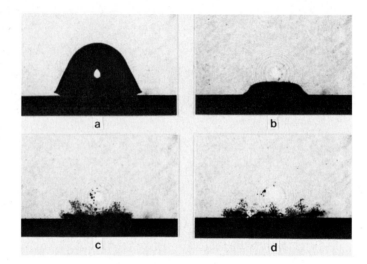

Fig. 6.   Stages of bubble implosion with formation of cap-shaped cavity and disintegration of cavity for $\gamma= 0.55$.
(a) $t/t_1=0.77$, (b) $t/t_1=0.96$,
(c) $t/t_1=1.35$, (d) $t/t_1=1.43$.

implosion and a vortex cavity is created afterwards. A number of characteristic photographs is presented in Figure 4. At about $t/t_1=0.93$ a counterjet begins to develop. An example is shown in Figure 4a ($t/t_1=0.96$). In Figure 4b ($t/t_1=1.03$) the bubble is attached to the wall and a distinct counterjet is observed. Next, in Figure 4c ($t/t_1=1.12$) the bubble has emerged into a vortex cavity with a toroidal shape, which develops further (Figure 4d, $t/t_1=1.60$). Finally, in Figure 4e ($t/t_2=1.11$), the vortex cavity begins to disintegrate. Also for $\gamma=1.00$ the bubble attaches to the wall shortly after the first implosion and a vortex cavity is created afterwards. In Figure 5a ($t/t_1=0.95$) a rather flat bubble is shown at the instant of counterjet inception. The counterjet appears to be weak. In Figure 5b ($t/t_1=1.09$) the vortex cavity structure and the counterjet are clearly distinguished. Remains of the vortex cavity of toroidal shape are shown in Figure 5c ($t/t_2=1.00$).

For $\gamma=0.55$ the bubble attaches to the wall during the initial stage of growth, develops into a cap-shaped cavity and implodes on the wall. A second implosion does not occur. Figure 6a ($t/t_1=0.77$) shows the cap-shaped cavity attached to the wall. In Figure 6b ($t/t_1=0.96$) the bubble has flattened, whereas the beginning of a weak counterjet can be observed. During subsequent stages the counterjet remains weak. In Figure 6c ($t/t_1=1.35$) the bubble begins to disintegrate. A further stage of disintegration is shown in Figure 6d ($t/t_1=1.43$). Finally, a cloud of bubbles in the vicinity of the wall remains.

The present results and those by Vogel [17] stress upon the fact that the implosion behaviour of a bubble near a boundary is strongly dependent on the relative proximity of the bubble to the boundary, expressed by the parameter γ. As shown in [20], this dependency is hardly affected by real flow conditions.

3.   ASSOCIATED PHYSICAL PHENOMENA

The collapse of bubbles, clusters of bubbles or large cavities may produce physical phenomena such as pressure pulses or light emission. Pressure pulses may lead to harmful effects such as erosion, noise or vibrations. One of the main problems in the applied fields of naval hydrodynamics and fluid machinery is the assessment of the intensity of cavitation, leading to either one of these effects. Full-scale predictions of these effects are based on computations of pressure fluctuations or on model test results. Erosion measurements applying so-called soft surface techniques (21, 22), noise measurements by means of hydrophones (23 e.g.) and pressure fluctuation measurements by means of pressure transducers (24, 25 e.g.) are all methods to obtain model values of the relevant cavitation effect. Scaling problems, however, may interfere with predictions. In particular, the quantitative prediction of full-scale erosion is still considered to be a too complex problem.

Besides pressure pulses, bubble collapse may also be accompanied by the emission of light, or luminescence. This phenomenon, widely recognized in the field of acoustic cavitation, has received little or no attention in the field of hydrodynamic cavitation, until recent years. Luminescence was discovered by Frenzel and Schultes (26) in 1935, when observing water subjected to a 500 kHz sound field. Since then, many investigations were made to explain the origin of this light emission. Kuttruff (27) performed a detailed experimental study on the relationship between the emission of flashes of luminescence light and shock waves, due to acoustic cavitation. According to Meyer and Kuttruff (28) the time duration of these light flashes is smaller than 10 ns. The experiments by Kuttruff (27) clearly showed that light emission occurs during the final stage of the implosion of bubbles. However, the intensity of the light flashes did not correlate with the intensity of the shock waves. Additional experiments were made in mercury, where luminescence was observed by means of a glass rod oscillating at a frequency of 25 kHz. These experiments eliminated the theory that luminescence would arise from discharges between electrical charges on the surface of cavitation bubbles. Flynn (29) made a thorough analysis of existing work on luminescence and concluded that the origin of luminescence is thermal: the temperature in the collapsing cavity increases to such an extent that the cavity contents are brought to incandescence. The computations and experiments performed by Hickling (30, 31) and Young (32) lend further support to the thermal theory. It was shown that the luminous intensity from a gas dissolved in water has an inverse relation with the thermal conductivity of the gas. For water saturated with Xenon gas, which had the lowest thermal conductivity of all gases investigated, the luminous intensity appeared to be 52 times higher than that of water saturated with air.

In an attempt to find an accessible manner to quantify the intensity of hydrodynamic cavitation, Van der Meulen (33, 35, 36) and Van der Meulen and Nakashima (34) studied the feasibility of employing the measurement of the intensity of luminescence due to hydrodynamic cavitation. Some of the results of this study are presented in the next section.

4.   RELATIONSHIP BETWEEN PHYSICAL PHENOMENA

In his classic treatise on acoustic cavitation, Flynn (29) introduced the concept of cavitation activity measure. He distinguished five different measures of cavitation activity, namely: (1) visual observation, (2) luminescence, (3) chemical reactions, (4) erosion and (5) radiated pressure waves (noise). Cavitation measures provide quantitative information about the effects of cavitation events in a cavitation field. When this concept is applied to hydrodynamic cavitation it may be more appropriate to use the definition: measure of cavitation intensity. Although observations of hydrodynamic cavitation are often used as a measure of cavitation intensity, it should be noted that observations have to be regarded in a qualitative sense only, and even then they may be quite misleading. Useful measures of cavitation intensity are erosion and noise. If hydrodynamic cavitation is accompanied by luminescence, then this phenomenon could also be considered as a measure of cavitation intensity.

The work at MARIN (33-36) was concerned with the relationship between luminescence, noise and erosion from cavitation on a hydrofoil. In particular,

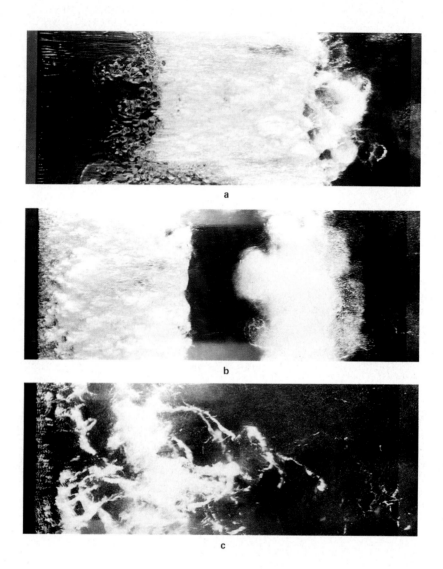

Fig. 7.   Photographs of (a) sheet cavitation ($\alpha=4^\circ$, $\sigma=1.00$), (b)
           sheet-cloud cavitation ($\alpha=7^\circ$, $\sigma=1.70$) and (c) vortex-cloud
           cavitation ($\alpha=12^\circ$, $\sigma=2.91$) on NACA 16-012 hydrofoil for
           $V_0=15$ m/s.

the study was aimed at finding the velocity dependence for these measures of
cavitation intensity. The experiments were made in the high speed water tunnel
with a test section having a rectangular cross-sectional area of 40 mm x 80
mm. The body was a two-dimensional, modified NACA 16-012 hydrofoil with a
chord length of 70 mm and a span of 40 mm. For the luminescence and noise
measurements a stainless steel hydrofoil was used, whereas the erosion
measurements were made with a series of pure copper (99.9 per cent Cu)
hydrofoils. Previous experiments with a modified NACA 16-012 hydrofoil (37)
had shown that, depending on the angle of attack α, cavitation could be
classified into four characteristic types: bubble, sheet, sheet-cloud and
vortex-cloud cavitation. Bubble cavitation led to anomalous results (35) and

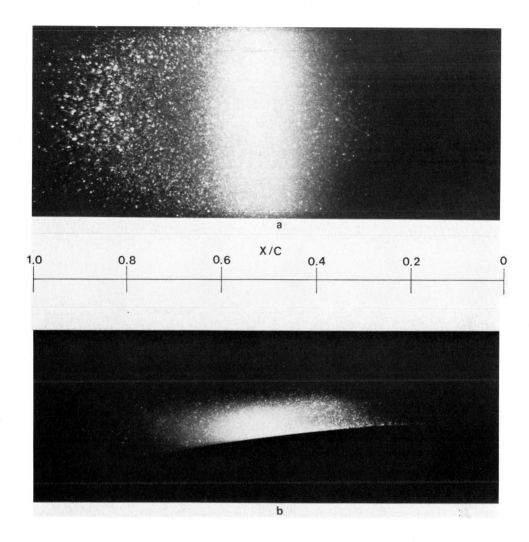

X/C

| 1.0 | 0.8 | 0.6 | 0.4 | 0.2 | 0 |

Fig. 8.   Photographic recordings of luminescence due to sheet-cloud
cavitation on NACA 16-012 hydrofoil at $\alpha=7^\circ$, $V_o=15$ m/s and
$\alpha=1.69$. The photographs (a) and (b) refer to top and side view
respectively. The flow direction is from right to left. The
chordwise position (x/c=0-1.0) is indicated.

will not be discussed here. In Figure 7, photographs are presented of sheet,
sheet-cloud and vortex-cloud cavitation. These photographs were all taken at a
velocity $V_o=15$ m/s, where V_o is the velocity in the test section ahead of
the hydrofoil. Figure 7a refers to sheet cavitation for $\alpha=4^\circ$ and $\sigma=1.00$,
where σ is the cavitation number. The sheet cavitation can be divided into a
sheet cavity proper, consisting of a part with a smooth surface and a part
with a cloudy surface, and trailing clouds. Figure 7b refers to sheet-cloud
cavitation ($\alpha=7^\circ$, $\sigma=1.70$). This type of cavitation is characterized by the
cyclic generation of a cavitation cloud. This cloud can be observed in the
photograph; it is preceded by an attached cavity with a cloudy appearance. Its
behaviour shows a strong resemblence with the observations of developed
cavitation on an axisymmetric body by Knapp (38) and with observations of
developed cavitation on an oscillating or still hydrofoil by Shen and Peterson
(39). Figure 7c refers to vortex-cloud cavitation ($\alpha=12^\circ$, $\sigma=2.91$). This type
of cavitation is typically unattached. It consists of both cavitation clouds
and cavitating vortices.

Initial visual observations of the possible occurrence of luminescence was made at the following test condition: $\alpha=10°$, $V_o=15$ m/s, a cavity length of 30 mm, a water temperature of 13 C and a gas content of 9 cm^3/l. Even with dark adapted eyes not a single trace of luminescence was observed. However, the addition of a small amount of Xenon gas to the tunnel water did result in observable light emission. The following procedure was used in all subsequent luminescence tests. First, the water was deaerated till a gas content of 3 cm^3/l was reached. Next, one liter of Xenon gas at a pressure of 98 kPa and a temperature of 18 C was released in the tunnel and allowed to dissolve. In this way the gas content raised to about 18 cm^3/l. The water temperature during the tests was kept at 13 C ($\pm$ 2 C). Besides visual observations, luminescence was recorded on film. For this purpose Kodak Tri-X-pan film with an ASA rating of 400 was used. Photographs of luminescence due to sheet-cloud cavitation ($\alpha=7°$, $V_o=15$ m/s, $\sigma=1.69$), taken with the camera above the test section and from the side of the test section, are presented in Figure 8a and 8b respectively. The exposure time amounted to 30 minutes. The direction of the flow is from right to left. A scale indicating the chordwise position (x/c=0-1.0) is included. In Figure 8a the maximum brightness occurs at about x/c=0.5. The structure of the light recordings in this area is very fine, indicating the occurrence of many weak light flashes. A course structure is found in the area around x/c=0.8, indicating the occurrence of a much smaller number of strong light flashes. In some of the recordings of the light flashes in this area the shape resembles that of a horseshoe, with the loose ends pointing in a downstream direction. Figure 8b displays the extent of the luminescence zone above the foil surface.

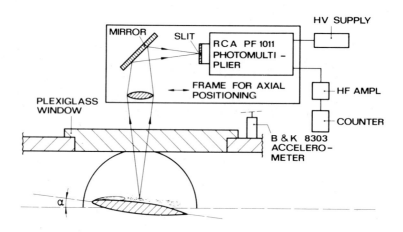

Fig. 9.   Schematic diagram of test set-up for luminescence measurements.

Although such photographs can provide useful qualitative information, they can not be used as a (quantitative) measure of cavitation intensity. Actual measurements of luminescence were made with a photomultiplier. A schematic diagram of the test set-up is presented in Figure 9. The light rays emitted by cavitation traversed a plexiglass window on top of the test section and via a lens, mirror and slit entered a RCA Type PF 1011 photomultiplier. The distance between the foil and the lens was the same as between the lens and the slit. The dimensions of the slit were 1.4 x 3.1 mm. Thus, light from the same area on the hydrofoil (1.4 mm along the chord and 3.1 mm along the span, in the central plane of the foil) was recorded by the photomultiplier. The lens, mirror, slit and photomultiplier were attached to a frame which could be moved in axial direction to enable measurement of the light intensity along the hydrofoil chord. The voltage applied to the photomultiplier was set at 2170 Volt. No attempt was made to measure the absolute light intensity. So,

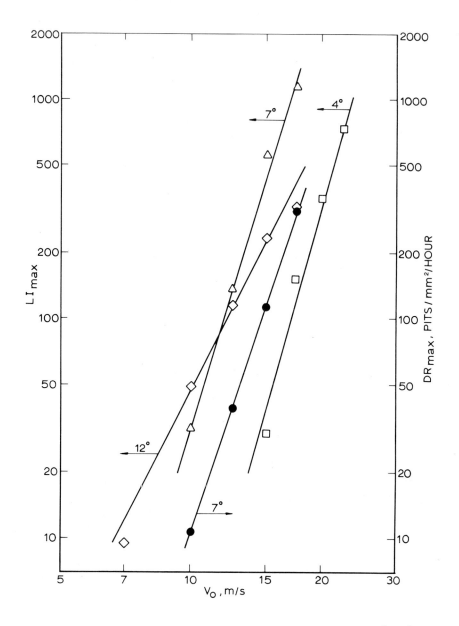

Fig. 10. Maximum luminous intensity levels LI_{max} for $\alpha=4°$, $7°$ and $12°$ and maximum damage rate DR_{max} for $\alpha=7°$ as a function of the velocity V_0.

only relative light intensities were measured by a pulse counter.

Measurements of the relative luminous intensity were performed at the following conditions: (a) $\alpha=4°$, $\sigma=1.00$; (b) $\alpha=7°$, $\sigma=1.68$ and (c) $\alpha=12°$, $\sigma=2.90$. For each condition, the cavitation number was kept constant whereas the velocity was varied. The results, in terms of the relative luminous intensity LI along the hydrofoil chord, are to be found in (35). In Figure 10, values of the maximum luminous intensity LI_{max} are plotted against the velocity V_0. It is found that the relationship between LI_{max} and V_0 can be represented by the power law

$$LI_{max} \sim V_0^{m}.$$

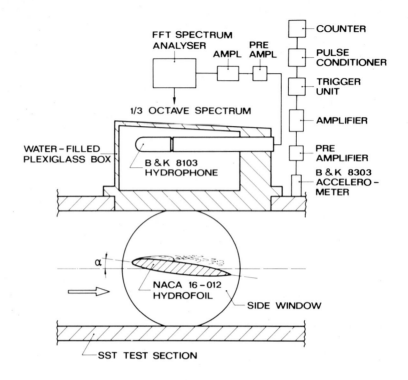

Fig. 11. Schematic diagram of test set-up for noise measurements.

The velocity exponents m derived from Figure 10 for $\alpha=4^{\circ}$, 7° and 12° appear to be 7.2, 6.5 and 3.9 respectively.

Noise measurements were made with a Brüel and Kjaer Type 8103 hydrophone, mounted in a water-filled plexiglass box on top of the test section. A schematic diagram of the test set-up is presented in Figure 11. The hydrophone signals were analysed in a Nicolet Scientific Type 444.A spectrum analyser, providing a 1/3 octave band sound pressure level spectrum in the frequency range 0.5-80 kHz. The measurements were performed at the following conditions: (a) $\alpha=4^{\circ}$, $\sigma=0.96$, (b) $\alpha=7^{\circ}$, $\sigma=1.69$ and (c) $\alpha=12^{\circ}$, $\sigma=2.89$. For each condition the cavitation number was kept constant, whereas the velocity was varied. The results, in terms of sound pressure levels in 1/3 octave bands, are to be found in (35). In Figure 12, broadband sound pressure levels $\bar{L}_p$ in the frequency range 0.5-80 kHz are plotted against the velocity V_o. $\bar{L}_p$ is a measure for the emitted acoustic energy and may be correlated to V_o by the power law

$$\bar{L}_p \sim V_o^{\,n}.$$

The velocity exponents n derived from Figure 12 for $\alpha=4^{\circ}$, 7° and 12° are found to be 5.9, 5.1 and 4.8 respectively.

Erosion measurements were made for sheet-cloud cavitation at $\alpha=7^{\circ}$ and $\sigma=1.69$, and at four different velocities V_o: 10, 12.5, 15 and 17.5 m/s. At these velocities the copper hydrofoils were subjected to cavitation for a period of 2 hours, 1 hour, 30 minutes and 15 minutes respectively. A pit counting method, described by Knapp (38), was used to analyse erosion. The erosion pits were classified in three pit diameter ranges: 10-35 µm, 35-70 µm and >70 µm. The analyses yielded damage rate curves, where the damage rate DR in (total) number of pits/mm^2/hour was plotted against the distance along the chord (36). In Figure 10, maximum values of the damage rate DR_{max} are plotted against the velocity V_o. It is found that the relationship between

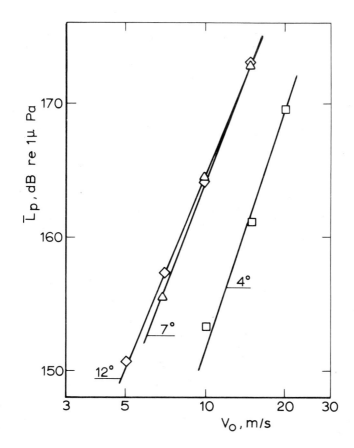

Fig. 12. Broadband sound pressure levels $\bar{L}_n$ as a function of velocity for $\alpha = 4°$, $7°$ and $12°$.

DR_{max} and V_o can be represented by the power law

$$DR_{max} \sim V_o^p,$$

where p appears to be 6.0. The percentage of large pits (d>35 μm) appeared to be strongly dependent on the location along the chord. For $V_o=15$ m/s this percentage was found to be 10 at the location of the maximum damage rate (x/c=0.43) and 43 at the location x/c=0.77. In a qualitative sense these findings are in agreement with the photographic recordings of luminescence (for the same condition) presented in Figure 8a.

Table 2. Velocity exponents for luminescence (m), noise (n) and erosion (p) generated by cavitation on NACA 16-012 hydrofoil.

α	m	n	p
4°	7.2	5.9	
7°	6.5	5.1	6
12°	3.9	4.8	

Velocity exponents for cavitation noise and erosion derived by other investigators are in general agreement with the above findings. Ramamurthy and Bhaskaran (40) reported on a velocity exponent of 5.45 for noise and erosion generated by a cavitating wedge, and Stinebring, Holl and Arndt (41) found a velocity exponent of 6 for erosion due to cavitation on a zero-caliber ogive. A survey of all velocity exponents for lminescence, noise and erosion, generated by cavitation on the NACA 16-012 hydrofoil is presented in Table 2. It would appear that the velocity exponent for sheet cavitation ($\alpha=4^O$) and sheet-cloud cavitation ($\alpha=7^O$) corresponds to about 6, whereas a velocity exponent of about 4 might be more appropriate for vortex-cloud cavitation ($\alpha=12^O$). The consistency of results obtained by the luminescence, noise and erosion measurements supports the feasibility of the concept of measure of cavitation intensity for hydrodynamic cavitation. Besides noise and erosion, luminescence is to be regarded as a genuine measure of cavitation intensity and may thus offer useful perspectives in future studies on the severity of hydrodynamic cavitation.

REFERENCES

1. Rayleigh, Lord, "On the Pressure Developed in a Liquid during the Collapse of a Spherical Cavity", Philosophical Magazine, Vol. 34, 1917, pp. 94-98.
2. Naudé, C.F., and Ellis, A.T., "On the Mechanism of Cavitation Damage by Nonhemispherical Cavities Collapsing in Contact With a Solid Boundary", ASME Journal of Basic Engineering, Vol. 83, No. 4, Dec. 1961, pp. 648-656.
3. Shutler, N.D., and Mesler, R.B., "A Photographic Study of the Dynamics and Damage Capabilities of Bubbles Collapsing Near Solid Boundaries", ASME Journal of Basic Engineering, Vol 87, June 1965, pp. 511-517.
4. Benjamin, T.B., and Ellis, A.T., "The Collapse of Cavitation Bubbles and the Pressures Thereby Produced Against Solid Boundaries", Philosophical Transactions of the Royal Society of London, A, Vol. 260, 1966, pp. 221-240.
5. Kling, C.L., and Hammitt, F.G., "A Photographic Study of Spark-Induced Cavitation Bubble Collapse", ASME Journal of Basic Engineering, Vol. 94, No. 4, Dec. 1972, pp. 825-833.
6. Gibson, D.C., "The Kinetic and Thermal Expansion of Vapor Bubbles", ASME Journal of Basic Engineering, Vol. 94, No. 1, March 1972, pp. 89-96.
7. Lauterborn, W., "Bubble Generation by Giant Laser Pulses and Resonance Curves of Gas Bubbles in Water", Proceedings of the IUTAM Symposium on Non-Steady Flow of Water at High Speeds, Leningrad, June 1971, pp. 267-275.
8. Lauterborn, W., "Kavitation durch Laserlicht", Acustica, Vol. 31, No. 2, 1974, pp. 51-78.
9. Lauterborn, W., and Bolle, H., "Experimental Investigations of Cavitation Bubble Collapse in the Neighbourhood of a Solid Boundary", Journal of Fluid Mechanics, Vol. 72, Part 2, 1975, pp. 391-399.
10. Van Wijngaarden, L., "On the Collective Collapse of a Large Number of Gas Bubbles in Water", Proceedings of the Eleventh International Congress of Applied Mechanics, Munich, 1964, pp. 854-861.
11. Hansson, I., and Mørch, K.A., "The Dynamics of Cavity Clusters in Ultrasonic (Vibratory) Cavitation Erosion", Journal of Applied Physics, Vol. 51, Sept. 1980, pp. 4651-4658.
12. Chahine, G.L., "Cloud Cavitation: Theory", Proceedings 14th Symposium on Naval Hydrodynamics, Ann Arbor, August 1982, pp. 165-194.
13. Tomita, Y., and Shima, A., "Mechanisms of Impulsive Pressure Generation and Damage Pit Formation by Bubble Collapse", Journal of Fluid Mechanics, Vol. 169, August 1986, pp. 535-564.
14. Blake, J.R., Taib, B.B., and Doherty, G., "Transient Cavities near Boundaries. Part 1. Rigid Boundary", Journal of Fluid Mechanics, Vol. 170, Sept. 1986, pp. 479-497.

15. Blake, J.R., Taib, B.B., and Doherty, G., "Transient Cavities near Boundaries. Part 2. Free Surface", Journal of Fluid Mechanics, Vol. 181, August 1987, pp. 197-212.

16. Blake, J.R., and Gibson, D.C., "Cavitation Bubbles Near Boundaries", Annual Reviews Fluid Mechanics, Vol. 19, 1987, pp. 99-123.

17. Vogel, A., "Optische und akustische Untersuchungen der Dynamik lasererzeugter Kavitationsblasen nahe fester Grenzflächen", Ph.D. Thesis, Göttingen, May 1987.

18. Dear, J.P., and Field, J.E., "A Study of the Collapse of Arrays of Cavities", Journal of Fluid Mechanics, Vol. 190, May 1988, pp. 409-425.

19. Van der Meulen, J.H.J., and Van Renesse, R.L., "A Study of the Collapse of Laser-Induced Bubbles in a Flow Near a Boundary", ASME Cavitation and Multiphase Flow Forum, Cincinnati, June 1987, pp. 21-25.

20. Van der Meulen, J.H.J., and Van Renesse, R.L., "The Collapse of Bubbles in a Flow Near a Boundary", To be presented at 17th Symposium on Naval Hydrodynamics, The Haque, August 1988.

21. Kato, H., Maeda, M., and Nakashima, Y., "A Comparison and Evaluation of Various Cavitation Erosion Test Methods", Proceedings ASME Symposium on Cavitation Erosion in Fluid Systems, Boulder, June 1981, pp. 83-94.

22. Ukon, Y., and Takei, Y., "An Investigation of the Effects of Blade Profile on Cavitation Erosion of Marine Propellers", Transactions West-Japan Society of Naval Architects, Vol. 61, March 1981, pp. 81-97.

23. Van der Kooij, J., and De Bruijn, A., "Acoustic Measurements in the NSMB Depressurized Towing Tank", International Shipbuilding Progress, Vol. 31, Jan. 1984, pp. 13-25.

24. Huse, E., "Cavitation Induced Hull Pressures, some recent Developments of Model Testing Techniques", Proceedings Symposium on High Powered Propulsion of Large Ships, Wageningen, Dec. 1974.

25. Van der Kooij, J., "Experimental Determination of Propeller Induced Hydrodynamic Hull Forces in the NSMB Depressurized Towing Tank", Proceedings RINA Symposium on Propeller Induced Ship Vibration, London, Dec. 1979, pp. 73-86.

26. Frenzel, H., and Schultes, H., "Luminescenz im ultraschallbeschickten Wasser", Zeitschrift Physikalische Chemie, Vol. 27, 1935, pp. 421-424.

27. Kuttruff, H., "Uber den Zusammenhang zwischen der Sonoluminescenz und der Schwingungskavitation in Flüssigkeiten", Akustica, Vol. 12, 1962, pp. 230-254.

28. Meyer, E., and Kuttruff, H., "Zur Phasenbeziehung zwischen Sonolumineszenz und Kavitationsvorgang bei periodischer Anregung", Zeitschrift für angewandte Physik, Vol. 11, No. 9, Sept, 1959, pp. 325-333.

29. Flynn, H.G., "Physics of Acoustic Cavitation in Liquids", Physical Acoustics, Vol. 1B, Academic Press, New York, 1964, pp. 57-172.

30. Hickling, R., "Effects of Thermal Conduction in Sonoluminescence", Journal of Acoustical Society of America, Vol. 35, No. 7, July 1963, pp. 967-974.

31. Hickling, R., "Some Physical Effects of Cavity Collapse in Liquids", Journal of Basic Engineering, Transactions ASME, Vol. 88, Series D, No. 1, March 1966, pp. 229-235.

32. Young, F.R., "Sonoluminescence from Water Containing Dissolved Gases", Journal of Acoustical Society of America, Vol. 60, No. 1, July 1976, pp. 100-104.

33. Van der Meulen, J.H.J., "The Use of Luminescence as a Measure of Hydrodynamic Cavitation Activity", ASME Cavitation and Multiphase Flow Forum, Houston, June 1983, pp. 51-53.

34. Van der Meulen, J.H.J., and Nakashima, Y., "A Study of the Relationship between Type of Cavitation, Erosion and Luminescence", Proceedings 2nd International Conference on Cavitation, Edinburgh, Sept. 1983, pp. 13-19.

35. Van der Meulen, J.H.J., "The Relation between Noise and Luminescence From Cavitation on a Hydrofoil", Proceedings ASME Symposium on Cavitation in Hydraulic Structures and Turbomachinery, Albuquerque, June 1985, pp. 149-159.

36. Van der Meulen, J.H.J., "On Correlating Erosion and Luminescence From Cavitation on a Hydrofoil", Proceedings International Symposium on Propeller and Cavitation, Wuxi, China, April 1986, pp. 261-267.
37. Van der Meulen, J.H.J., "Boundary Layer and Cavitation Studies of NACA 16-012 and NACA 4412 Hydrofoils", Proceedings 13th Symposium on Naval Hydrodynamics, Tokyo, Oct. 1980, pp. 195-219.
38. Knapp, R.T., "Recent Investigations of the Mechanics of Cavitation and Cavitation Damage", Transactions ASME, Vol. 77, Oct. 1955, pp. 1045-1054.
39. Shen, Y.T., and Peterson, F.B., "Unsteady Cavitation on an Oscillating Hydrofoil", Proceedings 12th Symposium on Naval Hydrodynamics, Washington D.C., June 1978, pp. 362-384.
40. Ramamurthy, A.S., and Bhaskaran, P., "Velocity Exponent for Erosion and Noise Due to Cavitation", ASME Journal of Fluids Engineering, Vol. 101, March 1979, pp. 69-75.
41. Stinebring, D.R., Holl, J.W., and Arndt, R.E.A., "Two Aspects of Cavitation Damage in the Incubation Zone: Scaling by Energy Considerations and Leading Edge Damage", ASME Journal of Fluids Engineering, Vol. 102, Dec. 1980, pp. 481-485.

Theoretical and Applied Mechanics
P. Germain, M. Piau and D. Caillerie (Editors)
Elsevier Science Publishers B.V. (North-Holland)
  IUTAM, 1989

FLOW OF BUBBLY LIQUIDS

L. VAN WIJNGAARDEN
University of Twente, Enschede, The Netherlands

1.   INTRODUCTION

Gas/liquid flows, the subject of my contribution, are interesting
scientifically and of great technological importance. These flows occur in
oil engineering, in distribution networks for natural.gas, in the cooling of
nuclear reactors, propeller hydrodynamics (cavitation) and many processes in
the petro-chemical industry. In many countries, for example here in France,
special programmes in two phase flows exist financed in co-operation between
industry, government and universities.

A first interesting phenomenon to mention is formed by the various
topologies exhibited by gas/liquid flows.

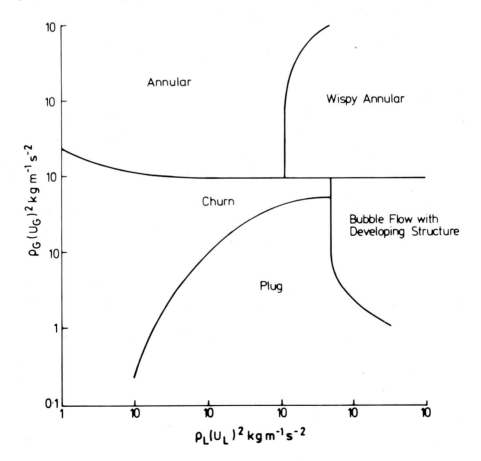

FIGURE 1
Flow map for two phase gas/liquid flows. From Hetsroni [23].

In figure 1 the occurrence of some of these is given as a function of the
volumetric gas and liquid flow in vertical pipe flow.

Flow maps of this and other kind suggest that transitions from one topology
to another depend on a few parameters viz. void fraction, gas velocity and
liquid velocity. It is unfortunately more complicated than that. Most
transitions are neither well documented nor well understood. We are, in fact,
a long way from being able to explain and predict those transitions. The
scientific approach to two phase flow problems is relatively young and the
field is in what Kuhn [1] calls the preparadigmatic stage. In this stage
there are few generally accepted ideas, abstractions and theories. To see
what I mean, take boundary layers. Things like stability, separation,
Tollmien-Schlichting waves, adverse pressure gradient have the same meaning
to everybody working in boundary layers, if not to every fluid dynamicist.
They form the paradigmas. Such paradigmas exist to a much lesser extent in
two phase flow. In consequence there are every day, so to speak, new theories
rejecting others and the field is very much sensitive for external influences.
There is an area in gas/liquid flow on which there is more or less agreement
about the principal effects and that concerns the propagation of pressure
waves. This seems therefore good to begin with, the more so while for the
newcomer to the field there are some unexpected phenomena.

2.   PROPAGATION OF PRESSURE WAVES OF SMALL AMPLITUDE

Consider a liquid in which gas is dispersed in lumps of various size and
shape. Let the concentration by volume be α and let ρ be the density of the
suspension. When the densities of gas and liquid are indicated with ρ_g and ρ_l
respectively, we have

$$\rho = \alpha\rho_g + (1-\alpha)\rho_l. \tag{2.1}$$

At a given temperature the gas has, associated with ρ_g, a certain pressure p
and if we take this to be also the local pressure in the mixture we might ask
for a relation between p and ρ giving us, like in single phase fluids, a
sound velocity c, defined by

$$c^2 = \left(\frac{dp}{d\rho}\right)_s \tag{2.2}$$

where the subscript means that the derivative needs to be taken at constant
entropy. Such a relation exists when we ignore possible velocity differences
between the two phases. Then the mass of gas in a unit mass of the mixture is
a constant,

$$\frac{\alpha\rho_g}{(1-\alpha)\rho_l + \alpha\rho_g} = \text{constant.} \tag{2.3}$$

From (2.1) - (2.3) together with the isentropic relation for the gas
$p = p_g \sim \rho_g^\gamma$, γ being the ratio of specific heats of the gas, one finds with α neither very close to zero nor to unity,

$$c^2 = \frac{\gamma p}{\rho_l \alpha(1-\alpha)} \, . \qquad (2.4)$$

This gives the interesting result that at 50% concentration by volume the sound speed is at atmospheric conditions only 23.7 m/s, much lower than the sound speed in water, but also much lower than the sound speed in air. In fact, the flow maps show that a bubbly flow cannot exist at such high concentrations. Take therefore $\alpha = 0.1$ in which case c equals 39.4 m/s which is still very small and illustrates the slow variation with α.

Now the densities of gas and liquid are quite different. The motion of the interface gives rise to inertia forces which in turn produce pressure differences. Minnaert [2] was conscious of the relation between bubble behaviour and the sound emitted by running streams and studied a simple system, one bubble produced at the mouth of a small glass tube , see fig. 2, in a liquid. He studied in particular free oscillations of air bubbles in water determining experimentally their frequency with help of a tuning fork.

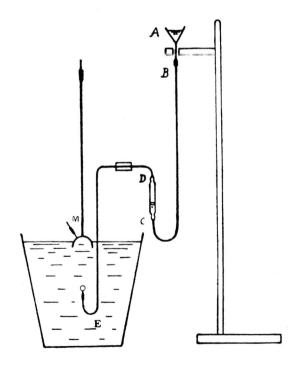

FIGURE 2

Experimental set up used by Minnaert [2]

Simple calculations based on the conservation of energy gave him for
the resonant angular frequency ω_B the expression, a being the bubble radius,

$$\omega_B^2 = \frac{1}{a^2} \frac{3\gamma p_o}{\rho_l} .$$
(2.5)

where p_o is the pressure, when undisturbed by the (small) oscillations.
For a bubble with radius a=1 mm the corresponding frequency is 3.2 kHz, which
is in the audible range. The sound of shallow running streams, for example in
the mountains is caused by bubbles being formed and breaking up. When placed
in a sound field of given frequency a bubble will, at small enough amplitude,
perform oscillations at this frequency. The incorporation of this in the
propagation of sound waves through a bubbly suspension was done originally
(Foldy [3]) by multiple scattering theory. The bubbles are considered as
point scatterers. Part of the scattered sound forms a coherent wave, part is
lost as incoherent wave. If we express the amplitude of the coherent wave as
proportional to expi(kx-ωt) where ω is real and k complex one finds

$$\left(\frac{k}{\omega}\right)^2 = \frac{1}{c_l^2} + \frac{1}{c^2} \frac{1 - (\omega/\omega_B)^2 + i\delta\omega/\omega_B}{\{1- (\frac{\omega}{\omega_B})^2\}^2 + \delta^2\omega^2/\omega_B^2} ,$$
(2.6)

where c_l is the sound speed in liquid, c is given in (2.4) and δ is a
coefficient allowing for attenuation.

The result (2.6) can also be found in a way perhaps appealing more to
fluid dynamicists, and which has the further advantage that it can be
extended to nonlinear amplitudes. For this we understand p, ρ and $\underset{\sim}{u}$ as local
averages of pressure, density and velocity. For these we can write down
equations of motion,

$$\frac{D\rho}{Dt} + \rho \nabla . \underset{\sim}{u} = 0,$$
(2.7)

$$\rho \frac{D\underset{\sim}{u}}{Dt} + \nabla p = 0,$$
(2.8)

ignoring viscous forces. The coupling between ρ, ρ_g, ρ_l and α is given by
(2.1), whereas the further coupling with bubble radius is given by

$$\alpha = \frac{4}{3} \pi n a^3,$$
(2.9)

and the conservation of bubbles, which makes the number density n subject to

$$\frac{Dn}{Dt} + n \nabla . \underset{\sim}{u} = 0.$$
(2.10)

Instead of the assumption of equal pressure in both phases we assume that the relation between the local average pressure p and the gas pressure p_g is the same as the relation between the pressure p_∞ at infinity and the internal pressure p_g when a bubble changes volume in an infinite liquid. The latter is the Rayleigh-Plesset equation

$$\rho_l \left\{ a \frac{d^2a}{dt^2} + \frac{3}{2} \left(\frac{da}{dt}\right)^2 \right\} = p_g + p_v - p_\infty - \frac{2T}{a} - \frac{4\nu_l}{a} \frac{da}{dt}. \qquad (2.11)$$

Apart from quantities already mentioned we have introduced here the vapour pressure p_v, the coefficient of surface tension T and the kinematic viscosity ν_l of the liquid.

In the original continuumdescription (Van Wijngaarden [4]) it was not specified how small α should be to make the replacing of p_∞ in (2.11) by the average pressure justified. This was done later in work by Caflisch e.o. [5]. Linearization of (2.7) - (2.11) and seeking solutions of the form expi(kx-ωt) gives again (2.6), for $c_l \to \infty$ and c^2, as given in (2.4) modified by surface tension into

$$c^2 = \frac{p-p_v}{\rho_l \alpha(1-\alpha)} + \frac{2}{3} (3\gamma-1) \frac{T}{\rho_l a} . \qquad (2.12)$$

At given pressure surface tension increases the sound speed of a bubbly mixture. This is, as (2.12) shows a negligible effect for bubbles larger than a few microns.

Properties expressed by (2.6) and (2.12) have some interesting effects on bubbly flows. Apart from the sound emission at resonant frequencies observed by Minnaert [2] a bubbly flow emits sound in another way. It is wellknown that in a single phase fluid part of the turbulent energy is radiated away as sound. Bubbles, present in such a fluid act as monopoles and magnify the sound production by a factor as large as $(c_l/c)^4$, c being given by (2.12) (Crighton & Ffowcs Williams [6]). This production is at the frequencies of the turbulence, and not as one is inclined to think, at bubble resonance. This, because turbulent pressure fluctuations are at such high frequencies not coherent over the surface of a bubble.

The upper layer of the oceans contain many bubbles. A knowledge of their size and number distributions is of interest, a.o. for underwater acoustics, and can be obtained with help of the just described properties.
Apart from the behaviour of the sound speed with frequency one makes, in what is called bubble spectrometry use of the attenuation of sound waves by bubbles. From (2.6) it follows that this is particularly strong at resonance where the imaginary part of k, k_i, is approximately given by

$$k_i \approx \frac{\omega_B}{c} \delta^{-1/2}.$$

The coefficient δ consists of, and is approximately equal to, the sum of three distinct contributions, an acoustical one due to the incoherent sound emission and of magnitude $\pi\omega a/c_1$, a thermal one due to heat conduction in the gas, of magnitude $3(\gamma-1)/2(D/\omega a^2)^{1/2}$, at high frequencies, D being the thermal diffusivity in the gas phase and finally a viscous contribution $4\nu_1/\omega_B a^2$. The thermal contribution dominates and is of order 10^{-2} for bubbles of 1 mm size. This may lead at resonance to tremendous sound attenuation, of the order of 100 db per meter. So, when sending a sound wave through a bubbly fluid one can safely ascribe the damping to bubbles of that resonant frequency. This type of considerations form the basis of bubble spectrometry at sea (Medwin [7]). Analysis like described above considers only one bubble size because a size distribution makes calculation complicated. This, however, doesn't make life easy for experimentalists who would like to make experimental verification. It is extremely difficult to produce bubbles of one size only. Silberman [8]) managed to do this. Figure 3 is from his paper and shows how the measured attenuation compares well with prediction.

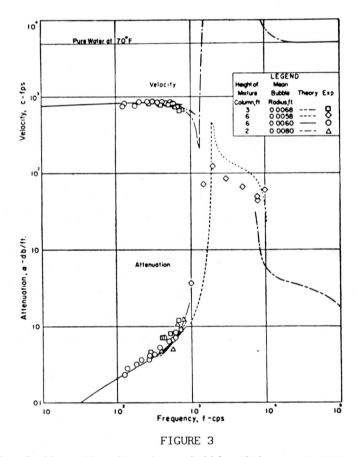

FIGURE 3

Sound speed and attenuation in water a bubble mixtures as measureed by Silberman [8]. α=2.2*10^{-3}. a$\approx$2x10^{-4}m. The solid lines represent the theoretical results. The experimental results are for various bubble radii as indicated in the figure.

3. PROPAGATION OF PRESSURE WAVES OF FINITE AMPLITUDE

When we now go on to extend the consideration of pressure waves to finite amplitude, we encounter new phenomena. The relation (2.4) shows that, locally, the speed of propagation increases with the pressure amplitude for low frequencies when dispersion is unimportant. This leads like in the theory of water waves of finite amplitude or of pressure waves in single phase gases to steepening of compression waves and flattening of expansion waves. On the other hand, for long waves,(2.6) allows the series expansion

$$\omega \approx kc - \frac{1}{2} \frac{k^3 c^3}{\omega_B^2} \ldots\ldots\ldots,$$

with which corresponds the differential equation $(k=-i\frac{\partial}{\partial x}, \omega=i\frac{\partial}{\partial t})$

$$(\frac{\partial}{\partial t} + c \frac{\partial}{\partial x} + \frac{1}{2} \frac{c^3}{\omega_B^2} \frac{\partial^3}{\partial x^3})p = o.$$

For weak dispersion and not too large amplitudes, analysis of the equations (2.7) – (2.11) leads for waves in one direction only, to a combination of Burgers's equation and the Korteweg-de Vries equation (Van Wijngaarden [4]). For example for waves propagating in the positive x-direction we have for the excursion $\tilde{p}$ from the undisturbed pressure p_o

$$\frac{\partial \tilde{p}}{\partial t} + c_o \frac{\partial \tilde{p}}{\partial x} + c_o \tilde{p} \frac{\partial \tilde{p}}{\partial x} + \frac{1}{2} \frac{c_o^3}{\omega_B^2} \frac{\partial^3 \tilde{p}}{\partial x^3} - \frac{1}{2} \frac{\delta c_o^2}{\omega_B} \frac{\partial^2 \tilde{p}}{\partial x^2} = o. \qquad (3.1)$$

Like with water waves on a fluid of finite depth (see e.g. Whitham [9]),we can have in a bubbly flow cnoidal waves, undular bores, solitons and similar phenomena. This prediction has found ample experimental verification (Noordzij & Van Wijngaarden [10], Kuznetsov, Nakoryakov, Pokusaev & Schreiber [11]).

Till thus far we have considered bubbles filled with insoluble gas, except for the allowance of a vapour pressure in (2.11). In many practical situations, notably the cooling of nuclear reactors bubbles consist of vapour and gas. A much larger transfer of heat can take place because of vaporization and condensation. While with gas filled bubbles heat transfer is limited by the conductivity of the gas here the limiting factor is the liquid conductivity. As a result of the large heat transfer the behaviour of weakly nonlinear waves is more monotonous than in a liquid with gas bubbles (Nigmatulin [12]). For strong oscillatory waves, however, vapour pressures may be reached in considerable excess of the liquid pressure. We have in mind here vapour bubbles at such temperatures as occur in nuclear plants.

The study of mixtures consisting of vapour and liquid has received much
stimulus in countries where a large part of the electric power is generated
in such plants. Especially at the Institute of Thermophysics at Novosibirsk
much valuable work has been done (Nakoryakov, Pokusaev & Schreiber [13].
Bubbles partly filled with vapour occur at room temperature in cavitation, a
phenomenon of great importance in shiphydrodynamics and in fluid machinery.

4. CAVITATION AND CAVITATION NOISE

When we work out Minnaert's resonance frequency including vapour content
and surface tension, we find e.g. from linearization of the Rayleigh-Plesset
equation (2.11)

$$\omega_B^2 = \frac{1}{a^2} \left\{ \frac{3\gamma(p_o - p_v)}{\rho_l} + \frac{2T(3\gamma-1)}{\rho_l a} \right\} \tag{4.1}$$

where p_v is, again, the pressure of the vapour inside the bubble. This
expression becomes negative indicating explosive growth at the cavitation
threshold

$$p_o = p_v - \frac{2T}{a} (\gamma - \frac{1}{3}).$$

This is the instability which causes, when the ambient pressure decreases
sufficiently below vapour pressure, microscopic bubbles, nuclei, to grow to
macroscopic size, 1 mm, say. From these bubbles other types of cavitation
like sheet cavitation or vortex cavitation may develop. We refer for more
information about cavitation to the sectional lecture by dr. J. van der Meulen
during this congres and we will deal here with one aspect, and that is the
sound emitted by cavitation. The sound is produced while eventually the
bubbles collapse in high pressure regions, for example in the flow along the
blades of a ship's propeller. The energy spectrum of cavitation noise looks
somewhat like shown in figure 4. This noise has traditionally been related to
collapse of single bubbles. Recently evidence has been obtained in various
ways (e.g. Arakeri [14]) that a considerable contribution comes from clouds
or clusters of collapsing cavitation bubbles. The study of such collapsing
clouds forms a still young subject but some interesting features may be
already mentioned. Consider a cloud with effective radius A consisting of
small bubbles in liquid. This cloud is hit by a sudden change of pressure.
One is interested in the sound radiated into the clear liquid which surrounds
the cloud.

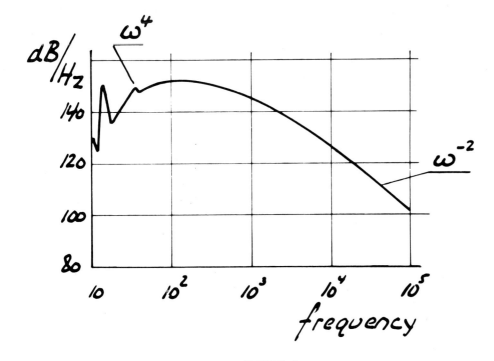

FIGURE 4

Global form of spectrum of cavitation noise.

Omta [15] found that at large distance, compared with the dimension of the cloud, the sound has primarily the frequency corresponding with the lowest natural frequency of the cloud

$$f_c \approx \frac{1}{4} \frac{\gamma^{1/2} c}{A} \; , \qquad\qquad (4.2)$$

where c is given by (2.4). With $A=5*10^{-2}$m, c=40 m/s this gives 237 Hz which corresponds, in order of magnitude, with the peak in the observed noise spectra. If we take a gas bubble containing all the gas present in the cloud this would have a radius $B=\alpha^{1/3}A$. Such a bubble, when immersed in an infinite liquid would have according to (2.5) a resonance frequency

$$f_B = \frac{(2\pi)^{-1}}{\alpha^{1/3}A} \left(\frac{3\gamma p}{\rho_l}\right)^{1/2} . \qquad\qquad (4.3)$$

Comparison with (4.2) shows that

$$f_c / f_B = \frac{\pi}{(12\alpha^{1/3})^{1/2}} \; ,$$

which is of unit order. This gives the rule of thumb that the lowest eigenfrequency of the cloud is approximately equal to the natural frequency of one bubble containing all the gas in the cloud. This frequency is obviously independent of the size of the individual bubbles.

Attenuation on the contrary, is not. Since the respresentative bubble, with natural frequency as in (4.3), is much larger than the individual bubbles making up the cloud, these are excited far from their resonance frequency and their thermal behaviour strongly depends on size.

When a fluid is irradiated by a sound wave of sufficient intensity, cavitation is provoked. This is known as acoustic cavitation. Long ago, in 1952 Esche found in the emitted sound a component $f_o/2$, f_o being the frequency of the primary beam. This phenomenon has puzzled cavitation specialists for a long time, see e.g. Plesset and Prosperetti [16]. It is now clear that this is the beginning of a Feigenbaum route to chaos. At higher intensity more subharmonics appear, on to chaos. This is found, both experimentally, Lauterborn and Cramer [17], and theoretically, Lauterborn and Suchla [18]. The latter by applying recently obtained techniques in nonlinear dynamics to the Rayleigh-Plesset equation (2.11). The increase in complexity which we see when we compare, for example, the analysis in Minnaert's work with that by Lauterborn on the Rayleigh-Plesset equation has its counterpart in experimental research. Next to Minnaert's experimental device shown in figure 2, used to study the natural oscillations of a bubble, the interested reader is advised to take a look at the, sometimes most ingenious instrumentation in gas/liquid flow collected in Delhaye and Cognet [19] .

5. GAS/LIQUID FLOW WITH RELATIVE VELOCITY

Till thus far we have ignored velocity differences between the two phases. Often such a difference occurs when the mixture is exposed to a sudden pressure gradient. The large density difference results in relative acceleration and hence in relative velocity. This type of situation has been intensively studied in many countries, especially in connection with the simulation of hazardous situations in nuclear plants, such as the Loss of Coolant Accident (LOCA). Incorporation of such relative velocities brings unexpected and severe problems. Let us first consider the relatively simple case provided by stratified flow where gas flows over the liquid, as in figure 5.

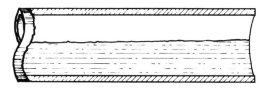

STRATIFIED

FIGURE 5

Stratified gas/liquid flow

In this case one can establish equations for mass, momentum and
energy conservation for each of the phases subjected to appropriate
continuity conditions at the interface, including, perhaps, heat and mass
transfer. In its simplest form fluid dynamicists know these equations in
connection with Kelvin-Helmholz instability. The formulation pertinent to
stratified flow has led, in particular in the development of computer codes
to predict transient flow behaviour, to the formulation of similar equations
for topologies in which the phases are arranged differently.

In such formulations average velocities u_l and u_g, say, are assigned to
liquid and gas respectively. Likewise with densities, pressures, stresses and
so on. The equations developed in this socalled two-fluid model have for mass
and momentum conservation, the form

$$\frac{\partial}{\partial t}\, \alpha\rho_g + \nabla\cdot\alpha\rho_g\, u_g = \Gamma \tag{5.1}$$

$$\frac{\partial}{\partial t}\, (1-\alpha)\rho_l + \nabla\cdot(1-\alpha)\rho_l u_l = -\Gamma \tag{5.2}$$

$$\frac{\partial}{\partial t}\, \alpha\rho_g u_g + \nabla\cdot\left\{\alpha p_g + \alpha\rho_g\, u_g u_g\right\} = \nabla\cdot\underset{\approx}{\sigma}_g + \underset{\sim}{B}_g \tag{5.3}$$

$$\frac{\partial}{\partial t}\, \alpha\rho u_l + \nabla\cdot\left\{(1-\alpha)(p_l + \rho_l u_l u_l)\right\} = \underset{\approx}{\nabla}\cdot\sigma_l + \underset{\sim}{B}_l . \tag{5.4}$$

Similar equations are contained in the two-fluid model for energies. The
quantity Γ represents mass transfer between the phases (condensation,
evaporization), $\underset{\sim}{B}_g$ and $\underset{\sim}{B}_l$ represent forces exerted by one phase on the other,
$\underset{\approx}{\sigma}_g$ and $\underset{\approx}{\sigma}_l$ are stresses in gas and liquid respectively.

Although much computation with equations of this type has been done, as
readers of the International Journal of Multiphase Flow will be aware of,
strong doubts must be expressed as to their validity for bubbly flows, among
others. Take, for example, vertical flow of bubbles and liquid as depicted in

figure 6. A point, marked $\underset{\sim}{x}$, will be part of the time in liquid and part of the time in gas. The quantities $\underset{\sim l}{u}$ and $\underset{\sim g}{u}$ could be defined as time averages over only those times in which $\underset{\sim}{x}$ is in the pertinent phase. It is however very hard to attach a meaning to quantities like $\underset{\approx g}{\sigma}$, $\underset{\approx l}{\sigma}, \underset{\sim g}{B}$ and $\underset{\sim l}{B}$. The equations are formulated as if the flow was like in figure 5 whereas in reality gas is dispersed as bubbles reacting to the liquid quite differently.

Another weakness of the two fluid equations is that they form a set for which the Cauchy initial value problem is ill posed in the sense of Hadamard. This can be seen as follows.

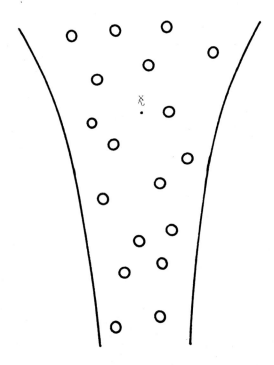

FIGURE 6
Bubbly flow in a vertical duct. A point x is sometimes in gas, sometimes in liquid.

Suppose the terms on the righthandside, other than pressure terms like $p_g \partial \alpha / \partial x$, are negligible or expressible in the dependent variables only, not in their derivatives. Then the equations form a set of quasilinear first order partial differential equations.

Inspection shows that these have no real characteristics. As a consequence solutions of the initial value problem do not depend on the initial data in a continuous way. Some people worry about this, others, like Spalding [20] argue that there is nothing to worry about on the ground that in practice numerical solutions of the equations can be computed. As said, the two-fluid model has been and is widely used by designers of computer codes but, today, the number of people who think it to be fit for all topologies, is decreasing.

The USA National Academy Press published in 1986 an interesting report called "Physics through the 1990's" [21] Panels composed of members of the National Academy of Science evaluate the state of affairs and express their expectations for various branches of physics. Two-phase gas/liquid flow is discussed among others and on the subject of the two-fluid model the report says "To date, it has been customary to consider separate differential equations for each phase and to use these as the basis for the computations. Unfortunately, it is quite likely that this aspect of research on multiphase flow is dangerously ahead of our basic knowledge".

6 THE USE OF STATISTICAL METHODS

In a flow like in figure 6 one is not interested in momentary values but in mean quantities. This can be a time average, volume average or other average. A convenient one is the ensemble average. Consider in figure 6 a volume with dimension large with respect to the bubble size but small with respect to the length over which mean quantities vary significantly. The scale of this volume is called mesoscale. When one would consider such a volume on many different times one would each time find roughly the same number N of bubbles but they would each time have different positions. Each configuration can be thought to have a certain probability to occur. This is expressed by the probability distribution density $P(\underset{\sim}{x}, \ldots \underset{\sim N}{x})$, or $P(c_N)$, for shorthand, defined in such a way that the probability of finding the first bubble in $\underset{\sim 1}{x}$, the second in $\underset{\sim 2}{x}$ and so on, is

$$P(\underset{\sim 1}{x} \ldots \ldots \underset{\sim N}{x}) d\underset{\sim 1}{x} \; d\underset{\sim 2}{x} d\underset{\sim 3}{x} \ldots d\underset{\sim N}{x},$$

or

$$P(c_N) dc_N.$$

The ensemble average of a quantity G, scalar, tensor or vector, at a location $\underset{\sim}{x}$ in the presence of N bubbles, then is

$$<G> = \frac{1}{N!} \int G(\underset{\sim}{x}, c_N) P(c_N) dc_N. \tag{6.1}$$

When a property belonging to the bubbles only, like bubble velocity, is involved one uses instead of $P(c_N)$ the conditional probability density, which expresses the probability of finding the first bubble at $\underset{\sim 1}{x}$ etc. given that there is an additional one in $\underset{\sim}{x}$. Then the average is

$$<G> = \frac{1}{N!} \int G(\underset{\sim}{x}, c_N) P(c_N / \underset{\sim}{x}) dc_N \tag{6.2}$$

In fixed media, like a porous medium, P is in principle known. The task is to find <G>, for example the average temperature in a polycristalline material subjected to a temperature gradient. In flowing media $P(C_N)$ and $P(C_N/\underset{\sim}{x})$ are not known at the outset but they are determined by the flow and therefore part of the problem. To determine the probability distribution it is necessary to study the hydrodynamic interactions, between bubbles or, in gas/solid or liquid/solid flow, between particles. Of course, it suggests itself to use here methods of statistical mechanics in particular the kinetic theory of gases. As in that area, multiple interactions are very hard to deal with and one has to restrict oneself almost completely to dilute mixtures. In the lowest approximation one treats the test particle or bubble, as if it is unaware of the presence of others.

In that case the average velocity of rise of a bubble is equal to the terminal velocity U_∞, say, of a bubble in an unbounded liquid. In the next approximation the configuration around the test particle consists of one other particle. Allowing for these binary encounters gives, analogeous to binary collisions in the kinetic theory of gases, results which are accurate in the first order of the concentration α, in the absence of long range order. This can be made plausible in the following way. Suppose in a unit volume of the flow are n bubbles. Then, recalling that $\alpha = 4/3\ \pi na^3$, the probability of finding a bubble in a sphere with radius λa, λ of unit order, is of order α, the probability of finding two bubbles there is of order α^2, and so on. Bubbles of 1 mm size rise in water with a velocity of about 25 cm/s giving a Reynolds number of 250. This is large enough for the relative motion to be dominated by inertia. When the influence of surface active material can be neglected and only the continuity of stresses must be required at the interface there is at Reynolds numbers of order 10^2 no appreciable wake behind the bubbles. This makes calculations possible. Before application to suspensions dominated by inertia, statistical methods were used to study sedimentation of small particles in liquid at small Reynolds numbers. Both at small and at large Reynolds numbers there is a particular difficulty of the kind with which plasma physics was confronted in the 1950 's. In a plasma the interaction between particles is mainly through Coulomb forces. The Coulomb potential falls off like r^{-2}, r being the distance from a particle centre, which is not fast enough to make certain integrals occurring in the averaging process uniformly convergent.

In hydrodynamics a similar difficulty occurs. The velocity induced by a moving bubble at distance r varies in potential flow (high Reynolds numbers) like r^{-3} and in low Reynolds flow even slower, like r^{-1} (Stokeslet). This behaviour makes integrals representing average velocities and similar averages either divergent or, at best not uniformly convergent.

There are two ways to cheat here. The first is to impose a periodic array on the suspension.

By the periodicity divergence problems are avoided but the answer which one gets is wrong. The first correction to U_∞, to stay with our example of the rising bubble cloud, is with a periodic array of order $\alpha^{1/3}$ instead of α, what it should be. The second way is to take, on the mesoscale, a finite volume instead of allowing the dimensions to become infinite ($r\to\infty$). This leads to socalled cell models. They give answers but the difficulty is that the accuracy of these is not known. The correct way to deal with the problem of nonconvergent integrals has been given by Batchelor [22] and amended by others. When restriction to two particle interaction is made the expression for the average velocity becomes, starting from (6.2) and taking for G the particle velocity $u(\underset{\sim}{x}, \underset{\sim}{x}+\underset{\sim}{r})$ in the presence of one other particle

$$<\underset{\sim}{u}> = \int \underset{\sim}{u}(\underset{\sim}{x}, \underset{\sim}{x}+\underset{\sim}{r}) \ P(\underset{\sim}{x}/\underset{\sim}{x}+\underset{\sim}{r}) d\underset{\sim}{r}$$

Once the problem of the nonconvergent integral is solved, the task of hydrodynamics is to find P from the dynamics of the interaction. This problem plays a role in suspensions at low relative Reynolds number as well as at high Reynolds numbers. It also manifests itself in the study of the flow of granular materials, which form the subject of Prof. Savage's lecture at this congress.

In the case of suspensions till thus far only two-bubble interactions have been considered. It is unlikely that in any general way the above sketched method could be extended to higher interactions. In some cases asymptotic results can be obtained for high concentrations. This leaves a gap for concentrations of moderate magnitude like 20-30%, say. This forms a challenging problem for the future. In the remainder of the lecture, I would like to discuss voidage waves as an illustration of how interactions affect phenomena in gas/liquid.

7. VOIDAGE WAVES IN BUBBLY LIQUIDS

Consider homogeneous flow of a bubbly suspension, in a vertical pipe. Such a flow can be realized at low void fractions α. In our laboratory we have been studying such flows over the years in several contexts. A remarkable feature is that in vertical direction no appreciable change in bubble size can be observed, apart from a small increase in size of the bubbles due to hydrostatic pressure decrease. This indicates that either coalescence is unimportant, or takes place at the same rate as breakup of bubbles. As is indicated in the flow map shown at the beginning of this lecture bubbly flow cannot persist when the void fraction increases but is replaced by plug flow. The usual explanation (see e.g. Hetsroni [23]) is that at increasing

void fraction big bubbles are formed by coalescence which then eventually
must lead to the formation of plugs. I find that hard to accept given the
absence, or annihilation, of coalescence at lower void fraction than the 25%
indicated in the literature as the threshold for transition. Moreover
practice shows that transition can be postponed to void fractions till 40%.
The question therefore is, if not coalescence, what triggers the transition
to plug flow. It has been suggested (Biesheuvel [24], Matuszkiewicz, Flamand
& Bouré [25]) that instability of concentration waves, also called voidage
waves, has something to do with this transition. These are waves at low
velocity, of the order of the mean gas velocity and therefore at constant
density, along wich disturbances in the void fraction are propagated. For the
analysis of such a wave, let us take them in x direction with a uniform
distribution normal to x.

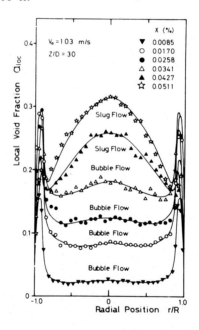

FIGURE 7

Void fraction profiles for vertical gas/liquid flow. From Serizawa [26]

In reality, α distributions exhibit in vertical up going pipe flow a profile
like in figure 7, flat in the central region and with peaks near the wall.
Indicating the mean gas velocity with U_g, conservation of volume for the gas
(remember that compressibility effects can be neglected) requires

$$\frac{\partial \alpha}{\partial t} + U_g \frac{\partial \alpha}{\partial x} + \alpha \frac{\partial U_g}{\partial x} = 0. \tag{7.1}$$

When there would be a one to one relationship between U_g and α, through a balance between frictional force and buoyancy, this could be written as

$$\frac{\partial \alpha}{\partial t} + (U_g + \alpha \frac{dU_g}{d\alpha}) \frac{\partial \alpha}{\partial x} = 0, \qquad (7.2)$$

which defines a propagation velocity, the kinematic wave velocity

$$c(\alpha) = U_g + \alpha \frac{dU_g}{d\alpha} . \qquad (7.3)$$

of a disturbance in α. Without interactions, that is for $\alpha \to 0$, the gas velocity with respect to the liquid velocity equals U_∞, the socalled terminal velocity. Interactions lead to dependence on α. The vertical motion, however, is governed by more forces than buoyancy and friction. First of all, there is an inertia force which represents the reaction of the fluid when the bubble accelerates. This force is expressed as $-D/Dt\{m(U_g-U_o)\}$ where m is the added mass of a bubble and U_o is the volume velocity

$$U_o = \alpha U_g + (1-\alpha)U_l .$$

At low void fractions it suffices, as explained earlier, to consider motion of pairs. Then the added mass is a function of the separation vector $\underset{\sim}{R}$. Knowledge of the pair probability distribution is needed to obtain a mean value for m

$$m(\alpha) = \int m(\underset{\sim}{R})P(\underset{\sim}{R})d\underset{\sim}{R} \qquad (7.4)$$

Then there are forces on the bubbles due to Reynolds stresses in the liquid. They may be due to higher interactions (in the case of pairs) and to turbulence in the liquid. Since bubbles cannot, their mass is zero, support any net force, the sum of all forces must be zero. For the mean motion in a pair this gives an equation of the type

$$\frac{D}{Dt}\left\{m(\underset{\sim}{R})(U_g-U_o)\right\} + 12\pi\mu a \; f(\underset{\sim}{R}) \; (U_g-U_o) = \rho g V + \rho_l V \frac{DU_o}{Dt} + \text{Reynolds stresses}$$

$$(7.5)$$

In this equation V is the volume of a bubble, in the righthand side forces due to acceleration of the volume flow U_o are allowed for apart from gravity. The functions $m(\underset{\sim}{R})$ and $f(\underset{\sim}{R})$ can be calculated with help of potential flow theory. Together with a momentum equation for the suspension solution for U_g-U_o can be obtained from (7.5). With knowledge of the distribution density $P(\underset{\sim}{R})$ mean quantities can be calculated.

For knowledge of $P(\underset{\sim}{R})$ one has to study the relative motion $\underset{\sim}{R}$ in a pair, as opposed to the common velocity dealt with in (7.5). This because P obeys an equation, expressing the conservation of bubbles

$$\frac{\partial P}{\partial t} + \nabla_{\underset{\sim}{R}} \cdot (P\underset{\sim}{\dot{R}}) = o.$$

Averaging of equation (7.5), using (assumed) knowledge about $P(\underset{\sim}{R})$ then leads to an equation for the average of $U_g - U_o$. Together with (7.1) this equation determines the behaviour of voidage waves. In our laboratory we carry out experimental and theoretical work in this field. The interaction between turbulence and bubbles is a subject in itself on which research has only recently started. We will, I think, hear more about that in the contribution by Bataille and Lance to this minisymposium.

Let us return to the discussion of voidage waves. They are governed by (7.1) and an equation which in the most simple form, that is leaving out Reynolds stresses and taking $DU_o/Dt=o$, looks like

$$\frac{D}{Dt}\left\{ m(\alpha)(U_g - U_o) \right\} + f(\alpha)(U_g - U_o) = U_\infty. \tag{7.6}$$

It should be mentioned that for many situations terms involving Reynolds stresses and diffusion should be added (To obtain insight in such terms is an important problem in this field). From (7.6) together with (7.1) two important quantities with the dimension of speed can be deduced. Take first the inertial term zero. Inserting the value following from (7.6) for that case in (7.3) gives

$$c_o = U_g\left\{ 1 - \alpha\, \frac{U_g - U_o}{U_g}\, \frac{f^{\,\prime}(\alpha)}{f(\alpha)} \right\}. \tag{7.7}$$

Secondly, assume that the inertia term dominates over the others. Then the quantity $m(\alpha)(U_g - U_o)$, the socalled impulse, is conserved. This gives, when inserted in (7.3) another wave speed

$$c_- = U_g\left\{ 1 - \alpha\, \frac{U_g - U_o}{U_g}\, \frac{m^{\,\prime}(\alpha)}{m(\alpha)} \right\}. \tag{7.8}$$

In the limit $\alpha \to o$, $U_g - U_o = U_\infty$, $dU_g/d\alpha = o$ and hence the wave speed $c(\alpha)$ is

$$c_+ = U_g. \tag{7.9}$$

Analysis (Van Wijngaarden and Biesheuvel [27]) of voidage waves at low concentration shows that, under these simplified circumstances these waves are stable when c_o lies in between c and c_f, which requires, from (7.7) - (7.9)

$$\frac{1}{m}\frac{dm}{d\alpha} > \frac{1}{f}\frac{df}{d\alpha} .$$

(7.10)

Void fraction waves in bubbly flows are therefore affected by inertia. This in contrast to similar waves in solid/gas flow, since there the added mass is unimportant. This would mean that concentration waves in gas/particle flows are always unstable. This, however, is not the case. Both in gas/particle and fluid/bubble flows there are several stabilizing mechanisms. The study of these form an important subject for research at present.

What can one learn from calculations and investigations as briefly described here ? First, for low α, one can design experiments to verify the results, for example by measuring mean gas speed, wave speeds etc. Such experiments are far from easy. The theory presupposes bubbles of uniform size, which are hard to produce experimentally. Unless much precaution is taken one gets too large bubbles, which are nonspherical and of too much variation in size. Experiments for which theory can be verified are therefore rare.

Secondly, especially for larger void fractions, where exact calculations are not possible the analysis with help of statistical methods may learn us what kind of experiments e.g. with voidage waves should be undertaken in order to measure certain functions or coefficients.

In the third place the study of voidage waves may be significant for the study of transition between regimes in the flow map.

8 CONCLUSION

In the foregoing I have discussed a number of topics, mainly selected from phenomena in bubbly flows. Similar problems occur with other topologies, in particular problems of transition.

It should be admitted that, at present, predictions for large scale transport lines of gas/liquid flow, based on small scale experiments, are hard to make. For this reason, industrial research is in this area often conducted in full scale experiments, e.g. for the measurement of pressure drops. Working on small scale, special problems and experiments is a good subject for university laboratories. There is a lot to learn for students. The subject of two phase flow offers possibilities to apply almost all aspect of fluid dynamics.

REFERENCES
[1] Kuhn, T.S., The structure of Scientific Revolutions (University of
     Chicago Press, 1962).
[2] Minnaert, M., Phil. Mag. 16 (1933) 235.
[3] Foldy, L.L., Phys. Rev. 67 (1945) 107.
[4] Wijngaarden, L. van, J. Fluid Mech. 33 (1968) 465.
[5] Caflisch, R.E., Miksis M.J., Papanicolaou G.C. and Ting L., J. Fluid
     Mech. 153 (1985) 259.
[6] Crighton, D.G., and Ffowcs Williams, J.E., J. Fluid Mech. 36 (1969) 585.
[7] Medwin, H., Acoustical bubble spectrometry at sea. in: Lauterborn, W.
     (ed) Cavitation and inhomogeneities in underwater acoustics (Springer
     Verlag 1980).
[8] Silberman, E., J. Acoust. Soc. Am. 29 (1957) 925.
[9] Whitham, G.B., Linear and nonlinear waves (Wiley & Sons, New York 1974).
[10] Noordzij, L. and Wijngaarden, L. van, J. Fluid Mech. 66 (1974) 115.
[11] Kuznetsov, V.V., Nakoryakov, V.E., Pokusaev, G. and Schreiber I.R.,
     J.Fluid Mech. 85 (1978) 85.
[12] Nigmatulin, R.I., Kabheev N.S. and Zuong Ngok Hai, J. Fluid Mech. 186
     (1988) 85.
[13] Nakoryakof V.Y., Pokusaev, B.G. and Schreiber I.R., Wave propagation in
     gas- and vapour-liquid media. (Institute of Thermophysics Novosibirsk
     1983) in Russian.
[14] Arakeri, V.H. and Shanmuganathan V.J., J. Fluid Mech.159 (1985) 131.
[15] Omta, R., J. Acoust. Soc. Am. 82 (1987) 1018.
[16] Plesset, M.S. and Prosperetti, A. Ann. Rev. Fluid Mech. (1977).
[17] Lauterborn, W. and Cramer, E., Phys. Rev. Lett. 47 (1981) 1445.
[18] Lauterborn, W. and Suchla, E., Phys. Rev. Lett. 53 (1984) 2304.
[19] Delhaye, J.M. and Cognet G. (ed), Measuring techniques in gas/liquid
     two-phase flows (Springer-Verlag, 1984).
[20] Spalding, D.B., Int. J. Multiphase Flow 6 (1980) 157.
[21] Physics through the 1990's. Plasmas and Fluids,(USA National Academy
     Press, 1986).
[22] Batchelor, G.K., J. Fluid Mech. 52 (1972) 245.
[23] Hetsroni Gad (ed)., Handbook of Multiphase Systems (Hemisphere
     Publ. corp.1982).
[24] Biesheuvel, A., On void fraction waves in dilute mixtures of liquid and
     gas bubbles. Thesis, University of Twente 1984.
[25] Matuszkiewicz, A., Flamand, J.C. and Bouré, J.A., Int. J. Multiphase
     Flow 13 (1987) 199.
[26] Serizawa, A. Kataoka I and Michiyoshi, I., Int.J.Multiphase Flow 2
     (1975) 235.
[27] Wijngaarden, L. van and Biesheuvel, A. Voidage Waves in bubbly flows.
     In: Proc. Int. Seminar on Transient Phenomena in Multiphase Flow.
     Ed.N. Afgan (Hemisphere 1988).

Theoretical and Applied Mechanics
P. Germain, M. Piau and D. Caillerie (Editors)
Elsevier Science Publishers B.V. (North-Holland)
© IUTAM, 1989

SOME PROBLEMS OF MECHANICS IN TECTONIC ANALYSIS

Ren WANG

Department of Mechanics
Peking University
Beijing, China

Problems using the continuum theory of solid mechanics to analyze the
tectonic movements are reviewed. The difference between these problems
and those of ordinary mechanics problems are discussed. The main difference is that the problems here are inverse ones, from the surficial
manifestation of tectonic movements to model the interior processes,
from the present information to trace their histories. Information gathered by Earth scientists are to be used to guide the inversion. The
establishment of proper constitutive relation and rupture criterion
is of paramount importance in the numerical simulation. Challenging
problems in solid mechanics are mentioned.

1. INTRODUCTION

The earth is under constant motion. Her motion around the Sun is perceived by
the four seasons, her own rotation by the 24 hours. The motion inside the Earth
manifests itself by volcanic eruptions and earthquakes. Some motions are so
small and so slow which one can hardly felt such as the ground under us is moving up and down twice daily about 20-30 cm, it is the Earth tide and requires
delicate instruments to measure. The continents move relative to each other
somewhere around a few centimeters per year which we are now able to detect by
space technology. The monitoring of surface deformation especially near the
seismic area is one of the major means in predicting oncoming earthquake or
other catastrophic events.

Tectonic movements are usually very slow processes. The rocks one sees at the
mountainside nearby were deposited on the sea bottom long time ago. It took several million years for sediments to form rock and to rise and become a mountain.
When geologists talk about something is very young, they may mean that it is
only a few million years old. Within such a slow process, there may be numerous
episodes of sudden ruptures known to us as earthquakes which may take place in
fractions of a second only. so the scale involved in Earth movements has a tremendous range which bridges at least 16 to 17 orders of magnitude.

In order to study the motion, we have to establish an Earth model. The commonly
known model is determined by the science of seismology using the elastic theory
of wave propagation. It is roughly composed of three layers: viz. crust, mantle
and core. A finer classification yields upper crust, lower crust, upper mantle,
etc. The crust together with the upper mantle is also termed the Lithosphere.
In the theory of plate tectonics, lithosphere is divided into plates, it is
around 100 km. thick (50-200km) and rests on a layer called asthenosphere in
which materials can flow laterally. In this rough model the lateral inhomogeneity
is neglected and the model is a spherically symmetric one. Further investigation
shows not only the uppermost layer is obviously not laterally homogeneous, but
even the core-mantle boundary is also not homogeneous which has a peak of 11 km
high directly under Hawaii, i.e. higher than Mt. Everest.

The subject of direct concern to mankind is the tectonic structure in the crust,

its driving mechanism, evolution process and results of tectonic movements.
These are directly related to resource exploration as well as to the prevention
and mitigation of natural disasters.

2. MANIFESTATION OF TECTONIC MOVEMENTS

We shall first take a look at the different manifestations of tectonic movements.

2.1 Plate Tectonics

The theory of plate tectonics concerning the global structure was advanced in
midsixties. If the earthquake foci are plotted for a few years over the whole
world, one can get a clear outline of the relatively rigid plates which are
moving against each other along these boundaries. From the paleomagnetism preserved in the rock, earth scientists try to locate the plate position in the
past. It is advocated that 200 ma (million year) ago there was a connected continent, Pangaea, which was broken up later on. The Indian subcontinent moved
Northeastward toward the present position. Figure 1 shows the position of Indian
plate some 65 ma ago, there was a Tethy's sea between Indian plate and the Eurasian plate. It disappeared some 20 ma ago as the two plates collided. It is not
improbable that Mediterranean today lying between the Northward moving African
plate and Eurasian plate would disappear some million years later (not long
geologically) and our meeting place today would be on top of a second "Tibetan"
plateau.

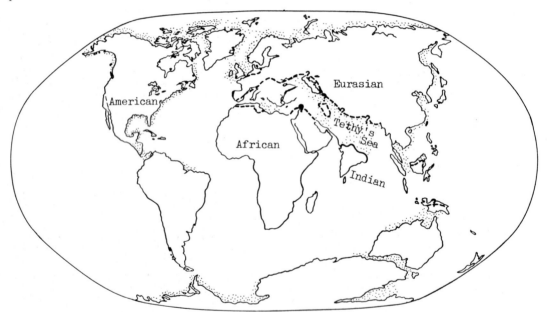

Fig. 1 Geographical arrangement of the continents in Palaeocene times.
(around 65 ma ago)(taken from The Evolving Earth, ed. L.R.M.Cocks
p. 222)

In the last decade, another major development in the theory of plate tectonics
is the discovery of amalgamation terrane. It was found that along the West coast
of U.S.A. there are geologically unrelated blocks of land lying side by side
across some vertical fault planes. Geological and paleomagnetic evidences show
that they may have been broken fragments from an originally intact continent
carried over a long distance by a mantle driven conveyer belt and attached themselves to the American plate. Since then, many amalgamation terranes have been
discovered in different continents.

These are tectonic movements on the global scale.

2.2 Folding

Folding is the principal mountain building process especially at the continental margin where two plates collide or when the amalgamation terrane meets the plate.

Folding is perhaps the most common obvious manifestation of ductile deformation of rocks. It looks very much like the bending of beams or the flexure of plates and sometimes it even has a wavy form like a ribbon. It may have been deformed by loads perpendicular to the rock layer, but most of the time it is due to the compressive stress in the plane of the layer. So it is one of the best studied process by solid mechanics using buckling theory. Here the buckled layer or layers are usually embedded in some more easily deformable media. Sometimes it is the buckling of the surface layer only, while the material below it is under simple compression. There are many different shapes of folds which are discussed fully in [1].

In tectonic analysis, it is the post buckling shape that is more interesting. Folds may undergo very large deformation and sometimes they overturned before fractures occur, see Figs. 2a and 2b.

Diagrammatic illustration of how an overturned field might evolve as a result of an increasing horizontal force. In the bottom sketch the formation has actually ruptured.

Fig. 2a A recumbent fold, Greenland. From top of cliff to valley bottom 800m. (from Press & Siever: Earth)

Fig. 2b A similar fold, Dahuichang Beijing. Note the ribbon-like fold at top left corner.

As rock is brittle near the surface of the earth, it becomes ductile only under the high temperature and pressure conditions at depth, folds must have been formed in the deeper part of the crust and then brought to the surface afterward and it must have been a very slow process.

On top of an earlier folded area, very often as a result of a change in direction of the compressive stress, there forms another set of folds, the interference of these folds produces arrays of domes and basins.

2.3 Faulting

A fault is defined as a planar discontinuity between blocks of rock that have been displaced past one another, in a direction parallel to the discontinuity, [2]. A fault zone is a region of many sub-parallel faults. These are products of brittle deformation. There is another phenomenon having similar appearance as a fault zone yet the materials inside undergo severe shear without macroscopic discontinuity, it is called a shear zone and is a product of localized ductile deformation.

Faulting is another most commonly seen tectonic process. Most of the faults seen are vertical or inclined. Reverse fault is formed under horizontal compressive stress so the rock mass is shortened. Normal fault is formed under gravity loading which overcomes the horizontal stress, so the rock mass is lengthened. Strike slip fault is formed under pure shear so it is usually vertical. Rivers usually follow tectonic faults, so earth scientists use the river course to deduce the fault pattern and get the regional stress orientation. In recent years reflection seismic exploration has discovered many sub-horizontal discontinuities in the crust which separate the severely folded layers or imbricate listric faults on the upper part of the crust from the practically undeformed rock mass at the bottom, they are called decóllement zone or detachment surface.

Faulting is usually accompanied by earthquake since it is a brittle rupture process and releases energy abruptly. In a tectonic process, the rock strata may undergo creep but after certain amount of creeping deformation, it may also end up in creep rupture.

The ground motion caused by an earthquake is a very complicated one. The horizontal ground motion induced is usually considered to be the most dangerous component, Fig.3a, however, near the epicenter, vertical ground motion can be very large. Weichert et al [3] have reported recently that a strong-motion accelerograph they installed during aftershock surveys of an M_S 6.6 earthquake in the Nahanni area, in the NW territories of Canada, captured very strong ground motions within a few km (less than 6 km estimated from other data) of a second earthquake of M_S 6.9. On one accelerograph the peak horizontal acceleration reached 1.25 g, but the peak vertical acceleration was offscale and must have exceeded 2 g in the upward direction. During Tangshan earthquake, it was reported that a mother grabbed her child and found herself sat on the roof of their home three meters off the ground, I would estimate it to be around 6 g. Brick chimneys were broken into sections which may be attributed to the transmission of stress waves travelling upward. The most interesting phenomenon is the twisting motion exhibited in these broken sections as shown in Fig.3b. The energy released in an earthquake is so large that a single earthquake can cause tens of kilometers of mountain chain to offset along a fault a distance over 10 meters.

Fig. 3a Ground faulting N29W during Haicheng earthquake 2,4,1975, M_S 7.3, crack 24 cm wide, left lateral offset 8cm, right block upheaves 8 cm. (taken by Geological Res. Inst. S.S.B.)

Fig. 3b A square section chimney broken into 4 sections, the top section has obvious right lateral twisting, during Haicheng earthquake. same as Fig. 3a.

Fault systems provide a channel for mineral bearing fluid to migrate and concentrate, it is also important in the utilization of geothermal energy, in regional planning for construction, in nuclear waste disposal, etc. so the investigation of fault system has been intensive and friutful.

3. BRIEF HISTORICAL SKETCH IN THE APPLICATION OF CONTINUUM MECHANICS TO TECTONIC ANALYSIS

Historically, the early literature on using continuum mechanics to analyze the Earth deformation problem can go back to Poisson, 1829, who gave the solution to the problem of free radial vibrations of a solid sphere. Lamé, 1854, discussed the general series solution for a homogeneous elastic spherical shell subjected to arbitrary forces on its surface. Sir W. Thomson, 1863, had dealt with the estimation of the rigidity of the Earth. G.H. Darwin, 1879, had dealt with the theory of elastic tide and the stresses in the Earth interior by the weight of continents and mountains. Lord Rayleigh, 1887, had investigated the wave propagated over the surface of an isotropic elastic solid body. J.H.Jeans, 1903, applied seismic wave propagation theory to determine the Earth modulus. A.E.H. Love, 1911,[4], in his prize winning treatise "Some Problems in Geodynamics" was the first one to use the term geodynamics, in which he made an excellent study on crustal isostasy, Earth tide, latitude variation, compressibility effect of the Earth, gravitational instability. Love number in the tidal theory and Love wave in seismology were named after him in his honor. So far those above had used solid sphere as the Earth model. L.S. Leibenson, 1915, [5], had calculated the global stress distribution due to the tilting of rotation axis and due to the variation of rotation speed using a two layer sphere. For early historical work the reader may be referred to Love [4] and Leibenson [5]. Other works on layered sphere will be discussed below.

As to the analysis of structural elements, A. Nadai has a chapter in his text [6] analyzing the bending of beam to simulate folding. M. Biot [7] had analyzed the folding process in great detail and developed the idea of dominant wave length. In recent years, there are many more mechanics people working in this direction and had contributed much to the understanding of tectonic movement and earthquake mechanics, however, much remains to be done.

4. DISTINCTIVE FEATURES IN THE FORMULATION OF MECHANICAL ANALYSIS OF TECTONICS

In applying continuum mechanics to analyze those tectonic movements introduced above we run into many difficulties. One of these is that the movements are the end products on the Earth surface seen today. To solve for the driving mechanism and the deformation process one faces a two-way inversion problem, viz. spatially to invert the process in the interior of the Earth and temporally to invert the process in the past. Besides, the material property is highly non-linear, the deformation is very large so it is a highly non-linear inversion problem.

In order to solve an inverse problem, we have to know how to solve the forward one. The proper formulation of a mechanics problem requires to know the following four conditions:

1, The structure of the body under consideration, its composition and its boundary.
2, Physical and mechanical properties of the materials, including the constitutive relations and the rupture criteria, the thermal properties, etc.
3, External action on the body, including body forces, boundary tractions or boundary displacements, thermal conditions.
4, Initial conditions.

For the tectonic problems these conditions can be very complicated and most of the data are not accessible. We have to depend on models based on correct understanding and reasonable assumptions. Thus it requires us to work closely with earth scientists. They use every technical means to collect data and specimens from the field. With the help of satellite they will be able to determine the relative movement of the plates. With the help of deep hole drilling they can obtain specimens over 10 km under the surface. From the world wide seismological network, the reflection seismograms, they are getting better understanding of the structure in the Earth interior, etc. The laboratory equipments for testing rock properties can now provide a statically maintained pressure of 550 GPa and temperature greater than 5000 K by laser heating in diamond cell, i.e. reaching the conditions in the core of the Earth. The electronic microscope is used to study the microscopic deformation mechanisms which in turn give constraints on the environmental conditions and the stress required for such deformation. From the fold and fault patterns down to the deformation of fossils one can deduce the stress directions and the strain tensor they had been subjected to and also the timing, [1,2].

Let us now discuss the work done related to these four conditions in a little more detail.

4.1. Structural Model

The preliminary model mentioned in the first section was obtained from seismic wave inversion, it is good for short term movements. Refinements to it had been done by taking into account the constraints from the global free oscillations recorded after large earthquakes, the perturbation of satellite orbits, etc.

For tectonic analysis, long-term movements are involved, the motion is so slow that viscosity of the Earth interior can usually be neglected, so the Earth can be regarded as a thin elastic (or visco-elastic) shell with an invisid, incompressible fluid interior. Since the lateral inhomogeneity is comparable to the thickness of the shell, the axi-symmetric model is no longer applicable. Presently, earth scientists are working on the determination of lateral inhomogeneity, only after this is done a good Earth model for tectonic analysis can be set up.

For regional tectonic analysis, the structural model is usually taken to be a plate with thickness of the lithosphere and the curvature of the Earth being neglected. A plane stress approximation may probably be better, but sometime plane strain approximation is also used. The problem here is to specify the lateral boundaries which depends on how does one specify the connections with neighboring regions.

4.2. Physical and Mechanical Properties of the Earth Media

With the advent of high speed computers and the development of numerical techniques such as finite element method, boundary element method, etc. the key problem in tectonic analysis is now the correct representation of constitutive relations and fracture criteria of the media. Earth media is very heterogeneous and subjected to highly diverse environmental conditions. Among these factors, temperature is probably the most important one which causes the constitutive relation to depend on the time scale under consideration. For tectonic movements in general, one has to use rheological models, moreover, it is found that they usually would be non-Newtonian ones. However, for the upper layers we may still use elastic models, but the thickness of the elastic layer will be a function of load duration because the creep rate is strongly temperature dependent. Melosh [8] advocated that for very short term loads of less than a few years the entire mantle is elastic, for a time scale of 10^4 years, the elastic layer is about 100 km thick and for loads lasting 10^6 years or more, the elastic layer may be only 15 to 30 km thick.

There are many experimental work done on the creep of rocks under high temperature and pressure, due to the difficulties in maintaining the conditions constant for a long time, most of the work done under high temperature and pressure lasted a few hours, the strain rate attained is between $10^{-7}/\text{sec}$ to $10^{-8}/\text{sec}$. To extrapolate them to $10^{-14}/\text{sec}$ or slower for the case of tectonic movement, one has to investigate the microscopic deformation mechanisms, only the behavior of the same deformation mechanism can be extrapolated. Thus the deformation mechanism for rocks has been investigated intensively. There are four main deformation mechanisms as discussed by Ashby and Verrall [9] and others, see Fig. 4.

a. Cataclastic Flow

This is the mechanism related to creep motion in a fault zone. The shear deformation is accomplished by repeated tensile fracture and the rolling of the cataclastics. So the grain size continues to decrease with the deformation, as the grain size distribution reaches a certain ratio, the stationary flow will persist. The shear resistance to slip increases with confining pressure and the volume increases with deformation -- causing dilatancy.

Related to the slip along a fault, frictional constitutive laws have been proposed based on the friction experiment of Dieterich (see Tullis [10] for a review)

$$\tau = \sigma[\mu_0 + \sum_i b_i \psi_i + a \ln(V/V^*)] \tag{1}$$

where τ is the shear stress, σ is the normal stress, the square bracket is the coefficient of friction, in which the first term represents the Coulomb's friction, the second terms represent the internal variables ψ_i that characterize the state of the sliding surface, V in the third term is the sliding velocity, V^* is a reference velocity. This law have been used to relate earthquake precursory phenomena and earthquake mechanism.

b. Dislocation Glide

Under low temperature or high stress, the plastic flow in a crystalline solid is accomplished by dislocation glide in the slip plane. Confining pressure does not have much effect on the rock strength. The rock hardens as the dislocations accumulate, this corresponds to primary creep. As the effects of diffusion and recrystallization come in to cause relaxation and balance off the hardening, the rock undergoes steady state creep, the steady state creep rate takes the form:

$$\dot{\varepsilon}_s = A \exp(-Q_e/RT) \sin(B\sigma) \tag{2}$$

where A,B are constants, Q_e is the effective activation energy, R is the gas constant and T is the temperature in $^\circ$K.

c. Dislocation Creep or Dislocation Climb and Polygonization -- Power Law Creep

This is the mechanism under medium temperature ($T > 0.5\ T_m$) and medium stress, the steady state creep rate is given by

$$\dot{\varepsilon}_s = A \exp(-Q_e/RT) \sigma^n \tag{3}$$

where T_m is the melting temperature, n usually lies between 3 - 5. The detail micromechanism is very complicated in this case.

d. Diffusion Creep under Higher Temperature

This is the mechanism under higher temperature and low stress with lower strain rate, the steady state strain rate for volumetric diffusion (Nabarro-Herring creep) is

$$\dot{\varepsilon} = a \ (D_S\Omega\sigma/RT)/d^2 \tag{4}$$

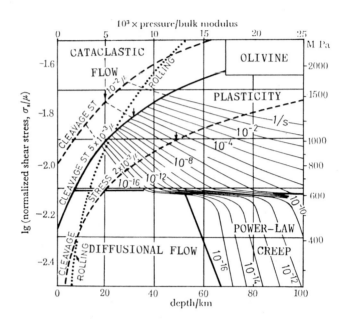

Fig. 4   Deformation map for olivine showing transition from
         cataclastic flow to plastic flow ([9])

where D_S is the volumetric diffusion coefficient, Ω is the atomic volume, d is the grain diameter. For grain boundary diffusion at lower temperature it is called Coble creep, we have,

$$\dot{\varepsilon} = a' \ (D_{gb}\Omega w\sigma/RT)/d^3 \tag{5}$$

where D_{gb} is the grain boundary diffusion coefficient, w is the average width of the boundary. In these relations

$$D = D_0 \ \exp \ (-Q/RT) \tag{6}$$

These relations have the form of Newtonian viscosity, but since the grain size of a deformed rock may itself be a function of stress, there is some question whether even these diffusion creep mechanism are really Newtonian, [8].

There is another important deformation mechanism in the tectonic process through liquid phase transportation, it is called pressure solution. For a rock under load, the stresses at the contact points between grains are higher than those at the non-contact points, there is a difference in chemical potential of ions to cause them to dissolve under high stress and through transportation to redeposit at the low stress region.

Since rock is more or less permeable, the role of pore fluids in the tectonic process is very important. When the pore pressure increases the effective stress is decreased, the shear required for rupture decreases so the rock becomes brittle. Otherwise when the pore pressure decreases the rock may act ductilely. In this manner the stress and deformation are coupled to fluid flow. Thermal expansion and pressurization of fluids further coupled the stress and fluid fields to the temperature field. The fully coupled hydro-thermo-mechanical aspect of tectonic processes is an important problem to be tackled. The hydrolytic weakening effect of some minerals is also important in analyzing reservoir induced earthquakes.

Another distinctive feature in the constitutive relation of rock is its strain-softening characteristics. It is a property especially important in dealing with instabilities of geological processes. There are several strain softening mechanisms: 1. it may be caused by the direct effect of grain boundary migration associated with recrystallization or the growth of new phases which wipe out the previously formed dislocation that rendered the hardening effect. 2. it stemmed from grain size reduction which facilitates the diffusion processes shown in Eqs.(5), (6) and Fig.5. 3. the rotation of grains to more favorable orientations for further intracrystalline slip. Besides the pore fluid weakening effect mentioned before, there are also thermal softening effect, metamorphic softening effect, etc, [11].

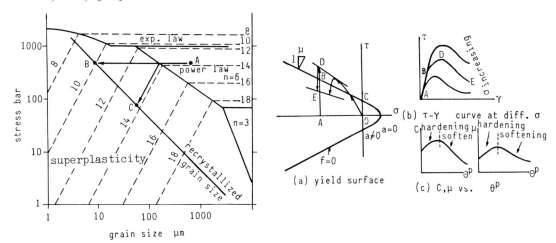

Fig. 5 Deformation map for calcite at 400°C, softening due to grain size reduction. From A to B under constant stress, strain rate increases. From A to C under constant strain rate, stress drops. (From S.M.Schmid in "Mountain Building Process" Ed. by K.J.Hsu, 1982, p. 108)

Fig.6. Work softening model based on Drucker-Prager yield condition. C -- cohesion strength and μ-- coefficient of internal friction both are functions of θ^p. (Yin[12]).

There are several constitutive representations for work softening materials. In Eq.(1) when $b_i > a$, it will represent velocity softening. Yin et al had put forth a simple work-softening relation based on Drucker-Prager's yield condition

$$f = \alpha I_1 + (J_2 + a^2 K^2)^{\frac{1}{2}} - K = 0 \qquad (7)$$

with the plastic potential

$$g = \bar{\alpha} I_1 + (J_2 + a^2 K^2)^{\frac{1}{2}} \qquad (8)$$

where I_1 is the average normal stress
    J_2 is the second invariant of the stress deviator,
    $\alpha, \bar{\alpha}$ and K are related to the cohesion strength C and internal friction μ, they are functions of the internal variable θ^p, the plastic volumetric strain. These functions may represent both the work hardening and the work softening behavior in accordance to test curve, see Fig.6.
    a is a small parameter to avoid singularity at $\tau=0$ (Fig.6).

From the flow rule associated with the plastic potential g, one gets the constitutive relation. It has been incorporated into a finite element program and used successfully to simulate the rupture process of a slotted marble plate under uniaxial compression (see section 5.3 below) and many other geotechnical problems, [12,40].

4.3. Driving Forces and Boundary Conditions

Among the many forces acting on the Earth, it is the long term forces that are important for tectonic movements. Those short term forces such as tidal attractions, forces due to polar wandering and annual change of rotation speed, etc. may have triggering effect only.

The fundamental force that is everpresent is the gravitational force and the centrifugal force due to the Earth rotation. The latter has a long term angular deceleration averaging 5×10^{-22} rad/sec over the last 600 ma. However, it is suspected that during this period the angular speed may have several sizable fluctuations although the overall trend is deceleration. These fluctuations might have been related to the global tectonic cycles occurring every 100–200 ma. Lee [13] had tried to correlate the two but no conclusive result was obtained.

From the consideration of energy source, one naturally thinks of the thermal energy in the Earth interior. The thermal driven convective motion in the mantle has been confirmed by the measured spreading rate of oceanic ridge, but the probable convective pattern is still a subject under debate. As to the driving force for the plate movement, many investigations are interested in the relative contribution from the downward pull of subducting slab under the continental margin, the sidewise push from the uprising ridge and the tangential force of mantle convection acting on the bottom of the plate, and whether the latter is an active or a resisting force. Further complication is that the whole process may vary with time.

As regards to the boundary conditions, on the Earth surface one may be able to measure the stress variation as well as the displacement over a certain time period. In this case then, we have redundant boundary conditions on this surface, while the boundary conditions over the remainnig surfaces are quite inadequate. From in-situ stress measurements in underground mines and hydrofracturing in oil wells, from the analysis of the earthquake focal mechanism, earth scientists are gathering more and more data for the present stress distribution over different parts of the world. There are many modern techniques to measure the surface displacements reaching an accuracy better than 1 in 10^7 or 10^8. The advancements will improve our working model and will also provide better constraints on the choice of boundary conditions.

4.4. Initial conditions

For quasi-static problems in engineering one usually neglect the initial stress state in the body i.e. to start from a virgin state. However, for the case of Earth structure, the residual stress can be preserved for a long time with quite a large magnitude. For example it was reported that a 18 cm x 18 cm basaltic column from a magma intrusion with all its sides free for over ten thousand years has a maximum tensile residual stress reaching 12.6 MP_a and a maximum compressive stress reaching 15.2 MPa. The Earth crust has undergone numerous tectonic processes, Turcotte et al [14] has advocated that the stress in the crust may take over 10^8 years to relax, so in analyzing tectonic movements, especially in predicting earthquakes, it is important to know the residual stress state already there. In this respect, one of the endeavors of earth scientists is to estimate the stress that a rock had undergone e.g. from the preserved micromechanisms of deformation. This is the so called paleopiezometer, see e.g. [15].

5. FORWARD AND INVERSE PROBLEMS

With all these conditions given or reasonably assumed, one can go ahead to solve the forward mechanics problem. As mentioned before, a good solution depends on having a good model. Earth scientists often do some qualitative experiments to help them decide the correct structural model. For example, Tapponnier and others

1982, [16], have used a test on a block of plasticine which is pushed from one side to represent the action of Indian peninsula on Tibet. From the test they wished to decide whether the Eastern boundary of the Eurasian plate should be fixed or be left free. They picked the latter because it showed a fault in the model that simulated well with the major fault in Western China. For this purpose, photoviscoelastic modelling, clay modelling using Moire techniques, etc. have been used, they help greatly in setting up a model. Of course, there is a question about the similarity rule, but for getting a qualitative picture, they may still be useful.

As to inverse problems, one simple algorithm is to solve a forward problem with several unknown parameters contained in any of the above conditions, they will be determined by comparing the final results with the measured data. Since the data can never be completely satisfied there can be no unique solution, one can only expect to pick out the optimum solution under the assumed condition. The result will show a direction to modify the original model so that one can go on with a better approximation. For the linear case, we can make use of the super-position principle and use the method of least square to get an optimum solution. For the non-linear case, there seems to be no other way than the trial and error method, when the parameters are few in number, we had tried to use orthogonal design to guide the choice of parameters.

5.1. Tectonic Stress Field

The deformation of the Earth due to Solar-Lunar attraction and tidal loading has long been subjects of mechanical analysis, most of them had treated the Earth as a solid sphere or a sphere composed of only a few layers. By using the solutions for axi-symmetric shell from three-dimensional elasticity, Wang et al [17] have obtained analytic solution for the spherically symmetric elastic model composed of any number of layers under tidal attraction and rotational accelera-tion. The displacement field and stress field were used to evaluate seismic trig-gering effect. Pan, Ding et al [18,19] had further calculated the displacement and stress fields for the latest PREM (a short-term) model suggested by IUGS in 1981. By expanding both the body force and the surface load potentials in the same spherical harmonic form, they are able to transform the basic equations of gravitating elastic sphere

$$\nabla \cdot \sigma - \nabla(\rho g \, U \cdot e_r) + g \, \nabla \cdot (\rho U) \, e_r - \rho \nabla \psi_1 = 0 \tag{9}$$

$$\nabla^2 \psi_1 = -4\pi G \nabla \cdot \rho U \tag{10}$$

into a set of six simultaneous first order linear differential equations and the related boundary conditions. In the above equations ρ is the density, g is the gravitational acceleration, U is the displacement vector, σ is the stress tensor, e_r is the unit vector along the radius and ψ_1 is the perturbation potential. The use of propagating matrices in the numerical treatment greatly simplifies the computation for the multilayer model. The time varying displacement fields and stress fields in the Earth interior are computed. The influence of ocean tide on the Love number compares favorably to the observed values and is better than those obtained by using other models.

For tectonic movements, Vening Meinesz, 1947, [20], had solved the stress field for a 30 km thick spherical symmetric elastic shell when the rotational speed had decreased 14.7%, equivalent to the change of ellipticity from 1/210 some 1600 ma ago to the present 1/297 suggested by Jeffreys [21]. He also solved the case of polar wandering. He compared the calculated stress fields with the shear pattern on the globe and concluded a 70° polar wandering, i.e. if the pole had wandered from somewhere near Calcutta to the present position it will give a better fit to the shear pattern than due to the change of rotational speed. This is the first trial on a long term Earth model. Wang et al [17] had also tried several layered elastic shell of different thicknesses with incompressible fluid

interior solving for the Earth's response to the long term rotational acceleration.

Richardson et al [22] has used finite element method to solve for the relative
importance of the three types of forces acting on the plates mentioned in the
last section. The model they used is a elastic shell of uniform thickness, 100km,
they solved 32 different ratios between these forces and used earthquake focal
plane solutions, in-situ stress measurements and strikes of stress-sensitive
geological features as criteria for choosing the optimum solution. Solomon et al
[23] had later extended the work to estimate stress magnitude. So far no long-
term model with lateral inhomogeneity has been studied, although the tidal loading used in [18] does vary over the Earth surface.

For regional stress field, Eastern Asia is probably the most popular place to
work on. Tapponnier et al [24] made use of plane strain slip line field theory
for the perfectly plastic body to investigate the displacement field resulting
from the collision of the Indian peninsula on the Eurasian plate. They used the
trend distributions and the relative movements of the faults to be the main criteria in judging the correctness of their assumptions. S.Y.Wang et al [25] have
investigated the same problem in the elastic case by using the finite element
method. They put in the stress boundary conditions along the East China coast
instead of leaving it free and added the seismicity and the fault-plane solutions in North China region to the distinguishing criteria and arrived at a ratio
between the tractions from the Pacific, Phillipine and Indian plates. Villote
et al [26] considered the same problem as the indentation of a rigid die into
an incompressible viscoplastic medium by the finite element method, both in the
plane strain and plane stress cases. Otsuki [27] treated this problem using 24
models in his computation.

Wang and Liang [28] made use of the principle of superposition to lessen the
computing work and used the method of least square to give a better way to pick
out the optimum solution. The traction boundary in this case was divided into
segments (Fig.7), unit load along x (or y) direction is applied on segment m and
solved for the displacement u_i^{mx} and stress σ_{ij}^{mx} fields. (superscript x will also
stand for y, in the equations below they are not written out but should be so
understood). By the principle of superposition, multiply the solution by a constant factor α_m^x and sum the displacement or the stress components over all the
segments at any interior point, it should be equal to the measured value at that
point, say at point n, i.e.

$$\sum_m \alpha_m (u_i^{mn}) = \bar{u}_i^{-n} \tag{11}$$

As the stress data are usually given by earthquake focal plane solution which
gives the direction $\bar{\theta}$ only for the maximum principal stress, so the corresponding equation is a homogeneous one:

$$F^n = 2 \sum_m \alpha_m (\tau_{xy}^{mn}) - \sum_m \alpha_m (\sigma_x^{mn} - \sigma_y^{mn}) \tan 2\bar{\theta}^n = 0 \tag{12}$$

rewrite Eq.(11) into:

$$D_i^{\,n} = \sum_m \alpha_m u_i^{mn} - \bar{u}_i^{\,n} \tag{13}$$

Formulate

$$f(\alpha_m) = \sum_n [(D_i^{\,n})^2 + (F^n)^2] \tag{14}$$

From $\partial f / \partial \alpha_m = 0$, we shall get 2m equations for obtaining the optimum values of
α_m. They used 5 slip displacements along 5 major faults and 9 focal plane solutions distributed over the region (see Fig.7) and obtained a finer distribution
of tractions along the boundaries than before. Jia [29] has further considered
the tangential traction applied at the bottom of the plate distributed according
to an inversion from satellite gravity measurement. He found that the mantle pro-

vided a resistance instead of a driving force as originally thought, so the
boundary forces along the Eastern boundaries were diminished. From the above
scheme, one notice that the accuracy of the absolute magnitudes of the displacement and stress depends very much on those of the measured $\bar{u}_i$.

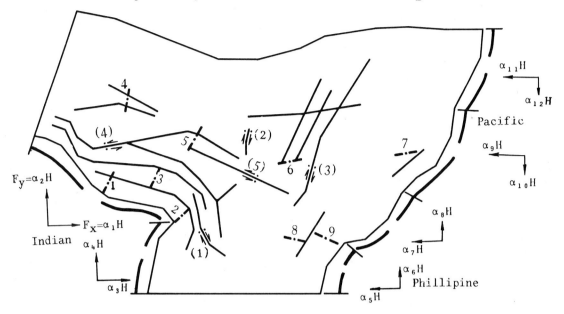

Fig. 7 Illustrating the division of traction boundary into segments.
Numerals with bracket shows where the relative fault slip is known,
Numerals without bracket shows where the focal-plane solution is given.

5.2. Folding

As mentioned before, folding is mainly analyzed as a buckling process. For a
more viscous layer (called competent layer in geology terms) embedded in a less
viscous matrix, Biot [7] and also Ramberg [30] have used a Newtonian model and
developed the concept of dominant wave length with beam theory.

$$L_d = a\pi h \ (\eta_1/6\eta_2)^{1/3} \tag{15}$$

where h is the thickness of the competent layer, η_1 and η_2 are the viscosities
of the competent layer and the matrix respectively. For n layers simply multiply η_1 by n. These equations have been modified by different people after comparison with those found in the field.

Dieterich and Onat [31] are probably the first to introduce finite element method
in analyzing geological structure. They used it to analyze folding mechanism. As
finite element method is further developed and applied, many simulations are made
for the large deformation process of folding either symmetric or asymmetric,
elasto-plastic or Newtonian. Lan et al [32] have simulated a case of an overturned fold and found that it had been formed in two stages, first by finite symmetric deformation and then sheared to become overturned.

Smith [33] used a unified hydrodynamic stability theory to treat the onset of
folding, boudinage and mullion structures and found that the Newtonian theory
fails to predict dominant wave length to thickness ratio between 4 to 6 which
appears most commonly in the field. He suggested that a power law Non-Newtonian
fluid has to be used and so obtained satisfactory results. In Fig.8.

$$m = \frac{\text{viscosity of layer}}{\text{viscosity of matrix}}$$

both viscosities are functions of strain rate. In 8a, for folding, strain rate
softening layer increases the growth rates but decreases the wave length. In 8b,
for boudinage, strain rate softening layer is necessary for growth. In 8c, for
inverse folding, extreme strain rate softening is necessary even for weak growth.
In 8d, for mullions or inverse boudinage, strong strain rate softening matrix
is necessary for significant growth.

The buckling of the surface layer of a
elastic half space under compression was
first investigated by Biot for the case
of Mooney material [7]. The problem was
studied later by many others for hypere-
lastic Hadamard material and for the combination of a layer over a half space,
etc. Wu and Cao [34, 35] recently have
examined the surface instability of an
incompressible half space under biaxial
loading and the axisymmetric surface
buckling for a standard material. Cao
has further extend the case to have the
upper surface also loaded by hydrostatic
pressure.

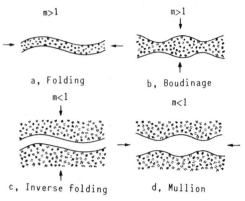

Fig.8 Four cases of single-layer
shearing instabilities (Smith [33]).

As to regional fold pattern, Huang and
Hu Haichang, [36], had analyzed analytically the onset of en-echelon fold pat-
tren within two straight boundaries moving parallelly relative to each other.
Zen [37] had treated analytically the fold pattern due to two vertical columns
in the crust approaching each other and used the corresponding principle between
linear viscoelastic and elastic materials to get the solution for the linear
viscoelastic case from which the evolution process of the structure can be deduced. The pattern compares well with the structure somewhere in Xinjiang au-
tonomous region where he had started the modelling. Finite element simulation
has also been used to simulate the logarithmic spiral fold pattern within two
concentric circles due to the relative rotation of the two boundaries, which
is of use in oil field exploration.

Numerous finite element simulations have been done on the uprising of Tibetan
plateau and on the subduction of slabs, etc. These we shall not discuss further.

5.3. Faulting

Faulting is another feature that has been intensively studied in mechanics because of its importance in the study of earthquakes, in geothermal energy and
oil extraction, in nuclear waste disposal etc. As mentioned before, fault system
is closely related to the migration of minerals, hence the fault zones especially
at their intersections are often favorable places for geological prospecting.

The study of fracture mechanics in Earth media differs from that in machine parts
by that the Earth media are under compression so that usually the two crack surfaces are in contact. However, there still may be regions of local tensile fractures at a depth of several kilometers, especially under high interstitial fluid
pressure.

The crack extension of a slotted rock specimen under uni-axial compression was
first demonstrated by Brace et al [38] in 1963 using photoelasticity. Under sufficient load two wing cracks of tensile fracture arise near the tips of the slot
which curve gradually toward the compressive direction and become stabilized.
The specimen can usually stand a further 30-50% increase of load before it ruptures. This phenomenon has been repeatedly demonstrated by experiments done on
glass, photoelastic materials, plexiglass, gypsum, monocrystalline quartz, etc.

However, for rocks, the process may be
different, as suggested by the results
of Ingraffea et al [39] for Indiana li-
mestone and St. Cloud charcoal granodio-
rite. For these rocks when the angle, θ,
between the slot and the compressive axis
is larger than 30°, and after the wing
cracks become stable, a second set of
symmetrically displaced cracks appeared
with a smaller angle, β (Fig.9). These
cracks called secondary cracks by the
authors, propagate and lead to final
rupture of the specimen. They used the
theory of mixed mode fracture mechanics
together with a finite element scheme
for elastic material to compute the
stress intensity factors and predicted
the trace of crack propagation which
agrees with the experimental observations.
Wang et al [40] studied the fracture process in the case of Fangshan marble. They
found that in the case of a slot, X-type
shear fractures in the direction opposite
to the primary and secondary cracks cause

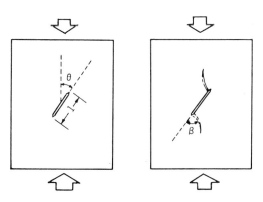

Fig.9 Thin marble plate with an
inclined slot under uniaxial compression and the initial tensile
cracks. See text for further explanation.

the final rupture of the specimen. The laser holographic interferometry technique was used. The wing cracks did appear, but as the load increases, local
microfractures increase around the tips of the slot. After the peak load, X-
type shear fractures in the conjugate directions become more and more evident,
sometime it is tensile fractures in these directions and followed by shear fractures. It was clear that the fractures all begin near the tips of the slot. They
also used the strain-softening constitutive relation mentioned in section 4.2 to
simulate the fracture process by finite element method and obtained good results.

We notice from the above experiment that the fracture behavior differs for marble
from that of homogeneous materials. The polycrystalline brittle rock materials
around the tips deteriorate and cause the stress field there to change. So in
the case of brittle rock-like materials, it is suggested that damage parameters
have to be used in the constitutive relation to describe the strain softening
characteristics. In other words their fracture behaviors are to be studied by
means of damage mechanics.

A little over a decade ago the reflection seismic exploration had shown that
many of the faults near the continental margin originally thought to have cut
vertically downward through the Moho discontinuity actually bend quickly and
form a horizontal detachment surface at the bottom. They are called listric
faults. Chen [41] have analyzed these faults analytically by a modified slip
line theory associated with the parabolic yield condition. Listric normal fault
begins at the top as tensile fracture, as it goes downward it turns into extensional shear, pure shear and then compressive shear. It agrees well with the
flow rule associated with the parabolic yield condition which takes the following form given by Wang et al in [42] (Fig.10).

$$\tau_n^2 = k^2 + (\sigma_c - 2k)\sigma_n \tag{16}$$

where τ_n, σ_n are the shear stress and normal stress on the fracture surface respectively,
$\quad \sigma_c \quad$ is the uniaxial compressive strength,
$\quad k \quad$ is the cohesion strength.

From Eq.(16) the tensile strength can be deduced to be $\sigma_t = k^2/(\sigma_c - 2k)$. The fracture shear angle is:

$$\mu = \tfrac{1}{2} \ \text{arctan}[\frac{\sqrt{2}}{\sigma_c - 2k} \sqrt{2k^2 + 2(\sigma_c - 2k)P - (\sigma_c - 2k)^2} \ ] \qquad (17)$$

where $P = \tfrac{1}{2}(\sigma_1 + \sigma_3)$, Using the concept of effective stress, the angle between the two families of slip line is 2μ, it varies between $0-90^{\circ}$, at 0°, it is tensile fracture. The two families of slip lines are:

$$\alpha\text{-line} \qquad \frac{dz}{dx} = \tan(\theta + \mu)$$

$$\sin 2\mu \ \frac{\partial P}{\partial S_\alpha} + 2R\frac{\partial \theta}{\partial S_\alpha} + (\frac{\partial G}{\partial x}\frac{\partial z}{\partial S_\alpha} - \frac{\partial G}{\partial z}\frac{\partial x}{\partial S_\alpha})$$

$$= \gamma[\cos(2\mu - \eta)\frac{\partial x}{\partial S_\alpha} + \sin(2\mu - \eta)\frac{\partial z}{\partial S_\alpha}]$$

$$\beta\text{-line} \qquad \frac{dz}{dx} = \tan(\theta - \mu)$$

$$\sin 2\mu \frac{\partial P}{\partial S_\beta} - 2R\frac{\partial \theta}{\partial S_\beta} - (\frac{\partial G}{\partial x}\frac{\partial z}{\partial S_\beta} - \frac{\partial G}{\partial z}\frac{\partial x}{\partial S_\beta})$$

$$= \gamma[-\cos(2\mu + \eta)\frac{\partial x}{\partial S_\beta} + \sin(2\mu + \eta)\frac{\partial z}{\partial S_\beta}]$$

where
$\quad \theta \quad$ angle between max. principle stress and x-direction
$\quad \eta \quad$ angle between body force and z-direction
$R=G(P,x,z)$ fracture criterion of the non-homogeneous body
$\quad \gamma \quad$ specific weight of the saturated rock layer

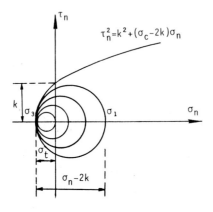

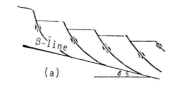

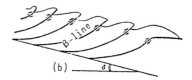

Fig.10. Limiting stress circles tangent to the parabolic yield curve at $\tau_n = 0$ (Chen [41])

Fig.11. Slip-line field in glide system under pore fluid pressure (a) extensional (b) contractional. (Chen [41].)

These slip lines can explain listric faults very well (Fig.11). Wang et al [42] had also used this fracture criterion to analyze the conjugate fault pattern, λ-type fracture along a main fault, en-echelon fracture, etc and calculated the accompanying tectonic stress fields.

Koide et al [43,44] have used tensile fracture criterion to study the relation between igneous intrusion and rift valley formation to see how the former produces tension above it and separates the two sides of the rift. As mentioned before pore pressure causes the rock to become brittle, around igneous intrusions interstitial fluid pressure are generally very high, so they have taken the ten-

sile fracture criterion in the analysis even at depth. The igneous body is assumed to be a thin vertical oblate fracture containing the pressurized fluid.
From an exact solution of three dimensional theory of elasticity, they obtained
the stress distribution around such an oblate fracture in an infinite and homogeneous elastic rock body. The maximum normal stress at the apex of the intrusion is:

$$\sigma_{max} = \frac{4S}{\pi} (\sigma_1 - P) + \sigma_1 \tag{18}$$

where P is the fluid pressure in the magma body, S is the aspect ratio of the
intrusion, σ_1 is the lithospheric stress. If S is high, e.g. equals 10, the stress
concentration factor is generally very high, the excess magma pressure may produce tensile stress around the apex of the intrusion and propagate straight upward, further on, the fractures branch outward which induces normal faults and
form a graben immediately above the intrusion. Finite element computation is
also done. When the top of the intrusive body is at a depth of 10 to 30 km, for
S larger than 10 the width of the rift valley will be 20 to 60 km consistent
from observation. When S is close to unity, however, it is the upheaval of the
ground surface to form a dome first before the fracture. This result is also
used in studying the premonitory motion of volcano activity. Formerly it was
thought only the upheaval is the promonitary motion and did not pay attention
to ground subsidence. The above result shows that the latter is possible for a
different type of magma intrusive body, [45].

Pore pressure effect on fracture is a very important subject in tectonic analysis
and seismology, it has been intensively studied by Rice [46], Carroll [47] and
many others, Its role in tectonic processes is reviewed recently by Mase et al
[48].

5.4 A non-linear inversion used in earthquake prediction and regional planning

For the non-linear case we have resorted to the trial and error method. In the
Summer of 1976 after the great Tangshan earthquake, people were anxious to know
where would the next big earthquake be. We think the important thing in predicting an earthquake site is to know its present (residual) stress state, and to
predict the future we should first be able to reproduce the past. Making use of
the long historical record that China has, we set up the scheme shown in Fig.12,
see [46].

A series of big earthquakes with magnitude larger than 7 in the last 700 years
in North China was used in the simulation. A parameter indicating the safety
against earthquake is used in accordance to Coulomb's shear fracture criterion:

$$G = \frac{\text{frictional resistance} - \text{tectonic shear stress}}{\text{frictional resistance}} = \frac{\mu\sigma_n - |\tau|}{\mu\sigma_n}$$

A decrease in G means increase of seismic risk. Earthquake is simulated by lo-
wering the coefficient of friction from a static one to a dynamic one. The media
are treated as elastic outside the fault zones and elastic-perfectly plastic
(obeying Coulomb's yield condition and associated flow rule) in the fault zone.
The result is compared to the magnitude of the earthquake, the aftershock area
and the smaller earthquakes in subsequent years. Each earthquake produces a residual stress distribution due to the locking of the fault slip. The earthquakes
are taken in succession. If the parameter has to be adjusted, the entire sequence
of events is to start from the beginning, since there is no reason for the parameter to change in such a short time. In the past decade, there were several magnitude 6 earthquakes in this area, they all fell in the seismic risky area in
the simulation. Of course, there are many things to be improved especially the
earthquake mechanism.

The above scheme can also be used in inverting tectonic movements provided the

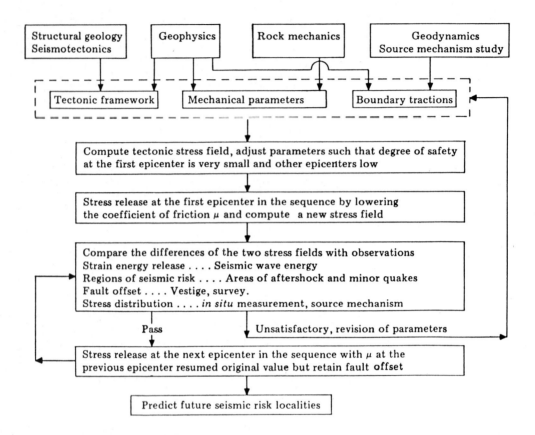

Fig. 12 Block diagram of a mathematical simulation of earthquake sequence

sequence of tectonic events are known and there are enough data for checking. In this respect, earth scientists try to deduce information from the past, besides those mentioned before, another example is they examine the disturbances in the sedimentary layers to look for the traces of old earthquake and to decide their repetitive intervals. This will also constrain the above simulation.

6. FURTHER STUDY NEEDED IN MECHANICS RELATED TO TECTONIC ANALYSIS

As mentioned above, the main problems in tectonic analysis are the driving mechanism of plate motion, the coupling of plate motion with mantle motion as related to the evolution of regional structures, the earthquake mechanism and the surface motion. From the mechanics point of view, the constitutive relations and the rupture criteria are probably the most important issue, their variations under different environmental conditions and the time factor in these relations. The next problem is a good scheme for non-linear inversion in tectonic analysis, related to this is the question on how does one evaluate a geophysical model as discussed by Guttorp in [51]. Another is to correctly interpret and further exploit the measured data in mechanics terms, e.g. from deep hole drilling, in-situ stress measurement, getting stress magnitudes from seismic records, studies of microtectonics in terms of deformation mechanism and paleopiezometer, etc. The problems mentioned above is far from complete, there are many other interesting problems and many unknowns in the field of tectonic analysis. Earth sciences is developing rapidly and mechanics can contribute a lot, we would like to see more mechanics people to work in this field.

REFERENCES

[1] Ramsay, J.G., 1967. Folding and Fracturing of Rocks. McGraw-Hill N.Y.

[2] Hobbs, B.E., Means, W.D. and Williams, P.F., 1976. An Outline of Structural Geology. John Wiley & Sons, N.Y.

[3] Weichert, D.H., Horner, R.B. and Baldwin, R., 1987. Nahanni strong motion records. IUGG XIX General Assembly Aug. 9-22, 1987, Vancouver, Abstract 1: 304.

[4] Love, A.E.H., Mathematical Theory of Elasticity. Dover Pub. New York, 1944.

[5] Leibenson, L.S., Deformation of elastic sphere in connection with the problem of Earth's structure, Collected work vol. IV, 1955 Acad. Sci. USSR, 186-266 (in Russian).

[6] Nadai, A., 1931. Plasticity. McGraw-Hill Inc. N.Y.

[7] Biot, M.A., 1965. Mechanics of Incremental Deformations, John Wiley & Sons, N.Y.

[8] Melosh, H.J. 1980. Rheology of the Earth: theory and observation. In "Physics of the Earth's Interior," by Soc. Italiana di Fisica-Bologna-Italy, 318-336.

[9] Ashby, M.F. and Verrall R.A.,1977 Micromechanisms of flow and fracture, and their relevance to the rheology of the upper mantle. Phil. Trans. R.Soc. Lond. A. 288: 59-95.

[10] Tullis, T.E. and Weeks, J.D., 1986. Constitutive behavior and stability of frictional sliding of granite, PAGEOPH 124 (3): 383-414.

[11] White, S.H., Burrows, S.E., Carreras, J., Shaw, N.D. and Humphreys, F.J. 1980. On mylonites in ductile shear zones. J. Structural Geology, 2: 175-187.

[12] Yin, Y.Q., Introduction to Non-linear Finite Element in Solid Mechanics. 1987. Peking Univ. Press, and Tsinghua Univ. Press, Beijing, (in Chinese).

[13] Lee, J.S., 1973. Crustal structure and Crustal movement. Scientia Sinica, 16 (4): 519-559.

[14] Turcotte, D.L., and Oxburgh, E.R., 1976. Stress accumulation in the Lithosphere. Tectonophysics, 35: 183-199.

[15] Tullis, T.E., 1980. The use of mechanical twinning in minerals as a measure of shear stress magnitudes. J.Geophys. Res. 85(B11): 6263-6268.

[16] Tapponnier, P., Peltzer, G., LeDain, A.Y. Armijo, R. and Cobbold, P., 1982.   Propagating extrusion tectonics in Asia, new insights from simple experiments with plasticine. Geology, 10:611-616.

[17] Wang, R. and Ding, Z.Y., 1979. On the Global Tectonic stress field due to the variation of Earth's rotation and tidal stress, Proc.Astrogeodynamics, Shanghai Observatory, 8-21, (in Chinese).

[18] Pan, E.N., Ding, Z.Y. and Wang, R., 1986. The response of a spherically stratified Earth model to body force and surface potentials' load, Acta Scien. Naturalium Univ. Pekin. 1986 No. 4:66-80, (English Abstract).

[19] Ding, Z.Y. and Wang, R., 1986. Global displacement and stress fields due to tidal attraction. Acta Geophy. Sinica 29(5): 578-596, (English abstract).

[20] Vening Meinesz, F.A., 1947. Shear pattern of the Earth's crust. EOS, Am. Geoph. Union, 28(1).

[21] Jeffreys, H., The Earth. Camb. Univ. Press, London, 3rd. ed., 1952.

[22] Richardson, R.M., Solomon, S.C. and Sleep, N.H., 1979. Tectonic stress in the plates, Rev. Geophys. Space Phys., 17:981-1019.

[23] Solomon, S.C., Richardson, R.M. and Bergman, E.A., 1980. Tectonic stress: models and magnitudes, J. Geophys. Res., 85(B11): 6086-6092.

[24] Tapponnier, P. and Molnar, P., 1976. Slip line field theory and large scale continental tectonics, Nature, 264: 319-324.

[25] Wang, S.Y. and Chen, P.S., 1980. A numerical simulation of the present tectonic stress field of China and its vicinity. Acta Geophy. Sinica, 23(1): 45-53 (English abstract)

[26] Villote, J.P. and Daignieres, M., 1982. Numerical modelling of intraplate deformation: simple mechanical models of continental collision, J. Geophys. Res. 87(B13): 10709-10728.

[27] Otsuki, K., 1982. Earth 4(10): 7-14, (in Japanese).

[28] Wang, R. and Liang, H.H., 1984. Inverting the stress field of Eastern Asia by the method of superposition, Collected works of Chinese scolars to 27th Inter, Geo. Cong., Geology Press, Beijing, 29–36, (English Abstract).

[29] Jia, J.K., 1987. Relative importance of driving forces in the stress field of Eastern Asia, M.S. thesis, Dept. of Geology, Peking University.

[30] Ramberg. H., 1964. Selective buckling of composite layers with contrasted rheological properties, Tectonophysics 1:307–341.

[31] Dieterich, J.H. and Onat, E.T. 1969. Slow finite deformation of viscous solids, J. Geophys. Res. 74:2081–2088.

[32] Lan, L. and Wang, R., 1987. Finite element analysis of an overfold using the viscous fluid model, Tectonophysics, 139(4): 309–314.

[33] Smith, R.B., 1977. Formation of folds, boudinages and mullions in Non-Newtonian materials, Geol. Soc. Amer. Bull. 88:312–320.

[34] Wu, C.H. and Cao G.Z., 1983 Buckling of an axially compressed incompressible half space, J. Struct. Mech., 11(1): 37–48.

[35] Cao, G.Z., 1986. Surface instability of an incompressible elastic half space subjected to biaxial loading, Acta Mech. Solida Sin. 1986(3):272–277

[36] Huang, Q.H., 1974. Analytic and experimental investigation of en echelon folding structure. Scientia Sinica, 1974(5): 492–500.

[37] Zeng, Z.X., 1982. Mechanical process study of a elliptic-S type structure in Xinjiang. M.S. Thesis in Wuhan Institute of Geology, Wuhan.

[38] Brace, W.F. and Bombolakis, E.G., 1963. A note on brittle crack growth in compression. J.Geophys. Res., 68: 3709–3717.

[39] Ingraffea, A.R., and Heuze, F.E., 1980. Finite element models for rock fracture mechanics. Int. J.Numer. Anal. Methods Geomech., 4: 25–43.

[40] Wang, R., Zhao, Y., Chen, Y., Yan, H., Yin, Y.Q., Yao, C.Y. and Zhang, H. 1987. Experimental and finite element simulation of X-type shear fracture from a crack in Marble. Tectonophysics, 144: 141–150.

[41] Chen, J., 1986. Mechanics analysis of Listric faults. Seismology and Geology, 8(3): 11–21 (in Chinese).

[42] Wang, W. and Han, Y., 1978. Mechanical analysis of Conjugate type structure. Collected works in Geomechanics, Vol. 4, Geology Press, Beijing, (in Chinese)

[43] Koide, H. and Bhattacharji, S., 1975. Mechanistic interpretation of rift valley formation. Science, 189: 791–793.

[44] Kouda, R. and Koide, H., 1978, Ring structures, resurgent cauldron, and ore deposits in the Hokuroku volcanic field, Northern Akita, Japan. Mining Geology, 28: 233–244.

[45] Koide, H., Hamajima, R. and Kawai, T., 1987. Analysis of Tectonic deformation by magma intrusion and its application to prediction of volcanic eruption. Proc. 7th National Sym. on Rock Mech., Japan, (English abstract).

[46] Rice, J.R. and Rudnicki, J.W., 1979. Earthquake precursory effects due to pore fluid stabilization of a weakening fault zone. J.Geophys. Res. 84(B5): 2177–2193.

[47] Carroll, M.M.,1980 Mechanical response of fluid saturated porous materials. IUTAM, 1980: 251–262.

[48] Mase, C.W. and Smith, L., 1987. The role of pore fluids in tectonic processes. Reviews of Geophysics, 25(6): 1348–1358.

[49] Wang, R., Sun, X.Y. and Cai, Y.E., 1983. A mathematical simulation of Earthquake sequence in North China in the last 700 years. Scientia Sinica, B26 (1): 103–112. (in English).

[50] Sun, X.Y., Liu, J.Y. and Wang, R., 1987. The inversion of postseismic deformation due to Tangshan earthquake, 1976. IUGG XIX General Assembly Aug. 9–22, 1987 Vancouver, Abstract 1: 337.

[51] Guttorp, P. and Walden, A., 1987. On the evaluation of geophysical models. Geophys. J.R. astr. Soc. 91:201–210.

---

* (English Abstract) means (in Chinese or Japanese with English abstract)

LIST OF CONTRIBUTED PAPERS
PRESENTED AT THE CONGRESS

AAZIZOU, K. - BURLET, H. : Cyclic elasto-plastic and creep models in finite element code.

ABDELMOUL,A R. - DAMIL, N. - POTIER-FERRY, M. : Cellular instabilities. Application to plate and shell buckling.

ABEYARATNE, R. - KNOWLES, J.K. : On the dissipative response due to discontinuous strains in bars of unstable elastic material.

ABOU-ELALA, N. : See Megahed, S..

ABRAHAMS, I.D. - WICKHAM, G.R. : On the scattering of sound by two semi-infinite parallel staggered plates : explicit matrix Wiener-Hopf factorisation.

ACHENBACH, J.D. : Crack characterization by ultrasonic scattering methods.

ACKER, P. : See Torrenti, J.M..

ADEROGBA, K. : Transmission through a thick layer between two media.

ADJEDJ, G. - AUBRY, D. - MODERASSI, A. - OZANAM, O. : Création adaptative de discontinuités cinématiques dans les géomatériaux.

ADLER, P.M. : See Bekki, S..

AEBLI, F. : See Thomann, H..

AFANASJEV, K.E. : See Terentjev, A.G..

AFANASJEVA, M.M. : See Terentjev, A.G..

ALAVYOON, F. - BARK, F.H. : On free convection in electrochemical systems.

ALEMANY, A. : See Lahjomri, J..

ALICI, E. - ALKU, O.Z. - DOST, S. : Design of prostheses for vertebral body replacement.

ALKU, O.Z. : See Alici, E..

ALLAIN, C. - SALOME, L. - LIMAT, L. : Mechanical properties near the gelation threshold.

ALTENBACH, J. : Modelling and analysis of thinwalled structures.

AMAZIGO, J.C. - BUDIANSKY, B. : Interaction of particulate and transformation thoughening.

ANDERECK, C.D. : See Mutabazi, I..

ANSOURIAN, P. : See Huang, C..

ANTORANZ, J.C. : See de la Torre, M..

ARBOCZ, J. : Koiter's stability theory in a computer-aided engineering (CAE) environment.

ARGOUL, P. - JEZEQUEL, L. : Une méthode d'identification non-paramétrique des structures vibratoires.

ARISTOV, V.V. - TCHEREMISSINE, F.G. : Investigation of nonequilibrium phenomena in a gas on the basis of the Boltzmann equation.

ASHIDA, F. : See Noda, N..

ASHRAF ALI, M. : See Sridharan, S..

ASSENHEIMER, M. : See Lorenzen, A..

ATANACKOVIC, T.M. : See Djukic, D.J..

ATLURI, S.N. : See Rajiyah, H..

AUBRY, E. - RENNER, M. : Dynamic aspects in unconventional spinning.

AUBRY, D. : See Adjedj, G..

AUDIBERT, S. : See Bruneau, M..

AURIAULT, J.L. : See Lebaigue, O..

AVILA-SEGURA, F. - CRUZ-MENA, J. - SERRANIA, F. - MENA, B. : On the slow motion of viscoelastic drops in two-phase flow.

AXELRAD, D.R. : Stochastic analysis of transient changes in clustered structures.

AYDEMIR, N.U. - VERNART, J.E.S. - SOLLOWS, K. : Free convective boundary layer of a thermomicropolar fluid past an isothermal flat plate.

AYOUB, E.F. : See Leissa, A.W..

BABESHKO, V.A. : See Obraztsov, I.P..

BAKKER, P.G. : Bifurcations in steady viscous flow patterns.

BAMMANN, D.J. : A micro-mechanically based model of finite deformation plasticity.

BANIOTOPOULOS, C.C. : See Panagiotopoulos, P..

BANKS, W.H.H. : See Zaturska, M.B..

BANKS-SILLS, L. - MARMUR, I. : Decrease of fracture thoughness induced by autofrettage.

BAPTISTE, D. : See Dan, W..

BARBI, C. : See Maresca, C..

BARGMANN, H.W. : The general problem of wear by liquid and solid impact.

BARK, F.H. : See Alavyoon, F..

BARNIER, B. : See Hua, B.L..

BARQUINS, M. : See Petit, J.P..

BARTAK, J. : A study of the rapid depressurization of hot water and the dynamics of vapor bubble generation in strongly superheated water.

BARTHES-BIESEL, D. : See Li, X.Z..

BATRA, R.C. - KIM, C.H. : Adiabatic shear banding in elastic-viscoplastic materials.

BAUER, J. : See Gutkowski, W..

BAZANT, Z.P. - TABBARA, M. - KAZEMI : Thermodynamics analysis of stable paths of structures with damage, fracture or plasticity.

BAZHLEKOV, I. : See Zapryanov, Z..

BEDA, T. : See Chevalier, Y..

BEKKI, S. - NAKACHE, E. - VIGNES-ADLER, M. - ADLER, P.M. : Solutal Marangoni effect and dissolution.

BELTZER, A.I. : Dynamic response of random viscoelastic composites.

BELYAEV, Y.N. - YAVORSKAYA, I.M. : Chaos in sperical Couette flow.

BEN HADID, H. : See Laure, P..

BEN OUEZDOU, M. : Crack-microcrack array interaction : a semi-empirical approach.

BEN-HAIM, Y. - ELISHAKOFF, I. : Non-probabilistic models of uncertainty in the buckling of shells with general imperfections : theoretical derivation of the knockdown factor.

BENABBAS, F. - DAUBE, O. - LOC, T.P. - MADANI, K. - DOFFIN, J. - PERRAULT, R. : Etude numérique de l'écoulement stationnaire et instationnaire autour d'une sphère aux grands nombres de Reynolds.

BENALLAL, A. - MARQUIS, D. : Thermodynamical modelling of the coupling between hardening, damage and ageing in metallic structures.

BENEDETTINI, F. - REGA, G. : Regular and chaotic dynamics of suspended cables under harmonic excitation.

BENNEY, D.J. : See Maslowe, S.A..

BERGAMASCHI, S. - CIARDO, R. : Tether assisted space station attitude stabilization.

BERGER, M.A. : The random walk winding number problem.

BERNADOU, M. - LALANNE, B. : On the approximation of free vibration modes of a general thin shell application to turbine blades.

BERNARD, M. : See Chollet, H..

BERTRAND, F. : See Tanguy, P.A..

BERVEILLER, M. : See Krier, J. and Patoor, E..

BESSIS, D. - FOURNIER, J.D. - SERVIZI, G. - SMITH, L.A. - SPIEGEL, E.A. - TURCHETTI, G. - VAIENTI, S. : Lacunarity of fractals, entropies of dynamical systems and intermittency of fluid turbulence.

BEVILACQUA, L. : See Pamplona, D..

BIELSKI, W.R. - TELEGA, J.J. : A contribution to the complementary energy principle, existence and uniqueness of solutions in geometrically nonlinear elasticity.

BILLOET, J.L. - GACHON, H. : A finite element analysis including shear deformations for highly anisotropic multilayered plates.

BINDER, G. : See Tardu, S..

BJÖRKMAN, G. : The solution of non linear frictionless contact problems using a sequence of linear complementarity problem.

BLANC-BENON, P. : See Karweit, M.J..

BLUME, J.A. - SHIH, C.F. - ORTIZ, M. : The singular behavior of an elastic-plastic bimaterial strip : infinitesimal and finite deformation analyses.

BODNER, S.R. - NAVEH, M. : Viscoplastic shell buckling.

BOEHLER, J.P. - EL AOUFI, L. : On experimental investigations of anisotropic solids.

BOGACZ, R. - POPP, K. : Modelling of dynamic interaction problems for transportation systems.

BOGY, D.B. : See Zhu, L.Y..

BOLT, H.M. : See Hunt, G.W..

BOMPARD, P. : See Dan, W..

BONNET, G. : See Lebaigue, O..

BORGHI, R. : See Chollet, J.P..

BORINO, G. : See Symonds, P.S..

BOSZNAY, A. : Obtaining an elastic continuum with prescribred eigenfrequencies - a new approach.

BOTSIS, J. : See Chudnovsky, A..

BOUC, R. : See Karray, M.A..

BOUCHERIT, A. : See Dambrine, B..

BOULON ,M. : Soil-structure interface behaviour and numerical solution of boundary values problems involving contact with friction.

BOURE, J.A. : Propriétés des ondes cinématiques dans les écoulements diphasiques liquide-gaz en conduite.

BOUSGARBIES, J.L. - GERON, M. - KHELIF, K. : Etude statistique de la structure tourbillonaire de grande échelle au sein d'une couche de mélange plane.

BOUSSAA, D. : See Labbé, P..

BRANCHER, J.P. - HENROT, A. - PIERRE, M. : Magnetic shaping of liquid metal : the direct and inverse problems.

BRAUCHLI, H. : See Sofer, M..

BREMAND, F. - LAGARDE, A. : New methods of large and small deformations measurements on a small size area.

BREMICKER, M. : See Eschenauer, H.A..

BRIANCON-MARJOLLET, L. - FRANC, J.P. - MICHEL, J.M. : Attached cavitation and travelling bubble cavitation.

BROCK, L.M. : Transient studies of dislocation emission and fracture under dynamic loading.

BROWAND, F. - PROST-DOMASKY, S. : Vortex patterns in high Reynolds number shear flow.

BRUCE, P.M. - HUPPERT, H.E. : Thermal control of basaltic eruptions

BRUN, L. : The self-sustained detonation as a partially reactive sonic shock-wave.

BRUNEAU, M. - HERZOG, P. - BRUNEAU, A.M. - COSNARD, C. - KERGOMARD, J. - AUDIBERT, S. - DENOIZE, P. : Acoustic field inside resonant gas-filled cavities - Theory and applications.

BUCKMASTER, J. : Detonation stability : selection of a critical wavelength.

BUDIANSKY, B. : See Amazigo, J.C..

BURCHULADZE, T.V. : Non-stationnary problems of generalized elastothermodiffusion for inhomogeneous media.

BURLET, H. : See Aazizou, K..

BUSSE, F.H. : See Umemura, A..

BYSKOV, E. : Smooth postbuckling stresses by a modified finite element method.

CAILLETAUD, G. : A "macro-micro" approach of the inelastic behaviour of CFC metals : macro/micro comparison.

CALLADINE, C.R. : Stability of the endeavour balloon.

CALTAGIRONE, J.P. - FABRIE, P. : Convection naturelle en milieu poreux à grands nombres de Rayleigh.

CAMBOU, B. - JAFARI, K. - SIDOROFF, F. : An elastic-plastic model for cohesionless soils.

CAMOTIM, D. - ROORDA, J. : Plastic buckling with residual stresses - Postbifurcation behaviour.

CANALES, S.M. : See Letelier, M.F..

CANOVA, G.R. - MOLINARI, A. - WENK, H.R. : Effects of anisotropy, grain shapes and grain-to-grain interaction in texture predictions.

CAO, H.L. : See Teodosiu, C..

CAPECCHI, D. - VESTRONI, F. : Stability of steady state oscillations of hysteretic systems.

CAPERAN, P. - SPEDDING, G. - MAXWORTHY, T. : Evolution of a freely decaying turbulence in a shallow layer of stratified fluid, for low Froude number.- See also Lahjomri, J. and Nguyen Duc, J.M..

CARMONA, C. : See Power, H..

CARRUTHERS, D.J. : See Hunt, J.C.R..

CASSOT, F. - RIEU, R. - MORVAN, D. - PELISSIER, R. : Ultrasonic velocity measurements for unsteady flows in elastic tubes and cavities simulating the cardiovascular system.
CAUGHEY, T.K. : See Masri, S.F..
CAUMES, P. : See Valentin, G..
CEDERBAUM, G. : See Elishakoff, I..
CEPTUREANU, G.H. : See Mangeron, D..
CERCIGNANI, C. - LAMPIS, M. : Variational calculation of the slip coefficent and the temperature jump for arbitrary gas-surface interactions.
CESCOTTO, S. - GROBER, H. - PREDELEANU, M. : Damage evaluation in wire drawing.
CHAI, L. : See Sun, Y.D..
CHAI, Y.S. : See Liechti, K.M..
CHANDRA, A. - MUKHERJEE, S. : Stress and deformation analysis of axisymmetric metal forming problems by the boundary element method.
CHANG, C.O. - CHOU, C.S. : Dynamic analysis of a viscous ring damper for a freely - precessing gyro.
CHAO, C.C. - CHERN, Y.C. - TUNG, T.P. : Dynamics of polar orthotropic hemispherical shells subjected to suddenly applied loads.
CHEN, C.F. : See Chen, F..
CHEN, F. - CHEN, C.F. : Onset of thermal convection in a horizontal porous layer underlying a fluid layer.
CHEN, H.L. : See Oshima, Y..
CHEN, X.J. : See Zhang, J..
CHEN, Y.N. - YANG, X. : Numerical studies for gas solid two phase steady mixed convection problems with phase change
CHENG, C.J. - SPENCER, A.J.M. - LUI, X.A. : Elastic instability of annular plates in shearing.
CHERENKOV, A.P. : On control by single switching.
CHERN, Y.C. : See Chao, C.C..
CHERNYI, G.G. : On a mechanism of body motion in solid media with anomalously low drag.
CHERY, J. : See Vilotte, J.P..
CHEUNG, E. - TENNYSON, R.C. : Buckling of composite sandwich cylinders under axial compression.
CHEVALIER, Y. - BEDA, T. - MERDRIGNAC, J. : Détermination des modules dynamiques complexes de matériaux viscoélastiques par la méthode d'identification.
CHIANG, C.R. : Yield criterion of a polycrystal under combined loadings.
CHIDAMBARRAO, D. : See Havner, K.S..
CHIEN, L.K. : See Li, J.C..
CHIRIACESCU, S.T. : See Mangeron, D..
CHOLLET, H. - MAUPU, J.L. - BERNARD, M. : Essieu à roues indépendantes orienté magnétiquement, conception, modélisation et essais à l'échelle 1/4.
CHOLLET, J.P. - PICART, A. - BORGHI, R. : Numerical simulations of reactive trubulent homogeneous flows and mixing layers.
CHOMAZ, J.M. - HUERRE, P. - REDEKOPP, L.G. : Global frequencies selection mechanisms in free shear flows.
CHONA, R. - NIGAM, H. - SHUKLA, A. : Characterizing non-singular stress. effects in fracture test specimens.- See also Shukla, A..

CHOU, C.S. : See Chang, C.O..

CHRISTOV, C.I. : The random point approximation for mechanical systems with stochastic behaviour.

CHRYSOCHOOS, A. : Stored energy measurements for behaviour laws thermomechanical analysis.

CHU, C.C. : See Li, J.C..

CHU, S.S. : See Lee, K.D..

CHUDNOVSKY, A. - BOTSIS, J. : On recent development of the crack layer model.

CHUNG, K. : See Germain, Y..

CIARDO, R. : See Bergamaschi, S..

CIMETIERE, A. : Elastic-plastic buckling of plates - A discussion of Hill's stability criterion.

CINQUINI, C. - GOBETTI, A. - ROVATI, M. : Two-dimensional structural optimization in presence of static and dynamic load conditions.

CLEGHORN, W.L. : See Tabarrok, B..

COELHO, S.L.V. : Vorticity dynamics of entraining jets in cross-flows;

COENE, R. : Aerodynamics of cyclogyro wing systems with concentrator effects.

COGNARD, J.Y. - LADEVEZE, P. - POSS, M. - ROUGEE, P. : A new approach in non-linear mechanics : the large time increment method.

COGNET, G. : See Lusseyran, F..

COHEN, D.S. - COX, R.W. - WHITE, A.B. : Stress-driven diffusion in polymers.

COINTE R. : Toward the realization of a numerical wave tank.

COMTE, P. : See Lesieur, M..

COMTE-BELLOT, G. : See Karweit, M.J..

CORONA, E. : See Kyriakides, S..

COSNARD, C. : See Bruneau, M..

COULLET, P. - LEGA, J. : Topological turbulence in wave patterns.

COURAGE, W.M.G. - SCHREURS, P.J.G. - JANSSEN, J.D. : Calculating parametervalues for effective mechanical behaviour of composites.

COX, R.W. : See Cohen, D.S..

CRAIK, A.D.D. : The Floquet instability of unbounded elliptical-vortex flows subject to a Coriolis force.

CRANDALL, S.H. : Material creep effects in rotordynamics.

CRUZ-MENA, J. : See Avila-Segura, F..

D'HUMIERES, D. : See Qian, Y.H..

DAHLKILD, A.A. - GREENSPAN, H.P. : On the flow of a rotating mixture in a sliced cylinder.

DAI, S. : See Zhang, S..

DAIGNIERES, M. : See Vilotte, J.P..

DALZIEL, S.B. : The hydraulics of two-layer rotating fluids.

DAMBRINE, B. - BOUCHERIT, A. - MASCARELL, J.P. : About the interest of using unified viscoplastic models in engine hot components life prediction.

DAMIL, N. : See Abdelmoula, R..

DAN, W. - BAPTISTE, D. - BOMPARD, P. - FRANCOIS, F. : Damage micromechanics modelisation of an heterogeneous material.

DAS, P. : See Narasimha, R..

DAUBE, O. - HUBERSON, S. : Onset of plumes and stagnant layer in natural convection. - See also Benabbas, F..

EBERHARDT, A. : See Patoor, E..

ECKELMANN, H. : See Detemple-Laake, E..

EDLUND, U. : Analysis of elastic and elastic-plastic adhesive joints using a mathematical programming approach.

EHLERS, W. : A thermodynamical development of liquid-saturated elasto-plastic porous media.

EHRLACHER, A. : See El Hawa, E..

EL AOUFI, L. : See Boehler, J.P..

EL HAWA, E. - EHRLACHER, A. : On the prediction of the evolution of damaged zones by finite elements.

EL MOUATASSIM, M. - TOUZOT, G. : Modélisation des problèmes de grandes déformations . Formulation et techniques de résolutions.

ELISHAKOFF, I. - CEDERBAUM, G. - LIBRESCU, L. : Random vibrations of moderately thick laminated shells excited by a ring loading.- See also Ben-Haim, Y..

ELLYIN, F. - KUJAWSKI, D. : Effects of different singularity fields and R-ratio on fatigue crack growth.

ELPERIN, T. : See Kitron, A..

ENESCU, I. : See Mangeron, D..

ENGBLOM, J.J. : See Fuehne, J.P..

ERBAY, H.A. - SUHUBI, E.S. : An asymptotic theory of thin nonlinear elastic plates.

ERBAY, S. - SUHUBI, E.S. : Nonlinear wave propagation in micropolar media.

ERDOGAN, F. - JOSEPH, P.F. : Surface cracks in plates and shells under mixed-mode loading conditions.

ERINGEN, C. : See Inan, E..

ESCHENAUER, H.A. - BREMICKER, M. : On strategies within the scope of a structural optimization process.

ETAY, J. : See Julliard, P..

ETHIER R. : Creeping flow through two-component fibrous materials.

EVANS, K.E. : See Zhang, W..

FABRIE, P. : See Caltagirone, J.P..

FAN, T.Y. - HAHN, H.G. : Spreading fault at high speed in fluid-infiltrated porous material.

FARGE, M. : Subgrid-scale parametrization of two-dimensional divergent flows.

FAUTRELLE, Y. : See Galpin, J.M..

FAVIER, D. : See Maresca, C..

FERRARI, M. - JOHNSON, G.C. : Evaluation of the effective. thermo-electro-mechanical properties of polycrystals exhibiting texture.

FISCHER, F.D. - MITTER, W. : A micromechanical explanation of transformation induced plasticity (TRIP).

FISZDON, W. : See Vogel, H..

FLECK, N.A. : Ductile fracture under remote shear.

FOLIAS, E.S. : A general 3D analytical solution for the equilibrium of linear elastic plates.

FORTIN, A. : See Tanguy, P.A..

FOURNIER, J.D. : See Bessis, D..

FRANC, J.P. : See Briançon-Marjollet, L..

FRANCOIS, F. : See Dan, W..

FREMOND, M. : Alliage à mémoire de forme.

HSIEH, R.K.T. - ZHOU, S.A. : Electromagnetic-Mechanical nondestructive evaluation of materials with defects.

HU, Y. - GUO, S. - HUANG, Y. : Mechanism of oil entrapment in real rock pore system

HU, C.H. : See Wang, K.C..

HUA, B.L. - LE PROVOST, C. - BARNIER, B. - VERRON, J. : Very high-resolution, multi-layer, quasi-geostrophic numerical simulations to investigate the general circulation of a mid-latitude ocean gyre.

HUANG, T.C. - HUANG, X.L. : A substructural modal perturbation method.

HUANG, X.L. : See Huang, T.C..

HUANG, Y. : See Hu, Y..

HUANG, C. - ANSOURIAN, P. : Steel tanks and large deformations.

HUBERSON, S. : See Daube, O..

HUERRE, P. : See Chomaz, J.M..

HULSE, D.A. : See Hogan, H.A..

HUNT, G.W. - BOLT, H.M. - THOMPSON, J.M.T. : Structural localization phenomena and the dynamical phase-space analogy.

HUNT, J.C.R. - CARRUTHERS, D.J. : Vorticity dynamics of idealised coherent structures.

HUO, Y.Z. - GUO, Z.H. : The Langrangean field theory of finite micropolar elasticity.

HUPPERT, H.E. : See Woods, A.W. and Bruce, P.M..

HUSSAIN, F. : See Hayakawa, M. ; Ishii, K. ; Ken, R.M. and Metcalfe, R.W..

HYCA, M. : Predicting shear-center location in open cross-sections of shear-deformable thin-walled beams.

IBRAHIM, R.A. : See Li, W..

INAN, E. - ERINGEN, C. : Nonlocal theory of wave propagation in thermoelastic plates.

INOUE, S. : See Funakoshi, M..

IOOSS, G. - LAURE, P. - ROSSI, M. : The linear stability of a spherical gas bubble in an incompressible viscous fluid

IOOSS, G. - LOS, J. : Bifurcations to quasiperiodic flows with many frequencies.

IRGENS, F. : A continuum model of granular media and simulation of snow avalanche flow in run-out zones.

IRSCHIK, H. : See Ziegler, F..

ISHII, K. - HUSSAIN, F. - KUWAHARA, K. - LIU, C.H. : The dynamics of an elliptic vortex ring.

ISHII, Y. : See Oshima, K..

ISOMÄKI, H. - VON BOEHM, J. - RÄTY, R. : Bifurcations, chaos and fractals in the motions of an impacting body.

IWANOW, Z. : See Gutkowski, W..

IZUTSU, N. : See Oshima, Y..

JAFARI, K. : See Cambou, B..

JANSSEN, J.D. : See Paas, M.H.J.W. and Courage, W.M.G..

JEFFREY, D.J. : Low-Reynolds number interactions between unequal rigid spheres.

JEZEQUEL, L. : See Argoul, P..

JIANG, J. : See Xu, J..

JOHNSON, G.C. - DIKE, J.J. : A method for the complete nondestructive evaluation of plane states of residual stress.- See also Ferrari, M..

JOHNSON, R.E. : Coating flows in rotational molding.

JOSEPH, P.F. : See Erdogan, F..

JOUSSELLIN, F. : See Morioka, S..

JULLIARD, P. - ETAY, J. : Experimental study of the impact of a sheet of liquid on the rim of a rotating wheel.

JUVE, D. : See Karweit, M.J..

KALAMKAROV, A.L. : See Parton, V.Z..

KALIFA, P. - DUVAL, P. : Crack growth in polycrystalline ice under triaxial compression.

KAM, T.Y. : Fracture analysis of brittle ceramics subjected to thermal shocks and random loads.

KAMBE, T. - KARATSU, M. - UMEKI, M. : Parametric and internal resonances of surface waves on a cylindrical fluid layer.

KANDA, H. : See Oshima, K..

KAN,T R. - DECKERT, K.L. : Laser induced heating of a multilayered medium resting on a half space. Part II - Moving source.

KAPITANIAK, T. : Stochastic chaotic processes in noisy mechanical systems.

KARAGIOZOVA, D. - HADJIKOV, L. - MANOACH, E. : On the inverse problems in dynamics of layered systems.

KARATSU, M. : See Kambe, T..

KARIHALOO, B.L. : A damage model for plain concrete.

KARLSSON, L. - NASSTROM, M. - TROIVE, L. - WEBSTER, P. - LOW, K.S. : Residual stresses and deformations in an inconel welded component.

KARRAY, M.A. - BOUC, R. : Etude de l'efficacité d'un système anti-sismique avec amortissement par plasticité.

KARWEIT, M.J. - BLANC-BENON, P. - JUVE, D. - COMTE-BELLOT, G. : Simulation of the propagation of an acoustic wave through a turbulent velocity field.

KATSIKADELIS, J. : See Sapountzakis, E..

KATSUBE, N. : See Liu, Q..

KAWATA, K. : Mechanics of high velocity brittleness and high velocity ductility of solids and related problems.

KAZEMI, : See Bazant, Z.P..

KEANE, A.J. - PRICE, W.G. : Statistical energy analysis of periodic structures.

KERGOMARD, J. : See Bruneau, M..

KERR, R.C. - LISTER, J.R. : The gravitational instability of a buoyant region of viscous fluid, applied to island arc and mid-ocean ridge volcanism. See also Worster, M.G..

KERR, R.M. - HUSSAIN, F. : Simulations of reconnection and helicity.

KESTIN ,J. - HERRMANN, G. : The fundamental elements of an exact thermodynamic theory of damage in elastic solids.

KHALAYINI, K. : See Petit, J.P..

KHELIF, K. : See Bousgarbiès, J.L..

KHELIL, A. - ROTH, J.C. : Etat des contraintes et champ de vitesse de matériaux granulaires dans les silos métalliques - Etudes théoriques et expérimentales sur modèles réduits et silos réels.

KHUSNUTDINOVA, N.V. : The problems of thermal boundary layer separation.

MAKINOUCHI, A. - LIU, S.D. : A finite element simulation of contact problems at finite elasto-plastic deformation.

MAKOWSKI, J. - STUMPF, H. : Buckling equations for elastic shells with rotational degrees of freedom undergoing finite strain deformation.

MALIER, Y. : See Torrenti, J.M..

MANGERON, D. - CHIRIACESCU, S.T. - CEPTUREANU, G.H. - ENESCU, I. - LIXANDROIU, D. : Optimization problem for elastic bodies in the Hertzian contact.

MANOACH, E. : See Karagiozova, D..

MAO, R. : See Ling, F..

MARANDI, S.R. - MODI, V.J. : Attitude stability of a rigid satellite through normalized Hamiltonian.

MARCELIN, J.L. : See Trompette, P..

MARESCA, C. - FAVIER, D. - BARBI, C. : Aérodynamique instationnaire d'un profil d'aile. Applications.

MARIGO, J.J. : See Geymonat, G..

MARKENSCOFF, X. - NI, L. - PAPADIMITRIOU, C.H. : Complete restraint of rigid bodies.- See also Dundurs, J..

MARKOV, K. : Functional series approach in mechanics of composite materials.

MARKUS, S. : Forced coupled wave motion in periodic, layered structures.

MARMUR, I. : See Banks-Sills, L..

MARQUIS, D. : See Benallal, A..

MARSHALL, J.S. - NAGHDI, P.M. : Wave reflection and transmission by steps and rectangular obstacles in channels of finite depth.

MASCARELL, J.P. : See Dambrine, B..

MASLOWE, S.A. - BENNEY ,D.J. : Wave packet critical layers in shear flows.

MASRI, S.F. - MILLER, R.K. - TRAINA, M.I. - CAUGHEY, T.K. - RUBIN, S. : Time-domain nonlinear system identification of bearing friction characteristics.

MATSUMOTO, Y. - NISHIKAWA, H. - OHASHI, H. : Numerical simulation of wave phenomena in bubbly liquid.

MAUGIN, G.A. : See Sabir, M..

MAUPU, J.L. : See Chollet, H..

MAXWORTHY, T. : See Caperan, P..

McCOY, J. : One-way theories for linearly elastic propagation media.

McDILL, J.M.J. : See Oddy, A.S..

MEGAHED, S. - ABOU-ELALA, N. : Forces analysis of robot manipulators.

MEIER, G.E.A. : See Lorenzen, A..

MELVILLE, W.K. - RENOUARD, D. - ZHANG, X. : Resonant generation of nonlinear internal Kelvin waves in a rotating channel.

MENA, B. : See Avila-Segura, F..

MENON, S. : See Metcalfe, R.W..

MERDRIGNAC, J. : See Chevalier, Y..

MERIC, R.A. : Identifying a surface with an integro-differential condition for a biharmonic structure.

METAIS, O. : Large eddy simulations of stably-stratified turbulence.

METCALFE, R.W. - MENON, S. - HUSSAIN, F. : Coherent structures in reacting free shear flows.

MEYER, M. : See Sayir, M..

MICHEL, J.M. : See Briançon-Marjollet, L..

MIEHE, C. - STEIN, E. : A quadratically counvergent procedure for thermomechanical coupling based on spatial formulation.
MIETTINEN, A. : See Parland, H..
MIHOVSKY, I.M. : See Herrmann, K.P..
MIKSAD, R.W. - SOLIS, R.S. - POWERS, E.J. : Nonlinear characteristics of an unsteady wake.
MILLER, R.K. : See Masri, S.F..
MILLIKEN, W.J. : See Powell, R.L..
MINEV, P. : See Zapryanov, Z..
MISHIMA, T. : See Mura, T..
MISHIRO, Y. : See Yamamoto, Y..
MITTER, W. : See Fischer, F.D..
MIYAMOTO, H. : A study on the ductile fracture -Al-alloys 7075 & 2017.
MIYAZAKI, T. - FUKUMOTO, Y. - HASIMOTO, H. : Long bending waves on a vortex.
MIZUSHIMA, J. : See Fujimura, K..
MOBERG ,H. : Sedimentation between conical discs in a centrifugal separator.
MODERASSI, A. : See Adjedj, G..
MODI, V.J. : See Marandi, S.R..
MOHRING, W. : See Holtfort, J..
MOLINARI, A. : See Canova, G.R. and Klepaczko, J..
MONDY, L.A. : See Powell, R.L..
MONJI, H. : See Morioka, S..
MOON, F.C. : See Paidoussis, M.P..
MOON, Y.J. : See Holt, M..
MORIOKA, S. - JOUSSELLIN, F. - MONJI, H. : Flow pattern transition in riser due to instability of voidage wave.
MOROZ, I.M. : See Leibovich, S..
MORVAN, D. : See Cassot, F..
MROZ, Z. : Development of damage in brittle-plastic solids with account for propagating damage interfaces.- See also Dems, K..
MUHE, H. : Turbulence modelling of the swirling radial free jet flow.
MUKHERJEE, S. : See Chandra, A..
MULLER, I. : Swelling and collapse of polyelectrolyste gels.
MURA, T. - TANAKA, K. - YAMASHITA, N. - MISHIMA, T. : A dislocation model for hardness indentation.
MUTABAZI, I. - PEERHOSSAINI, H. - WESFREID, J.E. - NORMAND, C. - ANDERECK, C.D. - HEGSETH, J. : Instabilities in the flow between two horizontal coaxial rotating cylinders with a partially filled gap
NAGATA, M. : Nonlinear properties of three-dimensional solutions in plane Couette flow.
NAGHDI, P.M. : See Marshall, J.S..
NAJAR, J. : Failure at cyclic loading : a CDM formulation at thermodynamic constraints.
NAKACHE, E. : See Bekki, S..
NARASIMHA, R. - DAS, P. : A spectral method for the Boltzmann equation.
NASSTROM, M. : See Karlsson, L..
NASTASE, A. : The optimum-optimorum theory and its application to the design of space shuttle.

PARTON, V.Z. - KALAMKAROV, A.L. - KUDRYAVTSEV, B.A. : Homogenization method in mechanics of deformable solids with regular microstructure.

PARVIN, M. : See Knauss, W.G..

PASCAL, M. : Dynamic analysis of a system of hinge-connected flexible bodies.

PATOOR, E. - EBERHARDT, A. - BERVEILLER, M. : Modélisation du comportement thermomécanique associé aux transformations martensitiques thermoélastiques.

PAYNE, E.R. : Turbulent boundary layer wall pressure fluctuation reduction via water injection.

PECHERSKI, R.B. : See Klepaczko, J..

PEDERSEN, P.T. : The stability of buried, heated, imperfect, pipelines.

PEERHOSSAINI, H. : See Mutabazi, I..

PELISSIER, R. : See Cassot, F..

PELLEGRINO, S. : Static response of pre-stressed kinematically indeterminate assemblies.

PENCE, T.J. : See Horgan, C.O..

PENG, M. : See Sridharan, S..

PEREGO, U. : See Symonds, P.S..

PEREGRINE, D.H. : See Teles da Silva, A.F. and Popat, N.R..

PERRAULT, R. : See Benabbas, F..

PERZYNA, P. : Induced anisotropy effects in shear band localization failure in inelastic solids.

PETIT, J.P. - KHALAYINI, K. - BARQUINS, M. : Propagation de fissures en milieu confiné dans le PMMA : modèle analogique de l'extension des failles en Mode II.

PETRYK, H. : Energy criteria of instability of plastic flow.

PFEIFFER, F. : Rattling in gearboxes - A chaotic example of the real world.

PHILIP, J.R. : Mechanics of water entry from unsaturated downward seepage into underground cavities and tunnels.

PIAN, T.H.H. - WU, C.C. : A rational approach for choosing stresses for hybrid finite element formulations.

PICART, A. : See Chollet, J.P..

PIERRE, C. : Weak and strong vibration localization in disordered structures : a statistical investigation.

PIERRE, M. : See Brancher, J.P..

PLATTEN, J.K. : See Lhost, O..

PLAUT, R. : Optimal beam and plate foundations with respect to deflection and buckling.

PONTE CASTANEDA, P. : On the overall properties of nonlinearly elastic composites.

POPAT, N.R. - PEREGRINE, D.H. : The propagation of finite amplitude capillary waves on steep gravity waves.

POPP, K. : See Bogacz, R..

POSS, M. : See Cognard, J.Y..

POTIER-FERRY, M. : See Abdelmoula, R..

POWELL, R.L. - MILLIKEN, W.J. - MONDY, L.A. - GRAHAM, A.L. - GOTTLIEB, M. : Rheology of suspensions of rods using falling ball rheometry.

POWER, H. - VILLEGAS, M. - CARMONA, C. - TRALLERO, J.L. : Slightly eccentric core-annular flow.
POWERS, E.J. : See Miksad, R.W..
PRASKOVSKY, A.A. : See Kuznetsov, V.R..
PREDELEANU, M. : See Cescotto, S. and Gelin, J.C..
PRICE, W.G. : See Keane, A.J..
PROCTOR, M.R.E. : Spatial resonance and the breakdown of steady convection.
PROST-DOMASKY, S. : See Browand, F..
QIAN, Y.H. - D'HUMIERES, D. - LALLEMAND, P. : One-dimensional lattice gas models.
RAJIYAH, H. - ZHONG, H.Q. - ATLURI, S.N : An embedded elliptical crack, in an infinite solid of transversely isotropic material, subjected to arbitrary crack-face tractions.
RAMMERSTORFER, F.G. : Dynamic instability criteria in fem.
RASZILLIER, H. : See Durst, F..
RÄTY, R. : Vibrations of one-dimensional systems carrying dynamic. elements with discrete degrees of freedom.- See also Isomäki, H..
REDEKOPP, L.G. : See Chomaz, J.M..
REGA, G. : See Benedettini, F..
RENNER, M. : See Aubry, E..
RENOUARD, D. : See Melville, W.K..
REY, C. : See Dubois, M..
RIBEIRO, G.S. - FROTA, M.N. - DOS SANTOS VARGAS, A. : Pressure drop induced by solid particles in non-Newtonian flows.
RIEU, R. : See Cassot, F..
RIEUTORD, M. : On the onset of convection in rotating spherical shells : the influence of surface gravity waves.
RIMROTT, F. : A floating frame for deforming axisymmetric gyros.
RIVET, J.P. : Three dimensionnal lattice gas hydrodynamics.
ROBERT, R. : See Duchon, J..
ROGALSKA, E. : See Litewka, A..
ROMANO, A. : See Kosinski, W..
ROORDA, J. : See Camotim, D..
ROSSI, M. : See Iooss, G..
ROTH, J.C. : See Khelil, A..
ROUGEE, P. : See Cognard, J.Y..
ROUX, B. : See Laure, P..
ROVATI, M. : See Cinquini, C..
ROY, A. : Effective elastic moduli and attenuation in material containing elliptic cracks.
ROZVANY, G. - ONG, T.G. - SZETO, W.T. : Optimal design of composite plates.
RUBIN, S. : See Masri, S.F..
RUBINSTEIN, A.A. : Effect of a curvilinear crackpath trajectory on material toughness.
RUBIO, M.A. : See de la Torre, M..
RUMJANTSEV, V.V. : On stability and stabilization with respect to a part of variables.
SAANOUNI, K. - LESNE, P.M. : On creep crack initiation and growth by a non local damage model.

SAASEN, A. - TYVAND, P.A. : Linear surface waves on a Maxwell fluid.
SABAG, M. : See Stavsky, Y..
SABELNIKOV, V.A. : See Kuznetsov, V.R..
SABIR, M. - MAUGIN, G.A. : Influence of stresses on ferromagnetic hysteresis and its use as a means of nondestructive testing.
SAITO, H. : See Sato, H..
SALOME, L. : See Allain, C..
SANCHEZ-PALENCIA, E. : See Leguillon, D..
SANO, O. : Thermal convection in a vertical torus in a uniform vertical temperature gradient.
SAPOUNTZAKIS, E. - KATSIKADELIS, J. : Unilaterally supported plates on elastic foundation by the boundary element method.
SARLET, W. : Adjoint symmetries in classical mechanics.
SATO, H. - SAITO, H. : Detailed mechanism of randomization process in the laminar-turbulent transition of a two-dimensional wake.
SAYIR, M. - MEYER, M. : Singular perturbations in plasticity : matched solutions in the plastic and elastic zones around an elliptic hole in a plate loaded in its plane.- See also Dual, J..
SCHNEIDER, W. : Asymptotic analysis of bubbles and slugs in fluidized beds.
SCHOLL, H. : See Sulem, P.L..
SCHREURS, P.J.G. : See Courage, W.M.G..
SCHREYER, H.L. : See Yazdani, S..
SCHUETZ, L.S. : See Kleinman, R..
SERRANIA, F. : See Avila-Segura, F..
SERVIZI, G. : See Bessis, D..
SHALOM, A.L. : See Shapiro, M..
SHAPIRO, M. - SHALOM, A.L. : Particle buildup on a single rod in magnetic capture from axisymmetric flows.
SHAW, P.K. : See Kyriakides, S..
SHE, Z.S. : See Sulem, P.L..
SHEMER, L. - KIT, E. : Study of the neutral stability and long-time evolution of cross-waves in the presence of dissipation.
SHIH, C.F. : See Blume, J.A..
SHIRRON, J. : See Kleinman, R..
SHOPOV, P. : See Zapryanov, Z..
SHRIVASTAVA, S.C. : See Neale, K.W..
SHUKLA, A. - CHONA, R. - ZHU, C. : On the preferred direction of crack extension in dynamic situations.- See also Chona, R..
SIDOROFF, F. - TEODOSIU, C. : Plastic spin and texture evolution during large deformation of polycristalline materials.- See also Cambou, B..
SINOPOLI, A. : Impact effects on the dynamical evolution by earthquake excitations of multi-block structures.
SMITH, L.A. : See Bessis, D..
SNYDER, J.M. - WILSON, J.F. : Dynamics of the end-loaded elastica.
SOFER, M. - BRAUCHLI, H. : Nonlocal elastic materials and regular dynamics.
SOLIS, R.S. : See Miksad, R.W..
SOLLOWS, K. : See Aydemir, N.U..
SOMMERIA, J. : See Nguyen Duc, J.M..

TURCHETTI, G. : See Bessis, D..

TYVAND, P.A. : See Saasen, A..

UENG, T.S. : See Lee, C.J..

UMEKI, M. : See Kambe, T..

UMEMURA, A. - BUSSE, F.H. : Axisymmetric convection at large Prandtl and Rayleigh numbers.

URBANEK, A. - WITTBRODT, E. : Reduction of support vibration and flexible rotor deflections and stresses by optimal balancing.

URSELL, F. : On trapping modes in the theory of water waves.

VAIENTI, S. : See Bessis, D..

VALENTIN, G. - CAUMES, P. : Critère de propagation de fissure dans le bois sous sollicitations mixtes.

VAN BEEK, P. : Small concentration expansion for the effective heat conductivity of a random disperse two-component material.

VAN DE VEN, A.A.F. : See van Lieshout, P.H..

VAN DEN BROECK, J.M. : Surfing on solitary waves.

VAN LIESHOUT, P.H. - VAN DE VEN, A.A.F. : Magneto-elastic buckling of superconducting structural systems.

VASIL'EV, D.G. : See Gol'denveizer, A.L..

VEIGA, M.F. : See Trabucho, L..

VERNART, J.E.S. : See Aydemir, N.U..

VERRON, J. : See Hua, B.L..

VESTRONI, F. : See Capecchi, D..

VIGNES-ADLER, M. : See Bekki, S..

VILLEGAS, M. : See Power, H..

VILOTTE, J.P. - CHERY, J. - DAIGNIERES, M. : Towards a thermo-mechanical approach of intra-continental deformations : mechanical and numerical aspects.

VINEY, B. : See Luu, T.S..

VIRGIN, L.N. : Nonlinear rolling motion of a ship prior to capsize : the role of approximate analytical methods.

VOGEL, H. - FISZDON, W. : On conditions for the existence of a counterflow in superfluid helium.- See also Holtfort, J..

VOLDOIRE, F. : Effect of a micro-crack array on the behaviour of a pressurized thin cylinder.

VON BOEHM, J. : See Isomäki, H..

VREENEGOOR, A.J. - WILDERS, P. - GEURST, J.A. : A numerical study of nonlinear wave-interactions in bubbly two-phase flow.

VUJICIC, V. : The modification of analitical dynamics of rheonomous systems.

WAGONER, R.H. : See Germain, Y..

WALLASCHEK, J. : Integral covariance analysis for random vibrations of linear continuous mechanical systems.

WANG, K.C. - ZHOU, H.C. - HU, C.H. : 3-D separated flow patterns over prolate spheroids.

WANG, M. : See Yan, G..

WANG, W. : See Ling, F..

WANG, W.L. : See Zhang, J..

WEBSTER, P. : See Karlsson, L..

WEDIG, W. : Lyapunov exponents and rotation numbers - A generalized Hermite analysis.

WEICHERT, D. : Shakedown-theory in the light of geometrically non-linear mechanics.

WEITSMAN, Y. : A viscoelastic continuum damage model.

WENG, G.J. - ZHU, Z.Q. : High temperature deformation of particle-reinforced metals.

WENK, H.R. : See Canova, G.R..

WESFREID, J.E. : See Mutabazi, I..

WHITE, A.B. : See Cohen, D.S..

WHITTAKER, A. : Numerical models of subduction

WICKHAM, G.R. : See Abrahams, I.D..

WILDERS, P. : See Vreenegoor, A.J..

WILSON, J.F. : See Snyder, J.M..

WINDRICH, H. : Time-dependent statistical linearization with application to randomly disturbed limit - Cycle systems.

WITTBRODT, E. : See Urbanek, A..

WOODS, A.W. - HUPPERT, H.E. : The generation of igneous layering by solidification above a horizontal boundary.

WOODS, B.A. : An approximate theory for viscous gravity currents with sources, applied to the lava dome problem.

WORSTER, M.G. - KERR, R.C. - WOODS, A.W. - HUPPERT, H.E. : Compositional variations in igneous rocks formed by cooling magma from above.

WU, C.C. : See Pian, T.H.H..

WU, X.Q. : See Lin, T.H..

XIA, R.W. : See Zhou, M..

XU, J. - JIANG, J. : Bifurcation and chaos in the forced Van der Pol oscillator.

YAMAMOTO, Y. - HOMMA, Y. - OSHIMA, K. - MISHIRO, Y. - TERADA, H. - MORIHANA, H. - YAMAUCHI, Y. - TAKENAKA, N. : Buckling of stiffened cylindrical shells under external pressure.

YAMASHITA, A. : See Oshima, Y..

YAMASHITA, N. : See Mura, T..

YAMAUCHI, Y. : See Yamamoto, Y..

YAN, G. - WANG, M. : The complex-variable method in Stokes flow and its applications.

YANG, G.S. - LU, Y. : The micro-cracks damage theory for fiber composites.

YANG, Y.B. - YAU, J.D. : Stability of pretwisted bars with various end torques.

YANG, X. : See Chen, Y.N..

YAU, J.D. : See Yang, Y.B..

YAVORSKAYA, I.M. : See Belyaev, Y.N..

YAZDANI, S. - SCHREYER, H.L. : A unified plasticity and damage model for plain concrete.

YIN, X.C. - LI, S. - TENG, C. - LI, H. : An experimental investigation on extension of non-penetrating crack in rocks and other brittle solids.

YOUNGDAHL, C.K. : The interaction between pulse shape and strain-hardening in dynamic plastic response.

ZAOUI, A. : Modelling the influence of morphological characteristics overall plastic behaviour of two-phase materials.- See also Dubois, M..

ZAPRYANOV, Z. - SHOPOV, P. - MINEV, P. - BAZHLEKOV, I. : The evolution of the deformation of fluid-fluid interfaces in viscous flows.

ALPHABETIC LIST OF TOPICS
CONTRIBUTED PAPERS AND MINI-SYMPOSIA

Acoustic and waves

Bifurcation and chaos

Biomechanics

Boundary layers

Ceramic materials

Chaos *(Solids & Fluids)*

Composites

Compressible flows

Computational fluid dynamics

Contact and friction

Control problems

Convection

Crack mechanics

Crustal and seismic problems

Damage

Diffusive transport

Elasticity

Experimental methods

Flow oscillations and unsteady flows

Flow in porous media

Flow in tubes

Fracture

Geotechnics

Identification problems

Inhomogeneous materials

Jets, wakes and shear layers

Kinetic theory

Liquid films

Low Reynolds number fluid mechanics

Mechanics of robots

Micromechanics

Non-newtonian fluid mechanics

Numerical methods

Optimization

Plasticity

Plates and shells

Space structures

Stability *(Fluids)*

Stability of structures

Structural dynamics

Stratified and rotating flows

Structural vibrations

Surface waves

Turbulence

Two-phase flows

Variational principles

Vibrations and waves in solids

Viscoelasticity and creep

Vortex dynamics

Waves

Waves in solids

Wave structure interaction

"Mechanics of large deformation and damage"

- Introductory lectures   Tuesday 23
- Sectional lecture : B. STORAKERS   Tuesday 23
- Contributed papers - see Damage *and* Plasticity   Tu 23 to Sa 27

"The dynamics of two-phase flow"

"Mechanics of the earth's crust"